Geometric Properties for Incomplete Data

Computational Imaging and Vision

Managing Editor

MAX A. VIERGEVER
Utrecht University, Utrecht, The Netherlands

Volume 31

Geometric Properties for Incomplete Data

Edited by

Reinhard Klette

The University of Auckland, Auckland, New Zealand

Ryszard Kozera

The University of Western Australia, Perth, Australia

Lyle Noakes

The University of Western Australia, Perth, Australia

and

Joachim Weickert

Saarland University, Saarbrücken, Germany

Springer

A C.I.P. Catalogue record for this book is available from the Library of Congress.

ISBN-13 978-90-481-6982-5
ISBN-10 1-4020-3858-5 (e-book)
ISBN-13 978-1-4020-3858-7 (e-book)

Published by Springer,
P.O. Box 17, 3300 AA Dordrecht, The Netherlands.

www.springeronline.com

Printed on acid-free paper

Printed in the Netherlands.

Contents

Contributors

Mirosław Bober
Visual Information Laboratory
Mitsubishi Electric
Guildford, United Kingdom

Andrés Bruhn
Mathematical Image Analysis Group, Faculty of Mathematics and Computer Science
Saarland University, Building 27
D–66 041 Saarbrücken, Germany

Isabelle Debled-Rennesson
Equipe Adage
LORIA Nancy
Nancy, France

Helene Dörksen-Reiter
Faculty of Mathematics
Hamburg University, Bundesstrasse 55
D–20 146 Hamburg, Germany

Ulrich Eckhardt
Faculty of Mathematics
Hamburg University, Bundesstrasse 55
D–20 146 Hamburg, Germany

Rida T. Farouki
Dept. of Mechanical and Aeronautical Engineering
University of California
Davis, CA 95616, USA

Markus Fenn
Dept. of Mathematics and Computer Science
University of Mannheim
D–68 131 Mannheim, Germany

Wolfgang Förstner
Institute of Photogrammetry
University Bonn, Nußallee 15
D–53 115 Bonn, Germany

Chang Yong Han
Dept. of Mechanical and Aeronautical Engineering
University of California
Davis, CA 95616, USA

Edwin R. Hancock
Dept. of Computer Science
University of York
York YO1 5DD, United Kingdom

Joachim Hornegger
Dept. of Computer Science
Friedrich-Alexander University Erlangen-Nürnberg
D–91 058 Erlangen, Germany

Martin N. Huxley
School of Mathematics
Cardiff University, 23 Senghennydd Road
Cardiff CF 24 4YH, United Kingdom

Atsushi Imiya
IMIT
Chiba University, Yayoi-cho 1-33, Inage-ku
Chiba 263-8522, Japan

Kenichi Kanatani
Dept. of Information Technology
Okayama University
Okayama 700-8530, Japan

Reinhard Klette
Dept. of Computer Science and CITR
The University of Auckland
Auckland, New Zealand

Ryszard Kozera
School of Computer Science and Software Engineering
The University of Western Australia
Perth, Australia

Krzysztof Kucharski
Institute of Radioelectronics
Warsaw University of Technology
Warsaw, Poland

Truong Kieu Linh
School of Science and Technology
Chiba University, Yayoi-cho 1-33, Inage-ku
Chiba 263-8522, Japan

Pavel Mrázek
Institute for Information Theory and Automation
Academy of Sciences of the Czech Republic, Pod vodárenskou věží 4
18208 Praha 8, Czech Republic (mrazek@mia.uni-saarland.de)

Matthias Mühlich
Institut für Angewandte Physik, Computer Vision Group
J.W. Goethe University
D–60 054 Frankfurt am Main, Germany

Lyle Noakes
School of Mathematics and Statistics
The University of Western Australia
Perth, Australia

Andreas Obereder
Dept. of Computer Science
University of Innsbruck, Technikerstr. 25
A–6020 Innsbruck, Austria

Stanley Osher
Mathematics Department
University of California
Los Angeles, CA 90095-1555, USA

Christian Perwass
Institute of Computer Science
Christian-Albrechts-University
D–24 118 Kiel, Germany

Antonio Robles-Kelly
National ICT Australia
Canberra Laboratory
Canberra ACT 2601, Australia

Bodo Rosenhahn
Dept. of Computer Science and CITR
The University of Auckland
Auckland, New Zealand

Otmar Scherzer
Dept. of Computer Science
University of Innsbruck, Technikerstr 25
A–6020 Innsbruck, Austria

Christoph Schnörr
Dept. M&CS, CVGPR-Group
University of Mannheim
D–68 131 Mannheim, Germany

Thomas Schüle
Dept. M&CS, CVGPR-Group
University of Mannheim
D–68 131 Mannheim, Germany

Władysław Skarbek
Institute of Radioelectronics
Warsaw University of Technology
Warsaw, Poland

Nir Sochen
School of Mathematics
Tel-Aviv University
Tel-Aviv, 69978 Israel

Gerald Sommer
Institute of Computer Science
Christian-Albrechts-University
D–24 118 Kiel, Germany

Gabriele Steidl
Dept. of Mathematics and Computer Science
University of Mannheim
D–68 131 Mannheim, Germany

Yasuyuki Sugaya
Dept. of Information Technology
Okayama University
Okayama 700-8530, Japan

Ariel Tankus
School of Computer Science
Tel-Aviv University
Tel-Aviv, 69978 Israel

Akihiko Torii
School of Science and Technology
Chiba University, Yayoi-cho 1-33, Inage-ku
Chiba 263-8522, Japan

Stefan Weber
Dept. M&CS, CVGPR-Group
University of Mannheim
D–68 131 Mannheim, Germany

Joachim Weickert
Mathematical Image Analysis Group, Faculty of Mathematics and Computer Science
Saarland University, Building 27
D–66 041 Saarbrücken, Germany

Yehezkel Yeshurun
School of Computer Science
Tel-Aviv University
Tel-Aviv, 69978 Israel

Joviša Žunić
Computer Science Dept.
Exeter University, Harrison Building
Exeter EX4 4QF, United Kingdom

Preface

Computer vision and image analysis requires interdisciplinary collaboration between mathematics and engineering. This book addresses the area of high-accuracy measurements of length, curvature, motion parameters and other geometrical quantities from acquired image data. It is a common problem that these measurements are incomplete or noisy, such that considerable efforts are necessary to regularise the data, to fill in missing information, and to judge the accuracy and reliability of these results. This monograph brings together contributions from researchers in computer vision, engineering and mathematics who are working in this area. A number of the chapters are expanded from invited lectures presented at the international computer science center at Dagstuhl/Germany which is funded by the German federal and two state governments. The editors gratefully acknowledge this support.

The invited authors are well-known experts in their area and have been encouraged to stress the survey character of their contributions. These are fundamental work, directed to applications in computer vision and engineering. Each paper was reviewed by at least three independent referees, and we thank all referees (see list below) for their efforts. Their comments were very valuable in order to improve the quality of this research monograph. The help of Joviša Žunić in finalizing the Latex files is very much appreciated.

By its nature, it was difficult to categorize the chapters in this book. We decided for a division into three parts, which should be of help to demonstrate the themes running through this volume.

Although geometry underlies most contributions in the present volume, it is possible to draw distinctions. Contributions in Part I are mainly concerned with continuous geometry, including algebraic tools designed for geometry, such as Clifford algebras in the chapters by Sommer and Perwass, and quaternions in the contribution of Farouki. The chapters by Noakes and Kozera concern geometrical aspects of approximation and interpolation. A somewhat more specific application of geometry to computer vision is given in the chapter by Robles-Kelly.

Discrete geometry in computer vision (Part II), including the important emerging area of digitization, has a different flavour. Even more than with other contributions to this volume, incomplete information is ever-present. The two chapters by Eckhardt, and by Doerksen-Reiter and Debled-Rennesson discuss segmentations of borders in a digital image into convex and concave parts. Linh, Imiya and Torii consider polygonalization and polyhedrization algorithms for approximating borders in 2D and 3D images.

Binary tomography (i.e., reconstruction of binary images from projections) is the subject of the chapter by Weber, Schnörr, Schule and Hornegger. Linear discriminant analysis for features derived from digital images is applied by Skarbek, Kucharski and Bober for face recognition. Huxley, Klette and Žunić study the accuracy of approximating real moments based on data available in digital images. 3D shape recovery based on digital images is the subject of the following two chapters; Tankus, Sochen and Yeshurun use shading models, and Imiya considers shadows.

Part III is concerned with approximation and regularisation methods that can be interpreted in a statistical or deterministic way. Typical applications include robust denoising of signals and images, the reliable estimation of model parameters, and motion estimation in image sequences. This area is characterised by transparent mathematical models, optimality results and performance evaluation. It includes two contributions on motion analysis: The chapter by Bruhn and Weickert evaluates a novel confidence measure for variational optic flow methods, while Kanatani and Sugaya review their contributions on feature point tracking in video sequences. The subsequent chapters deal with approximation methods: Fenn and Steidl establish connections between robust local estimation methods in image processing and approximation theoretical techniques for scattered data. The contribution by Mrázek et al. presents a unified framework for edge-preserving denoising and interpolation, while Mühlich's work deals with data fitting to geometric manifolds. The contribution by Obereder et al. concludes this category by presenting an analysis of higher order bounded variation regularisation in terms of generalised G norms.

Finally, it has to be said that the majority of contributions transcend the categories we have (sometimes arguably) assigned. This reflects the interdisciplinary nature of the work. We hope that the reader will enjoy an exciting journey.

Reinhard Klette
Ryszard Kozera
Lyle Noakes
Joachim Weickert

Auckland, Perth, Saarbrücken, March 2005.

List of referees for submissions to this book

Valentin Brimkov
Andres Bruhn
Thomas Bülow
Simon Colling
Alain Daurat
Isabell Debled-Rennesson
Patrice Delmas
Helene Dörksen
Leo Dorst
Ulrich Eckhardt
Rida Farouki
Michael Felsberg
Fabian Feschet
Vojtech Franc
Georgy Gimel'farb
Richard Hartley
Gabor T. Herman
Michael Hofer
Atsushi Imiya
Kenichi Kanatani
Antonio Kelly
Ron Kimmel
Nahum Kiryati
Gisela Klette
Reinhard Klette
Ryszard Kozera
Norbert Krüger
Attila Kuba
Wayne M. Lawton
Hrushikesh N. Mhaskar
Knut M. Moerken
Pavel Mrázek
Matthias Mühlich
Mila Nikolova
Lyle Noakes
Christian Perwass
Massi Pontil
Emmanuel Prados
Otmar Scherzer
Christoph Schnörr
Władysław Skarbek
Nir Sochen
Gerald Sommer
Gabriele Steidl
Mohamed Tajine
Rein van den Boomgaard
Joachim Weickert
Joviša Žunić

Part I

Continuous Geometry

THE TWIST REPRESENTATION OF FREE-FORM OBJECTS*

GERALD SOMMER
Institute of Computer Science, Christian-Albrechts-University, Kiel, Germany

BODO ROSENHAHN
Centre for Image Technology and Robotics, University of Auckland, Auckland, New Zealand

CHRISTIAN PERWASS
Institute of Computer Science, Christian-Albrechts-University, Kiel, Germany

Abstract. We give a contribution to the representation problem of free-form curves and surfaces. Our proposal is an operational or kinematic approach based on the Lie group $SE(3)$. While in Euclidean space the modelling of shape as orbit of a point under the action of $SE(3)$ is limited, we are embedding our problem into the conformal geometric algebra $\mathbb{R}_{4,1}$ of the Euclidean space $\mathbb{R}^3$. This embedding results in a number of advantages which makes the proposed method a universal and flexible one with respect to applications. It makes possible the robust and fast estimation of the pose of 3D objects from incomplete and noisy image data. Especially advantagous is the equivalence of the proposed shape model to that of the Fourier representations.

Key words: shape representation, conformal geometric algebra, Lie algebra, Fourier transform, free-form curves, free-form surfaces, motor, twist

1.1. Introduction

Two objects can be said to have the same shape if they are similar in the sense of Euclidean geometry. By leaving out the property of scale invariance, we can define the shape of an object as that geometric concept that is invariant under the special Euclidean group. Furthermore, we allow our

* This work has been partially supported (G.S. and C.P.) by EC Grant IST-2001-3422 (VISATEC), by DFG Grant RO 2497/1-1 (B.R.), and by DFG Graduiertenkolleg No. 357 (B.R. and C.P.).

R. Klette et al. (eds.), Geometric Properties for Incomplete Data, 3-22.

objects to change their shape in a well-defined manner under the action of some external forces.

The literature on shape modelling and applications is vast. May it be visualization and animation in computer graphics or shape and motion recognition in computer vision. The central problem for the usefulness in either field is the chosen representation of shape.

Here we present a new approach to the modelling of free-form shape of curves and surfaces which has some features that make it especially attractive for computer vision and computer graphics. In our applications of pose estimation of 3D objects we could easily handle incomplete and noisy image data for numerically stable estimations with nearly video real-time capability.

That new representation results from the fusion of two concepts:

1) Free-form curves and surfaces are modelled as the orbit of a point under the action of the Lie group $SE(3)$, caused by a set of coupled infinitesimal generators of the group, called twists (Murray et al., 1994).
2) These object models are embedded in the conformal geometric algebra (CGA) of the Euclidean space $\mathbb{R}^3$ (Li et al., 2001), that is $\mathbb{R}_{4,1}$. Only in conformal geometry the above mentioned modelling of shape unfolds its rich set of useful features.

The concept of fusing a local with a global algebraic framework has been proposed already in (Sommer, 1997). But only the pioneering work in (Li et al., 2001) made it feasible to consider the Lie algebra $se(3)$, the space of tangents to an object, embedded in $\mathbb{R}_{4,1}$, as the source of our shape model instead of using $se(3)$ in $\mathbb{R}^3$.

The tight relations of geometry and kinematics are known to the mathematicians for centuries, see e.g. (Farouki, 2000). But in contrast to most applications in mechanical engineering we are not restricted in our approach by physically feasible motions nor will we get problems in generating spatial curves or surfaces.

By embedding our design method into CGA, both primitive geometric entities as points or objects on the one side and actions on the other side will have algebraic representations in one single framework. Furthermore, objects are defined by actions, and also actions can take on the role of operands.

Our proposed kinematic definition of shape uses infinitesimal actions to generate global patterns of low intrinsic dimension. This phenomen corresponds to the interpretation of the special Euclidean group in CGA, $SE(3)$, as a Lie group, where an element $g \in SE(3)$ performs a transformation of an entity $\underline{\boldsymbol{u}} \in \mathbb{R}_{4,1}$,

$$\underline{\boldsymbol{u}}' = \underline{\boldsymbol{u}}(\theta) = g\{\underline{\boldsymbol{u}}(0)\} \tag{1}$$

with respect to the parameter θ of g. Any special $g \in SE(3)$ that represents a general rotation in CGA corresponds to a Lie group operator $\boldsymbol{M} \in \mathbb{R}^{+}_{4,1}$ which is called a motor and which is applied by the bilinear spinor product

$$\underline{\boldsymbol{u}}' = \boldsymbol{M}\underline{\boldsymbol{u}}\widetilde{\boldsymbol{M}}, \tag{2}$$

where $\widetilde{\boldsymbol{M}}$ is the reverse of $\boldsymbol{M}$. This product indicates that $\boldsymbol{M}$ is an orthogonal operator. If g is an element of the Lie group $SE(3)$, than its infinitesimal generator, ξ, is defined in the corresponding Lie algebra, that is $\xi \in se(3)$. That Lie algebra element of the rigid body motion is geometrically interpreted as the rotation axis $\underline{\boldsymbol{l}}$ in conformal space. Then the motor $\boldsymbol{M}$ results from the exponential map of the generator $\underline{\boldsymbol{l}}$ of the group element, which is called a twist:

$$\boldsymbol{M} = \exp\left(-\frac{\theta}{2}\underline{\boldsymbol{l}}\right). \tag{3}$$

While θ is the rotation angle as the parameter of the motor, its generator is defined by the five degrees of freedom of a line $\underline{\boldsymbol{l}}$ in space.

In our approach, the motor $\boldsymbol{M}$ is the effective operator which causes arbitrarily complex object shape. This operator may result from the multiplicative coupling of a set of primitive motors $\{\boldsymbol{M}_i | i = n, ..., 1\}$,

$$\boldsymbol{M} = \boldsymbol{M}_n\boldsymbol{M}_{n-1}...\boldsymbol{M}_2\boldsymbol{M}_1. \tag{4}$$

Each of these motors $\boldsymbol{M}_i$ is representing a circular motion of a point around its own axis.

Based on that approach rather complex free-form objects can be designed which behave as algebraic entities. That means, they can be transformed by motors in a covariant and linear way. To handle complete objects in that way as unique entities makes sense from both a cognitive and a numeric point of view.

The conformal geometric algebra $\mathbb{R}_{4,1}$ makes this possible. This is caused by two essential facts. First, the representation of the special Euclidean group $SE(3)$ in $\mathbb{R}_{4,1}$ as a subgroup of the conformal group $C(3)$ is isomorphic to the special orthogonal group $SO^{+}(4,1)$. Hence, rigid body motion can be performed as rotation in CGA and therefore has a covariant representation. Second, the basic geometric entity of the conformal geometric algebra of the Euclidean space is the sphere. All geometric entities derived by incidence operations from the sphere can be transformed in CGA by an element $g \in SE(3)$, that is a motor $\boldsymbol{M} \in \mathbb{R}^{+}_{4,1}$, in the same linear way, just as a point in the homogeneous Euclidean space $\mathbb{R}^4$. Because there exists a dual representation of a sphere (and of all derived entities) in CGA, which considers points as the basic geometric entity of the Euclidean space in the conformal space, all the known concepts from Euclidean space can be transformed to the conformal one.

Finally, we can take advantage of the stratification of spaces by CGA. Since the seminal paper (Faugeras, 1995) the purposive use of stratified geometries became an important design principle of vision systems. This means that an observer in dependence of its possibilities and needs can have access to different geometries as projective, affine or metric ones. So far this could hardly be realized. In CGA we have quite another situation.

The CGA $\mathbb{R}_{4,1}$ is a linear space of dimension 32. This mighty space represents not only conformal geometry but also affine geometry. Note that the special Euclidean group is a special affine group. Because $\mathbb{R}_{4,1}$ is derived from the Euclidean space $\mathbb{R}^3$, it encloses also Euclidean geometry, which is represented by the geometric algebra $\mathbb{R}_{3,0}$. In addition, the projective geometric algebra $\mathbb{R}_{3,1}$ is enclosed in $\mathbb{R}_{4,1}$. Thus, we have the stratification of the geometric algebras $\mathbb{R}_{3,0} \subset \mathbb{R}_{3,1} \subset \mathbb{R}_{4,1}$. This enables to consider metric (Euclidean), projective and kinematic (affine) problems in one single algebraic framework.

1.2. Rigid Body Motion in Conformal Geometric Algebra

After giving a bird's eye view on the construction of a geometric algebra and on the features of the conformal geometric algebra, we will present the possibilities of representing the rigid body motion in CGA.

1.2.1. SOME CONSTRUCTIVE PRINCIPLES OF A GEOMETRIC ALGEBRA

A geometric algebra (GA) $\mathbb{R}_{p,q,r}$ is a linear space of dimension 2^n, $n = p+q+r$, which results from a vector space $\mathbb{R}^{p,q,r}$. We call (p,q,r) the signature of the vector space of dimension n. This indicates that there are $p/q/r$ unit vectors $\boldsymbol{e}_i$ which square to $+1/-1/0$, respectively. While $n = p$ in case of the Euclidean space $\mathbb{R}^3$, $\mathbb{R}^{p,q,r}$ indicates a vector space with a metric different than the Euclidean one. In the case of $r \neq 0$ there is a degenerate metric. We will omit the signature indexes from right if the interpretation is unique, as in the case of $\mathbb{R}^3$.

The basic product of a GA is the geometric product, indicated by juxtaposition of the operands. This product is associative and anticommutative. There can be used a lot of other product forms in CA too, as the outer product $(\wedge)$ and the inner product $(\cdot)$.

The space $\mathbb{R}_{p,q,r}$ is spanned by a set of 2^n linear subspaces of different grade called blades. Giving the blades a geometric interpretation makes the difference of a GA from a Clifford algebra. A blade of grade k, a k-blade $\boldsymbol{B}_{\langle k\rangle}$, results from the outer product of k independent vectors $\{\boldsymbol{a}_1, ..., \boldsymbol{a}_k\} \in \mathbb{R}^{p,q,r} \equiv \langle \mathbb{R}_{p,q,r}\rangle_1$,

$$\boldsymbol{B}_{\langle k\rangle} = \boldsymbol{a}_1 \wedge ... \wedge \boldsymbol{a}_k = \langle \boldsymbol{a}_1 ... \boldsymbol{a}_k\rangle_k, \tag{5}$$

where $\langle\cdot\rangle$ is the grade operator. There are $l_k = \binom{n}{k}$ different blades of grade k, $\boldsymbol{B}_{\langle k\rangle j}, j = 1, ..., l_k$. If $\boldsymbol{e}_0 \in \mathbb{R}_{p,q,r}$, $\boldsymbol{e}_0 \equiv 1$, is the unit scalar element and $\boldsymbol{e}_{1...n} \in \mathbb{R}_{p,q,r}$, $\boldsymbol{e}_{1...n} \equiv \boldsymbol{e}_1 ... \boldsymbol{e}_n \equiv \boldsymbol{I}$, is the unit pseudoscalar element of the GA, then $\boldsymbol{B}_{\langle 0\rangle}$ is the scalar blade and $\boldsymbol{B}_{\langle n\rangle} \equiv \boldsymbol{I}$ is the pseudoscalar blade. Hence, $\sum_{k=0}^{n} l_k = 2^n$ is the dimension of the GA. Blades are directed numbers, thus $\boldsymbol{I}_{\langle k\rangle} = \boldsymbol{e}_{i_1} \wedge ... \wedge \boldsymbol{e}_{i_k}$ gives the direction of a blade. Any linear combination

$$\boldsymbol{A}_k = \sum_{j=1}^{l^*} \alpha_j \boldsymbol{B}_{\langle k\rangle j}, \; l^* \leq l_k \;, \; \alpha_j \in \mathbb{R} \tag{6}$$

is called a k-vector, $\boldsymbol{A}_k \in \langle \mathbb{R}_{p,q,r}\rangle_k$. This rich structure of a GA can be further increased by the linear combination of k-vectors,

$$\boldsymbol{A} = \sum_{k=k_*}^{k^*} \beta_k \boldsymbol{A}_k, \; 0 \leq k_* < k^* \leq n \;, \; \beta_k \in \mathbb{R} \tag{7}$$

Here $\boldsymbol{A}$ is called a (general) multivector. It is composed of components of different grade. The multivector may result from the geometric product of an r-vector $\boldsymbol{A}_r$ with an s-vector $\boldsymbol{B}_s$,

$$\boldsymbol{A} = \boldsymbol{A}_r \boldsymbol{B}_s = \langle \boldsymbol{A}_r \boldsymbol{B}_r\rangle_{|r-s|} + \langle \boldsymbol{A}_r \boldsymbol{B}_s\rangle_{|r-s|+2} + \cdots + \langle \boldsymbol{A}_r \boldsymbol{B}_s\rangle_{r+s} \tag{8}$$

with the pure inner product

$$\boldsymbol{A}_r \cdot \boldsymbol{B}_s = \langle \boldsymbol{A}_r \boldsymbol{B}_s\rangle_{|r-s|} \tag{9}$$

and the pure outer product

$$\boldsymbol{A}_r \wedge \boldsymbol{B}_s = \langle \boldsymbol{A}_r \boldsymbol{B}_s\rangle_{r+s}. \tag{10}$$

All other components of $\boldsymbol{A}$ result from a mixture of inner and outer products. The product of two multivectors, $\boldsymbol{A}$ and $\boldsymbol{B}$, can always be decomposed in the sum of an even and an odd component,

$$\boldsymbol{AB} = \frac{1}{2}(\boldsymbol{AB} + \boldsymbol{BA}) + \frac{1}{2}(\boldsymbol{AB} - \boldsymbol{BA}). \tag{11}$$

In the case of the product of two vectors, $\boldsymbol{a}$ and $\boldsymbol{b}$, $\boldsymbol{a}, \boldsymbol{b} \in \langle \mathbb{R}_{p,q,r} \rangle_1$, we get

$$\begin{aligned} \boldsymbol{ab} &= \frac{1}{2}(\boldsymbol{ab} + \boldsymbol{ba}) + \frac{1}{2}(\boldsymbol{ab} - \boldsymbol{ba}) = \boldsymbol{a} \cdot \boldsymbol{b} + \boldsymbol{a} \wedge \boldsymbol{b} && (12) \\ &= \langle \boldsymbol{ab} \rangle_0 + \langle \boldsymbol{ab} \rangle_2 = \alpha + \boldsymbol{A}_2 && (13) \end{aligned}$$

with $\alpha \in \langle \mathbb{R}_{p,q,r} \rangle_0$ and $\boldsymbol{A}_2 \in \langle \mathbb{R}_{p,q,r} \rangle_2$.

An important concept of a GA is that of duality. This means that it is possible to change the blade base of a multivector $\boldsymbol{A} \in \mathbb{R}_{p,q,r}$. Its dual is written as $\boldsymbol{A}^*$ and is defined as

$$\boldsymbol{A}^* = \boldsymbol{A} \cdot \boldsymbol{I}^{-1}, \tag{14}$$

where $\boldsymbol{I}$ is the unit pseudoscalar of $\mathbb{R}_{p,q,r}$. In the case where $\boldsymbol{A}_k \in \langle \mathbb{R}_{p,q,r} \rangle_k$ the dual is given by $\boldsymbol{A}_k^* = \boldsymbol{A}_{n-k} \in \langle \mathbb{R}_{p,q,r} \rangle_{n-k}$. The duality expresses the relations between the inner product null space, IPNS, and the outer product null space, OPNS, of a multivector, see (Perwass and Hildenbrand, 2003). The OPNS defines a collinear subspace of dimension k to a k-blade $\boldsymbol{B}_{\langle k \rangle} \subset \mathbb{R}_{p,q,r}$ which is given by all $\boldsymbol{x} \in \mathbb{R}^{p,q,r}$ so that

$$\boldsymbol{x} \wedge \boldsymbol{B}_{\langle k \rangle} = 0. \tag{15}$$

The IPNS defines a subspace of $\mathbb{R}_{p,q,r}$ which is orthogonal to a k-blade $\boldsymbol{B}_{\langle k \rangle} \subset \mathbb{R}_{p,q,r}$ and, hence

$$\boldsymbol{x} \cdot \boldsymbol{B}_{\langle k \rangle} = 0. \tag{16}$$

1.2.2. CGA OF THE EUCLIDEAN SPACE

The conformal geometry of Euclidean and non-Euclidean spaces is known for a long time (Yaglom, 1988) without giving strong impact on the modelling in engineering with the exception of electrical engineering. There are different representations of the conformal geometry. Most disseminated is a complex formulation (Needham, 1997). Based on an idea in (Hestenes, 1984), in (Li et al., 2001) and in two other papers of the same authors in (Sommer, 2001), the conformal geometries of the Euclidean, spherical and hyperbolic spaces have been worked out in the framework of GA.

The basic approach is that a conformal geometric algebra (CGA) $\mathbb{R}_{p+1,q+1}$ is built from a pseudo-Euclidean space $\mathbb{R}^{p+1,q+1}$. If we start with an Euclidean space $\mathbb{R}^n$, the construction $\mathbb{R}^{n+1,1} = \mathbb{R}^n \oplus \mathbb{R}^{1,1}$, $\oplus$ being the direct sum, uses a plane with Minkowski signature for augmenting the basis of $\mathbb{R}^n$ by the additional basis vectors $\{\boldsymbol{e}_+, \boldsymbol{e}_-\}$ with $\boldsymbol{e}_+^2 = 1$ and $\boldsymbol{e}_-^2 = -1$. Because that model can be interpreted as a homogeneous stereographic projection of all points $\boldsymbol{x} \in \mathbb{R}^n$ to points $\underline{\boldsymbol{x}} \in \mathbb{R}^{n+1,1}$, this space is called the homogeneous model of $\mathbb{R}^n$. Furthermore, by replacing the basis $\{e_+, e_-\}$ with the basis $\{e, e_0\}$, the homogeneous stereographic representation will become a representation of null vectors. This is caused by the properties $e^2 = e_0^2 = 0$ and $e \cdot e_0 = -1$. The relation between the null basis $\{e, e_0\}$ and the basis $\{e_+, e_-\}$ is given by

$$e := (e_- + e_+) \text{ and } e_0 := \frac{1}{2}(e_- - e_+). \tag{17}$$

Any point $x \in \mathbb{R}^n$ transforms to a point $\underline{x} \in \mathbb{R}^{n+1,1}$ according to

$$\underline{x} = x + \frac{1}{2}x^2 e + e_0 \tag{18}$$

with $\underline{x}^2 = 0$. In fact, any point $\underline{x} \in \mathbb{R}^{n+1,1}$ is lying on an n-dimensional subspace $N_e^n \subset \mathbb{R}^{n+1,1}$, called horosphere (Li et al., 2001). The horosphere is a non-Euclidean model of the Euclidean space $\mathbb{R}^n$.

It must be mentioned that the basis vectors e and e_0 have a geometric interpretation. In fact, e corresponds the north pole and e_0 corresponds the south pole of the hypersphere of the stereographic projection, embedded in $\mathbb{R}^{n+1,1}$. Thus, e is representing the points at infinity and e_0 is representing the origin of $\mathbb{R}^n$ in the space $\mathbb{R}^{n+1,1}$.

By setting apart these two points from all others of the $\mathbb{R}^n$ makes $\mathbb{R}^{n+1,1}$ a homogeneous space in the sense that each $\underline{x} \in \mathbb{R}^{n+1,1}$ is a homogeneous null vector without having reference to the origin. This enables coordinate-free computing to a large extent. Hence, $\underline{x} \in N_e^n$ constitutes an equivalence class $\{\lambda\underline{x}, \lambda \in \mathbb{R}\}$ on the horosphere. The reduction of that equivalence class to a unique entity with metrical equivalence to the point $x \in \mathbb{R}^n$ needs a normalization.

The CGA $\mathbb{R}_{4,1}$, derived from the Euclidean space $\mathbb{R}^3$, offers 32 blades as basis of that linear space. This rich structure enables one to represent low order geometric entities in a hierarchy of grades. These entities can be derived as solutions of either the IPNS or the OPNS depending on what we assume as the basis geometric entity of the conformal space, see (Perwass and Hildenbrand, 2003). So far we only considered the mapping of an Euclidean point $x \in \mathbb{R}^3$ to a point $\underline{x} \in N_e^3 \subset \mathbb{R}^{4,1}$. But the null vectors on the horosphere are only a special subset of all the vectors of $\mathbb{R}^{4,1}$. All the vectors of $\mathbb{R}^{4,1}$ are representing spheres as the basic entities of the conformal space. A sphere $\underline{s} \in \mathbb{R}^{4,1}$ is defined by its center position, $c \in \mathbb{R}^3$, and its radius $\rho \in \mathbb{R}$ according to

$$\underline{s} = c + \frac{1}{2}(c - \rho)^2 e + e_0. \tag{19}$$

And because $\underline{s}^2 = \rho^2 > 0$, it must be a non-null vector. A point $\underline{x} \in N_e^3$ can be considered as a degenerate sphere of radius zero. Hence, spheres $\underline{s}$

and points $\underline{\boldsymbol{x}}$ are entities of grade 1. By taking the outer product of spheres $\underline{\boldsymbol{s}}_i$, other entities of higher grade can be constructed. So we get a circle $\underline{\boldsymbol{z}}$ (grade 2), which exists outside the null cone in $\mathbb{R}^{4,1}$,

$$\underline{\boldsymbol{z}} = \underline{\boldsymbol{s}}_1 \wedge \underline{\boldsymbol{s}}_2 \tag{20}$$

as solution of the IPNS. If we consider the OPNS on the other hand, we are starting with points $\underline{\boldsymbol{x}}_i \in N_e^3$ and can proceed similarly to define a circle $\underline{\boldsymbol{Z}}$ and a sphere $\underline{\boldsymbol{S}}$ as entities of grade 3 and 4 derived from points $\underline{\boldsymbol{x}}_i$ on the null cone of $\mathbb{R}_{4,1}$ according to

$$\underline{\boldsymbol{Z}} = \underline{\boldsymbol{x}}_1 \wedge \underline{\boldsymbol{x}}_2 \wedge \underline{\boldsymbol{x}}_3 \tag{21}$$
$$\underline{\boldsymbol{S}} = \underline{\boldsymbol{x}}_1 \wedge \underline{\boldsymbol{x}}_2 \wedge \underline{\boldsymbol{x}}_3 \wedge \underline{\boldsymbol{x}}_4. \tag{22}$$

These sets of entities are obviously related by the duality $\underline{\boldsymbol{u}}^* = \underline{\boldsymbol{U}}$. Finally,

$$\underline{\boldsymbol{X}} = e \wedge \underline{\boldsymbol{x}}$$

is called the affine representation of a point (Li et al., 2001). This representation of a point is used if the interplay of the projective with the conformal representation is of interest in applications as in (Rosenhahn, 2003). With respect to lines $\underline{\boldsymbol{l}}$ and planes $\underline{\boldsymbol{p}}$ or $\underline{\boldsymbol{L}}$ and $\underline{\boldsymbol{P}}$ we refer the reader to (Sommer et al., 2004).

Let us come back to the stratification of spaces mentioned in Section 1. Let be $\boldsymbol{x} \in \mathbb{R}^n$ a point of the Euclidean space, $\boldsymbol{X} \in \mathbb{R}^{n,1}$ a point of the projective space and $\underline{\boldsymbol{X}} \in \mathbb{R}^{n+1,1}$ a point of the conformal space. Then the operations which transform the representation between the spaces are for $\mathbb{R}_3 \longrightarrow \mathbb{R}_{3,1} \longrightarrow \mathbb{R}_{4,1}$

$$\underline{\boldsymbol{X}} = e \wedge \boldsymbol{X} = e \wedge (\boldsymbol{x} + e_-), \tag{23}$$

and for $\mathbb{R}_{4,1} \longrightarrow \mathbb{R}_{3,1} \longrightarrow \mathbb{R}_3$

$$\boldsymbol{x} = -\frac{\boldsymbol{X}}{\boldsymbol{X} \cdot e_-} = \frac{((e_+ \cdot \underline{\boldsymbol{X}}) \wedge e_-) \cdot e_-}{(e_+ \cdot \underline{\boldsymbol{X}}) \cdot e_-}. \tag{24}$$

1.2.3. THE SPECIAL EUCLIDEAN GROUP IN CGA

A geometry is defined by its basic entity, the geometric transformation group which is acting in a linear and covariant manner on all the entities which are constructed from the basic entity by incidence operations, and the resulting invariances with respect to that group. The search for such a geometry was motivated in Section 1. Next we want to specify the required features of the special Euclidean group in CGA.

To make a geometry a proper one, we have to require that any action A of that group on an entity, say $\boldsymbol{u}$, is grade preserving, or in other words structure preserving. This makes it necessary that the operator $\boldsymbol{A}$ applies as versor product (Perwass and Sommer, 2002)

$$A\{\boldsymbol{u}\} = \boldsymbol{A}\boldsymbol{u}\boldsymbol{A}^{-1}. \tag{25}$$

This means that the entity $\boldsymbol{u}$ should transform covariantly (Dorst and Fontijne, 2004). If $\boldsymbol{u}$ is composed by e.g. two representants $\boldsymbol{u}_1$ and $\boldsymbol{u}_2$ of the basis entities of the geometry, then $\boldsymbol{u}$ should transform according to

$$A\{\boldsymbol{u}\} = A\{\boldsymbol{u}_1 \circ \boldsymbol{u}_2\} = (\boldsymbol{A}\boldsymbol{u}_1\boldsymbol{A}^{-1}) \circ (\boldsymbol{A}\boldsymbol{u}_2\boldsymbol{A}^{-1}) = \boldsymbol{A}\boldsymbol{u}\boldsymbol{A}^{-1}. \tag{26}$$

The invariants of the conformal group $C(3)$ in $\mathbb{R}^3$ are angles. The conformal group $C(3)$ is mighty (Needham, 1997), but other than (25) and (26) it is nonlinear and transforms not covariantly in $\mathbb{R}^3$. Besides, in $\mathbb{R}^3$ there exist no entities other than points which could be transformed.

As we have shown in Section 2.2, in $\mathbb{R}_{4,1}$ the situation is quite different because all the geometric entities derived there can be seen also as algebraic entities in the sense of Section 1. Not only the elements of the null cone transform covariantly but also those of the dual space of $\mathbb{R}_{4,1}$. Furthermore, the representation of the conformal group $C(3)$ in $\mathbb{R}_{4,1}$ has the required properties of (25) and (26), see (Li et al., 2001). All vectors with positive signature in $\mathbb{R}_{4,1}$, that is a sphere, a plane as well as the components inversion and reflection of $C(3)$ compose a multiplicative group. That is called the versor representation of $C(3)$. This group is isomorphic to the Lorentz group of $\mathbb{R}_{4,1}$. The subgroup, which is composed by products of an even number of these vectors, is the spin group $Spin^+(4,1)$, that is the spin representation of $O^+(4,1)$. To that group belong the subgroups of rotation, translation, dilatation, and transversion of $C(3)$. They are applied as a spinor $\boldsymbol{S}$, $\boldsymbol{S} \in \mathbb{R}^+_{4,1}$ and $\boldsymbol{S}\widetilde{\boldsymbol{S}} = |\boldsymbol{S}|^2$. A rotor $\boldsymbol{R}, \boldsymbol{R} \in \langle\mathbb{R}_{4,1}\rangle_2$ and $\boldsymbol{R}\boldsymbol{R}^2 = 1$, is a special spinor. Rotation and translation are represented in $\mathbb{R}_{4,1}$ as rotors.

The special Euclidean group $SE(3)$ is defined by $SE(3) = SO(3) \oplus \mathbb{R}^3$. Therefore, the rigid body motion $g = (R, \boldsymbol{t})$, $g \in SE(3)$ of a point $\boldsymbol{x} \in \mathbb{R}^3$ writes in Euclidean space

$$\boldsymbol{x}' = g\{\boldsymbol{x}\} = R\boldsymbol{x} + \boldsymbol{t}. \tag{27}$$

Here R is a rotation matrix and $\boldsymbol{t}$ is a translation vector. Because $SE(3) \subset C(3)$, in our choice of a special rigid body motion the representation of $SE(3)$ in CGA is isomorphic to the special orthogonal group, $SO^+(4,1)$. Hence, such $g \in SE(3)$ does not represent the full screw, but a general rotation in $\mathbb{R}_{4,1}$, that is the rotation axis in $\mathbb{R}^3$ is shifted out of the origin by the translation vector $\boldsymbol{t}$.

That transformation $g \in SE(3)$ is represented in CGA by a special rotor $\boldsymbol{M}$, called a motor, $\boldsymbol{M} \in \langle \mathbb{R}_{4,1} \rangle_2$. The motor may be written as in equation (3). To specify the line $\underline{\boldsymbol{l}} \in \langle \mathbb{R}_{4,1} \rangle_2$ by the rotation and translation in $\mathbb{R}^3$, the motor has to be decomposed into its rotation and translation components. The normal rotation in CGA is given by the rotor

$$\boldsymbol{R} = \exp\left(-\frac{\theta}{2}\boldsymbol{l}\right) \tag{28}$$

with $\boldsymbol{l} \in \langle \mathbb{R}_3 \rangle_2$ indicating the rotation plane which passes the origin. The translation in CGA is given by a special rotor, called a translator,

$$\boldsymbol{T} = \exp\left(\frac{\boldsymbol{et}}{2}\right) \tag{29}$$

with $\boldsymbol{t} \in \langle \mathbb{R}_3 \rangle_1$ as the translation vector. Rotors constitute a multiplicative group. If we interprete the rotor $\boldsymbol{R}$ as that entity of $\mathbb{R}_{4,1}$ which should be transformed by translation in a covariant manner, then

$$\boldsymbol{M} = \boldsymbol{T}\boldsymbol{R}\widetilde{\boldsymbol{T}}. \tag{30}$$

We call this special motor representation the twist representation. Its exponential form is given by

$$\boldsymbol{M} = \exp\left(\frac{1}{2}\boldsymbol{et}\right)\exp\left(-\frac{\theta}{2}\boldsymbol{l}\right)\exp\left(-\frac{1}{2}\boldsymbol{et}\right). \tag{31}$$

This equation expresses the shift of the rotation axis $\boldsymbol{l}^*$ in the plane $\boldsymbol{l}$ by the vector $\boldsymbol{t}$ to perform the normal rotation and finally shifting back the axis.

Because $SE(3)$ is a Lie group, the line $\underline{\boldsymbol{l}} \in \langle \mathbb{R}_{4,1} \rangle_2$ is the representation of the infinitesimal generator of $\boldsymbol{M}$, $\xi \in se(3)$. We call the generator representation a twist because it represents rigid body motion as general rotation. It is parameterized by the position and orientation of $\underline{\boldsymbol{l}}$ which are the Plücker coordinates, represented by the rotation plane $\boldsymbol{l}$ and the inner product $(\boldsymbol{t} \cdot \boldsymbol{l})$, (Rosenhahn, 2003),

$$\underline{\boldsymbol{l}} = \boldsymbol{l} + \boldsymbol{e}(\boldsymbol{t} \cdot \boldsymbol{l}). \tag{32}$$

The twist model of the rigid body motion, equation (30), is that one we are using in that paper. The most general formulation of the rigid body motion is the screw motion (Rooney, 1978). But instead of presenting that in detail, we refer the reader to the report (Sommer et al., 2004).

A motor $\boldsymbol{M}$ transforms covariantly any entity $\underline{\boldsymbol{u}} \in \mathbb{R}_{4,1}$ according to

$$\underline{\boldsymbol{u}}' = \boldsymbol{M}\underline{\boldsymbol{u}}\widetilde{\boldsymbol{M}} \tag{33}$$

with $\underline{\boldsymbol{u}}' \in \mathbb{R}_{4,1}$. An equivalent equation is valid for the dual entity $\underline{\boldsymbol{U}} \in \mathbb{R}_{4,1}$. Because motors concatenate multiplicatively, a multiple-motor transformation of $\underline{\boldsymbol{u}}$ resolves recursively. Let be $\boldsymbol{M} = \boldsymbol{M}_2\boldsymbol{M}_1$, then

$$\underline{\boldsymbol{u}}'' = \boldsymbol{M}\underline{\boldsymbol{u}}\widetilde{\boldsymbol{M}} = \boldsymbol{M}_2\boldsymbol{M}_1\underline{\boldsymbol{u}}\widetilde{\boldsymbol{M}}_1\widetilde{\boldsymbol{M}}_2 = \boldsymbol{M}_2\underline{\boldsymbol{u}}'\widetilde{\boldsymbol{M}}_2. \tag{34}$$

It is a feature of any GA that also composed entities, which are built by the outer product of other ones, transform covariantly by a linear transformation. This is called outermorphism (Hestenes, 1991) and it means the preservation of the outer product under linear transformations. Following Section 1, this is an important feature of the chosen algebraic embedding that will be demonstrated in Section 3.

1.3. Shape Models from Coupled Twists

In this section we will approach step by step the kinematic design of algebraic and transcendental curves and surfaces by coupling a certain set of twists as generators of a multiple-parameter Lie group action.

1.3.1. THE KINEMATIC CHAIN AS MODEL OF CONSTRAINED MOTION

In the preceding section we argued that each entity $\underline{\boldsymbol{u}}_i$ contributing to the rigid model of another entity $\underline{\boldsymbol{u}}$ is performing the same transformation, represented by the motor $\boldsymbol{M}$. Now we assume an ordered set of non-rigidly coupled rigid components of an object. Such model is called a kinematic chain (Murray et al., 1994). In a kinematic chain the task is to formulate the net movement of the end-effector at the n-th joint by movements of the j-th joints, $j = 1, ..., n-1$, if the 0-th joint is fixed coupled with a world coordinate system. These movements are discribed by the motors $\boldsymbol{M}_j$. Let T_j be the transformation of an attached joint j with respect to the base coordinate system, then for $j = 1, ..., n$ the point $\underline{\boldsymbol{x}}_{j,i_j}$, $i_j = 1, ..., m_j$, transforms according to

$$T_j(\underline{\boldsymbol{x}}_{j,i_j}, \boldsymbol{M}_j) = \boldsymbol{M}_1 ... \boldsymbol{M}_j \underline{\boldsymbol{x}}_{j,i_j} \widetilde{\boldsymbol{M}}_j ... \widetilde{\boldsymbol{M}}_1 \tag{35}$$

and

$$T_0(\underline{\boldsymbol{x}}_{0,i_0}) = \underline{\boldsymbol{x}}_{0,i_0}. \tag{36}$$

The motors $\boldsymbol{M}_j$ are representing the flexible geometry of the kinematic chain very efficiently. This results in an object model O defined by a kinematic chain with n segments and described by any geometric entity $\underline{\boldsymbol{u}}_{j,i_j} \in \mathbb{R}_{4,1}$ attached to the j-th segment,

$$O = \left\{T_0(\underline{\boldsymbol{u}}_{0,i_0}), T_1(\underline{\boldsymbol{u}}_{1,i_1}, \boldsymbol{M}_1), ..., T_n(\underline{\boldsymbol{u}}_{n,i_n}, \boldsymbol{M}_n) | n, i_0, ..., i_n \in \mathbb{N}\right\}. \tag{37}$$

If $\underline{\boldsymbol{u}}_{j,i_j}$ is performing a motion caused by the motor $\boldsymbol{M}$, then

$$\underline{\boldsymbol{u}}'_{j,i_j} = \boldsymbol{M}\left(T_j(\underline{\boldsymbol{u}}_{j,i_j}, \boldsymbol{M}_j)\right)\widetilde{\boldsymbol{M}} \tag{38}$$

$$= \boldsymbol{M}(\boldsymbol{M}_1 ... \boldsymbol{M}_j \underline{\boldsymbol{u}}_{j,i_j} \widetilde{\boldsymbol{M}}_j ... \widetilde{\boldsymbol{M}}_1)\widetilde{\boldsymbol{M}}. \tag{39}$$

1.3.2. THE OPERATIONAL MODEL OF SHAPE

We will now introduce another type of constrained motion, which can be realized by physical systems only in special cases but should be understood as a generalization of a kinematic chain. This is our proposed model of operational or kinematic shape (Rosenhahn, 2003). An operational shape means that a shape results from the net effect, that is the orbit, of a point under the action of a set of coupled operators. So the operators at the end are the representations of the shape. A kinematic shape means the shape for which these operators are the motors as representations of $SE(3)$ in $\mathbb{R}_{4,1}$. The principle is simple. It goes back to the interpretation of any $g \in SE(3)$ as a Lie group action (Murray et al., 1994), see equation (1). But only in $\mathbb{R}_{4,1}$ we can take advantage of its representation as rotation around the axis $\underline{\boldsymbol{l}}$, see equations (3), (30) and (31).

In Section 2.2 we introduced the sphere and the circle from IPNS and OPNS, respectively. We call these definitions the canonical ones. On the other hand, a circle has an operational definition which is given by the following. Let $\underline{\boldsymbol{x}}_\phi$ be a point which is a mapping of another point $\underline{\boldsymbol{x}}_0$ by $g \in SE(3)$ in $\mathbb{R}_{4,1}$. This may be written as

$$\underline{\boldsymbol{x}}_\phi = \boldsymbol{M}_\phi \underline{\boldsymbol{x}}_0 \widetilde{\boldsymbol{M}}_\phi \tag{40}$$

with $\boldsymbol{M}_\phi$ being the motor which rotates $\underline{\boldsymbol{x}}_0$ by an angle ϕ,

$$\boldsymbol{M}_\phi = \exp\left(-\frac{\phi}{2}\Psi\right). \tag{41}$$

Here again is Ψ the twist as a generator of the rotation around the axis $\underline{\boldsymbol{l}}$, see equation (3). Note that $\Psi = \alpha \boldsymbol{l}, \alpha \in \mathbb{R}$. If ϕ covers densely the whole span $[0, ..., 2\pi]$, then the generated set of points $\{\underline{\boldsymbol{x}}_\phi\}$ is also dense. The infinite set $\{\underline{\boldsymbol{x}}_\phi\}$ is the orbit of a rotation caused by the infinite set $\{\boldsymbol{M}_\phi\}$, which has the shape of a circle in $\mathbb{R}^3$. The set $\{\boldsymbol{x}_\phi\}$ represents the well-known subset concept in a vector space of geometric objects in analytic geometry. In fact, that circle is on the horosphere N_e^3 because it is composed only by points. We will write for the circle $\underline{\boldsymbol{z}}_{\{1\}}$ instead of $\{\underline{\boldsymbol{x}}_\phi\}$ to indicate the different nature of that circle in comparison to either $\underline{\boldsymbol{z}}$ or $\underline{\boldsymbol{Z}}$ of Section

2.2. The index $\{1\}$ means that the circle is generated by one twist from a continuous argument ϕ. So the circle, embedded in $\mathbb{R}_{4,1}$, is defined by

$$\underline{\boldsymbol{z}}_{\{1\}} = \left\{\underline{\boldsymbol{x}}_{\phi} \middle| \text{ for all } \phi \in [0, ..., 2\pi]\right\}. \tag{42}$$

Its radius is given by the distance of the chosen point $\underline{\boldsymbol{x}}_0$ to the axis $\underline{\boldsymbol{l}}$ whose orientation and position in space depends on the parameterization of $\underline{\boldsymbol{l}}$. That $\underline{\boldsymbol{z}}_{\{1\}}$ is defined by an infinite set of arguments is no real problem in the case of computational geometry or applications where only discretized shape is of interest. More interesting is the fact that in the canonical definitions of Section 2.2 the geometric entities are all derived from either spheres or points. In the case of the operational definition of shape, the circle is the basic geometric entity instead, respectively rotation is the basic operation.

A sphere results from the coupling of two motors, $\boldsymbol{M}_{\phi_1}$ and $\boldsymbol{M}_{\phi_2}$, whose twist axes meet at the center of the sphere and which are perpendicularly arranged.

The resulting constrained motion of a point $\underline{\boldsymbol{x}}_{0,0}$ performs a rotation on a sphere given by $\phi_1 \in [0, ..., 2\pi]$ and $\phi_2 \in [0, ..., \pi]$,

$$\underline{\boldsymbol{x}}_{\phi_1,\phi_2} = \boldsymbol{M}_{\phi_2}\boldsymbol{M}_{\phi_1}\underline{\boldsymbol{x}}_{0,0}\widetilde{\boldsymbol{M}}_{\phi_1}\widetilde{\boldsymbol{M}}_{\phi_2}. \tag{43}$$

The complete orbit of a sphere is given by

$$\underline{\boldsymbol{s}}_{\{2\}} = \left\{\underline{\boldsymbol{x}}_{\phi_1,\phi_2} \middle| \text{ for all } \phi_1 \in [0, ..., 2\pi], \phi_2 \in [0, ..., \pi]\right\}. \tag{44}$$

Let us come back to the point of generalization of the well-known kinematic chains. These models of linked bar mechanisms have to be physically feasible. Instead, our model of coupled twists is not limited by that constraint. Therefore, the sphere expresses a virtual coupling of twists. This includes both location and orientation in space, and the possibility of fixating several twists at the same location, for any dimension of the space $\mathbb{R}^n$. There are several extensions of the introduced kinematic model which are only possible in CGA.

First, while the group $SE(3)$ can only act on points, its representation in $\mathbb{R}_{4,1}$ may act in the same way on any entity $\underline{\boldsymbol{u}} \in \mathbb{R}_{4,1}$ derived from either points or spheres. This results in high complex free-form shapes caused from the motion of relatively simple generating entities and low order sets of coupled twists.

Second, only by coupling a certain set of twists, high complex free-form shapes may be generated from a complex enough constrained motion of a point.

Let $\underline{\boldsymbol{u}}_{\{n\}}$ be the shape generated by n motors $\boldsymbol{M}_{\phi_1}, ..., \boldsymbol{M}_{\phi_n}$. We call it the n-twist model,

$$\underline{\boldsymbol{u}}_{\{n\}} = \left\{\underline{\boldsymbol{x}}_{\phi_1,...,\phi_n} \middle| \text{ for all } \phi_1, ..., \phi_n \in [0, ..., 2\pi]\right\} \tag{45}$$

with

$$\underline{\boldsymbol{x}}_{\phi_1,\ldots,\phi_n} = \boldsymbol{M}_{\phi_n}\ldots\boldsymbol{M}_{\phi_1}\underline{\boldsymbol{x}}_{0,\ldots,0}\widetilde{\boldsymbol{M}}_{\phi_1}\ldots\widetilde{\boldsymbol{M}}_{\phi_n}. \tag{46}$$

1.3.3. FREE-FORM OBJECTS

There are a lot of more degrees of freedom to design free-form objects embedded in $\mathbb{R}_{4,1}$ by the motion of a point caused by coupled twists. While a single rotation-like motor generates a circle, a single translation-like motor generates a line as a root of non-curved objects. Of course, several of both variants can be mixed. Other degrees of freedom of the design result from the following extensions:

- Introducing an individual angular frequency λ_i to the motor $\boldsymbol{M}_{\phi_i}$ also influences the synchronization of the rotation angles ϕ_i.
- Rotation within limited angular segments $\phi_i \in [\alpha_{i_1}, ..., \alpha_{i_2}]$ with $0 \leq \alpha_{i_1} < \alpha_{i_2} \leq 2\pi$ is possible.

Let us consider the simple example of a 2-twist model of shape,

$$\underline{\boldsymbol{u}}_{\{2\}} = \{\underline{\boldsymbol{x}}_{\phi_1,\phi_2} | \text{ for all } \phi_1, \phi_2 \in [0, ..., 2\pi]\} \tag{47}$$

with

$$\underline{\boldsymbol{x}}_{\phi_1,\phi_2} = \boldsymbol{M}_{\lambda_2\phi_2}\boldsymbol{M}_{\lambda_1\phi_1}\underline{\boldsymbol{x}}_0\widetilde{\boldsymbol{M}}_{\lambda_1\phi_1}\widetilde{\boldsymbol{M}}_{\lambda_2\phi_2}, \tag{48}$$

$\lambda_1, \lambda_2 \in \mathbb{R}$ and $\phi_1 = \phi_2 = \phi \in [0, ..., 2\pi]$.

That model can generate not only a sphere, but an ellipse ($\lambda_1 = -2, \lambda_2 = 1$), several well-known algebraic curves (in space), see (Rosenhahn, 2003), such as cardioid, nephroid or deltoid, transcendental curves like a spiral, or surfaces. For the list of examples see Table 1.1.

Interestingly, the order of nonlinearity of algebraic curves grows faster than the number of the generating motors.

1.3.4. EXTENSIONS OF THE CONCEPTS

By replacing the initial point $\underline{\boldsymbol{x}}_0$ by any other geometric entity, $\underline{\boldsymbol{u}}_0$, built from either points or spheres by applying the outer product, the concepts remain the same. This makes the kinematic object model in conformal space a recursive one.

The infinite set of arguments ϕ_i of the motor $\boldsymbol{M}_{\phi_i}$ to generate the entity $\underline{\boldsymbol{u}}_{\{n\}}$ will in practice reduce to a finite one, which results in a discrete entity $\underline{\boldsymbol{u}}_{[n]}$. The index $[n]$ indicates that n twists are used with a finite set of arguments $\{\phi_{i,j_i} | j_i \in \{0, ..., m_i\}\}$.

TABLE 1.1. Simple geometric entities generated from up to three twists

Entity	Generation	Class
point	twist axis intersected with a point	0twist curve
circle	twist axis non-collinear with a point	1twist curve
line	twist axis is at infinity	1twist curve
conic	2 parallel non-collinear twists	2twist curve $\lambda_1 = 1, \lambda_2 = -2$
line segment	2 twists, building a degenerate conic	2twist curve $\lambda_1 = 1, \lambda_2 = -2$
cardioid	2 parallel non-collinear twists	2twist curve $\lambda_1 = 1, \lambda_2 = 1$
nephroid	2 parallel non-collinear twists	2twist curve $\lambda_1 = 1, \lambda_2 = 2$
rose	2 parallel non-collinear twists, j loops	2twist curve $\lambda_1 = 1, \lambda_2 = -j$
spiral	1 finite and 1 infinite twist	2twist curve $\lambda_1 = 1, \lambda_2 = 1$
sphere	2 perpendicular twists	2twist surface $\lambda_1 = 1, \lambda_2 = 1$
plane	2 parallel twists at infinity	2twist surface
cylinder	2 twists, one at infinity	2twist surface
cone	2 twists, one at infinity	2twist surface
quadric	a conic rotated with a third twist	3twist surface

The previous formulations of free-form shape did assume a rigid model. As in the case of the kinematic chain, the model can be made flexible. This happens by encapsulating the entity $\underline{\boldsymbol{u}}_{[n]}$ into a set of motors $\{\boldsymbol{M}_j^d | j = J, ..., 1\}$, which results in a deformation of the object.

$$\underline{\boldsymbol{u}}_{[n]}^d = \boldsymbol{M}_J^d ... \boldsymbol{M}_1^d \underline{\boldsymbol{u}}_{[n]} \widetilde{\boldsymbol{M}}_1^d ... \widetilde{\boldsymbol{M}}_J^d \tag{49}$$

Finally, the entity $\underline{\boldsymbol{u}}_{[n]}^d$ may perform a motion under the action of a motor $\boldsymbol{M}$, which itself may be composed by a set of motors $\{\boldsymbol{M}_i | i = I, ..., 1\}$ according to equation (4),

$$\underline{\boldsymbol{u}}_{[n]}^{d'} = \boldsymbol{M} u_{[n]}^d \widetilde{\boldsymbol{M}}. \tag{50}$$

But a twist is not only an operator but it may play in CGA also the role of an operand,

$$\Psi' = \boldsymbol{M} \Psi \boldsymbol{M}. \tag{51}$$

This causes a dynamic shape model as an alternative to (49).

So far, the entity $\boldsymbol{u}_{\{n\}}$ was embedded in the Euclidean space. Lifting up the entity to the conformal space, $\underline{\boldsymbol{u}}_{\{n\}} \in \mathbb{R}_{4,1}$, is simply done by

$$\underline{\boldsymbol{u}}_{\{n\}} = \boldsymbol{e} \wedge \left(\boldsymbol{u}_{\{n\}} + \boldsymbol{e}_-\right) = \boldsymbol{e} \wedge \boldsymbol{U}_{\{n\}} \tag{52}$$

with $\boldsymbol{U}_{\{n\}}$ being the shape in the projective space $\mathbb{R}_{3,1}$.

1.4. Twist Models and Fourier Representations

The message of the last subsection is the following. A finite set of coupled twist (or nested motors) performs a constrained motion of any set of geometric entities, whose orbit uniquely represents either a curve, a surface or a volume of arbitrary complexity. This needs a parameterized model of the generators of the shape. In some applications the reverse problem may be of interest. That is to find a parameterized twist model for a given shape. That task can be solved: Any curve, surface or volume of arbitrary complexity can be mapped to a finite set of coupled twists, but in a non-unique manner. That means, that there are different models which generate the same shape.

We will show here that there is a direct and intuitive relation between the twist model of shape and the Fourier representations. The Fourier series decomposition and the Fourier transforms in their different representations are well-known techniques of signal analysis and image processing. The interesting fact that this equivalence of representations results in a fusion of concepts from geometry, kinematics, and signal theory is of great importance in engineering. Furthermore, because the presented modelling of shape is embedded in a conformal space, there is also a single access for embedding the Fourier representations in either conformal or projective geometry. This is quite different from the recent publication (Turski, 2004).

1.4.1. THE CASE OF A CLOSED PLANAR CURVE

Let us consider a closed curve $\boldsymbol{c} \in \mathbb{R}^2$ in a parametric representation with $t \in \mathbb{R}$. Then its Fourier series representation is given by

$$\boldsymbol{c}(t) = \sum_{\nu=-\infty}^{\infty} \gamma_\nu \exp\left(\frac{j2\pi\nu t}{T}\right) \tag{53}$$

with the Fourier coefficients γ_ν, $\nu \in \mathbb{Z}$ as frequency and j, $j^2 = -1$, as the imaginary unit and T as the curve length.

This model of a curve has been used for a long time in image processing for shape analysis by Fourier descriptors (these are the Fourier coefficients) (Zahn and Roskies, 1972).

We will translate this spectral representation into the model of an infinite number of coupled twists by following the method presented in (Rosenhahn et al., 2004). Because equation (53) is valid in an Euclidean space, the twist model has to be reformulated accordingly. This will be shown for the case of a 2-twist curve $\underline{c}_{\{2\}}$ based on equation (27). Then equation (48) can be written in $\mathbb{R}_3$ for $\phi_1 = \phi_2 = \phi$ as

$$\begin{aligned} \boldsymbol{x}_\phi &= \boldsymbol{R}_{\lambda_2\phi}\left((\boldsymbol{R}_{\lambda_1\phi}(\boldsymbol{x}_0 - \boldsymbol{t}_1)\widetilde{\boldsymbol{R}}_{\lambda_1\phi} + \boldsymbol{t}_1) - \boldsymbol{t}_2\right)\widetilde{\boldsymbol{R}}_{\lambda_2\phi} + \boldsymbol{t}_2 && (54)\\ &= \boldsymbol{p}_0 + \boldsymbol{V}_{1,\phi}\boldsymbol{p}_1\widetilde{\boldsymbol{V}}_{1,\phi} + \boldsymbol{V}_{2,\phi}\boldsymbol{p}_2\widetilde{\boldsymbol{V}}_{2,\phi}. && (55)\end{aligned}$$

Here the translation vectors have been absorbed by the vectors $\boldsymbol{p}_i$ and the $\boldsymbol{V}_i$ are built by certain products of the rotors $\boldsymbol{R}_{\lambda_i\phi}$. We call the $\boldsymbol{p}_i$ the phase vectors. Next, for the aim of interpreting that equation as a Fourier series expansion, we rewrite the Fourier basis functions as rotors of an angular frequency $i \in \mathbb{Z}$, in the plane $\boldsymbol{l} \in \mathbb{R}_2$, $\boldsymbol{l}^2 = -1$,

$$\boldsymbol{R}_{\lambda_i\phi} = \exp\left(-\frac{\lambda_i\phi}{2}\boldsymbol{l}\right) = \exp\left(-\frac{\pi i\phi}{T}\boldsymbol{l}\right). \qquad (56)$$

All rotors of a planar curve lie in the same plane as the phase vectors $\boldsymbol{p}_i$. After some algebra, see (Rosenhahn et al., 2004), we get for the transformed point

$$\boldsymbol{x}_\phi = \sum_{i=0}^{2} \boldsymbol{p}_i \exp\left(\frac{2\pi i\phi}{T}\boldsymbol{l}\right) \qquad (57)$$

and for the curve as subspace of $\mathbb{R}^3$ the infinite set of points

$$\boldsymbol{c}_{\{2\}} = \{\boldsymbol{x}_\phi | \text{ for all } \phi \in [0, ..., 2\pi] \text{ and for all } i \in \{0,1,2\}\}. \qquad (58)$$

A general (planar) curve is given by

$$\boldsymbol{c}_{\{\infty\}} = \{\boldsymbol{x}_\phi | \text{ for all } \phi \in [0, ..., 2\pi] \text{ and for all } i \in \mathbb{Z}\}, \qquad (59)$$

respectively as Fourier series expansion, written in the language of kinematics

$$\begin{aligned} \boldsymbol{c}_{\{\infty\}} &= \left\{\lim_{n\longrightarrow\infty} \sum_{i=-n}^{n} \boldsymbol{p}_i \exp\left(\frac{2\pi i\phi}{T}\boldsymbol{l}\right)\right\} && (60)\\ &= \left\{\lim_{n\longrightarrow\infty} \sum_{i=-n}^{n} \boldsymbol{R}_{\lambda_i\phi}\boldsymbol{p}_i\widetilde{\boldsymbol{R}}_{\lambda_i\phi}\right\}. && (61)\end{aligned}$$

A discretized curve is called a contour. In that case equation (60) has to consider a finite model of n twists and the Fourier series expansion becomes the inverse discrete Fourier transform. Hence, a planar contour is given by the finite sequence $\boldsymbol{c}_{[n]}$ with the contour points $c_k, -n \leq k \leq n$, in parametric representation

$$c_k = \sum_{i=-n}^{n} \boldsymbol{p}_i \exp\left(\frac{2\pi ik}{2n+1}\boldsymbol{l}\right), \qquad (62)$$

and the phase vectors are computed as a discrete Fourier transform of the contour

$$\boldsymbol{p}_i = \frac{1}{2n+1} \sum_{k=-n}^{n} c_k \exp\left(-\frac{2\pi ik}{2n+1}\boldsymbol{l}\right). \tag{63}$$

These equations imply that the angular argument ϕ_k is replaced by k.

1.4.2. EXTENSIONS OF THE CONCEPTS

The extension of the modelling of a planar curve, embedded in $\mathbb{R}^3$, to a 3D curve is easily done. This happens by taking its projections to either $\boldsymbol{e}_{12}$, $\boldsymbol{e}_{23}$, or $\boldsymbol{e}_{31}$ as periodic planar curves. Hence, we get the superposition of these three components. Let $\boldsymbol{c}^j_{[n]}$ be these components in the case of a 3D contour with the rotation axes $\boldsymbol{l}^*_j$ perpendicular to the rotation planes $\boldsymbol{l}_j$. Then

$$\boldsymbol{c}_{[n]} = \sum_{j=1}^{3} \boldsymbol{c}^j_{[n]} \tag{64}$$

with the contour points of the projections c^j_k, $j = 1, 2, 3$ and $-n \leq k \leq n$,

$$c^j_k = \sum_{i=-n}^{n} \boldsymbol{p}^j_i \exp\left(\frac{2\pi ik}{2n+1}\boldsymbol{l}_j\right). \tag{65}$$

Another useful extension is with respect to surface representations, see (Rosenhahn et al., 2004). If this surface is a 2D function orthogonal to a plane spanned by the bivectors $\boldsymbol{e}_{ij}$, then the twist model corresponds to the 2D inverse FT. In the case of an arbitrary orientation of the rotation planes $\boldsymbol{l}_j$ instead, or in the case of the surface of a 3D object, the procedure is comparable to that of equation (65). The surface is represented as a two-parametric surface $\boldsymbol{s}(t_1, t_2)$ as superposition of the three projections $\boldsymbol{s}^j(t_1, t_2)$.

In the case of a discrete surface in a two-parametric representation we have the finite surface representation $\boldsymbol{s}_{[n_1,n_2]}$,

$$\boldsymbol{s}_{[n_1,n_2]} = \sum_{j=1}^{3} \boldsymbol{s}^j_{[n_1,n_2]} \tag{66}$$

with the surface points of the projections $s^j_{k_1,k_2}$, $j = 1, 2, 3$ and $-n_1 \leq k_1 \leq n_1$, $-n_2 \leq k_2 \leq n_2$,

$$s^j_{k_1,k_2} = \sum_{i_1=-n_1}^{n_1} \sum_{i_2=-n_2}^{n_2} \boldsymbol{p}^j_{i_1,i_2} \exp\left(\frac{2\pi i_1 k_1}{2n_1+1}\boldsymbol{l}_j\right) \exp\left(\frac{2\pi i_2 k_2}{2n_2+1}\boldsymbol{l}_j\right) \tag{67}$$

and the phase vectors

$$\boldsymbol{p}^j_{i_1,i_2} = \frac{1}{2n_1+1}\frac{1}{2n_2+1}\boldsymbol{p}^{j'}_{i_1,i_2} \tag{68}$$

$$\boldsymbol{p}^{j'}_{i_1,i_2} = \sum_{k_1=-n_1}^{n_1}\sum_{k_2=-n_2}^{n_2} s^j_{k_1,k_2} \exp\left(\frac{2\pi i_1 k_1}{2n_1+1}\boldsymbol{l}_j\right)\exp\left(\frac{2\pi i_2 k_2}{2n_2+1}\boldsymbol{l}_j\right) \tag{69}$$

Finally, we will give the hint to an alternative model of a curve $\underline{\boldsymbol{c}} \in \mathbb{R}_{4,1}$, see (Rosenhahn, 2003). While equation (60) expresses the additive superposition of rotated phase vectors in Euclidean space, the multiplicative coupling of the twists directly in conformal space is possible.

The discussed equivalence of the twist model and the Fourier representation has several advantages in practical use of the model. The most important may be the applicability to low-frequency approximations of the shape. For instance in pose estimation (Rosenhahn, 2003) the estimations of the motion parameters of non-convex objects can be regularized efficiently in that way. Instead of estimating motors, the parameters of the twists are estimated because of numeric reasons.

1.5. Summary and Conclusions

We presented an operational or kinematic model of shape in $\mathbb{R}^3$. This model is based on the Lie group $SE(3)$, embedded in the conformal geometric algebra $\mathbb{R}_{4,1}$ of the Euclidean space. While the modelling of shape in $\mathbb{R}^3$ caused by actions of $SE(3)$ is limited, a lot of advantages result from the chosen algebraic embedding in real applications. As one of these the possibility of conformal (and projective) shape models should be mentioned. We did not discuss any applications in detail. Instead, we refer the reader to the website http://www.ks.informatik.uni-kiel.de with respect to the problem of pose estimation. In that work we could show that the pose estimation based on the presented shape model can cope with incomplete and noisy data. In addition to that robustness the pose estimation is numerically stable and fast.

Because the chosen twist model is equivalent to the Fourier representation (in some aspects it overcomes that), the proposed shape representation unifies geometry, kinematics, and signal theory. It can be expected that this will have a great impact on both theory and practice in computer vision, computer graphics and modelling of mechanisms.

An extended version of this paper can be found as report (Sommer et al., 2004).

Acknowledgement: The authors thank the anonymous reviewers for their valuable suggestions for improving the quality of this paper.

References

Dorst, L. and D. Fontijne: An algebraic foundation for object-oriented Euclidean geometry. In Proc. *Innovative Teaching in Mathematics with Geometric Algebra* (E. Hitzer and R. Nagaoka, editors), pages 138–153, Kyoto. Research Institute for Mathematics, Kyoto University, 2004.

Farouki, R.: Curves from motion, motion from curves. In *Curve and Surface Design* (P.-J. Laurent, P. Sablonniere, and L. Schumaker, editors), pages 63–90, Vanderbilt University Press, Nashville, TN, 2000.

Faugeras, O.: Stratification of three-dimensional vision: projective, affine and metric representations. *J. Optical Soc. of America,* **12**:465–484, 1995.

Hestenes, D.: The design of linear algebra and geometry. *Acta Appl. Math.,* **23**:65–93, 1991,

Hestenes, D. and G. Sobczyk: *Clifford Algebra to Geometric Calculus.* D. Reidel Publ. Comp., Dordrecht, 1984.

Li, H., D. Hestenes, and A. Rockwood: Generalized homogeneous coordinates for computational geometry. In *Geometric Computing with Clifford Algebras* (G. Sommer, editor), pages 27–59, Springer, Heidelberg, 2001.

Murray, R., Z. Li, and S. Sastry: *A Mathematical Introduction to Robotic Manipulation.* CRC Press, Boca Raton, 1994.

Needham, T.: *Visual Complex Analysis.* Clarendon Press, Oxford, 1997.

Perwass, C. and D. Hildenbrand: Aspects of geometric algebra in Euclidean, projective and conformal space. TR 0310, Kiel University, Institut für Informatik und Praktische Mathematik, 2003.

Perwass, C. and G. Sommer: Numerical evaluation of versors with Clifford algebra. In *Applications of Geometric Algebra in Computer Science and Engineering* (L. Dorst, C. Doran, and J. Lasenby, editors), pages 341–350, Birkhäuser, Boston, 2002.

Rooney, J.: A comparison of representations of general screw displacements. *Environment and Planning,* **5**:45–88, 1978.

Rosenhahn, B.: Pose estimation revisited. TR 0308, PhD thesis, Kiel University, Institut für Informatik und Praktische Mathematik, 2003.

Rosenhahn, B., C. Perwass, and G. Sommer: Free-form pose estimation by using twist representations. *Algorithmica,* **38**:91–113, 2004.

Sommer, G.: Algebraic aspects of designing behavior based systems. In *Algebraic Frames for the Perception-Action Cycle* (G. Sommer and J. Koenderink, editors), LNCS 1315, pages 1–28, Springer, Berlin, 1997.

Sommer, G.: *Geometric Computing with Clifford Algebras.* Springer, Heidelberg, 2001.

Sommer, G., Rosenhahn B., and Perwass C.: The twist representation of shape. TR 0407, Kiel University, Institut für Informatik und Praktische Mathematik, 2004.

Turski, J.: Geometric Fourier analysis of the conformal camera for active vision. *SIAM Review,* **46**:230–255, 2004.

Yaglom, M.: *Felix Klein and Sophus Lie.* Birkhäuser, Boston, 1988.

Zahn, C. and R. Roskies: Fourier descriptors for plane closed curves. *IEEE Trans. Computers,* **21**:269–281, 1972.

UNCERTAIN GEOMETRY WITH CIRCLES, SPHERES AND CONICS

CHRISTIAN PERWASS
Christian-Albrechts-University Kiel, Institute of Computer Science, Olshausenstr. 40, D-24098 Kiel, Germany

WOLFGANG FÖRSTNER
University Bonn, Institute of Photogrammetry, Nußallee 15, D-53115 Bonn, Germany

Abstract. In this text the description of uncertain geometric entities is extended from points, lines and planes to circles, spheres and 2D-conic sections. While the former has been treated previously by Kanatani (Kanatani, 1996), Förstner (Förstner et al., 2000) and Heuel (Heuel, 2004) in matrix spaces, the latter can be treated advantageously in a multilinear setting using Clifford algebra. It is shown how error propagation can be applied to Clifford algebra operations in general, and specifically for the construction of circles, spheres and 2D-conic sections. While circles and spheres are treated in the Clifford algebra of conformal space (Hestenes, 1991; Li et al., 2001), the construction of uncertain 2D-conic sections is treated in the Clifford algebra of a specially developed vector space. Some results on synthetic data are presented.

Key words: Clifford algebra, error propagation, conformal space, conic sections

2.1. Introduction

Spatial reasoning is one of the central tasks in Computer Vision. It always has to deal with uncertain data. Projective geometry has become the working horse for modelling multiple view geometry, while modelling uncertainty with statistical tools has become a standard. Geometric reasoning in projective geometry with uncertain geometric elements has been advocated by Kanatani in the early 90's, and recently made transparent and generalized to basic entities in projective geometry including transformations by Förstner and Heuel, exploiting the multilinearity of nearly all relations, such as incidence and identity, which results from the underlying

R. Klette et al. (eds.), Geometric Properties for Incomplete Data, 23-41.

Grassmann-Cayley algebra; see (White, 1995; Faugeras and Papadopoulo, 1998; Faugeras and Luong, 1998).

This paper generalizes geometric reasoning under uncertainty towards circles and spheres, which play a role in many computer vision applications. The basic step is to embed all entities into a more general algebra, namely the Clifford algebra of conformal space as proposed by Hestenes et al. (Hestenes, 1991; Li et al., 2001). The basic elements in conformal algebra are spheres in any dimension, including points, straight lines, planes but also point pairs, i. e. spheres in $\mathbb{E}^1$ and circles, i. e. spheres in $\mathbb{E}^2$. We also introduce the Clifford algebra over the vector space of 2D-conics, which, to the best of our knowledge, has not yet been discussed in the literature. This allows us to model 2D-conics and their intersections and thus also apply geometric reasoning under uncertainty to these entities. Clifford algebra is similar to Grassmann-Cayley algebra and also covers projective geometry (Hestenes and Ziegler, 1991).

Modelling uncertainty of uncertain homogeneous entities is not straight forward (cf. (Collins, 1993)). In case of good relative accuracy, i. e. directional errors of less than 1 %, the representation with covariance matrices has been widely accepted (cf. e. g. (Kanatani, 1996; Criminisi, 2001)). A direct integration into projective geometry has been proposed by Förstner (Förstner et al., 2000). For the simple case of the join $\mathbf{l} = \mathbf{x} \times \mathbf{y} = \mathsf{S}(\mathbf{x})\mathbf{y} = -\mathsf{S}(\mathbf{y})\mathbf{x}$, where $\mathsf{S}(\mathbf{x}) = [\mathbf{x}]_\times$ is the skew matrix induced by the 3-vector $\mathbf{x}$, we obtain:

$$\Sigma_{l,l} = \mathsf{S}(\mathbf{y})\Sigma_{x,x}\mathsf{S}^\mathsf{T}(\mathrm{y}) + \mathsf{S}(\mathbf{x})\Sigma_{y,y}\mathsf{S}^\mathsf{T}(\mathbf{x})$$

for independent 2D points with covariance matrices $\Sigma_{x,x}$ and $\Sigma_{y,y}$, e. g.

$$\Sigma_{x,x} = \begin{bmatrix} \sigma_x^2 & \sigma_{xy} & 0 \\ \sigma_{xy} & \sigma_y^2 & 0 \\ 0 & 0 & 0 \end{bmatrix}$$

This type of uncertainty representation and propagation can be extended to all types of geometric entities and also transformations within projective geometry, in case the expressions are multilinear in the given entities.

The paper generalizes these developments towards circles, spheres and conic sections by embedding all entities in a more general algebra. The paper is organized as follows: Sect. 2.2 presents the basic concepts of Clifford algebra making the multilinearity of the expressions explicit. Sect. 2.3 describes the embedding of n-spheres into Clifford algebra via the special instance of conformal algebra and the versatility of the concept. Sect. 2.4 introduces the embedding of 2D-conics in a 6D-vector space and the Clifford algebra over this vector space. Based on the statistical error propagation in sect. 2.5 the uncertainty propagation in conformal algebra and the algebra of conics is demonstrated for 3D circles and 2D conics.

2.2. Clifford Algebra

Without explaining exactly what it is, we will denote a Clifford algebra on $\mathbb{R}^n$ by $\mathcal{C}\ell(\mathbb{R}^n)$, or simply $\mathcal{C}\ell_n$ if it is clear that we are forming the Clifford algebra over the reals. The latter will in fact be the case for the whole of this text. For more detailed introductions to Clifford algebra see e.g. (Hestenes and Sobczyk, 1984; Porteous, 1995; Lounesto, 1997; Dorst,, 2001; Perwass and Hildenbrand, 2003). A Clifford algebra $\mathcal{C}\ell_n$ over a vector space $\mathbb{R}^n$ has dimension 2^n. An algebraic basis of $\mathcal{C}\ell_n$ may therefore be denoted by a set $\{E_i\}_{i=1}^{2^n}$ of so called *basis blades*. It may be shown that these basis blades satisfy a number of constraints with respect to the algebra product which is also called the *geometric* or *Clifford* product. This product will simply be denoted by juxtaposition, i.e. the geometric product of two elements $A, B \in \mathcal{C}\ell_n$ is written as AB. The basis blades of $\mathcal{C}\ell_n$ have the following properties:

$$\begin{aligned} &\exists E_1 \text{ such that } E_i E_1 = E_1 E_i = E_i,\ \forall i \in \{1, \ldots, 2^n\}, \\ &E_i E_i = \lambda_i\, E_1,\ \lambda_i \in \{-1, 1\},\ \forall i \in \{1, \ldots, 2^n\}, \\ &E_i E_j = \textstyle\sum_{k=1}^{2^n}\, g^k{}_{ij}\, E_k,\ \forall i, j \in \{1, \ldots, 2^n\}. \end{aligned} \tag{1}$$

The last condition basically says that the geometric product of basis blades is invertible. For example, given indices (i, j, k) such that $E_i E_j = E_k$, we find that

$$E_i E_j = E_k \iff E_i E_j E_j = E_k E_j \iff E_k E_j = \lambda_j\, E_i,$$

and thus $g^i{}_{kj} = \lambda_j$.

A general element of $\mathcal{C}\ell_n$ is called *multivector*. In terms of basis blades a general multivector $A \in \mathcal{C}\ell_n$ may be given by $A = \sum_{i=1}^{2^n} \alpha^i\, E_i$. In the following we will use the Einstein summation convention, that a superscript index repeated within a product as a subscript index is implicitly summed over its range. That is, a multivector may be written as $A = \alpha^i\, E_i$, if it is clear that $i \in \{1, \ldots, 2^n\}$. The geometric product of two multivectors $A, B \in \mathcal{C}\ell_n$, with $A = \alpha^i E_i$ and $B = \beta^i E_i$, is then given by

$$AB = (\alpha^i\, E_i)\,(\beta^j\, E_j) = \alpha^i\, \beta^j\, E_i E_j = \alpha^i\, \beta^j\, g^k{}_{ij}\, E_k. \tag{2}$$

Writing the result multivector $M \in \mathcal{C}\ell_n$ of $M = AB$ as $M = \mu^i E_i$ then gives

$$M = AB \iff \mu^k E_k = \alpha^i\, \beta^j\, g^k{}_{ij}\, E_k \iff \mu^k = \alpha^i\, \beta^j\, g^k{}_{ij}\ \forall k. \tag{3}$$

This shows that if multivectors in $\mathcal{C}\ell_n$ are expressed as vectors in $\mathbb{R}^{2^n}$, the geometric product between them becomes a bilinear function. Therefore, if we want to discuss error propagation in Clifford algebra, we can look at the error propagation of bilinear functions. Note that other products available in Clifford algebra like the inner and outer product, which will be discussed in the following, may also be expressed in this way.

As an example for an $\{E_i\}_{i=1}^{2^n}$ basis, consider the space $\mathbb{R}^4$ with orthonormal basis $\{\mathsf{e}_1, \mathsf{e}_2, \mathsf{e}_3, \mathsf{e}_4\}$. A basis for the Clifford algebra $\mathcal{C}\ell(\mathbb{R}^4)$ is then be given by

$$\begin{aligned}\{1,\ &\mathsf{e}_1,\ \mathsf{e}_2,\ \mathsf{e}_3,\ \mathsf{e}_4,\ \mathsf{e}_2\mathsf{e}_3,\ \mathsf{e}_3\mathsf{e}_1,\ \mathsf{e}_1\mathsf{e}_2,\ \mathsf{e}_4\mathsf{e}_1,\ \mathsf{e}_4\mathsf{e}_2,\ \mathsf{e}_4\mathsf{e}_3,\\ &\mathsf{e}_2\mathsf{e}_3\mathsf{e}_4,\ \mathsf{e}_3\mathsf{e}_1\mathsf{e}_4,\ \mathsf{e}_1\mathsf{e}_2\mathsf{e}_4,\ \mathsf{e}_1\mathsf{e}_2\mathsf{e}_3,\ \mathsf{e}_1\mathsf{e}_2\mathsf{e}_3\mathsf{e}_4\}\end{aligned} \tag{4}$$

Each of the elements of this basis may now be denoted by one E_i. From the associativity of the algebra product $(\mathsf{e}_1(\mathsf{e}_2\mathsf{e}_3) = (\mathsf{e}_1\mathsf{e}_2)\mathsf{e}_3)$ and the signature of the vector space, in this case $\mathsf{e}_i\mathsf{e}_i = 1$, the particular values of the tensor $g^k{}_{ij}$ follow. For example, using the above basis we can define $E_1 := 1$, $E_2 := \mathsf{e}_1$, $E_3 := \mathsf{e}_2$, …, $E_8 := \mathsf{e}_1\mathsf{e}_2$, …, $E_{16} := \mathsf{e}_1\mathsf{e}_2\mathsf{e}_3\mathsf{e}_4$. We then clearly have $E_2\, E_3 = E_8$. Hence, $g^8{}_{23} = 1$ and $g^k{}_{23} = 0\, \forall\, k \neq 8$.

The representation of algebra products in the form of equation (3) allows us to apply standard error propagation directly to Clifford algebra, as will be seen later on. However, this representation is not particularly enlightening when it comes to the description of geometry. Geometry is in fact represented through the null-spaces of algebraic entities with respect to particular algebra products. In Clifford algebra these are the inner and the outer product (Hestenes and Sobczyk, 1984) and in Grassmann-Cayley algebra the meet and join (Faugeras and Luong, 1998).

The outer product is a special operation defined within Clifford algebra and is denoted by $\wedge$. It is, in fact, equivalent to the exterior product of Grassmann algebra. The outer product is associative and distributive. For vectors $\mathbf{x}, \mathbf{y} \in \mathbb{E}^n$ it is also anti-commutative, i. e. $\mathbf{x} \wedge \mathbf{y} = -\mathbf{y} \wedge \mathbf{x}$. Another important property is that if $\mathbf{x} \wedge \mathbf{y} = 0$, then $\mathbf{x}$ and $\mathbf{y}$ are linearly dependent. More generally, for a set $\{\mathbf{x}_1, \ldots, \mathbf{x}_k\} \subset \mathbb{R}^n$ of $k \leq n$ mutually linearly independent vectors, $(\mathbf{x}_1 \wedge \mathbf{x}_2 \wedge \ldots \wedge \mathbf{x}_k) \wedge \mathbf{y} = 0$ if and only if $\mathbf{y}$ is linearly dependent on $\{\mathbf{x}_1, \ldots, \mathbf{x}_k\}$. The outer product of a number of vectors is also called a *blade*. The *grade* of a blade is simply the number of vectors that "wedged" together give the blade. Hence, the outer product of k linearly independent vectors gives a blade of grade k, a k-blade.

The set of vectors in $\mathbb{R}^n$ whose outer product with the k-blade gives zero, spans a k-dimensional subspace. The null space of a k-blade in some $\mathcal{C}\ell(\mathbb{R}^n)$ with respect to the outer product, i. e. the *outer product null space* of a k-blade, is therefore a k-dimensional subspace of $\mathbb{R}^n$. Geometrically this means for Euclidean space $\mathbb{E}^3$ that a vector represents a line through the origin, a 2-blade (or *bivector*) a plane through the origin, and a 3-blade (or *trivector*) the whole $\mathbb{E}^3$. For more details see (Hestenes and Ziegler, 1991).

Instead of looking at the null space of algebraic entities with respect to the outer product, we can do the same for the *inner product* of Clifford algebra. The inner product will be denoted by $\cdot$. For vectors $\mathbf{x}, \mathbf{y} \in \mathbb{R}^n$, their inner product is just the same as their scalar product denoted by $*$. That is, $\mathbf{x} \cdot \mathbf{y} = \mathbf{x} * \mathbf{y} \quad \in \mathbb{R}$. This may be called the "metric" property of the inner product, since the result of the scalar product of two vectors depends on the metric of the vector space they lie in. However, the inner product also has some purely algebraic properties for elements in $\mathcal{C}\ell(\mathbb{R}^n)$, which are independent of the metric of the vector space $\mathbb{R}^n$. For example, let $\mathbf{x}, \mathbf{a}, \mathbf{b} \in \mathbb{R}^n$, then the inner product of $\mathbf{x}$ with $\mathbf{a} \wedge \mathbf{b}$ gives,

$$\mathbf{x} \cdot (\mathbf{a} \wedge \mathbf{b}) = (\mathbf{x} \cdot \mathbf{a})\, \mathbf{b} - (\mathbf{x} \cdot \mathbf{b})\, \mathbf{a}. \tag{5}$$

Since $(\mathbf{x} \cdot \mathbf{a})$ and $(\mathbf{x} \cdot \mathbf{b})$ are scalars, we see that the inner product of a vector with a bivector results in a vector. In terms of the null space of entities with respect to the inner product, this formula shows that vector $\mathbf{x}$ lies in the *inner product null space* of $\mathbf{a} \wedge \mathbf{b}$ if and only if $\mathbf{x}$ lies in the inner product null space of $\mathbf{a}$ *and* $\mathbf{b}$. That is, the inner product null space of $\mathbf{a} \wedge \mathbf{b}$ is the intersection of the inner product null spaces of $\mathbf{a}$ and $\mathbf{b}$. For example, in the Clifford algebra of projective space $\mathcal{C}\ell(\mathbb{PE}^3)$, vectors $\mathbf{a}$ and $\mathbf{b}$ may represent planes w.r.t. their inner product null space. Hence, the bivector $\mathbf{a} \wedge \mathbf{b}$ then represents the intersection line of the two planes.

2.3. Conformal Space

In the previous section it was shown how Clifford algebra can be used to represent geometric entities like lines and planes through the origin in $\mathcal{C}\ell(\mathbb{E}^3)$. Conformal space extends this idea by embedding a n - dimensional Euclidean space in a nonlinear manner in a $(n+2)$-dimensional space. Conformal space takes its name from the fact that certain types of reflections in conformal space represent inversion in Euclidean space and conformal transformations can be represented by combinations of inversions. See (Needham, 199; Li et al., 2001) for more details. In this text we cannot go into all the details relating to conformal space and the Clifford algebra over this space. We can only state the important formulae and give a basic idea of how we can use conformal space to work with geometric entities.

As before we will denote vectors in a n-dimensional Euclidean vector space $\mathbb{E}^n$ by small, bold faced letters as in $\mathbf{x}$. Note that even though we will work in the following with the conformal space of 3-dimensional Euclidean space, all formulae extend directly to n dimensions. In order to obtain a conformal space, which we will denote by $\mathbb{PK}^n$, we extend the orthonormal basis $\{\mathsf{e}_i\}_{i=1}^n$ of $\mathbb{E}^n$ by two orthogonal basis vectors $\{\mathsf{e}_+, \mathsf{e}_-\}$ with $\mathsf{e}_+^2 = 1$ and $\mathsf{e}_-^2 = -1$. The embedding of a Euclidean vector $\mathbf{x}$ in conformal space is then given by

$$\mathbf{X} = \mathbf{x} + \tfrac{1}{2}\mathbf{x}^2\,\mathsf{e}_\infty + \mathsf{e}_o, \tag{6}$$

where $\mathsf{e}_\infty := \mathsf{e}_- + \mathsf{e}_+$ and $\mathsf{e}_o := \frac{1}{2}(\mathsf{e}_- - \mathsf{e}_+)$. The properties of e_∞ and e_o are therefore $\mathsf{e}_\infty^2 = \mathsf{e}_o^2 = 0$ and $\mathsf{e}_\infty \cdot \mathsf{e}_o = -1$. We use the null basis $\{\mathsf{e}_\infty, \mathsf{e}_o\}$ instead of the Minkowski basis $\{\mathsf{e}_+, \mathsf{e}_-\}$ since e_∞ and e_o have a clear semantic meaning as the point at infinity (there is only one) and the origin, respectively. We can now ask which Euclidean vectors $\mathbf{y} \in \mathbb{E}^3$ when embedded in conformal space, lie in the inner product null space of $\alpha\mathbf{X}$, with $\alpha \in \mathbb{R}$. Since we know that the embedding of $\mathbf{y}$ in conformal space is $\mathbf{Y} = \mathbf{y} + \frac{1}{2}\mathbf{y}^2\,\mathsf{e}_\infty + \mathsf{e}_o$, the question becomes for which $\mathbf{y}$ the inner product of $\mathbf{Y}$ and $\alpha\mathbf{X}$ becomes zero. We find that $\mathbf{Y} \cdot (\alpha\,\mathbf{X}) = \alpha\,(\mathbf{Y} \cdot \mathbf{X}) = \alpha\,(-\frac{1}{2}\,(\mathbf{y} - \mathbf{x})^2)$, which is clearly zero if and only if $\mathbf{y} = \mathbf{x}$. Similarly, we can ask what a vector of the form $\mathbf{S} = \mathbf{X} - \frac{1}{2}\,\rho^2\,\mathsf{e}_\infty$, with $\rho \in \mathbb{R}$, represents, where $\mathbf{X}$ is the same as above. We find that $\mathbf{Y}\cdot\mathbf{S} = -\frac{1}{2}\,(\mathbf{y}-\mathbf{x})^2 + \frac{1}{2}\,\rho^2$, which is zero if and only if $(\mathbf{x} - \mathbf{a})^2 = \rho^2$. Note that if $\mathbf{x}$ and $\mathbf{y}$ are elements of $\mathbb{E}^2$, $\mathbf{Y}\cdot\mathbf{S}$ is equivalent to this equation. That is, a vector of the form of $\mathbf{S}$ in $\mathbb{PK}^2$ represents a circle in $\mathbb{E}^2$ centered on $\mathbf{x}$ with radius ρ. In $\mathbb{PK}^3$, $\mathbf{S}$ represents a sphere centered on $\mathbf{x}$ with radius ρ and in even higher dimensional spaces it would represent a hypersphere. This shows that it is possible to represent circles and spheres in a linear manner in conformal space, which is of course due to the non-linear embedding of Euclidean vectors.

In fact, the embedding of a vector and the metric from equation (6) are equivalent to the embedding of Euclidean vectors in $\mathbb{PK}^2$ and the metric of the basis $\{\mathsf{e}_1, \mathsf{e}_2, \mathsf{e}_\infty, \mathsf{e}_o\}$ of $\mathbb{PK}^2$.

Since equation (5) holds in any Clifford algebra, it is also valid for $\mathcal{C}\ell(\mathbb{PK}^3)$. Given two vectors $\mathbf{S}_1, \mathbf{S}_2 \in \mathbb{PK}^3$ both representing spheres in $\mathbb{E}^3$, their outer product $\mathbf{S}_1 \wedge \mathbf{S}_2$ represents the intersection circle of the spheres with respect to the inner product null space of the bivector. That is, we can also represent circles in $\mathbb{E}^3$ in a linear manner in conformal space $\mathbb{PK}^3$.

While a circle is represented in the inner product null space by the intersection of two spheres, it may be shown that in terms of the outer product null space a circle through three points $\mathbf{x}, \mathbf{y}, \mathbf{z} \in \mathbb{E}^3$ can be represented by the outer product of the three corresponding conformal vectors $\mathbf{X}$, $\mathbf{Y}$ and $\mathbf{Z}$. For $\mathbb{E}^2$ this is then equivalent to the circle equation above, albeit not the same since $\mathbf{X}\wedge\mathbf{Y}\wedge\mathbf{Z}$ results in a trivector and not in a vector. Furthermore, four points $\mathbf{X}_1, \mathbf{X}_2, \mathbf{X}_3, \mathbf{X}_4 \in \mathbb{PK}^2$ are co-circular if $\mathbf{X}_1 \wedge \mathbf{X}_2 \wedge \mathbf{X}_3 \wedge \mathbf{X}_4 = 0$.

As it turns out, within the Clifford algebra over conformal space, the only geometric entity that can be represented is a sphere, albeit in any dimension and with any radius. For example, a sphere with infinite radius,

i. e. a plane, can be represented with finite components. A point, on the other hand, is a sphere with zero radius and a sphere in $\mathbb{E}^1$ is a point pair. The following list shows the geometric entities in $\mathbb{E}^3$ represented by blades of different grades in $\mathcal{C}\ell(\mathbb{PK}^3)$, in terms of their outer product null space. The $\{\mathbf{X}_i\} \subset \mathbb{PK}^3$ are assumed to be the conformal embeddings of Euclidean vectors $\{\mathbf{x}_i\} \subset \mathbb{E}^3$.

$$
\begin{array}{rl}
\mathbf{X}_1 & : \text{ Point } \mathbf{x}_1 \\
\mathbf{X}_1 \wedge \mathbf{X}_2 & : \text{ Point pair } (\mathbf{x}_1,\, \mathbf{x}_2) \\
\mathbf{X}_1 \wedge \mathsf{e}_\infty & : \text{ Point pair } (\mathbf{x}_1,\, \infty) \\
\mathbf{X}_1 \wedge \mathbf{X}_2 \wedge \mathbf{X}_3 & : \text{ Circle through } \mathbf{x}_1,\, \mathbf{x}_2,\, \mathbf{x}_3 \\
\mathbf{X}_1 \wedge \mathbf{X}_2 \wedge \mathsf{e}_\infty & : \text{ Line through } \mathbf{x}_1,\, \mathbf{x}_2 \\
\mathbf{X}_1 \wedge \mathbf{X}_2 \wedge \mathbf{X}_3 \wedge \mathbf{X}_4 & : \text{ Sphere through } \mathbf{x}_1,\, \mathbf{x}_2,\, \mathbf{x}_3, \mathbf{x}_4 \\
\mathbf{X}_1 \wedge \mathbf{X}_2 \wedge \mathbf{X}_3 \wedge \mathsf{e}_\infty & : \text{ Plane through } \mathbf{x}_1,\, \mathbf{x}_2,\, \mathbf{x}_3 \\
\mathbf{X}_1 \wedge \mathbf{X}_2 \wedge \mathbf{X}_3 \wedge \mathbf{X}_4 \wedge \mathbf{X}_5 & : \text{ The whole space } \mathbb{E}^3.
\end{array}
\tag{7}
$$

2.4. The Vector Space of Conic Sections

In conformal space we defined a particular embedding of Euclidean vectors in a higher dimensional vector space with particular properties. The Clifford algebra over this conformal vector space then allowed for the linear representation of circles, spheres, etc. A set of geometric entities of particular interest in computer vision are conic sections. It would therefore be advantageous to be able to form a Clifford algebra over a vector space such that conics and their intersections can be represented. This is indeed possible and may be done in the following way.

It is well known that given a symmetric 3×3 matrix A, the set of vectors $\mathsf{x} = (x, y, 1)^\mathsf{T}$ that satisfy $\mathsf{x}^\mathsf{T}\,\mathsf{A}\,\mathsf{x} = 0$, lie on a conic. This can also be written using the scalar product of matrices, denoted here by $*$, as $(\mathsf{x}\mathsf{x}^\mathsf{T}) * \mathsf{A} = 0$. It makes therefore sense to define a vector space of symmetric matrices in the following way. If a_{ij} denotes the component of matrix A at row i and column j, we can define a transformation $\mathcal{T}$ that maps elements of $\mathbb{R}^{3\times3}$ to $\mathbb{R}^6$ as

$$\mathcal{T} : \mathsf{A} \in \mathbb{R}^{3\times3} \mapsto \left(a_{13},\, a_{23},\, \tfrac{1}{\sqrt{2}}\, a_{33},\, \tfrac{1}{\sqrt{2}}\, a_{11},\, \tfrac{1}{\sqrt{2}}\, a_{22},\, a_{12}\right)^\mathsf{T} \in \mathbb{R}^6. \tag{8}$$

A vector $\mathsf{x} \in \mathbb{R}^3$ may now be embedded in the same six dimensional space via $\mathbf{x} := \mathcal{T}(\mathsf{x}\mathsf{x}^\mathsf{T})$. If we define $\mathbf{a} := \mathcal{T}(\mathsf{A})$, then $\mathsf{x}^\mathsf{T}\,\mathsf{A}\,\mathsf{x} = 0$ can be written as the scalar product

$$\mathbf{x} \cdot \mathbf{a} = 0 \iff x^2\, a_{11} + y^2\, a_{22} + 2xy\, a_{12} + 2x\, a_{13} + 2y\, a_{23} + a_{33} = 0. \tag{9}$$

Finding the vector **a** that best satisfies the above equation for a set of points is usually called the algebraic estimation of a conic, see e.g. (Bookstein, 1979).

We will denote the 6D-vector space in which 2D-conics may be represented by $\mathbb{D}^2 \equiv \mathbb{R}^6$. A 2D-vector $(x, y) \in \mathbb{R}^2$ is transformed to $\mathbb{D}^2$ by the function

$$\mathcal{D} : \quad (x, y) \in \mathbb{R}^2 \;\mapsto\; (x,\, y,\, \tfrac{1}{\sqrt{2}},\, \tfrac{1}{\sqrt{2}}\, x^2,\, \tfrac{1}{\sqrt{2}}\, y^2,\, xy) \in \mathbb{D}^2. \tag{10}$$

The Clifford Algebra $\mathcal{C}\ell(\mathbb{D}^2)$ has (algebra) dimension $2^6 = 64$. The inner product null space of a vector $\mathbf{A} \in \mathbb{D}^2$ is the set of all those vectors $\mathbf{X} \in \mathbb{D}^2$ that satisfy $\mathbf{X} \cdot \mathbf{A} = 0$. As was shown before, this null space is a (possibly degenerate) conic. Furthermore, the inner product null space of the outer product of two vectors $\mathbf{A}, \mathbf{B} \in \mathbb{D}^2$, $\mathbf{A} \wedge \mathbf{B}$, now has to represent the intersection of the conics represented by $\mathbf{A}$ and $\mathbf{B}$. Let $\mathbf{x}_i \in \mathbb{R}^2$ and let $\mathbf{X}_i \in \mathbb{D}^2$ be defined by $\mathbf{X}_i = \mathcal{D}(\mathbf{x}_i)\, \forall\, i$. Then the outer product null space of blades in $\mathcal{C}\ell(\mathbb{D}^2)$ may be shown to represent the following objects.

$$\begin{aligned} \mathbf{X}_1 &: \text{ Point } \mathbf{x}_1 \\ \mathbf{X}_1 \wedge \mathbf{X}_2 &: \text{ Point pair } (\mathbf{x}_1,\, \mathbf{x}_2) \\ \mathbf{X}_1 \wedge \mathbf{X}_2 \wedge \mathbf{X}_3 &: \text{ Point triplet } (\mathbf{x}_1,\, \mathbf{x}_2,\, \mathbf{x}_3) \\ \mathbf{X}_1 \wedge \mathbf{X}_2 \wedge \mathbf{X}_3 \wedge \mathbf{X}_4 &: \text{ Point quadruplet } (\mathbf{x}_1,\, \mathbf{x}_2,\, \mathbf{x}_3, \mathbf{x}_4) \\ \mathbf{X}_1 \wedge \mathbf{X}_2 \wedge \mathbf{X}_3 \wedge \mathbf{X}_4 \wedge \mathbf{X}_5 &: \text{ The conic through } \mathbf{x}_1,\, \mathbf{x}_2,\, \mathbf{x}_3,\, \mathbf{x}_4,\, \mathbf{x}_5. \end{aligned} \tag{11}$$

In particular, it can be shown that the outer product null space of $\mathbf{X}_1 \wedge \mathbf{X}_2 \wedge \mathbf{X}_3 \wedge \mathbf{X}_4 \wedge \mathbf{X}_5$ is the same as the inner product null space of its dual, which is a vector. Hence, this is also a simple way to construct the symmetric matrix that represents a conic through five points. Note that to the best of our knowledge the Clifford algebra $\mathcal{C}\ell(\mathbb{D}^2)$ has not yet been discussed in the literature. We believe that it offers an intuitive way to deal with 2D-conics and warrants further investigation.

2.5. Error Propagation in Clifford Algebra

It was shown previously that operations like the geometric, inner and outer product in Clifford algebra are basically bilinear functions. This implies that standard error propagation methods (cf. e. g. (Koch, 1997)) can be applied in the evaluation of these products. Therefore, we can, for example, evaluate the mean circle through three points, given the three points with corresponding covariance and cross-covariance matrices in conformal space. The same could be done, given two spheres with appropriate covariance and cross-covariance matrices. Before the details of such calculations are presented, error propagation in Clifford algebra is introduced from a somewhat more general point of view.

Let $\{E_i\}_{i=1}^{2^n}$ denote again the algebra basis of $\mathcal{C}\ell(\mathbb{R}^n)$. Given three multivectors $A, B, M \in \mathcal{C}\ell(\mathbb{R}^n)$, with $A = \alpha^i E_i$, $B = \beta^i E_i$ and $M = \mu^i E_i$, we may regard them as vectors in some $\mathbb{R}^m$, with orthonormal basis $\{\mathsf{e}_i\}_{i=1}^m$, where $m = 2^n$. In this vector space the multivectors may be written as column vectors $\mathsf{a} = [\alpha^1, \ldots, \alpha^m]^\mathsf{T}$, $\mathsf{b} = [\beta^1, \ldots, \beta^m]^\mathsf{T}$ and $\mathsf{m} = [\mu^1, \ldots, \mu^m]^\mathsf{T}$, respectively. We use here sans serif letters to denote vectors in $\mathbb{R}^m$ in order to distinguish them from (multi-)vectors in $\mathcal{C}\ell(\mathbb{R}^n)$. The relation between multivectors in $\mathcal{C}\ell(\mathbb{R}^n)$ and their representation in $\mathbb{R}^m$ may be regarded as an isomorphism Φ between these two spaces, whereby $\Phi(A \in \mathcal{C}\ell_n) = \mathsf{a} \in \mathbb{R}^m$ and $\Phi^{-1}(\mathsf{a}) = A$. This isomorphism also transforms Clifford algebra products to matrix products with special matrices. For example, if $M = A \wedge B$ then

$$\mathsf{m} = \Phi(M) = \Phi(A \wedge B) = \mathsf{U}(\Phi(A))\,\Phi(B) = \mathsf{U}(\mathsf{a})\,\mathsf{b},$$

where $\mathsf{U}(\mathsf{a})$ is a matrix whose entries depend on a. In the following all matrices will be written as capital sans-serif letters. The form of matrix U is derived through the following considerations. A product in $\mathcal{C}\ell(\mathbb{R}^n)$ between two multivectors can be expressed as a bilinear function g which is a map $\mathbb{R}^m \times \mathbb{R}^m \to \mathbb{R}^m$ and may be written as $\mathsf{g}(\mathsf{a}, \mathsf{b}) := \alpha^i \beta^j g^k{}_{ij}\, \mathsf{e}_k$, where again we have implicit sums over i, j and k. The object $g^k{}_{ij}$ is again the 3-valence tensor from equation (1). It encodes the relation between the basis blades of $\mathcal{C}\ell_n$ for a particular product. For example, if $g^k{}_{ij}$ encodes the outer product, then the equation $M = A \wedge B$ may be written in $\mathbb{R}^m$ as

$$\mathsf{m} = \mathsf{g}(\mathsf{a}, \mathsf{b}) \iff \mu^k = \alpha^i \beta^j g^k{}_{ij}\ \forall k. \tag{12}$$

If we now denote the matrix of derivatives of $\mathsf{g}(\mathsf{a}, \mathsf{b})$ with respect to the $\{\beta^j\}$ as $\mathsf{U}(\mathsf{a})$, and with respect to the $\{\alpha^i\}$ as $\mathsf{V}(\mathsf{b})$, we can write $M = A \wedge B$ equivalently in $\mathbb{R}^m$ as

$$\mathsf{m} = \mathsf{U}(\mathsf{a})\,\mathsf{b} = \mathsf{V}(\mathsf{b})\,\mathsf{a}. \tag{13}$$

Note that U and V are basically the Jacobi matrices of g.

2.5.1. ERROR PROPAGATION

Suppose now that multivectors A and B cannot be known exactly. Instead only their expectation value, covariance and cross-covariance matrices are known. The question is then how general Clifford algebra operations can be performed while propagating the covariances of the initial multivectors.

In the following we will denote random variables by underlining the variable name. That is, $\underline{A}$ and $\underline{B}$ denote two random multivector variables with an embedding in $\mathbb{R}^m$ as $\Phi(\underline{A}) = \underline{\mathsf{a}} = [\underline{\alpha}^i, \ldots, \underline{\alpha}^m]^\mathsf{T}$ and $\Phi(\underline{B}) = \underline{\mathsf{b}} = [\underline{\beta}^i, \ldots, \underline{\beta}^m]^\mathsf{T}$. The expectation value of a random variable will be denoted by overlining the variable name and the expectation value operator will be denoted by $\mathcal{E}$. The covariance matrix of a and b will be denoted by $\Sigma_{\mathsf{a},\mathsf{b}}$. Given the expectation values $\bar{\mathsf{a}}$, $\bar{\mathsf{b}}$, the covariance matrices $\Sigma_{\mathsf{a},\mathsf{a}}$, $\Sigma_{\mathsf{b},\mathsf{b}}$, and the cross-covariance $\Sigma_{\mathsf{a},\mathsf{b}}$ of $\underline{\mathsf{a}}$ and $\underline{\mathsf{b}}$, we ask what the expectation and covariance matrix of a bilinear function $\mathsf{g}(\underline{\mathsf{a}}, \underline{\mathsf{b}})$ as defined in equation (12) is. By expanding $\mathsf{g}(\underline{\mathsf{a}}, \underline{\mathsf{b}})$ with a Taylor expansion about the expectation values of $\underline{\mathsf{a}}$ and $\underline{\mathsf{b}}$, we find that

$$\bar{\mathsf{m}} = \mathcal{E}[\mathsf{g}(\underline{\mathsf{a}}, \underline{\mathsf{b}})] = \mathsf{U}(\bar{\mathsf{a}})\,\bar{\mathsf{b}} + \mathrm{tr}((\mathsf{H}^k)^\mathsf{T}\,\Sigma_{\mathsf{a},\mathsf{b}})\mathsf{e}_k, \tag{14}$$

where H^k is the Hessian matrix of the kth component of g, and $\mathrm{tr}(\mathsf{U})$ denotes the trace of a matrix U. In this case the Hessian matrix is simply $\mathsf{H}^k = {g^k}_{ij}$. Note that in most cases the term containing the Hessian matrix will be negligible. By using the same Taylor expansion of g as before, it may be shown that the covariance matrix of $\mathsf{g}(\underline{\mathsf{a}}, \underline{\mathsf{b}})$ is approximately given by

$$\begin{aligned}\Sigma_{\mathsf{m},\mathsf{m}} = \quad & \mathsf{V}(\bar{\mathsf{b}})\,\Sigma_{\mathsf{a},\mathsf{a},}\,\mathsf{V}^\mathsf{T}(\bar{\mathsf{b}}) + \mathsf{U}(\bar{\mathsf{a}})\,\Sigma_{\mathsf{b},\mathsf{b},}\,\mathsf{U}^\mathsf{T}(\bar{\mathsf{a}}) \\ + \; & \mathsf{V}(\bar{\mathsf{b}})\,\Sigma_{\mathsf{a},\mathsf{b},}\,\mathsf{U}^\mathsf{T}(\bar{\mathsf{a}}) + \mathsf{U}(\bar{\mathsf{a}})\,\Sigma_{\mathsf{b},\mathsf{a},}\,\mathsf{V}^\mathsf{T}(\bar{\mathsf{b}}),\end{aligned} \tag{15}$$

where we neglected an additional term $\mathrm{tr}((\mathsf{H}^r)^\mathsf{T}\;\Sigma_{\mathsf{a},\mathsf{b}})\,\mathrm{tr}((\mathsf{H}^s)^\mathsf{T}\;\Sigma_{\mathsf{a},\mathsf{b}})$ for each element $\Sigma^{rs}_{\mathsf{m},\mathsf{m}}$. For most applications it may be assumed that this is a good approximation. Furthermore, the cross-covariance matrix of $\mathsf{g}(\underline{\mathsf{a}}, \underline{\mathsf{b}})$ and another random multivector variable $\underline{\mathsf{c}} \in \mathbb{R}^m$ is given by

$$\Sigma_{\mathsf{m},\mathsf{c}} = \mathsf{U}(\bar{\mathsf{b}})\,\Sigma_{\mathsf{a},\mathsf{c}} + \mathsf{V}(\bar{\mathsf{a}})\,\Sigma_{\mathsf{b},\mathsf{c}}. \tag{16}$$

Note that in the previous two equations the matrices U and V are the Jacobean matrices of the bilinear function g. Equations (14), (15) and (16) complete describe error propagation for any combination of Clifford algebra operations.

2.5.2. CONFORMAL SPACE

For any expression we want to obtain in the Clifford algebra of conformal space $\mathcal{C}\!\ell(\mathbb{PK}^3)$, we can now use the equations presented in the previous section to implement error propagation. Nevertheless, the initial expectation values and covariance matrices will typically only be given for vectors in Euclidean space $\mathbb{E}^3$ and not for the corresponding embedded vectors in $\mathbb{PK}^3$. We therefore first have to do the error propagation for the embedding of a Euclidean random vector variable $\underline{\mathbf{x}} \in \mathbb{E}^3$ into conformal space, where we will denote the corresponding conformal random vector variable by $\underline{\mathbf{X}} \in \mathbb{PK}^3$. Note that while $\underline{\mathbf{x}}$ is 3-dimensional, $\underline{\mathbf{X}}$ is $(3+2)$-dimensional, and therefore also the corresponding covariance matrices will be of different dimensions. We find that the expectation value of $\underline{\mathbf{X}}$ is given by

$$\bar{\mathbf{X}} = \mathcal{E}[\mathcal{K}(\underline{\mathbf{x}})] = \bar{\mathbf{x}} + \tfrac{1}{2}\,\bar{\mathbf{x}}^2\,\mathsf{e}_\infty + \mathsf{e}_o + \tfrac{1}{2}\,\mathrm{tr}(\Sigma_{\mathsf{x},\mathsf{x}})\,\mathsf{e}_\infty, \tag{17}$$

where $\mathcal{K}$ is the function describing the embedding of a Euclidean vector in conformal space. This result is obtained by calculating the expectation value of the Taylor expansion of $\mathcal{K}(\underline{\mathbf{x}})$ about $\bar{\mathbf{x}}$. The term $\mathrm{tr}(\Sigma_{\mathbf{x},\mathbf{x}})$ is typically very small and may be neglected. If we denote by $\mathsf{J}_{\mathcal{K}}(\bar{\mathbf{x}})$ the Jacobi matrix of $\mathcal{K}$ evaluated at $\bar{\mathbf{x}}$, then the covariance matrix $\Sigma_{\mathbf{X},\mathbf{X}}$ of $\underline{\mathbf{X}}$ is given in terms of the covariance matrix $\Sigma_{\mathbf{x},\mathbf{x}}$ of $\underline{\mathbf{x}}$ as

$$\Sigma_{\mathbf{X},\mathbf{X}} = \mathsf{J}_{\mathcal{K}}(\bar{\mathbf{x}})\, \Sigma_{\mathbf{x},\mathbf{x}}\, \mathsf{J}_{\mathcal{K}}^{\mathsf{T}}(\bar{\mathbf{x}}). \tag{18}$$

Denoting the components of $\bar{\mathbf{x}}$ by $\{\bar{\xi^i}\}$, the Jacobi matrix is in fact given by

$$\mathsf{J}_{\mathcal{K}}(\bar{\mathbf{x}}) = \begin{bmatrix} 1 & 0 & 0 & \bar{\xi}^1 & 0 \\ 0 & 1 & 0 & \bar{\xi}^2 & 0 \\ 0 & 0 & 1 & \bar{\xi}^3 & 0 \end{bmatrix}^{\mathsf{T}} . \tag{19}$$

The cross-covariance $\Sigma_{\mathbf{X},\mathbf{Y}}$ is simply given in terms of $\Sigma_{\mathbf{x},\mathbf{y}}$ as

$$\Sigma_{\mathbf{X},\mathbf{Y}} = \mathsf{J}_{\mathcal{K}}(\bar{\mathbf{x}})\, \Sigma_{\mathbf{x},\mathbf{y}}\, \mathsf{J}_{\mathcal{K}}^{\mathsf{T}}(\bar{\mathbf{y}}). \tag{20}$$

2.5.3. EVALUATION OF CIRCLES

We mentioned earlier that in conformal space a circle may be represented by the outer product of three points, where a point is represented by a vector as given in equation (6). The problem we now want to discuss is, given three points in Euclidean space with associated covariance and cross-covariance matrices, what is the expected circle through these three points and what is its covariance matrix.

In order to somewhat simplify the formulas in the following, when we write a (multi-)vector we mean in fact the expectation value of a corresponding random (multi-)vector variable. For example, we will simply write $\mathbf{x}$ instead of $\bar{\mathbf{x}}$.

Let $\bar{\mathbf{x}}, \bar{\mathbf{y}}, \bar{\mathbf{z}} \in \mathbb{E}^3$ denote the expectation of three Euclidean vectors. Their corresponding covariance and cross-covariance matrices are $\Sigma_{\mathbf{x},\mathbf{x}}$, $\Sigma_{\mathbf{y},\mathbf{y}}$, $\Sigma_{\mathbf{z},\mathbf{z}}$, and $\Sigma_{\mathbf{x},\mathbf{y}}$, $\Sigma_{\mathbf{y},\mathbf{z}}$, $\Sigma_{\mathbf{z},\mathbf{x}}$. In section 2.5.2 we have shown how these three Euclidean vectors together with their covariance and cross-covariance matrices may be embedded in conformal space. The corresponding conformal vectors will be denoted by $\bar{\mathbf{X}}$, $\bar{\mathbf{Y}}$, $\bar{\mathbf{Z}}$ and the corresponding covariance and cross-covariance matrices likewise. Once this is done, we can use equations (14) and (15) to first evaluate $\bar{\mathbf{P}} = \mathcal{E}[\bar{\mathbf{X}} \wedge \bar{\mathbf{Y}}]$ and the corresponding $\Sigma_{\mathbf{P},\mathbf{P}}$. Then we use equation (16) to evaluate $\Sigma_{\mathbf{P},\mathbf{Z}}$. This then enables us to calculate $\bar{\mathbf{C}} = \mathcal{E}[\bar{\mathbf{P}} \wedge \bar{\mathbf{Z}}]$ and $\Sigma_{\mathbf{C},\mathbf{C}}$. We could, of course, also have evaluated the Jacobians directly for the trilinear product $\bar{\mathbf{X}} \wedge \bar{\mathbf{Y}} \wedge \bar{\mathbf{Z}}$ and then found the expectation and covariance. The former method is however useful since it

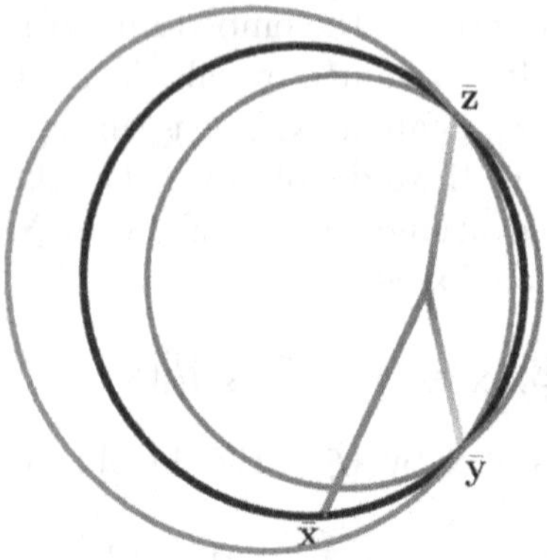

Figure 2.1. Standard deviation circles if vector $\bar{\mathbf{x}}$ has variance in only one component.

applies to any combination of products. Note that the statistical relation between the components of the trivector $\bar{\mathbf{C}}$ are linear, while the relation between the actual radius, center and normal of the circle need not be. That is, due to the conformal embedding not only the representation of a circle is linearized but also the statistical relationship between its embedded components.

This allows us to very easily evaluate the standard deviation circles of the mean circle $\bar{\mathbf{C}}$. We do this by evaluating a singular value decomposition (SVD) on $\Sigma_{\mathbf{C},\mathbf{C}}$. The singular vectors that correspond to non-zero singular values give the principal components of $\Sigma_{\mathbf{C},\mathbf{C}}$, while the singular values give the variances along them. If $\bar{\mathbf{C}}$ were a point, the principal components would give the axes of an ellipsoid which represents the surface of standard deviation about this point. In the present case, where $\bar{\mathbf{C}}$ represents a circle, we have to draw for each point on the ellipsoid a circle. Hence, if $\Sigma_{\mathbf{C},\mathbf{C}}$ only has one principal component, we obtain two standard deviation circles as shown in figure 2.1. Here points $\bar{\mathbf{y}}$ and $\bar{\mathbf{z}}$ were held fixed and only point $\bar{\mathbf{x}}$ was taken to have a variance along one dimension. The central black circle is the mean circle which goes through all three points. The two gray circles are the ones that will occur with a likelihood of $\exp(-\frac{1}{2})$, i. e. they give the standard deviation from the mean.

If we now only hold point $\bar{\mathbf{z}}$ fixed and assume that $\bar{\mathbf{x}}$ and $\bar{\mathbf{y}}$ each have a variance in one dimension, then $\Sigma_{\mathbf{C},\mathbf{C}}$ has two principal components that give the axes of an ellipse. If we draw for each point on the ellipse one circle, we obtain the surface shown in figure 2.2. That is, each circle on the surface has a probability of $\exp(-\frac{1}{2})$ to occur. How the actual circle parameter may be extracted from a trivector $\mathbf{C} \in \mathcal{C}\ell(\mathbb{PK}^3)$ that represents it, may be found in some detail, for example, in (Li et al., 2001). Only a

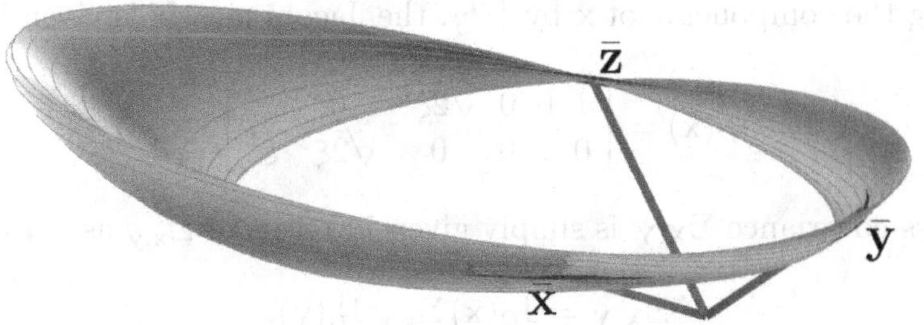

Figure 2.2. Standard deviation surface if vectors $\bar{\mathbf{x}}$ and $\bar{\mathbf{y}}$ have each a variance in only one component.

short overview will be given here.

First of all, evaluate $\mathbf{L} = \mathbf{C} \cdot \mathsf{e}_\infty$ and $\mathbf{A} = \mathbf{C} \wedge \mathsf{e}_\infty$. It turns out that $\mathbf{L}$ represents w.r.t. the inner product null space, the line through the center of the circle with direction perpendicular to the plane the circle lies in. $\mathbf{A}$ represents w.r.t. the outer product null space, the plane the circle lies in. The intersection of $\mathbf{A}$ and $\mathbf{L}$ may simply be evaluated by $\mathbf{P} = \mathbf{L} \cdot \mathbf{A}$, whence $\mathbf{P}$ is of the form $\mathbf{P} = \mathbf{X} \wedge \mathsf{e}_\infty$, if $\mathbf{X}$ gives the center of the circle. The normal of the plane the circle lies in is given by $\mathbf{N} = \mathbf{L} \cdot (\mathsf{e}_3 \wedge \mathsf{e}_2 \wedge \mathsf{e}_1)$. However, $\mathbf{N}$ still has to be normalized, since its magnitude is related to the radius of the circle. The radius r can simply be evaluated by $r^2 = -(\mathbf{C} \cdot \mathbf{C})/(\mathbf{A} \cdot \mathbf{A})$. In fact, $\mathbf{S} = \mathbf{C}/\mathbf{A}$ results in a vector of the same form as the vector representing a sphere in section 2.3. The center and radius of $\mathbf{S}$ are then the same as those of the circle $\mathbf{C}$. Note that error propagation can be applied to all of the above calculations, such that expectation values and covariance matrices are available for all of these properties.

2.5.4. EVALUATION OF CONICS

Constructing a conic from five uncertain points in $\mathbb{D}^2$ is very similar to constructing a circle from three uncertain points in conformal space $\mathbb{PK}^3$. We assume that we are given five points in $\mathbb{R}^2$, each with an associated covariance matrix. These are embedded in $\mathbb{D}^2$ using standard error propagation.

Let $\mathcal{D}$ again denote the function embedding vectors from $\mathbb{R}^2$ in $\mathbb{D}^2$. A random vector variable $\underline{\mathbf{x}} \in \mathbb{R}^2$ is embedded in $\mathbb{D}^2$ via $\underline{\mathbf{X}} = \mathcal{D}(\underline{\mathbf{x}})$. The expectation value of $\underline{\mathbf{X}}$ is then given by $\bar{\mathbf{X}} = \mathcal{E}[\mathcal{D}(\underline{\mathbf{x}})] \approx \mathcal{D}(\bar{\mathbf{x}})$. If we denote by $\mathsf{J}_\mathcal{D}(\bar{\mathbf{x}})$ the Jacobi matrix of $\mathcal{D}$ evaluated at $\bar{\mathbf{x}}$, then the covariance matrix $\Sigma_{\mathbf{X},\mathbf{X}}$ of $\underline{\mathbf{X}}$ is given in terms of the covariance matrix $\Sigma_{\mathbf{x},\mathbf{x}}$ of $\underline{\mathbf{x}}$ as

$$\Sigma_{\mathbf{X},\mathbf{X}} = \mathsf{J}_\mathcal{D}(\bar{\mathbf{x}})\, \Sigma_{\mathbf{x},\mathbf{x}}\, \mathsf{J}_\mathcal{D}^\mathsf{T}(\bar{\mathbf{x}}). \tag{21}$$

Denoting the components of $\bar{\mathbf{x}}$ by $\{\bar{\xi}^i\}$, the Jacobi matrix is given by

$$\mathsf{J}_{\mathcal{D}}(\bar{\mathbf{x}}) = \begin{bmatrix} 1 & 0 & 0 & \sqrt{2}\bar{\xi}^1 & 0 & \bar{\xi}^2 \\ 0 & 1 & 0 & 0 & \sqrt{2}\bar{\xi}^2 & \bar{\xi}^1 \end{bmatrix}^{\mathsf{T}}. \tag{22}$$

The cross-covariance $\Sigma_{\mathbf{X},\mathbf{Y}}$ is simply given in terms of $\Sigma_{\mathbf{x},\mathbf{y}}$ as

$$\Sigma_{\mathbf{X},\mathbf{Y}} = \mathsf{J}_{\mathcal{D}}(\bar{\mathbf{x}})\,\Sigma_{\mathbf{x},\mathbf{y}}\,\mathsf{J}_{\mathcal{D}}^{\mathsf{T}}(\bar{\mathbf{y}}). \tag{23}$$

Figure 2.3 shows an example for such a construction. Given are five points, of which two have a non-zero covariance matrix indicated by small black bars. Taking the outer product of these five points after having them embedded in $\mathbb{D}^2$, we can evaluate the mean conic, represented as black conic, and also the covariance matrix of the conic. In this case the covariance matrix is of rank 2, which generates a whole set of conics that have probability $\exp(-\frac{1}{2})$ of a occurring, represented by the gray conics. It can be seen that the area swept by this set of "standard deviation conics" has a highly non-linear shape. Nevertheless, this surface is represented by the covariance matrix of the conic in $\mathbb{D}^2$.

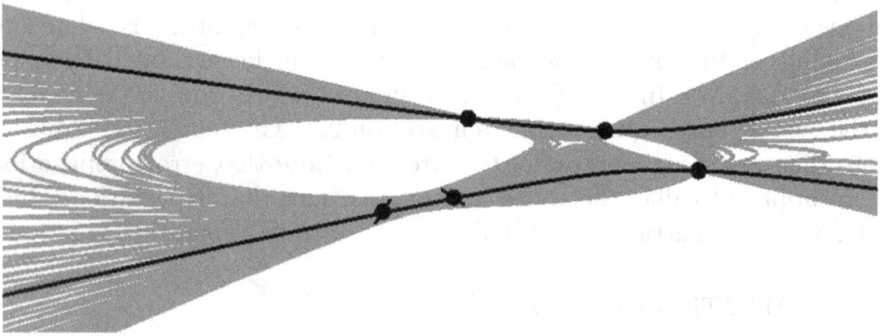

Figure 2.3. Standard deviation conics if two of the five points have rank 1 covariance matrices (indicated by small black bars).

2.6. Fitting of Circles and Conics to Data

There is also a linear solution to find the best circle that passes through a set of points in $\mathbb{E}^3$, or the best conic that passes through a set of points in $\mathbb{E}^2$, in a least squares sense. This follows directly from equation (13). In both cases the entities we would like to evaluate can be calculated from a set of linear constraint equations. In conformal space we can in this way extend the method given, for example, in (Delonge, 1972; Bookstein, 1979) for fitting circles in 2D-Euclidean space, to 3D-Euclidean space.

In the space of conics $\mathbb{D}^2$, it is not only possible to fit conics but also the intersection of conics to data by solving a linear system of equations. For example, if the data consists of four clusters of points, then fitting the intersection of two conics to the data will return a point quadruplet whose points specify the centers of the clusters.

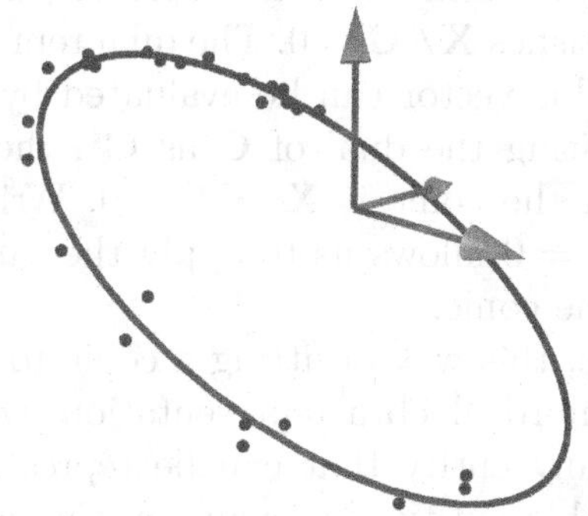

Figure 2.4. Result of linear fitting of a circle to a set of slightly scattered points in $\mathbb{E}^3$.

2.6.1. FITTING IN CONFORMAL SPACE

Here is a short description of how a circle may be fitted to a set of 3D-points using the Clifford algebra of conformal space. It was mentioned earlier that if a point $\mathbf{X}$ lies on a circle $\mathbf{C}$, then $\mathbf{X} \wedge \mathbf{C} = 0$. If we write $\mathsf{c} = \Phi(\mathbf{C})$ and $\mathsf{x} = \Phi(\mathbf{X})$, then this condition can be written as $\mathsf{U}(\mathsf{x})\,\mathsf{c} = \mathbf{0}$. That is, c lies in the null space of the matrix $\mathsf{U}(\mathsf{x})$. Given a set of points $\{\mathsf{x}_1, \ldots, \mathsf{x}_k\}$ that all have to lie on a circle, we can define a matrix W that contains the set of matrices $\{\mathsf{U}(\mathsf{x}_1), \ldots, \mathsf{U}(\mathsf{x}_k)\}$ stacked on top of each other. The condition a circle passing through all these points then has to satisfy becomes $\mathsf{W}\mathsf{c} = \mathbf{0}$. We could now simply find the null space of W using a SVD. However, this would give the subspace of multivectors and not trivectors that satisfy the constraint. Therefore, we first remove those columns from W that are not related to trivector components and only then find the null space. Since a SVD gives the best solution in a least squares sense, we should obtain a fairly good solution for the best circle fit, even though we have not taken the covariance matrix of the $\{\mathsf{x}_i\}$ into account. As discussed in (Chernov and Lesort, 2002), this simple method is therefore only likely to supply a good initial guess for an iterative algorithm (Gander et al., 1994). Figure 2.4 shows an example of a circle fitted to a set of artificially generated noisy

3D-points using this method. Of course, any linear regression method, as for example the Gauss-Helmert model may be applied here.

2.6.2. FITTING IN THE SPACE OF CONICS

In order to fit a conic to a set of points, the same method as above can be used, only this time in $\mathcal{C}\ell(\mathbb{D}^2)$. A conic $\mathbf{C} \in \mathcal{C}\ell(\mathbb{D}^2)$ is represented by the outer product of five points, and any point represented by a vector $\mathbf{X} \in \mathbb{D}^2$ that lies on the conic satisfies $\mathbf{X} \wedge \mathbf{C} = 0$. The dual representation of a conic in $\mathcal{C}\ell(\mathbb{D}^2)$ is a vector. This vector can be evaluated by the dual operation in Clifford algebra. Writing the dual of $\mathbf{C}$ as $\mathbf{C}^*$, the constraint a point satisfies when it lies on the conic is $\mathbf{X} \cdot \mathbf{C}^* = 0$. Writing this constraint again as $\mathsf{U}(\Phi(\mathbf{X}))\,\Phi(\mathbf{C}) = 0$ allows us to apply the same method we used for circles to evaluate the conic.

As mentioned before, this way of fitting a conic to data is well known. However, using the Clifford algebra representation, we can use the same linear approach to fit any entity that can be represented in the algebra to any other representable entity. For example, the outer product of two vectors $\mathbf{X}, \mathbf{Y} \in \mathbb{D}^2$ representing points in $\mathbb{R}^2$, represents this pair of points. Hence, we can also fit a conic to point pairs. Maybe more interestingly, we can also fit point pairs to a set of data points. Since the outer product of four vectors in $\mathcal{C}\ell(\mathbb{D}^2)$ represents a point quadruplet, it is also possible to fit a point quadruplet to a set of points. An example of this is shown in figure 2.5. In $\mathcal{C}\ell(\mathbb{D}^2)$ a point quadruplet can also be regarded as the intersection of two conics, since for every point quadruplet there exists a whole pencil of conics who all intersect in the same four points. For better visualization two conics of such a pencil are drawn in figure 2.5. As can be seen, the two conics intersect more or less in the centers of the four clusters. Hence, $\mathcal{C}\ell(\mathbb{D}^2)$ offers a simple, linear method to find the centers of up to four clusters in a set of data points.

Figure 2.6 shows the result of fitting point quadruplets to line segment structures, which are of particular interest in computer vision problems. It can be seen that the two conics drawn in each example intersect on the line structures in such a way that the structures may be further analyzed. This offers a method to distinguish between junctions and corners in images and also to evaluate the opening angle and orientation of corners. A detailed description of an algorithm based on this type of conic fitting can be found in (Perwass, 2004).

The Gauss-Helmert model (cf. e. g. (Koch, 1997)) is a linear model for a least-squares fitting of parameters to uncertain data, given constraints between the parameters and the data and constraints on the parameters alone. A Gaussian distribution of each data point is assumed, such that data

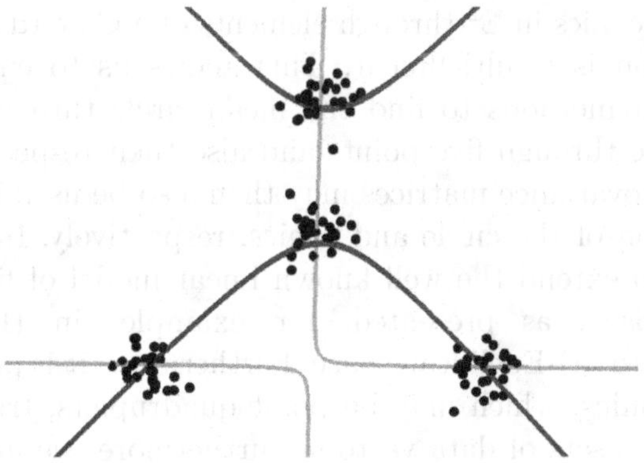

Figure 2.5. Result of linear fitting of a point quadruplet represented as the intersection of two conics to a set of scattered points in $\mathbb{E}^2$.

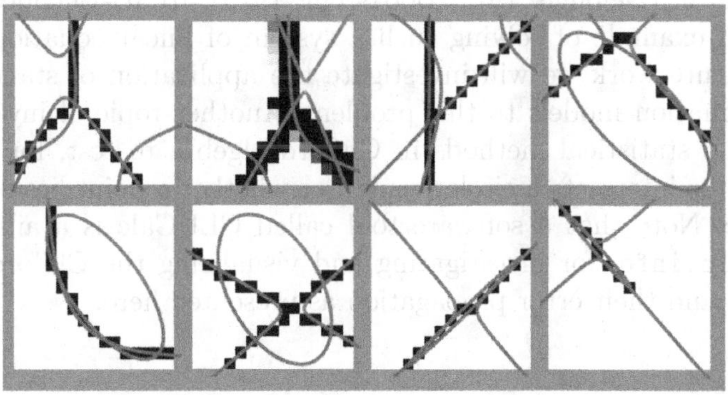

Figure 2.6. Fitting point quadruplets (represented as intersections of two conics) to line segment structures.

points are fully described by their mean value and covariance (matrix). We already saw in section 2.2 that operators in a Clifford algebra may be represented by bilinear function. In section 2.5 it was then shown how this fact may be used to apply error propagation to Clifford algebra. The representation of Clifford algebra products as bilinear functions using the Φ-isomorphism also allows us to apply linear regression models like the Gauss-Helmert model to Clifford algebra.

2.7. Conclusions

In this text we presented a method of constructing circles in 3D-Euclidean space and conics in 2D-Euclidean space from a number of uncertain points using error propagation methods. The main advantage of representing circles in $\mathbb{E}^3$ and conics in $\mathbb{E}^2$ through elements of a Clifford algebra, is that this representation is (multi-)linear. This allows us to employ standard error propagation methods to find the mean circle through three points or the mean conic through five points and also their respective covariance matrices. These covariance matrices may then also be used to visualize the standard deviation of the circle and conics, respectively. In this setting it is also possible to extend the well known linear model of fitting circles in 2D-Euclidean space, as presented, for example, in (Delonge, 1972; Bookstein, 1979), to 3D-Euclidean space. Furthermore, it is possible to fit the intersection of conics, which may be point quadruplets, triplets, doublets or single points, to sets of data vectors. Furthermore, we have shown that the constraint equations for fitting a circle to a set of scattered points in $\mathbb{E}^3$ can be given by a set of linear equations. This extends the well known linear model of fitting circles in 2D-Euclidean space, as presented, for example, in (Delonge, 1972; Bookstein, 1979), to 3D-Euclidean space. We gave an example of solving such a system of linear equations with a SVD. In future work we will investigate the application of standard statistical estimation models to this problem. Another topic of investigation is to develop statistical methods in Clifford algebra to test, for example, whether a line intersects a circle (conic), or whether a point lies on a circle (conic), etc. Note that a software tool called CLUCalc is available from `www.clucalc.info`, for investigating and visualizing the Clifford algebra expressions and their error propagation as presented here.

References

Bookstein, F.: Fitting conic sections to scattered data. *Computer Graphics Image Processing* **9**:56–71, 1979.

Chernov, N. and C. Lesort: Least squares fitting of circles and lines. *J. Mathematical Imaging Vision (submitted)*, 2002 (preprint available at `http://www.math.uab.edu/cl/cl1/`).

Collins, R. T.: Model acquisition using stochastic projective geometry. Ph.D. thesis, Dept. of Computer Science, University of Massachusetts, 1993.

Criminisi, A.: *Accurate Visual Metrology from Single and Multiple Uncalibrated Images.* Springer, Berlin, 2001.

Delonge, P.: Computer optimization of Deschamps' method and error cancellation in reflectometry. In Proc. *IMEKO-Symp. Microwave Measurement*, pages 117–123, 1972.

Dorst, L.: Honing geometric algebra for its use in the computer sciences. In *Geometric Computing with Clifford Algebra* (G. Sommer, editor), pages 127–151, Springer, Berlin, 2001.

Faugeras, O. and Q.-T. Luong: *The Geometry of Multiple Images.* MIT Press, Cambridge, MA, 2001.

Faugeras, O. and T. Papadopoulo: Grassmann-Cayley algebra for modelling systems of cameras and the algebraic equations of the manifold of trifocal tensors. *Phil. Trans. R. Soc. Lond. A*, **356**:1123–1152, 1998.

Förstner, W., A. Brunn, and S. Heuel: Statistically testing uncertain geometric relations. In *Mustererkennung 2000* (G. Sommer, N. Krüger, and C. Perwass, editors), pages 17–26, Springer, Berlin, 2000.

Gander, W., G. H. Golub, and R. Strebel: Fitting of circles and ellipses, least square solution. TR-217, Dept. of Computer Science, ETH Zürich, 1994.

Hestenes, D.: The design of linear algebra and geometry. *Acta Applicandae Mathematicae*, **23**:65–93, 1991.

Hestenes, D. and G. Sobczyk: *Clifford Algebra to Geometric Calculus: A Unified Language for Mathematics and Physics.* Kluwer, Dordrecht, 1984.

Hestenes, D. and R. Ziegler: Projective geometry with Clifford algebra. *Acta Applicandae Mathematicae*, **23**:25–63, 1991.

Heuel, S.: *Uncertain Projective Geometry*, LNCS 3008, Springer, Berlin, 2004.

Kanatani, K.: *Statistical Optimization for Geometric Computation: Theory and Practice.* Elsevier Science, Amsterdam, 1996.

Koch, K.-R.: *Parameter Estimation and Hypothesis Testing in Linear Models.* Springer, Berlin, 1997.

Li, H., D. Hestenes, and A. Rockwood: Generalized homogeneous coordinates for computational geometry. In *Geometric Computing with Clifford Algebra* (G. Sommer, editor), pages 27–59, Springer, Berlin, 2001.

Lounesto, P.: *Clifford Algebras and Spinors.* Cambridge University Press, Cambridge, 1997.

Needham, T.: *Visual Complex Analysis.* Oxford University Press, Oxford, 1997.

Perwass, C.: Analysis of local image structure using intersections of conics. TR 0403, Kiel University, Institut für Informatik und Praktische Mathematik, 2004.

Perwass, C. and D. Hildenbrand: Aspects of geometric algebra in Euclidean, projective and conformal space. TR 0310, Kiel University, Institut für Informatik und Praktische Mathematik, 2003.

Porteous, I. R.: *Clifford Algebras and the Classical Groups.* Cambridge University Press, Cambridge, 1995.

White, N. L.: A tutorial on Grassmann–Cayley algebra. In *Invariant Methods in Discrete and Computational Geometry* (N. L. White, editor), pages 93–106, Kluwer, Dordrecht, 1995.

ALGORITHMS FOR SPATIAL PYTHAGOREAN-HODOGRAPH CURVES

RIDA T. FAROUKI AND CHANG YONG HAN
Department of Mechanical and Aeronautical Engineering, University of California, Davis, CA 95616, USA.

Abstract. The quaternion representation for spatial Pythagorean-hodograph (PH) curves greatly facilitates the formulation of basic algorithms for their construction and manipulation, such as first-order Hermite interpolation, transformations between coordinate systems, and determination of rotation-minimizing frames. By virtue of their algebraic structures, PH curves offer unique computational advantages over "ordinary" polynomial curves in geometric design, graphics, path planning and motion control, computer vision, and similar applications. We survey some recent advances in theory, algorithms, and applications for spatial PH curves, and present new results on the unique determination of PH curves by the tangent indicatrix, and the use of generalized stereographic projection as a tool to obtain deeper insight into the basic structure and properties of spatial PH curves.

Key words: Pythagorean-hodograph curves, quaternion representation, rotation-minimizing frame, arc length, elastic energy, Hermite interpolation, tangent indicatrix

3.1. Introduction

Pythagorean-hodograph curves (Farouki and Sakkalis, 1990; Farouki, 2002) incorporate special algebraic structures that offer computational advantages in diverse application contexts, such as computer aided design, computer vision, computer graphics, robotics, and motion control. Planar PH curves are most conveniently expressed in terms of a complex variable model (Farouki, 1994), which facilitates key constructions (Albrecht and Farouki, 1996; Farouki, 1996; Farouki et al., 2001; Farouki and Neff, 1995; Jüttler, 2001; Moon et al., 2001). To achieve a necessary-and-sufficient characterization for the spatial PH curves, a quaternion model is required (Choi et al., 2002; Farouki et al., 2002). Our goal in this paper is to survey new algorithms that employ this representation (Choi and Han, 2002; Farouki, 2002; Farouki et al., 2002; Farouki and Han, 2003; Farouki et al., 2003; Jüttler

R. Klette et al. (eds.), Geometric Properties for Incomplete Data, 43-58.

and Mäurer, 1999); to describe new results on the unique correspondence between PH curves and tangent indicatrices; and to highlight important open problems concerning the theory, construction, and applications of spatial PH curves.

Among the key distinguishing features of any PH curve $\mathbf{r}(t)$—as distinct from an "ordinary" polynomial curve—we cite the following:

- The cumulative arc length $s(t)$ is a *polynomial* in the curve parameter t, and the total arc length can be computed *exactly* by rational arithmetic on the curve coefficients (Farouki, 1992).
- Integral shape measures, such as the *elastic energy*—the integral of the square of curvature—are amenable to exact evaluation (Farouki, 1996).
- PH curves admit *real-time interpolator algorithms* that allow computer numerical control (CNC) machines to accurately traverse curved paths with speeds dependent upon time, arc length, or curvature (Farouki et al., 1998; Farouki and Shah, 1996; Tsai et al., 2001).
- The *offsets* (or *parallels*) to any planar PH curve admit an exact rational parameterization—likewise for the "tubular" *canal surfaces* that have a given spatial PH curve as the "spine" curve (Farouki et al., 2002; Farouki and Sakkalis, 1990; Farouki and Sakkalis, 1994).
- An exact derivation of *rotation-minimizing frames* (which eliminate the "unnecessary" rotation of the Frenet frame in the curve normal plane) is possible for spatial PH curves (Farouki, 2002). These incur logarithmic terms—efficient rational approximations are available as an alternative (Farouki and Han, 2003).
- PH curves typically yield "fair" interpolants (with more even curvature distributions) to discrete data—as compared to "ordinary" polynomial splines or Hermite interpolants (Albrecht and Farouki, 1996; Farouki, 1996; Farouki and Neff, 1995; Farouki and Sakkalis, 1994; Moon et al., 2001).

Our plan for this paper is as follows. After reviewing basic properties of the quaternion formulation for spatial PH curves in Section 3.2, we briefly summarize in Section 3.3 the first-order Hermite interpolation problem using this form. Computation of rotation-minimizing frames on spatial PH curves is then discussed in Section 3.4. A remarkable property of PH curves is newly identified in Section 3.5: they are *uniquely determined* (modulo translation and uniform scaling) *by the tangent indicatrix*, the curve on the unit sphere describing the variation of the tangent vector. In Section 3.6 we discuss the *generalized stereographic projection* as a means to obtain deeper insight into the structure of the space of spatial PH curves. Throughout the paper we identify important open problems in the theory, algorithms, and applications of spatial PH curves that deserve further investigation. Finally, Section 3.7 summarizes the recent advances and makes some closing remarks.

3.2. Quaternion formulation of spatial PH curves

The defining characteristic of a Pythagorean-hodograph (PH) curve $\mathbf{r}(t)$ in $\mathbb{R}^n$ is the fact that the coordinate components of its derivative or *hodograph* $\mathbf{r}'(t)$ comprise a Pythagorean n-tuple of polynomials—i.e., the sum of their squares coincides with the perfect square of some polynomial $\sigma(t)$. Satisfaction of this condition requires the incorporation of a special algebraic structure in $\mathbf{r}'(t)$, dependent on the dimension n of the space[1] under consideration.

The polynomial $\sigma(t)$ defines the *parametric speed* of the curve $\mathbf{r}(t)$—i.e., the rate of change

$$\sigma = \frac{\mathrm{d}s}{\mathrm{d}t}$$

of its arc length s with respect to the curve parameter t. The fact that $\sigma(t)$ is a *polynomial* (rather than the square-root of a polynomial) in t is the source of the many advantageous properties of PH curves.

In the planar case ($n = 2$), a necessary-and-sufficient condition for $\mathbf{r}'(t) = (x'(t), y'(t))$ to be Pythagorean, with $\gcd(x'(t), y'(t)) = \text{constant}$, can be expressed (Farouki and Sakkalis, 1990) as

$$x'^2(t) + y'^2(t) = \sigma^2(t) \iff \begin{cases} x'(t) = u^2(t) - v^2(t) \\ y'(t) = 2\,u(t)v(t) \\ \sigma(t) = u^2(t) + v^2(t) \end{cases}$$

for some polynomials $u(t), v(t)$. The *complex-variable model* (Farouki, 1994) for planar PH curves succinctly embodies this condition: identifying the point (x, y) with the complex number $x + \mathrm{i}\,y$, the Pythagorean hodograph structure is ensured by writing $\mathbf{r}'(t) = \mathbf{w}^2(t)$ for any complex polynomial $\mathbf{w}(t) = u(t) + \mathrm{i}\,v(t)$ with $\gcd(u(t), v(t)) = \text{constant}$. The complex formulation simplifies many algorithms (Albrecht and Farouki, 1996; Farouki, 1996; Farouki et al., 2001; Farouki and Neff, 1995; Jüttler, 2001; Moon et al., 2001) for planar PH curves.

[1] PH curves have also been defined in the *Minkowski metric* of relativity theory (Choi et al., 2002; Moon, 1999): such "MPH curves" play a key role in reconstructing the boundary of a shape from its medial axis transform.

In the spatial case ($n = 3$), we need[2] *four* polynomials (Choi et al., 2002; Dietz et al., 1993) to characterize the Pythagorean nature of a hodograph $\mathbf{r}'(t) = (x'(t), y'(t), z'(t))$. Namely,

$$x'^2(t) + y'^2(t) + z'^2(t) = \sigma^2(t) \iff \begin{cases} x'(t) = u^2(t) + v^2(t) - p^2(t) - q^2(t) \\ y'(t) = 2\,[\,u(t)q(t) + v(t)p(t)\,] \\ z'(t) = 2\,[\,v(t)q(t) - u(t)p(t)\,] \\ \sigma(t) = u^2(t) + v^2(t) + p^2(t) + q^2(t) \end{cases}$$

for some polynomials $u(t)$, $v(t)$, $p(t)$, $q(t)$. The quaternion formulation for spatial Pythagorean hodographs, first introduced in (Choi et al., 2002), provides a very elegant and succinct embodiment of this structure.

Quaternions can be represented as pairs $\mathcal{A} = (a, \mathbf{a})$ and $\mathcal{B} = (b, \mathbf{b})$ where a, b are the *scalar parts* and $\mathbf{a} = a_x\mathbf{i} + a_y\mathbf{j} + a_z\mathbf{k}$, $\mathbf{b} = b_x\mathbf{i} + b_y\mathbf{j} + b_z\mathbf{k}$ are the *vector parts*. For brevity, we will often simply write a for the "pure scalar" quaternion $(a, \mathbf{0})$ and $\mathbf{a}$ for the "pure vector" quaternion $(0, \mathbf{a})$. The sum and product of $\mathcal{A}$, $\mathcal{B}$ are given by

$$\mathcal{A} + \mathcal{B} = (a + b, \mathbf{a} + \mathbf{b})\,, \quad \mathcal{A}\mathcal{B} = (ab - \mathbf{a}\cdot\mathbf{b}, a\,\mathbf{b} + b\,\mathbf{a} + \mathbf{a}\times\mathbf{b})\,.$$

Note that the product is non-commutative (i.e., $\mathcal{B}\mathcal{A} \neq \mathcal{A}\mathcal{B}$ in general).

Now if $\mathcal{A}(t) = u(t) + v(t)\mathbf{i} + p(t)\mathbf{j} + q(t)\mathbf{k}$ is a quaternion polynomial, and $\mathcal{A}^*(t) = u(t) - v(t)\mathbf{i} - p(t)\mathbf{j} - q(t)\mathbf{k}$ is its *conjugate*, the product

$$\begin{aligned} \mathbf{r}'(t) = \mathcal{A}(t)\,\mathbf{i}\,\mathcal{A}^*(t) \; &= \; [\,u^2(t) + v^2(t) - p^2(t) - q^2(t)\,]\,\mathbf{i} \\ &+ 2\,[\,u(t)q(t) + v(t)p(t)\,]\,\mathbf{j} \,+\, 2\,[\,v(t)q(t) - u(t)p(t)\,]\,\mathbf{k} \end{aligned} \tag{1}$$

generates the PH structure in $\mathbb{R}^3$ ($\mathbf{j}$ or $\mathbf{k}$ can be interposed between $\mathcal{A}(t)$, $\mathcal{A}^*(t)$ in place of $\mathbf{i}$, yielding a permutation of $u(t)$, $v(t)$, $p(t)$, $q(t)$). We may express (1) as $\mathbf{r}'(t) = |\mathcal{A}(t)|^2\,\mathcal{U}(t)\,\mathbf{i}\,\mathcal{U}^*(t)$, where $|\mathcal{A}(t)|^2 = \mathcal{A}(t)\mathcal{A}^*(t)$ and $\mathcal{U}(t) = (\cos\frac{1}{2}\theta(t), \sin\frac{1}{2}\theta(t)\,\mathbf{n}(t))$ defines a *unit quaternion*, expressed in terms of an angle $\theta(t)$ and a unit vector $\mathbf{n}(t)$. The product $\mathcal{U}(t)\,\mathbf{i}\,\mathcal{U}^*(t)$ defines a *spatial rotation* of the basis vector $\mathbf{i}$ by angle $\theta(t)$ about the axis vector $\mathbf{n}(t)$, while the factor $|\mathcal{A}(t)|^2$ imposes a *scaling* of this rotated vector. Thus, we can interpret the form (1) as generating a spatial hodograph through a continuous family of spatial rotations and scalings of the basis vector $\mathbf{i}$.

An important feature of the form (1) is its *structural invariance* (Farouki et al., 2002) under arbitrary spatial rotations of the coordinate system.[3]

---

[2] An earlier formulation (Farouki and Sakkalis, 1994) employing only three polynomials provides a sufficient, but not necessary, characterization of spatial Pythagorean hodographs (this form is not rotation-invariant).

[3] This is essential for a characterization of spatial Pythagorean hodographs to be sufficient and necessary, and distinguishes (1) from an earlier formulation (Farouki and Sakkalis, 1994), which is only sufficient.

Namely, when the coordinate system $\tilde{\mathbf{r}} = (\tilde{x}, \tilde{y}, \tilde{z})$ is obtained from $\mathbf{r} = (x, y, z)$ by a rotation through angle ϕ about the unit vector $\mathbf{n} = n_x\mathbf{i} + n_y\mathbf{j} + n_z\mathbf{k}$, the hodograph in the new coordinate system becomes $\tilde{\mathbf{r}}'(t) = \tilde{A}(t)\,\mathbf{i}\,\tilde{A}^*(t)$, where $\tilde{A}(t) = U\,A(t)$ with $U = (\cos\frac{1}{2}\phi, \sin\frac{1}{2}\phi\,\mathbf{n})$. The components $\tilde{u}$, $\tilde{v}$, $\tilde{p}$, $\tilde{q}$ of $\tilde{A}$ can be expressed in terms of those of A in matrix form as

$$\begin{bmatrix} \tilde{u} \\ \tilde{v} \\ \tilde{p} \\ \tilde{q} \end{bmatrix} = \begin{bmatrix} \cos\frac{1}{2}\phi & -n_x\sin\frac{1}{2}\phi & -n_y\sin\frac{1}{2}\phi & -n_z\sin\frac{1}{2}\phi \\ n_x\sin\frac{1}{2}\phi & \cos\frac{1}{2}\phi & -n_z\sin\frac{1}{2}\phi & n_y\sin\frac{1}{2}\phi \\ n_y\sin\frac{1}{2}\phi & n_z\sin\frac{1}{2}\phi & \cos\frac{1}{2}\phi & -n_x\sin\frac{1}{2}\phi \\ n_z\sin\frac{1}{2}\phi & -n_y\sin\frac{1}{2}\phi & n_x\sin\frac{1}{2}\phi & \cos\frac{1}{2}\phi \end{bmatrix} \begin{bmatrix} u \\ v \\ p \\ q \end{bmatrix}.$$

Adoption of the quaternion model for spatial PH curves greatly facilitates the formulation and solution of key problems in their construction and analysis, and offers new theoretical insights. Compared to the complex-number model for planar PH curves, however, it requires greater care and attention to detail in its use, due to the non-commutative nature of the quaternion product.

3.3. First-order spatial PH quintic Hermite interpolants

A basic algorithm (Farouki et al., 2002) in the construction of spatial PH curves is concerned with the problem of first-order Hermite interpolation, i.e., interpolation of given end points $\mathbf{p}_0$, $\mathbf{p}_1$ and derivatives $\mathbf{d}_0$, $\mathbf{d}_1$ by a spatial PH curve $\mathbf{r}(t)$ for $t \in [0, 1]$. The lowest-order PH curves capable of solving this problem for arbitrary data are—as with planar PH curves—quintics.

Whereas the planar PH quintic Hermite interpolation problem yields *four distinct solutions* (Farouki and Neff, 1995) in general, interpolation by spatial PH quintics incurs a *two-parameter family of solutions* (Farouki et al., 2002). The shape of these interpolants may depend rather sensitively on these two free parameters, and the question of choosing "optimal" values for them is still an open problem (one possibility is to impose an additional constraint, such as a helicity condition (Farouki et al., 2003), on the interpolants).

To construct spatial PH quintic Hermite interpolants, we begin by inserting a quadratic quaternion polynomial

$$A(t) = A_0(1-t)^2 + A_1 2(1-t)t + A_2 t^2$$

into the representation (1). Here the quaternion coefficients A_0, A_1, A_2 are to be determined by matching the Hermite data $\mathbf{p}_0$, $\mathbf{d}_0$ and $\mathbf{p}_1$, $\mathbf{d}_1$. The conditions $\mathbf{r}'(0) = \mathbf{d}_0$, $\mathbf{r}'(1) = \mathbf{d}_1$, and $\int_0^1 \mathbf{r}'(t)\,dt = \mathbf{p}_1 - \mathbf{p}_0$ thus yield (Farouki et al., 2002) the system of three equations

$$A_0\,\mathbf{i}\,A_0^* = \mathbf{d}_0\,, \quad A_2\,\mathbf{i}\,A_2^* = \mathbf{d}_1\,, \tag{2}$$

$$\begin{aligned}&(3A_0 + 4A_1 + 3A_2)\,\mathbf{i}\,(3A_0 + 4A_1 + 3A_2)^* \\ &\quad = 120(\mathbf{p}_1 - \mathbf{p}_0) - 15(\mathbf{d}_0 + \mathbf{d}_1) + 5(A_0\,\mathbf{i}\,A_2^* + A_2\,\mathbf{i}\,A_0^*)\end{aligned} \tag{3}$$

for A_0, A_1, A_2. This system may be solved by noting that the equation

$$A\,\mathbf{i}\,A^* = \mathbf{d} \tag{4}$$

for a given vector $\mathbf{d} = |\mathbf{d}|(\lambda, \mu, \nu)$ admits a one-parameter family of solutions

$$A(\phi) = \sqrt{\tfrac{1}{2}(1+\lambda)|\mathbf{d}|}\left(-\sin\phi\,, \ \frac{\cos\phi\,(\mathbf{i} + \mu\mathbf{j} + \nu\mathbf{k}) + \sin\phi\,(\nu\mathbf{j} - \mu\mathbf{k})}{1+\lambda}\right)$$

where ϕ is a free angular variable. The quaternion A serves to scale/rotate the basis vector $\mathbf{i}$ into the given vector $\mathbf{d}$—the appearance of a free parameter in the solution reflects the fact that, in $\mathbb{R}^3$, there is a *continuous family of spatial rotations* that will map one unit vector into another.

Equations (2) can be solved directly for A_0, A_2 using the known form of the solution to (4). These quaternions depend on free parameters, ϕ_0 and ϕ_2 say. Substituting them into (3) we may determine $3A_0 + 4A_1 + 3A_2$, and hence A_1, using the solution to (4). Again, this incurs a new free parameter—ϕ_1, say. Although the complete solution incurs three indeterminate angular variables ϕ_0, ϕ_1, ϕ_2, close inspection reveals (Farouki et al., 2002) that the Hermite interpolants depend only upon the *differences* of these angles. Hence, we may take $\phi_1 = 0$ without loss of generality, and the Hermite interpolants to $\mathbf{p}_0$, $\mathbf{p}_1$ and $\mathbf{d}_0$, $\mathbf{d}_1$ comprise a two-parameter family. Once A_0, A_1, A_2 are known, the Bezier control points of the interpolant are given by

$$\begin{aligned}
\mathbf{p}_1 &= \mathbf{p}_0 + \frac{1}{5}\,A_0\,\mathbf{i}\,A_0^*\,, \\
\mathbf{p}_2 &= \mathbf{p}_1 + \frac{1}{10}(A_0\,\mathbf{i}\,A_1^* + A_1\,\mathbf{i}\,A_0^*)\,, \\
\mathbf{p}_3 &= \mathbf{p}_2 + \frac{1}{30}(A_0\,\mathbf{i}\,A_2^* + 4\,A_1\,\mathbf{i}\,A_1^* + A_2\,\mathbf{i}\,A_0^*)\,, \\
\mathbf{p}_4 &= \mathbf{p}_3 + \frac{1}{10}(A_1\,\mathbf{i}\,A_2^* + A_2\,\mathbf{i}\,A_1^*)\,, \\
\mathbf{p}_5 &= \mathbf{p}_4 + \frac{1}{5}\,A_2\,\mathbf{i}\,A_2^*\,,
\end{aligned}$$

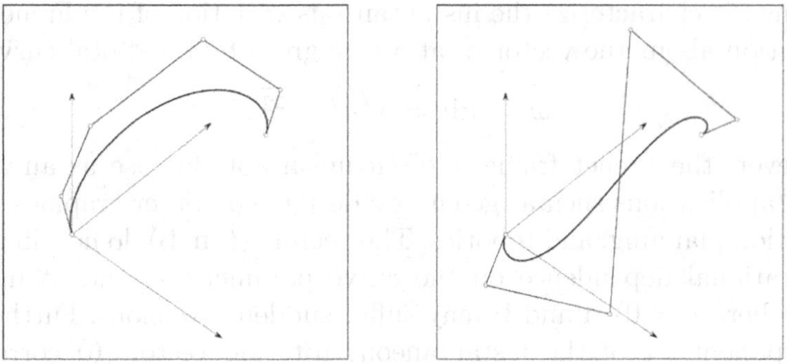

Figure 3.1. Examples of spatial PH quintics constructed as first-order Hermite interpolants.

with $\mathbf{p}_0$ being an arbitrary integration constant. Examples of spatial PH quintic Hermite interpolants, for specific ϕ_0, ϕ_2 values, are shown in Figure 3.1.

A challenging open problem is to generalize the formulation and solution of the two-point Hermite interpolation to the smooth interpolation of $N + 1$ points $\mathbf{p}_0, \ldots, \mathbf{p}_N$ in $\mathbb{R}^3$ by C^2 spatial PH quintic splines. In the planar case, the PH spline problem incurs solution of a "tridiagonal" system of N quadratic equations in N complex unknowns (Albrecht and Farouki, 1996). An analogous quaternion system can be formulated in the spatial case (Farouki et al., 2003). In solving it, one must account for the non-commutative nature of quaternion products, and the residual freedoms associated with each spline segment.

3.4. Rotation-minimizing frames on spatial PH curves

An *adapted frame* along a space curve $\mathbf{r}(t)$ is a right-handed system of three mutually orthogonal unit vectors $(\mathbf{t}, \mathbf{e}_1, \mathbf{e}_2)$ of which $\mathbf{t} = \mathbf{r}'/|\mathbf{r}'|$ is the tangent vector, and $\mathbf{e}_1$, $\mathbf{e}_2$ span the normal plane at each point such that $\mathbf{e}_1 \times \mathbf{e}_2 = \mathbf{t}$. The most familiar example is the *Frenet frame* $(\mathbf{t}, \mathbf{n}, \mathbf{b})$ comprising the tangent, normal $\mathbf{n}$ (pointing to the center of curvature), and binormal $\mathbf{b} = \mathbf{t} \times \mathbf{n}$. The variation of the Frenet frame with arc length is described (Kreyszig, 1959) by the equations

$$\frac{d\mathbf{t}}{ds} = \mathbf{d} \times \mathbf{t}\,, \qquad \frac{d\mathbf{n}}{ds} = \mathbf{d} \times \mathbf{n}\,, \qquad \frac{d\mathbf{b}}{ds} = \mathbf{d} \times \mathbf{b}\,, \tag{5}$$

where the *Darboux vector* is given in terms of the curvature κ and torsion τ by

$$\mathbf{d} = \kappa\,\mathbf{b} + \tau\,\mathbf{t}\,. \tag{6}$$

Equations (5) characterize the instantaneous variation of the Frenet frame as a rotation about the vector **d**, at a rate given by the "total curvature"

$$\omega = |\mathbf{d}| = \sqrt{\kappa^2 + \tau^2}\,.$$

However, the Frenet frame is often unsuitable for use as an adapted frame in applications such as geometric design, computer graphics, animation, motion planning, and robotics. The vectors $(\mathbf{t}, \mathbf{n}, \mathbf{b})$ do not, in general, have a rational dependence on the curve parameter t, and at inflection points (where $\kappa = 0$) **n** and **b** may suffer sudden inversions. Furthermore, the component $\tau\,\mathbf{t}$ of the instantaneous rotation vector (6) corresponds to an "unnecessary" rotation in the curve normal plane, that yields undesirable results in computer animation, swept surface constructions, and motion planning. Among the many adapted frames (Bishop, 1975) on a space curve, (Klok, 1986) has suggested the *rotation-minimizing frame* (RMF) as the most suitable for such applications. The RMF is defined so as to "cancel" the $\tau\,\mathbf{t}$ component of the rotation vector by setting

$$\begin{bmatrix} \mathbf{e}_2 \\ \mathbf{e}_3 \end{bmatrix} = \begin{bmatrix} \cos\theta & \sin\theta \\ -\sin\theta & \cos\theta \end{bmatrix} \begin{bmatrix} \mathbf{n} \\ \mathbf{b} \end{bmatrix},$$

where the angular function $\theta(t)$ is defined[4] (Guggenheimer, 1989) by

$$\theta(t) = \theta_0 - \int_0^t \tau(u)\,|\mathbf{r}'(u)|\,\mathrm{d}u\,. \tag{7}$$

Because this integral does not admit a closed-form reduction for "ordinary" polynomial and rational curves, schemes have been proposed to approximate RMFs or to approximate given curves by "simple" segments—e.g., circular arcs—with known RMFs (Jüttler, 1998; Jüttler and Mäurer, 1999; Jüttler and Mäurer, 1999; Wang and Joe, 1997).

For PH curves, the integrand in (7) is a *rational function*, and thus admits closed-form integration (Farouki, 2002). A simplification of this integral arises through the fact that PH curves exhibit the remarkable factorization

$$|\mathbf{r}' \times \mathbf{r}''|^2 = \sigma^2 \rho\,,$$

where $\sigma = u^2 + v^2 + p^2 + q^2$, and ρ is the polynomial defined by

$$\rho = 4\,[\,(up' - u'p)^2 + (uq' - u'q)^2 + (vp' - v'p)^2 + (vq' - v'q)^2 \\ + 2(uv' - u'v)(pq' - p'q)\,]\,.$$

Thus, for a PH curve, we have

$$\frac{\mathrm{d}\theta}{\mathrm{d}t} = -\,\frac{[\,\mathbf{r}'(t) \times \mathbf{r}''(t)\,] \cdot \mathbf{r}'''(t)}{\sigma(t)\,\rho(t)}\,.$$

[4] An incorrect sign before the integral is given in (Guggenheimer, 1989).

For PH quintics, $(\mathbf{r}' \times \mathbf{r}'') \cdot \mathbf{r}'''$ is of degree 6, while σ and ρ are both quartic in t. The latter must be factorized to perform a partial fraction decomposition of the integrand: this can be accomplished by Ferrari's method (Uspensky, 1948). Complete details on the closed-form integration of this equation may be found in (Farouki, 2002).

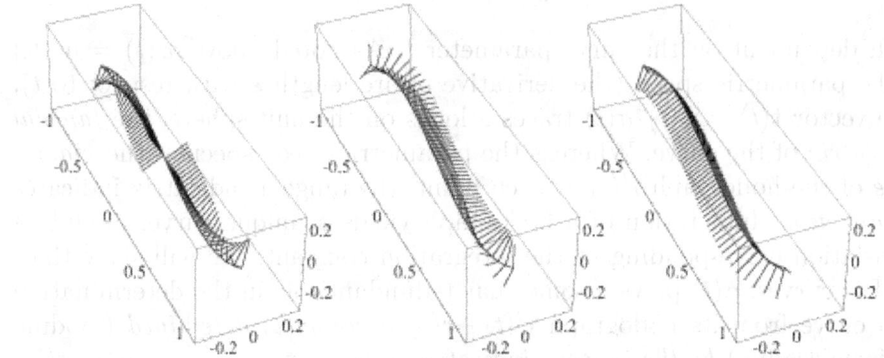

Figure 3.2. Comparison of Frenet frame (left), Euler-Rodrigues frame (center), and the rational approximation of rotation-minimizing frame (right) on a spatial PH quintic (for clarity, the tangent vector is omitted). Note the sudden reversal of the Frenet frame at the inflection point.

Since the integral in (7) involves rational and logarithmic terms, we describe in (Farouki and Han, 2003) an alternative rational approximation scheme, based on the equation

$$\frac{\mathrm{d}\theta}{\mathrm{d}t} = 2\,\frac{u'v - uv' - p'q + pq'}{u^2 + v^2 + p^2 + q^2}$$

characterizing the variation of the RMF relative to the *Euler-Rodrigues frame* (ERF), defined (Choi and Han, 2002) by

$$\mathbf{t}(t) = \frac{\mathcal{A}(t)\,\mathbf{i}\,\mathcal{A}^*(t)}{\mathcal{A}(t)\mathcal{A}^*(t)}\,, \quad \mathbf{u}(t) = \frac{\mathcal{A}(t)\,\mathbf{j}\,\mathcal{A}^*(t)}{\mathcal{A}(t)\mathcal{A}^*(t)}\,, \quad \mathbf{v}(t) = \frac{\mathcal{A}(t)\,\mathbf{k}\,\mathcal{A}^*(t)}{\mathcal{A}(t)\mathcal{A}^*(t)}\,.$$

The ERF is a rational adapted frame defined on spatial PH curves. In Figure 3.2 we compare the Frenet frame, ERF, and rational RMF approximation on an inflectional PH quintic—the superior behavior of the RMF is clearly apparent.

A natural question is when (or whether) one can have a *rational* RMF on a given non-planar polynomial curve. Note that the curve must be a PH curve to have a rational adapted frame, since only PH curves have rational unit tangents. A partial answer to this question can be given in terms of the ERF, defined above (Choi and Han, 2002): the minimum degree of non-planar PH curves that have rotation-minimizing ERFs is seven.

3.5. Tangent indicatrix uniquely determines PH curves

The hodograph $\mathbf{r}'(t)$ of a parametric curve can be expressed as the product of a scalar magnitude and a unit vector

$$\mathbf{r}'(t) \,=\, \sigma(t)\,\mathbf{t}(t)\,,$$

both dependent on the curve parameter t. As noted above, $\sigma(t) = |\mathbf{r}'(t)|$ is the parametric speed (the derivative of arc length s with respect to t). The vector $\mathbf{t}(t) = \mathbf{r}'(t)/\sigma(t)$ traces a locus on the unit sphere, the *tangent indicatrix* of the curve. Whereas the parametric speed specifies the *magnitude* of the hodograph $\mathbf{r}'(t)$ at each point, the tangent indicatrix indicates its *direction.* Integration of a hodograph yields a unique curve, modulo a translation corresponding to the integration constant. We will show that, for PH curves, $\sigma(t)$ plays a somewhat redundant role in the determination of a curve from its hodograph $\mathbf{r}'(t)$—$\mathbf{r}(t)$ *is uniquely detemined* (modulo uniform scaling) *by the tangent indicatrix only.*

This property distinguishes PH curves from "ordinary" polynomials curves, for which both the parametric speed and tangent indicatrix influence the shape of the curve $\mathbf{r}(t)$ obtained by integration of the hodograph $\mathbf{r}'(t) = \sigma(t)\,\mathbf{t}(t)$. Two ordinary polynomial curves with the same tangent indicatrix but different parametric speeds have, in general, quite different shapes. Note also that, for PH curves, the tangent indicatrix $\mathbf{t}(t) = \mathbf{r}'(t)/\sigma(t)$ is a *rational* curve on the unit sphere, since $\sigma(t)$ is a polynomial (whereas, for an ordinary polynomial curve, it is the square root of a polynomial, and hence $\mathbf{t}(t)$ is not rational).

PROPOSITION 3.1. *Let $\mathbf{r}(t)$, $\tilde{\mathbf{r}}(t)$ be two polynomial PH curves whose hodographs have relatively prime components. If these curves possess the same tangent indicatrix, they differ by at most a translation and uniform scaling, i.e., $\mathbf{r}'(t) = \gamma\,\tilde{\mathbf{r}}'(t)$ for some $\gamma \neq 0$.*

Proof Since $\mathbf{r}(t) = (x(t), y(t), z(t))$ and $\tilde{\mathbf{r}}(t) = (\tilde{x}(t), \tilde{y}(t), \tilde{z}(t))$ are both PH curves, polynomials $\sigma(t)$ and $\tilde{\sigma}(t)$ exist such that their hodographs

$$\mathbf{r}'(t) \,=\, (x'(t), y'(t), z'(t)) \quad \text{and} \quad \tilde{\mathbf{r}}'(t) \,=\, (\tilde{x}'(t), \tilde{y}'(t), \tilde{z}'(t))$$

satisfy

$$x'^2(t) + y'^2(t) + z'^2(t) = \sigma^2(t) \quad \text{and} \quad \tilde{x}'^2(t) + \tilde{y}'^2(t) + \tilde{z}'^2(t) \,=\, \tilde{\sigma}^2(t)\,.$$

Furthermore, since $x'(t), y'(t), z'(t)$ and $\tilde{x}'(t), \tilde{y}'(t), \tilde{z}'(t)$ are relatively prime, $\sigma(t)$ and $\tilde{\sigma}(t)$ never vanish, and we can assume $\sigma(t) > 0$ and $\tilde{\sigma}(t) > 0$ for all t. The tangent indicatrices of $\mathbf{r}(t)$ and $\tilde{\mathbf{r}}(t)$ are then $\mathbf{t}(t) = \mathbf{r}'(t)/\sigma(t)$ and $\tilde{\mathbf{t}}(t) = \tilde{\mathbf{r}}'(t)/\tilde{\sigma}(t)$, and if they are the same we must have

$$x'(t) = \frac{\sigma(t)\tilde{x}'(t)}{\tilde{\sigma}(t)}, \quad y'(t) = \frac{\sigma(t)\tilde{y}'(t)}{\tilde{\sigma}(t)}, \quad z'(t) = \frac{\sigma(t)\tilde{z}'(t)}{\tilde{\sigma}(t)}.$$

We claim that $\tilde{\sigma}(t)$ divides into $\sigma(t)$. The Proposition then follows, since we can swap the roles of $\sigma(t)$ and $\tilde{\sigma}(t)$, and thus $\sigma(t) = \gamma\,\tilde{\sigma}(t)$ for some $\gamma \neq 0$.

Consider the equation $x'(t) = \sigma(t)\tilde{x}'(t)/\tilde{\sigma}(t)$. Since $x'(t)$ is a polynomial, each non-constant factor of $\tilde{\sigma}(t)$ must divide into either $\sigma(t)$ or $\tilde{x}'(t)$, and an analogous statement holds for $y'(t)$ and $z'(t)$. However, if we postulate that a non-constant factor of $\tilde{\sigma}(t)$ divides into $\tilde{x}'(t)$, but not $\sigma(t)$, we must conclude that this factor also divides into $\tilde{y}'(t)$ and $\tilde{z}'(t)$, which contradicts the fact that $\gcd(\tilde{x}'(t), \tilde{y}'(t), \tilde{z}'(t)) = \text{constant}$. Hence, we may deduce that $\sigma(t) = \gamma\,\tilde{\sigma}(t)$ for some $\gamma \neq 0$. □

COROLLARY 3.1. *Given a rational curve* $\mathbf{q}(t) = (a(t), b(t), c(t))/\sigma(t)$ *on the unit sphere in* $\mathbb{R}^3$ *with* $\gcd(a(t), b(t), c(t)) = \text{constant}$, *there is—modulo scaling and translation—a unique polynomial PH curve* $\mathbf{r}(t) = (x(t), y(t), z(t))$ *with* $\gcd(x'(t), y'(t), z'(t)) = \text{constant}$ *that has* $\mathbf{q}(t)$ *as its tangent indicatrix.*

Proof Since $\mathbf{q}(t)$ lies on the unit sphere, we have $a^2(t)+b^2(t)+c^2(t) = \sigma^2(t)$. Since $\gcd(a(t), b(t), c(t)) = \text{constant}$, $\sigma(t)$ is never zero, and thus we may assume $\sigma(t) > 0$ for all t. Clearly, for any scalar factor $\gamma \neq 0$ and integration constant $\mathbf{r}_0$,

$$\mathbf{r}(t) = \gamma \int \sigma(t)\mathbf{q}(t)\,\mathrm{d}t + \mathbf{r}_0$$

defines a PH curve, and the uniqueness of $\mathbf{r}(t)$, modulo the uniform scaling γ and translation $\mathbf{r}_0$, follows from the Proposition. □

Although we have phrased the above results in terms of spatial PH curves, they obviously also apply to planar PH curves (for which the tangent indicatrix lies on the unit circle, rather than the unit sphere).

3.6. Inversion of spatial Pythagorean hodographs

In (Farouki, 1994) we addressed the question of "how many" planar PH curves exist by showing that, in the plane, the infinite sets of regular PH curves and of regular "ordinary" polynomial curves have the same cardinality—we can establish a *one-to-one correspondence* between their members (corresponding curves are of different degree). This was accomplished by invoking the map $\mathbf{z} \to \mathbf{z}^2$ and its inverse in the complex-variable model for planar PH curves.

In seeking an analogous result for spatial PH curves, we need the ability to invert the map $\mathcal{A}(t) \to \mathbf{r}'(t)$ defined by (1)—i.e., given a spatial Pythagorean hodograph $\mathbf{r}'(t)$, we wish to identify the pre-image curve(s) $\mathcal{A}(t)$ in quaternion space that generate it through expression (1). In Section 3.3 we gave the general solution to the analogous equation (4) for a fixed vector $\mathbf{d}$. However, this solution is not appropriate to the problem of inverting (1), since if we replace $\mathbf{d}$ by $\mathbf{r}'(t)$ it exhibits the factor $\sqrt{|\mathbf{r}'(t)|}$ and we require a *polynomial* pre-image $\mathcal{A}(t)$.

(Dietz et al., 1993) give a constructive proof for the existence of polynomials $u(t)$, $v(t)$, $p(t)$, $q(t)$ satisfying (1) for a given $\mathbf{r}'(t)$ by invoking the *generalized stereographic projection.* Since it is not convenient for actually computing the pre-image, our goal here is to re-work this proof into a practical algorithm.

GENERALIZED STEREOGRAPHIC PROJECTION

The notation in (Dietz et al., 1993) differs somewhat from our quaternion model: (x_0, x_1, x_2, x_3) denote homogeneous coordinates in real 3-dimensional projective space $\mathbb{RP}^3$, and the hodograph $\mathbf{r}'(t) = (x'(t), y'(t), z'(t))$ is mapped to its tangent indicatrix by the correspondence $x_0 = \sigma$, $x_1 = x'$, $x_2 = y'$, $x_3 = z'$. The generalized stereographic projection δ maps points

$$(p_0, p_1, p_2, p_3) \in \mathbb{RP}^3$$

to points (x_0, x_1, x_2, x_3) on the unit sphere (satisfying $x_1^2 + x_2^2 + x_3^2 = x_0^2$) according to

$$\begin{aligned} x_0 &= p_0^2 + p_1^2 + p_2^2 + p_3^2 \,, \\ x_1 &= 2p_0p_1 - 2p_2p_3 \,, \\ x_2 &= 2p_1p_3 + 2p_0p_2 \,, \\ x_3 &= p_1^2 + p_2^2 - p_0^2 - p_3^2 \,. \end{aligned}$$

This is equivalent to the quaternion formulation if we identify

$$(x', y', z', \sigma) = (x_3, x_2, x_1, x_0) \quad \text{and} \quad (u, v, p, q) = (p_2, p_1, p_3, p_0) \,.$$

Now suppose we are given a Pythagorean hodograph $\mathbf{r}'(t)$ with relatively prime coordinate components in $\mathbb{R}[t]$. Then the tetrad members x_0, x_1, x_2, x_3 are also relatively prime in $\mathbb{R}[t]$. We describe how to compute a pre-image p_0, p_1, p_2, p_3 under the above quadratic transformation.

1. Re-write the condition $x_1^2 + x_2^2 + x_3^2 = x_0^2$ as $(x_1 + \mathrm{i}x_2)(x_1 - \mathrm{i}x_2) = (x_0 + x_3)(x_0 - x_3)$.
2. Compute $d = \gcd(x_1, x_2)$ in $\mathbb{R}[t]$ such that $x_1 + \mathrm{i}x_2 = d(\tilde{x} + \mathrm{i}\tilde{x}_2)$ and $\gcd(\tilde{x}_1, \tilde{x}_2) = 1$ in $\mathbb{R}[t]$.
3. Factorize d in $\mathbb{R}[t]$ and $\tilde{x}_1 + \mathrm{i}\tilde{x}_2$ in $\mathbb{C}[t]$.
4. Let r be a factor of d in $\mathbb{R}[t]$. We claim that r divides into either $x_0 + x_3$ or $x_0 - x_3$, but not both. For, if r divides into both it must be a real common factor of x_0 and x_3, but r is already a real common factor of x_1 and x_2, which contradicts the fact that x_0, x_1, x_2, x_3 are relatively prime in $\mathbb{R}[t]$. Hence, we can write $d = r_1 \cdots r_m s_1 \cdots s_n$ such that the r_i's are factors of $x_0 + x_3$ and the s_i's are factors of $x_0 - x_3$.
5. Let $\gamma_1 \ldots \gamma_\kappa$ be a prime factor decomposition of $\tilde{x}_1 + \mathrm{i}\tilde{x}_2$ in $\mathbb{C}[t]$ (note that some factors may be repeated). Since we have already removed all the real factors, there is no pair $i \neq j$ such that $\gamma_i = \overline{\gamma}_j$. Since $\mathbb{C}[t]$ is a unique factorization domain, we can uniquely arrange the decomposition into groups $\alpha_1 \ldots \alpha_\rho$ and $\beta_1 \ldots \beta_\sigma$ such that the first divides into $x_0 + x_3$ and the second $x_0 - x_3$. Note that if γ_i divides into $x_0 + x_3$, so does $\overline{\gamma}_i$.
6. From the preceding two steps, we have $x_0 + x_3 = (r_1 \cdots r_m)^2 |\alpha_1 \cdots \alpha_\rho|^2$ and $x_0 - x_3 = (s_1 \cdots s_n)^2 |\beta_1 \cdots \beta_\sigma|^2$. Defining $\phi = r_1 \cdots r_m \alpha_1 \cdots \alpha_\rho$ and $\psi = s_1 \cdots s_n \beta_1 \cdots \beta_\sigma$, one can verify that $p_0 = \mathrm{Re}(\psi)$, $p_1 = \mathrm{Re}(\phi)$, $p_2 = \mathrm{Im}(\phi)$, $p_3 = \mathrm{Im}(\psi)$ constitute a solution (in fact $p_i/\sqrt{2}$ gives the exact form but we ignored the common constant multiple).

For example, consider the case

$$
\begin{aligned}
x_0 &= 7 - 8\,t + 13\,t^2 + 2\,t^3 + 4\,t^4\,, \\
x_1 &= -6 + 8\,t - 4\,t^2 + 10\,t^3\,, \\
x_2 &= 2 + 8\,t + 2\,t^3 + 4\,t^4\,, \\
x_3 &= 3 - 8\,t + t^2 + 2\,t^3\,.
\end{aligned}
$$

Then x_1 and x_2 are relatively prime in $\mathbb{R}[t]$, and we have

$$x_1 + \mathrm{i}x_2 = (t + 1 - 2\mathrm{i})(2t - 1 + \mathrm{i})(t - \mathrm{i}\tfrac{1}{2}(1 - \sqrt{5}))(2\mathrm{i}\,t + 1 + \sqrt{5})\,,$$

$$x_0 + x_3 = (t + 1 + 2\mathrm{i})(t + 1 - 2\mathrm{i})(2t - 1 + \mathrm{i})(2t - 1 - \mathrm{i})\,.$$

Hence, we can set $\alpha_1 = t + 1 - 2\mathrm{i}$, $\alpha_2 = 2t - 1 + \mathrm{i}$, $\beta_1 = 2\mathrm{i}\,t + 1 + \sqrt{5}$, $\beta_2 = t - \mathrm{i}(1 - \sqrt{5})/2$. Then $\phi = 1 + t + 2t^2 + \mathrm{i}3(1 - t)$ and $\psi = 2t + \mathrm{i}2(1 + t^2)$, giving $p_0 = \sqrt{2}t$, $p_1 = (1 + t + 2t^2)/\sqrt{2}$, $p_2 = 3(1 - t)/\sqrt{2}$, $p_3 = \sqrt{2}(1 + t^2)$.

However, this solution is not unique. In fact x_0, x_1, x_2, x_3 were initially constructed using $p_0 = 1 + t + t^2$, $p_1 = -1 + 2t + t^2$, $p_2 = 2 - t + t^2$, $p_3 = 1 - t + t^2$. The non-uniqueness has also been noted in the quaternion formulation (Farouki et al., 2002)—if we define

$$\begin{bmatrix} \hat{u}(t) \\ \hat{v}(t) \\ \hat{p}(t) \\ \hat{q}(t) \end{bmatrix} = \begin{bmatrix} \cos\xi & -\sin\xi & 0 & 0 \\ \sin\xi & \cos\xi & 0 & 0 \\ 0 & 0 & \cos\xi & \sin\xi \\ 0 & 0 & -\sin\xi & \cos\xi \end{bmatrix} \begin{bmatrix} u(t) \\ v(t) \\ p(t) \\ q(t) \end{bmatrix}$$

then $\hat{u}(t)$, $\hat{v}(t)$, $\hat{p}(t)$, $\hat{q}(t)$ and $u(t)$, $v(t)$, $p(t)$, $q(t)$ define the same PH curve.

This is also apparent from the generalized stereographic projection. The pre-image $\delta^{-1}(Q)$ of any point $Q = (x_0, x_1, x_2, x_3)$ on the sphere comprises the straight line $\lambda A + \mu B$ in $\mathbb{RP}^3$, where $A = (x_2, 0, x_0 + x_3, -x_1)$, $B = (x_1, x_0 + x_3, 0, x_2)$ and $\lambda, \mu \in \mathbb{R}$. In fact, one can verify that

$$\delta(\lambda A + \mu B) = 2(\lambda^2 + \mu^2)(x_0 + x_3)(x_0, x_1, x_2, x_3) .$$

Now suppose we have a degree-$2n$ curve $x(t) = (x_0(t), x_1(t), x_2(t), x_3(t))$ on the unit sphere such that $x_0(t), \ldots, x_3(t)$ are relatively prime in $\mathbb{R}[t]$. The pre-image of this curve under the generalized stereographic projection, i.e., the set $S = \{\delta^{-1}(x(t)) : t \in \mathbb{R}\}$, is clearly a ruled surface with base curves given by $A(t) = (x_2(t), 0, x_0(t) + x_3(t), -x_1(t))$ and $B(t) = (x_1(t), x_0(t) + x_3(t), 0, x_2(t))$. Note that these two base curves are also of degree $2n$.

However, a theorem in (Dietz et al., 1993) states that we can always find another pair $\{P(t), \tilde{P}(t)\}$ of base curves whose degree is just n. If $P(t) = (p_0(t), p_1(t), p_2(t), p_3(t))$, we can take $\tilde{P}(t) = (-p_3(t), p_2(t), -p_1(t), p_0(t))$. For each t, the four points $A(t)$, $B(t)$, $P(t)$, $\tilde{P}(t)$ are collinear.

3.7. Closure

The quaternion formulation for spatial PH curves, first introduced in (Choi et al., 2002), has paved the way for development of basic algorithms concerned with their construction, analysis, and applications. In this paper, we surveyed these new developments and identified a number of important open problems. Compared to planar PH curves, the construction of spatial PH curves typically incurs free parameters. A deeper theoretical understanding of the role of these parameters, and their optimal selection, remains to be achieved.

Acknowledgements: This work was supported in part by the National Science Foundation, under grants CCR-9902669, CCR-0202179, and DMS-0138411.

References

Albrecht, G. and R. T. Farouki: Construction of C^2 Pythagorean hodograph interpolating splines by the homotopy method. *Adv. Comp. Math.,* **5**:417–442, 1996.

Bishop, R. L.: There is more than one way to frame a curve. *Amer. Math. Monthly,* **82**:246–251, 1975.

Choi, H. I. and C. Y. Han: Euler-Rodrigues frames on spatial Pythagorean-hodograph curves. *Comput. Aided Geom. Design,* **19**:603–620, 2002.

Choi, H. I., D. S. Lee, and H. P. Moon: Clifford algebra, spin representation, and rational parameterization of curves and surfaces. *Adv. Comp. Math.,* **17**:5–48, 2002.

Dietz, R., J. Hoschek, and B. Jüttler: An algebraic approach to curves and surfaces on the sphere and on other quadrics. *Comput. Aided Geom. Design,* **10**:211–229, 1993.

Farouki, R. T.: Pythagorean-hodograph curves in practical use. In *Geometry Processing for Design and Manufacturing* (R. E. Barnhill, editor), pages 3-33, SIAM, 1992.

Farouki, R. T.: The conformal map $z \to z^2$ of the hodograph plane. *Comput. Aided Geom. Design,* **11**:363–390, 1994.

Farouki, R. T.: The elastic bending energy of Pythagorean hodograph curves. *Comput. Aided Geom. Design,* **13**:227–241, 1996.

Farouki, R. T.: Exact rotation-minimizing frames for spatial Pythagorean-hodograph curves. *Graph. Models,* **64**:382–395, 2002.

Farouki, R. T.: Pythagorean-hodograph curves. In *Handbook of Computer Aided Geometric Design* (G. Farin. J. Hoschek, and M-S. Kim, editors), pages 405–427, North Holland, 2002.

Farouki, R. T., M. al-Kandari, and T. Sakkalis: Structural invariance of spatial Pythagorean hodographs. *Comput. Aided Geom. Design,* **19**:395–407, 2002.

Farouki, R. T., M. al-Kandari, and T. Sakkalis: Hermite interpolation by rotation-invariant spatial Pythagorean-hodograph curves. *Adv. Comp. Math.,* **17**:369–383, 2002.

Farouki, R. T. and C. Y. Han: Rational approximation schemes for rotation-minimizing frames on Pythagorean-hodograph curves. *Comput. Aided Geom. Design,* **20**:435–454, 2003.

Farouki, R. T., C. Y. Han, C. Manni, and A. Sestini: Characterization and construction of helical Pythagorean-hodograph quintic space curves. *J. Comput. Applied Math.,* **162**:365–392, 2004.

Farouki, R. T., B. K. Kuspa, C. Manni, and A. Sestini: Efficient solution of the complex quadratic tridiagonal system for C^2 PH quintic splines. *Numer. Algor.,* **27**:35–60, 2001.

Farouki, R. T., J. Manjunathaiah, D. Nicholas, G.-F. Yuan, and S. Jee: Variable feedrate CNC interpolators for constant material removal rates along Pythagorean-hodograph curves. *Comput. Aided Design,* **30**:631–640, 1998.

Farouki, R. T., C. Manni, and A. Sestini: Spatial C^2 PH quintic splines. In *Curve and Surface Design: Saint Malo 2002* (T. Lyche, M.-L. Mazure, and L. L. Schumaker, editors), pages 147–156, Nashboro Press, 2003.

Farouki, R. T. and C. A. Neff: Hermite interpolation by Pythagorean hodograph quintics. *Math. Comp.,* **64**:1589–1609, 1995.

Farouki, R. T. and T. Sakkalis: Pythagorean hodographs. *IBM J. Res. Develop.,* **34**: 736–752, 1990.

Farouki, R. T. and T. Sakkalis: Pythagorean-hodograph space curves. *Adv. Comp. Math.,* **2**:41–66, 1994.

Farouki, R. T. and S. Shah: Real-time CNC interpolators for Pythagorean-hodograph curves. *Comput. Aided Geom. Design,* **13**:583–600, 1996.

Guggenheimer, H.: Computing frames along a trajectory. *Comput. Aided Geom. Design,* **6**:77–78, 1989.

Jüttler, B.: Generating rational frames of space curves via Hermite interpolation with Pythagorean hodograph cubic splines. In *Differential/Topological Techniques in Geometric Modeling and Processing '98* (D. P. Choi, H. I. Choi, M. S. Kim, and R. R. Martin, editors), pages 83–106, Bookplus Press, 1998.

Jüttler, B.: Hermite interpolation by Pythagorean hodograph curves of degree seven. *Math. Comp.,* **70**:1089–1111, 2001.

Jüttler, B. and C. Mäurer: Cubic Pythagorean hodograph spline curves and applications to sweep surface modelling. *Comput. Aided Design,* **31**:73–83, 1999.

Jüttler, B. and C. Mäurer: Rational approximation of rotation minimizing frames using Pythagorean-hodograph cubics. *J. Goem. Graphics,* **3**:141–159, 1999.

Klok, F.: Two moving coordinate frames for sweeping along a 3D trajectory. *Comput. Aided Geom. Design,* **3**:217–229, 1986.

Kreyszig, E.: *Differential Geometry.* University of Toronto Press, 1959.

Moon, H. P.: Minkowski Pythagorean hodographs. *Comput. Aided Geom. Design,* **16**:739–753, 1999.

Moon, H. P., R. T. Farouki, and H. I. Choi: Construction and shape analysis of PH quintic Hermite interpolants. *Comput. Aided Geom. Design,* **18**:93–115, 2001.

Tsai, Y.-F., R. T. Farouki, and B. Feldman: Performance analysis of CNC interpolators for time-dependent feedrates along PH curves. *Comput. Aided Geom. Design,* **18**: 245–265, 2001.

Uspensky, J. V.: *Theory of Equations.* McGraw-Hill, New York, 1948.

Wang, W. and B. Joe: Robust computation of the rotation minimizing frame for sweep surface modelling. *Comput. Aided Design,* **29**:379–391, 1997.

CUMULATIVE CHORDS, PIECEWISE-QUADRATICS AND PIECEWISE-CUBICS

LYLE NOAKES
School of Mathematics and Statistics
The University of Western Australia
35 Stirling Highway Crawley, WA 6009, Perth
Australia

RYSZARD KOZERA
School of Computer Science and Software Engineering
The University Western Australia
35 Stirling Highway, Crawley WA 6009, Perth
Australia

Abstract. Cumulative chord piecewise-quadratics and piecewise-cubics are examined in detail, and compared with other low degree interpolants for unparameterized data from regular curves in $\mathbb{R}^n$, especially piecewise-4-point quadratics. Orders of approximation are calculated and compared with numerical experiments. Good performance of the interpolant is also confirmed experimentally on sparse data. This work may be applicable in computer graphics and vision: image segmentation, medical image processing, or computer aided geometrical design.

Key words: interpolation, cumulative chord parameterization, length and trajectory estimation

4.1. Introduction

Let $\gamma : [0,T] \to \mathbb{R}^n$ be a smooth regular curve, namely γ is C^r for some $r \geq 1$ and $\dot{\gamma}(t) \neq \mathbf{0}$ for all $t \in [0,T]$. Our task is to estimate γ from *an ordered* $m+1$*-tuple* $Q = (q_0, q_1, \ldots, q_m)$ of points in $\mathbb{R}^n$, where $q_i = \gamma(t_i)$, $0 = t_0 < t_1 < \ldots < t_i < \ldots < t_m = T$, and m. Depending on what is known about the t_i the problem may be straightforward or unsolvable. Set $\delta = \max\{t_i - t_{i-1} : i = 1, 2, \ldots, m\}$. For *admissible samplings* we assume that $\delta \to 0$ as $m \to \infty$.

R. Klette et al. (eds.), Geometric Properties for Incomplete Data, 59-75.

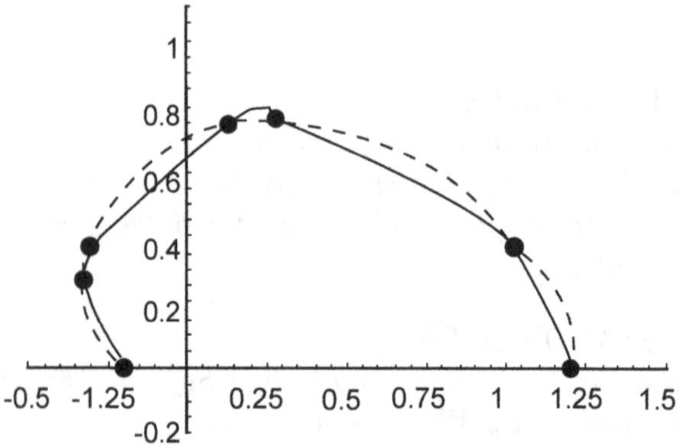

Figure 4.1. 7 data points on the spiral γ_{sp} (dashed) in Example 4.2, interpolated by a uniform piecewise-quadratic $\hat{\gamma}_{sp}$ (solid), with $d(\hat{\gamma}_{sp}) = d(\gamma_{sp}) + 2.422 \times 10^{-3}$.

EXAMPLE 4.1. (Kozera et al., 2003) *If the t_i are given then piecewise Lagrange interpolation through successive $k+1$-tuples $(q_i, q_{i+1}, q_{i+2}, \ldots, q_{i+k})$, $k \geq 1$ and $i = 0, k, 2k, 3k, \ldots$, approximates γ with uniform errors at most δ^{k+1}. Without real loss we may suppose m divisible by k.*

However in practice the t_i may not be given, so that Q is the only data available for approximation. Then γ can at most be approximated up to reparameterizations: we seek piecewise-quadratic and piecewise-cubic curves $\hat{\gamma} : [0, \hat{T}] \to \mathbb{R}^n$ with $\tilde{\gamma} \equiv \hat{\gamma} \circ \psi$ uniformly close to γ, for some piecewise-C^1 reparameterization $\psi : [0, T] \to [0, \hat{T}]$.

DEFINITION 4.1. *A family $\{f_\delta, \delta > 0\}$ of functions $f_\delta : [0, T] \to \mathbb{R}$ is said to be $O(\delta^p)$ when there is a constant $K > 0$ such that, for some $\delta_0 > 0$, $|f_\delta(t)| < K\delta^p$ for all $\delta \in (0, \delta_0)$ and all $t \in [0, T]$. In such a case write $f_\delta = O(\delta^p)$. For a family of vector-valued functions $F_\delta : [0, T] \to \mathbb{R}^n$, write $F_\delta = O(\delta^p)$ when $\|F_\delta\| = O(\delta^p)$, where $\|\cdot\|$ denotes the Euclidean norm.*

An approximation $\tilde{\gamma} : [0, T] \to \mathbb{R}^n$ to γ determined by Q is said to have *order p* when $\tilde{\gamma} - \gamma = O(\delta^p)$. Of course a larger order means the image of $\hat{\gamma}$ resembles $\gamma([0, T])$ more closely, at least in the limit as $\delta \to 0$. A different but related comparison can be made between the length

$$d(\gamma) \equiv \int_0^T \|\dot{\gamma}(t)\| \, dt$$

of γ and that of $\hat{\gamma}$. Piecewise-linear interpolation approximates γ to order 2, but piecewise-quadratic approximations can actually degrade estimates:

EXAMPLE 4.2. *If we guess* $t_i \approx \hat{t}_i = \frac{i}{m}$, *then the resulting* uniform piecewise-quadratic $\hat{\gamma} : [0,1] \to \mathbb{R}^n$ *is sometimes uninformative. For instance suppose* $t_i = \frac{(3i+(-1)^{i+1})}{3m}$, *take* $n = 2$, *and define a spiral* $\gamma_{sp} : [0,1] \to \mathbb{R}^2$ *by* $\gamma_{sp}(t) = (t + 0.2)(\cos(\pi(1-t)), \sin(\pi(1-t)))$. *Then* $d(\gamma_{sp}) = 2.452$. *When* m *is small* $\hat{\gamma}_{sp}$ *does not much resemble* γ_{sp}*: in Fig. 4.1,* $m = 6$ *and* $d(\hat{\gamma}_{sp}) = d(\gamma_{sp}) + 2.422 \times 10^{-3}$ *Errors in length tend to cancel and the curve is a worse approximation than these numbers suggest. When* m *is large* $\hat{\gamma}_{sp}$ *looks more like* γ_{sp}*: in Figure 4.2,* $m = 30$. *In this case however* $d(\hat{\gamma}_{sp}) = d(\gamma_{sp}) + 8.163 \times 10^{-2}$, *with nearly* 34 *times the error for* $m = 6$. *Even piecewise-linear interpolation with* 31 *points is better, with error* -3.622×10^{-3}.

Without knowing the t_i it seems difficult to match the order 3 achieved in Example 4.1 (and also to extend it to the same orders for length estimation), but it turns out that *higher* order approximations are achievable for many planar curves γ, and for fairly general sampling schemes, including *more-or-less uniform* sampling:

DEFINITION 4.2. *The* t_i *are said to be sampled* more-or-less uniformly *when there are constants* $0 < K_l < K_u$ *such that, for any sufficiently large integer* m, *and all* $1 \leq i \leq m$,

$$\frac{K_l}{m} \leq t_i - t_{i-1} \leq \frac{K_u}{m} .$$

Then $\frac{K_l}{m} \leq \delta \leq \frac{K_u}{m}$, *and increments between successive parameters are neither large nor small in proportion to* $\frac{T}{m}$.

In Example 4.2 sampling is more-or-less uniform, with $K_l = \frac{1}{3}, K_u = \frac{5}{3}$. On the other hand the samplings from Example 4.5, 4.6, and 4.9 are not more-or-less uniform. For $n = 2$, with $\gamma : [0,T] \to \mathbb{R}^2$ being C^4 and strictly convex, Theorem 4.1 below guarantees order 4 approximations by piecewise quadratics $\hat{\gamma}$, called *piecewise-4-point quadratics* because the quadratic arcs interpolate *quadruples* of points in Q rather than triples as in Example 4.1, 4.2. Implementations of such schemes, also studied in (de Boor et al., 1987), (Lachance and Schwartz, 1991), (Mørken, 1997), (Rababah, 1995), (Schaback, 1989), usually require solutions of systems of nonlinear equations.

THEOREM 4.1. (Noakes and Kozera, 2002), (Noakes and Kozera, 2003) *Let* $n = 2$. *Let* γ *be regular, strictly convex and* C^r *where* $r \geq 4$. *Let* Q *be sampled more-or-less uniformly. Then there is a piecewise-quadratic curve* $\hat{\gamma} : [0,1] \to \mathbb{R}^2$, *calculable in terms of* Q, *and a piecewise-*C^r *reparameterization* $\psi : [0,T] \to [0,1]$, *with* $\hat{\gamma} \circ \psi = \gamma + O(\delta^4)$, *and* $d(\hat{\gamma}) = d(\gamma) + O(\delta^4)$.

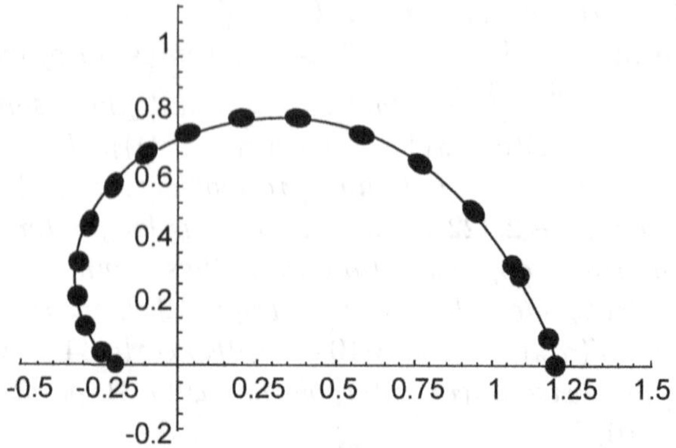

Figure 4.2. 31 points on the spiral in Example 4.2, interpolated by a uniform piecewise-quadratic $\hat{\gamma}_{sp}$ (solid), with $d(\hat{\gamma}_{sp}) = d(\gamma_{sp}) + 8.163 \times 10^{-2}$.

The requirement that γ be planar and strictly convex seems very restrictive. An alternative, noted in Chapter 11 of (Kvasow, 2000), is Lagrange interpolation based on *cumulative chord length parameterizations* (Epstein, 1976), (Lee, 1992). More precisely, set

$$\hat{t}_0 = 0 \quad \text{and} \quad \hat{t}_j = \hat{t}_{j-1} + \|q_j - q_{j-1}\| \,, \quad \text{for} \quad j = 1, 2, \ldots, m \,.$$

For k dividing m and $i = 0, k, 2k, \ldots, m-k$, let $\hat{\gamma}$ be the curve satisfying $\hat{\gamma}(\hat{t}_j) = q_j$ for all $j = 0, 1, 2, \ldots, m$, and whose restriction $\hat{\gamma}^i$ to each $[\hat{t}_i, \hat{t}_{i+k}]$ is polynomial of degree at most k. Call $\hat{\gamma} : [0, \hat{T}] \to \mathbb{R}^n$, where $\hat{T} = \sum_{j=1}^{m} \|q_j - q_{j-1}\|$, the *cumulative chord piecewise degree-k approximation* to γ defined by Q. Our *main result* is

THEOREM 4.2. *Suppose γ is a regular C^r curve in $\mathbb{R}^n$, where $r \geq k+1$ and k is 2 or 3. Let $\hat{\gamma} : [0, \hat{T}] \to \mathbb{R}^n$ be the cumulative chord piecewise degree-k approximation defined by Q. Then there is a piecewise-C^r reparameterization $\psi : [0, T] \to [0, \hat{T}]$, with $\hat{\gamma} \circ \psi = \gamma + O(\delta^{k+1})$, and if additionally $m\delta = O(1)$ then $d(\hat{\gamma}) = d(\gamma) + O(\delta^{k+1})$. Also, if $r \geq 4$, $k = 2$, and if, for some $\varepsilon \in (0, 1]$, we have the uniformity condition*

$$t_{i+1} - 2t_i + t_{i-1} = O(\delta^{1+\varepsilon}) \quad \textit{for} \quad i = 1, 3, 5, \ldots, m-1 \,, \tag{1}$$

then if $m\delta = O(1)$ we have $d(\hat{\gamma}) = d(\gamma) + O(\delta^{3+\varepsilon})$.

EXAMPLE 4.3. *Sampling is said to be ε-uniform when there is a C^∞ reparameterization $\phi : [0, T] \to [0, T]$ such that*

$$t_i = \phi(\frac{Ti}{m}) + O(\frac{1}{m^{1+\varepsilon}}) \,.$$

Then (1) *holds. For piecewise-quadratic interpolation based on guessed uniform grids* $\hat{t}_i = i$, $\hat{\gamma} = \gamma + O(\delta^{1+2\varepsilon})$, *and* $d(\hat{\gamma}) = d(\gamma) + O(\delta^{4\varepsilon})$ (Noakes et al., 2001), (Noakes et al., 2001). *Evidently, Theorem 4.2 asserts better orders of approximation except when* $\varepsilon = 1$. *Note here that any* ε*-uniform sampling arises from two types of perturbations of uniform samplings: first via a diffeomorphic distortion* $\phi : [0,T] \to [0,T]$ *combined subsequently with added extra distortion term* $O(1/m^{1+\varepsilon})$. *The perturbation via* ϕ *has no effect on both* $d(\gamma)$ *and geometrical representation of* γ. *The only potential nuisance stems from the second perturbation term* $O(1/m^{1+\varepsilon})$.

Unlike Theorem 4.1, Theorem 4.2 holds for any sufficiently smooth regular curve γ (not necessarily convex) in any Euclidean space $\mathbb{R}^n$, and is applicable even without tight conditions on sampling (for length estimation we need a weak constraint $m\delta = O(1)$ which is automatically satisfied by more-or-less and ε-uniform samplings). Cumulative chord piecewise-cubics approximate at least to order 4, as do the piecewise-4-point quadratics of Theorem 4.1. Notice also that cumulative chord piecewise-quadratics and piecewise-cubics approximate to the same order as in Example 4.1, where the t_i are given. Cumulative chord piecewise-quadratics also match length estimates for ε-uniform sampling where the t_i are given (Kozera et al., 2003).

A related work on cumulative chord piecewise-quartics and C^1 interpolation with cumulative chord cubics can be found in (Kozera, 2003), (Kozera, 2004), (Kozera and Noakes, 2004), and (Kozera and Noakes, 2005). After some preliminaries in Section 4.2, *the main result* i.e. Theorem 4.2 is proved in Section 4.3. In Section 4.4, Theorem 4.2 is illustrated by examples and compared with other results. In the last Section 4.5 the relevant conclusions are drawn and some possible applications are hinted.

4.2. Divided Differences and Cumulative Chords

First recall some facts about divided differences (de Boor, 2001): the *first divided difference* of γ at t_i is

$$\gamma[t_i, t_{i+1}] \equiv \frac{\gamma(t_{i+1}) - \gamma(t_i)}{t_{i+1} - t_i},$$

and, for $k = 2, 3, \ldots, m-i$, the kth *divided difference* is defined inductively as

$$\gamma[t_i, t_{i+1}, \ldots, t_{i+k}] \equiv \frac{\gamma[t_{i+1}, t_{i+2}, \ldots, t_{i+k}] - \gamma[t_i, t_{i+1}, \ldots, t_{i+k-1}]}{t_{i+k} - t_i}.$$

Newton's Interpolation Formula is $\gamma(t) = L + R$, where

$$L \equiv \gamma(t_i) + (t - t_i)\gamma[t_i, t_{i+1}] + (t - t_i)(t - t_{i+1})\gamma[t_i, t_{i+1}, t_{i+2}] +$$

$$\ldots + (t - t_i)(t - t_{i+1}) \ldots (t - t_{i+k-1})\gamma[t_i, t_{i+1}, \ldots, t_{i+k}]$$

is the polynomial of degree at most k interpolating γ at $t_i, t_{i+1}, \ldots, t_{i+k}$, and

$$R \equiv (t - t_i)(t - t_{i+1}) \ldots (t - t_{i+k})\gamma[t, t_i, t_{i+1}, \ldots, t_{i+k}] \,.$$

When γ is C^{k+1} and $t_i, t_{i+1}, t_{i+2}, \ldots, t_{i+k+1} \in (t - \tilde{\delta}, t + \tilde{\delta})$ where $\tilde{\delta} > 0$ then, for $j = 1, 2, \ldots, n$, the jth component of the $k+1$th divided difference is given by

$$\gamma[t, t_i, t_{i+1}, \ldots, t_{i+k}]_j = \frac{\gamma_j^{(k+1)}(\tilde{t}_j)}{(k+1)!} \,, \tag{2}$$

for some $\tilde{t}_j \in (t-\tilde{\delta}, t+\tilde{\delta})$. We now work with the hypotheses of Theorem 4.2. In particular γ is C^r and regular, where $r \geq k+1$ and k is 2 or 3. After a C^r reparameterization, as in (Klingenberg, 1978) Chapter 1, Proposition 1.1.5, we can assume γ is parameterized by arc-length, namely $\|\dot{\gamma}\|$ is identically 1. Then

$$\langle \dot{\gamma}, \dot{\gamma} \rangle \equiv 1 \,, \quad \langle \ddot{\gamma}, \dot{\gamma} \rangle \equiv 0 \,, \quad \langle \frac{d^3\gamma}{dt^3}, \dot{\gamma} \rangle \equiv -\langle \ddot{\gamma}(t), \ddot{\gamma}(t) \rangle = -\kappa(t)^2, \tag{3}$$

where $\kappa(t)$ is the curvature $\|\ddot{\gamma}(t)\|$ of γ at $t \in [0, T]$. For $i = 0, k, 2k, \ldots, m-k$, let $\psi^i : [t_i, t_{i+k}] \to [\hat{t}_i, \hat{t}_{i+k}]$ be the polynomial function of degree at most k satisfying $\psi^i(t_{i+j}) = \hat{t}_{i+j}$, for $0 \leq j \leq k$ [the analytical formula for ψ^i is given by Equation (10)]. Substituting for the $\hat{t}_{i+j}$ in the first divided difference, we find

$$\psi^i[t_{i+j}, t_{i+j+1}] = \|\gamma[t_{i+j}, t_{i+j+1}]\| \,, \tag{4}$$

for $j = 0, 1, \ldots, k-1$. Using (4) to substitute in the second divided difference,

$$\psi^i[t_i, t_{i+1}, t_{i+2}] = \frac{\|\gamma[t_{i+1}, t_{i+2}]\| - \|\gamma[t_i, t_{i+1}]\|}{t_{i+2} - t_i} \,, \quad \text{and so}$$

$$|\psi^i[t_i, t_{i+1}, t_{i+2}]| \leq \frac{\|\gamma[t_{i+1}, t_{i+2}] - \gamma[t_i, t_{i+1}]\|}{t_{i+2} - t_i} = \|\gamma[t_i, t_{i+1}, t_{i+2}]\| \,.$$

By (2) and because γ is C^2, the right hand side is bounded, and therefore $\psi^i[t_i, t_{i+1}, t_{i+2}] = O(1)$. The same can be said for the right hand side of (4). The assumption that γ is C^3 permits us to say more:

LEMMA 4.1. *$\psi^i[t_{i+j}, t_{i+j+1}] = 1+O(\delta^2)$, for $j = 0, 1$. Also $\psi^i[t_i, t_{i+1}, t_{i+2}] = O(\delta)$.*

Proof By (4) and Taylor's Theorem, $\psi^i[t_{i+j}, t_{i+j+1}] = \|\gamma[t_{i+j}, t_{i+j+1}]\|$

$$= \|\dot{\gamma}(t_{i+j}) + \frac{\ddot{\gamma}(t_{i+j})}{2}(t_{i+j+1} - t_{i+j}) + O((t_{i+j+1} - t_{i+j})^2)\| \,. \tag{5}$$

Because $\|\dot{\gamma}\| \equiv 1$ and $\langle \dot{\gamma}, \ddot{\gamma} \rangle \equiv 0$, the Binomial Theorem gives

$$\psi^i[t_{i+j}, t_{i+j+1}] = 1 + O((t_{i+j+1} - t_{i+j})^2) \,, \tag{6}$$

proving the first part. The second assertion follows by comparing (6) with the definition of the second divided difference, because t_i increases with i. □

The next lemma uses the sampling condition (1).

LEMMA 4.2. *Let γ be C^4. Then $\psi^i[t_{i+j}, t_{i+j+1}, t_{i+j+2}] = O(\delta^{1+\varepsilon})$ for $0 \le j \le k-1$. Also if $k = 3$, then $\psi^i[t_i, t_{i+1}, t_{i+2}, t_{i+3}] = O(1)$.*

Proof By Taylor's Theorem, $\|\gamma[t_{i+j}, t_{i+j+1}]\|$ is

$$\|\dot{\gamma} + \frac{\ddot{\gamma}}{2}(t_{i+j+1} - t_{i+j}) + \frac{d^3\gamma}{dt^3}\frac{(t_{i+j+1} - t_{i+j})^2}{6} + O((t_{i+j+1} - t_{i+j})^3)\| \,,$$

where $\dot{\gamma}$, $\ddot{\gamma}$, $\frac{d^3\gamma}{dt^3}$ are evaluated at t_{i+j}, and $0 \le j \le k$. Therefore, and by Equation (3), $\|\gamma[t_{i+j}, t_{i+j+1}]\|$ is

$$\begin{aligned}
&(1 - \frac{\kappa^2}{12}(t_{i+j+1} - t_{i+j})^2 + O((t_{i+j+1} - t_{i+j})^3))^{1/2} \\
&= 1 - \frac{\kappa^2}{24}(t_{i+j+1} - t_{i+j})^2 + O((t_{i+j+1} - t_{i+j})^3) \,,
\end{aligned} \tag{7}$$

by the Binomial Theorem, with κ evaluated at t_{i+j}. Therefore, for $0 \le j \le k-1$,

$$\begin{aligned}
\psi^i[t_{i+j}, t_{i+j+1}, t_{i+j+2}] &= -\kappa^2 \frac{(t_{i+j+2} - t_{i+j+1})^2 - (t_{i+j+1} - t_{i+j})^2}{24(t_{i+j+2} - t_{i+j})} \\
&\quad + O((t_{i+3} - t_i)^2) \\
&= -\frac{\kappa^2}{24}(t_{i+j+2} - 2t_{i+j+1} + t_{i+j}) \\
&\quad + O((t_{i+3} - t_i)^2) \,,
\end{aligned} \tag{8}$$

where κ is evaluated at t_{i+j}. This proves the first part. For $k = 3$ we then obtain

$$\begin{aligned}
\psi^i[t_i, t_{i+1}, t_{i+2}, t_{i+3}] &= -\frac{\kappa^2}{24}\frac{(t_{i+3} - 3t_{i+2} + 3t_{i+1} - t_i)}{t_{i+3} - t_i} + O(t_{i+3} - t_i) \\
&= O(1) \,,
\end{aligned} \tag{9}$$

where κ is evaluated at t_i. □

By Newton's Interpolation Formula, as ψ^i is polynomial of degree at most 3,

$$\begin{aligned}\psi^i(t) &= \psi^i(t_i) + (t-t_i)\psi^i[t_i,t_{i+1}] + (t-t_i)(t-t_{i+1})\psi^i[t_i,t_{i+1},t_{i+2}] \quad (10)\\ &\quad +(k-2)(t-t_i)(t-t_{i+1})(t-t_{i+2})\psi^i[t_i,t_{i+1},t_{i+2},t_{i+3}]\,,\end{aligned}$$

for $k = 2, 3$. Differentiating twice, for $k = 2, 3$, we first have

$$\dot{\psi}^i(t) = \psi^i[t_i,t_{i+1}] + (2t-t_i-t_{i+1})\psi^i[t_i,t_{i+1},t_{i+2}] + (k-2)\cdot$$

$$((t-t_{i+1})(t-t_{i+2})+(t-t_i)(t-t_{i+2})+ (t-t_i)(t-t_{i+1}))\psi^i[t_i,t_{i+1},t_{i+2},t_{i+3}]\,,$$

and then

$$\begin{aligned}\ddot{\psi}^i(t) &= 2\psi^i[t_i,t_{i+1},t_{i+2}]\\ &\quad +2(k-2)(3t-t_i-t_{i+1}-t_{i+2})\psi^i[t_i,t_{i+1},t_{i+2},t_{i+3}]\,.\end{aligned}$$

So, by Lemmas 4.1, 4.2,

LEMMA 4.3. *$\dot{\psi}^i = 1 + O(\delta^2)$ and $\ddot{\psi}^i = O(\delta)$. When γ is C^4 and $k = 3$, $\frac{d^3\psi^i}{dt^3} = O(1)$.*

In particular, ψ^i is a C^∞ diffeomorphism for δ small, which we assume from now on.

LEMMA 4.4. *For $k = 2, 3$, and $s \in [\hat{t}_i, \hat{t}_{i+k}]$, $\hat{\gamma}^i(s)$, $\frac{d\hat{\gamma}^i}{ds}$, $\frac{d^2\hat{\gamma}^i}{ds^2}$ are $O(1)$. For $k = 3$ and if γ is C^4, then $\frac{d^3\hat{\gamma}^i}{ds^3}$ is $O(1)$.*

Proof $\hat{\gamma}^i$ is the polynomial of degree at most k interpolating $\gamma \circ (\psi^i)^{-1}$ at $t_i, t_{i+1}, \ldots, t_{i+k}$, and the derivatives to order k of $\gamma \circ (\psi^i)^{-1}$ are $O(1)$ by Lemma 4.3. By (2) the divided differences to order k of $\gamma \circ (\psi^i)^{-1}$ are also $O(1)$. By Newton's Interpolation Formula, these are nonzero constant multiples of the derivatives of $\hat{\gamma}$ to order k. □

Remark. Let $f_i : [a_i, b_i] \to \mathbb{R}^n$ with $a_i < b_i$, be given for $0 \le i \le l$. By the *track-sum* of $\{f_i\}_{i=0}^{l}$ we understand the function $f : [0, \bar{T}] \to \mathbb{R}^n$, where $\bar{T} = \sum_{i=0}^{l}(b_i - a_i)$ satisfying:

$$\begin{aligned}f(t) &= f_0(t+a_0)\,,\ t \in [0,\bar{T}_0]\,,\ \bar{T}_0 = b_0 - a_0\,;\\ f(t) &= f_{k+1}(t+a_{k+1}-\bar{T}_k)\,,\ t \in [\bar{T}_k, \bar{T}_{k+1}]\,,\ \bar{T}_{k+1} = \bar{T}_k + b_{k+1} - a_{k+1}\,;\end{aligned}$$

for $0 \le k \le l-1$.

Let $\psi : [0, T] \to [0, \hat{T}]$ be the track-sum of the ψ^i defined in Equation (10).

4.3. Proof of Main Result - Theorem 4.2

By Lemmas 4.3, 4.4, all coefficients of the polynomials $\hat{\gamma}^i$ and ψ^i are $O(1)$, So the restriction $\hat{\gamma}^i \circ \psi^i$ of $\hat{\gamma} \circ \psi$ to $[t_i, t_{i+k}]$ is a polynomial of degree at most k^2 and with all coefficients $O(1)$. So all derivatives of the C^{k+1} function $f^i \equiv \hat{\gamma}^i \circ \psi^i - \gamma : [t_i, t_{i+k}] \to \mathbb{R}^n$ are $O(1)$. By Lemma 2.1 of Part I of (Milnor, 1963), because $f^i(t_{i+j}) = \mathbf{0}$ for $0 \le j \le k$,

$$f^i(t) = (t - t_i) \dots (t - t_{i+k-1}) g^i(t) \quad \text{where} \quad g^i(t) = (t - t_{i+k}) h^i(t) \,,$$

with $g^i, h^i : [t_i, t_{i+k}] \to \mathbb{R}^n$ C^1 and C^0 respectively. Also $h^i = O(\frac{d^{k+1} f^i}{dt^{k+1}}) = O(1)$, so that $g^i = O(\delta)$, and $\dot{g}^i = O(\frac{d^{k+1} f^i}{dt^{k+1}}) = O(1)$. So

$$f^i = O(\delta^{k+1}) \quad \text{and} \quad \dot{f}^i = O(\delta^k) \,. \tag{11}$$

In particular $\hat{\gamma} \circ \psi$ approximates γ uniformly with $O(\delta^{k+1})$ errors. To compare lengths, write $\hat{\gamma}^i \circ \psi^i \equiv \tilde{\gamma}^i$. Then $\dot{\tilde{\gamma}}^i(t) = (1 + \langle \dot{f}^i(t), \dot{\gamma}(t) \rangle) \dot{\gamma}(t) + v(t)$, where $v(t)$ is the projection of $\dot{f}^i(t)$ onto the line orthogonal to $\dot{\gamma}(t)$. By (11), $v = O(\delta^k)$ and, because $\|\dot{\gamma}\| \equiv 1$,

$$\|\dot{\tilde{\gamma}}^i(t)\| = (1 + \langle \dot{f}^i(t), \dot{\gamma}(t) \rangle) \|\dot{\gamma}(t)\| + O(\delta^{2k}) \,.$$

Then

$$\int_{t_i}^{t_{i+k}} (\|\dot{\tilde{\gamma}}^i(t)\| - \|\dot{\gamma}(t)\|) \, dt = \int_{t_i}^{t_{i+k}} \langle \dot{f}^i(t), \dot{\gamma}(t) \rangle \, dt + O(\delta^{2k+1})$$

which, on integration by parts, becomes $- \int_{t_i}^{t_{i+k}} \langle f^i(t), \ddot{\gamma}(t) \rangle \, dt + O(\delta^{2k+1})$. By (11) the right hand side is $O(\delta^{k+2})$, namely

$$\int_{t_i}^{t_{i+k}} \|\dot{\gamma}(t)\| \, dt - d(\tilde{\gamma}^i) = O(\delta^{k+2}) \,.$$

So as $m\delta = O(1)$ we arrive that

$$d(\hat{\gamma}) = \Sigma_{j=0}^{\frac{m}{k}-1} d(\tilde{\gamma}^{jk}) = d(\gamma) + O(\delta^{k+1}) \,.$$

When $r \ge 4$, $k = 2$, and (1) holds, we can say more. As before,

$$\int_{t_i}^{t_{i+2}} (\|\dot{\tilde{\gamma}}^i(t)\| - \|\dot{\gamma}(t)\|) \, dt = - \int_{t_i}^{t_{i+2}} \langle f^i(t), \ddot{\gamma}(t) \rangle \, dt + O(\delta^5) \,. \tag{12}$$

Now $h^i = O(1)$ and, since $r \ge 4$, h^i is C^1 with $\dot{h}^i = O(1)$ by Lemma 2.1 of Part I of (Milnor, 1963) again. Therefore $\langle h^i(t), \ddot{\gamma}(t) \rangle = a_i + O(\delta)$, where $a_i \equiv \langle h^i(t_i), \ddot{\gamma}(t_i) \rangle = O(1)$, and

$$\langle f^i(t), \ddot{\gamma}(t) \rangle \;=\; a_i (t - t_i)(t - t_{i+1})(t - t_{i+2}) + O(\delta^4) \,.$$

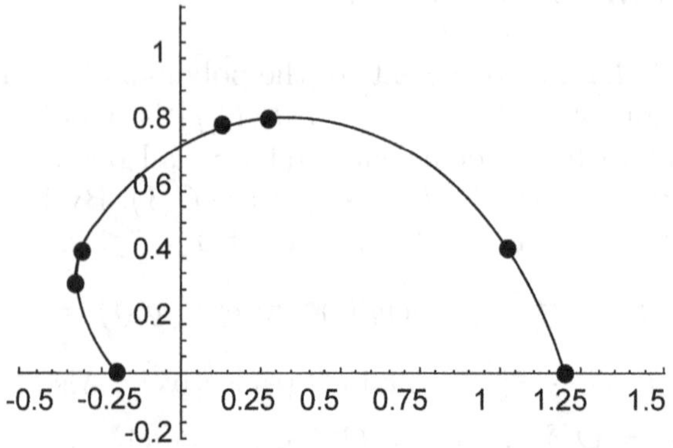

Figure 4.3. 7 points on the spiral γ_{sp} (dashed) in Example 4.2, 4.4, interpolated by a piecewise-4-point quadratic $\hat{\gamma}_{sp}$ (solid), with $d(\hat{\gamma}_{sp}) = d(\gamma_{sp}) - 1.155 \times 10^{-2}$.

Now by (1)

$$\int_{t_i}^{t_{i+2}} (t-t_i)(t-t_{i+1})(t-t_{i+2})\,dt = \frac{1}{12}(t_i-t_{i+2})^3(t_{i+2}-2t_{i+1}+t_i) = O(\delta^{4+\varepsilon}) .$$

So Equation (12) gives

$$\int_{t_i}^{t_{i+2}} (\|\dot{\tilde{\gamma}}^i(t)\| - \|\dot{\gamma}(t)\|)\,dt = O(\delta^{4+\varepsilon}) ,$$

and then as $m\delta = O(1)$

$$\Sigma_{j=0}^{\frac{m}{2}-1} d(\tilde{\gamma}^{2j}) = d(\gamma) + O(\delta^{3+\varepsilon}) .$$

This supplementary argument does not apply when $k = 3$. Indeed, in the next Section the orders of approximation in Theorem 4.2 for cumulative chord piecewise-cubics are seen to be best-possible, even when sampling is ε-uniform.

4.4. Numerical Experiments

Here are some experiments, using *Mathematica*, and admissible samplings from smooth regular curves in $\mathbb{R}^2$ and $\mathbb{R}^3$. First we verify Theorem 4.1, 4.2 in the situation of Example 4.2.

EXAMPLE 4.4. *Uniform piecewise-quadratics, piecewise-4-point quadratics, cumulative chord piecewise-quadratics and cumulative chord piecewise-cubics based on the 7-tuple Q of Example 4.2 are shown as solid curves in*

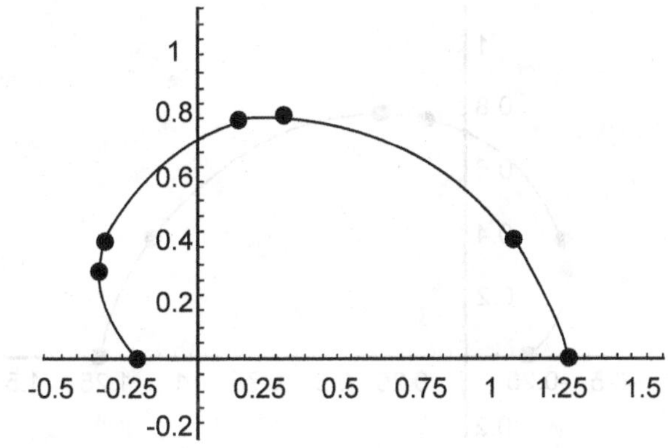

Figure 4.4. 7 points on the spiral γ_{sp} (dashed) in Example 4.4, interpolated by a cumulative chord piecewise-quadratic $\hat{\gamma}_{sp}$ (solid), with $d(\hat{\gamma}_{sp}) = d(\gamma_{sp}) - 2.459 \times 10^{-2}$.

Figure 4.1, 3, 4.4, and 4.5 respectively. Figure 4.3, 4.4 show markedly better approximations to γ_{sp} *than Figure 4.1, and in Figure 4.5 the cumulative chord piecewise-cubic is nearly indistinguishable from the dashed spiral. The respective errors in length estimates are* -1.155×10^{-2}, -2.459×10^{-2}, -1.048×10^{-3} *and* 2.422×10^{-3}. *For larger values of* m *differences in performance are more marked:*

- *Uniform piecewise-quadratics behave badly with respect to length estimates, as noted in Example 4.2.*
- *Piecewise-4-point quadratics with* $m = 30, 99, 198$ *yield errors* 5.499×10^{-5}, 5.127×10^{-7}, 3.200×10^{-8} *respectively in length estimates. The numerical estimate of order of convergence for length estimates, based on samples of up to* 199 *points is* 3.93.
- *Cumulative chord piecewise-quadratics with* $m = 30, 100, 200$ *yield errors* 1.738×10^{-4}, 4.332×10^{-6}, 5.302×10^{-7} *respectively in length estimates. The numerical estimate of order of convergence for length estimates, based on samples of up to* 201 *points is* 3.03.
- *Cumulative chord piecewise-cubics with* $m = 30, 99, 198$ *yield errors* 9.514×10^{-6}, 8.741×10^{-8}, 5.670×10^{-9} *respectively. The numerical estimate of order of convergence for length estimates, based on samples of up to* 199 *points is* 3.96.

So the orders of convergence for length estimates given in Theorem 4.2 *for cumulative chord piecewise-quadratics and piecewise-cubics are sharp. Note that condition* $m\delta = O(1)$ *holds for sampling from Example 4.2 used also here. Although* $\hat{\gamma}_{sp}$ *is not* C^1, *differences in left and right derivatives (at* t_3 *in Figure 4.3, 4.5 and* t_2, t_4 *in Figure 4.4) are hardly discernible for*

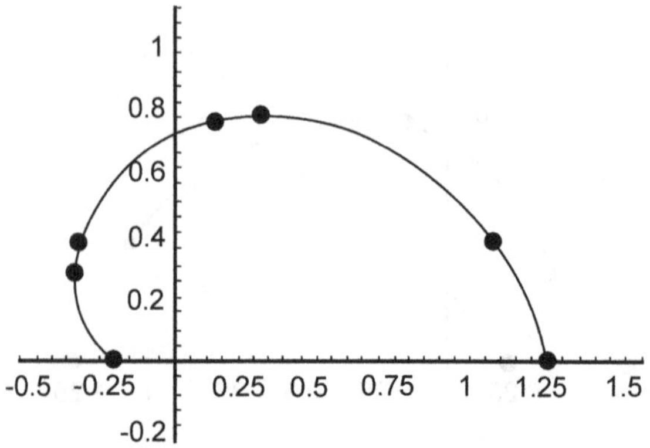

Figure 4.5. 7 points on the spiral γ_{sp} of Example 4.4, interpolated by a cumulative chord piecewise-cubic $\hat{\gamma}_{sp}$ (solid), with $d(\hat{\gamma}_{sp}) = d(\gamma_{sp}) - 1.048 \times 10^{-3}$.

piecewise-4-point-quadratics and cumulative chord piecewise-quadratics and piecewise-cubics. Such features are practically invisible when m is large.

Piecewise-4-point quadratics and cumulative chord piecewise-cubics have the same orders of convergence, but cumulative chord piecewise-cubics and piecewise-quadratics are more generally applicable: *curves need not be planar or convex and sampling need not be more-or-less uniform* for estimating orders of approximation.

EXAMPLE 4.5. *Let $\gamma_c : [0,1] \to \mathbb{R}^2$ be the cubic, with inflection point $(0,0)$, given by $\gamma_c(t) = (t - 0.5, 4(t - 0.5)^3)$. Given m, take t_i to be $\frac{i}{m}$ or $\frac{(i-1)}{m} + \frac{1}{m^2}$ according as i is even or odd. Then sampling is clearly not more-or-less uniform, though again $m\delta = O(1)$ holds. The plot for cumulative chord piecewise-quadratic interpolation of $-\log|d(\hat{\gamma}_c) - d(\gamma_c)|$ against $\log m$ in Figure 4.6, for $m = 4, 6, \ldots, 100$, appears almost linear, with least squares estimate of slope 3.86. Theorem 4.2 says the slope should be at least 3.*

EXAMPLE 4.6. *Figure 4.7 shows a cumulative chord piecewise-quadratic interpolant $\hat{\gamma}_h$ of 9 points on the elliptical helix $\gamma_h : [0, 2\pi] \to \mathbb{R}^3$, given by $\gamma_h(t) = (1.5\cos t, \sin t, t/4)$. The exact formula for the sampling is defined below. Although sampling is uneven, sparse, and not available for interpolation, $\hat{\gamma}$ seems very close to γ_h: $d(\gamma_h) = 8.090$ and $d(\hat{\gamma}_h) = 8.019$. For (not more-or-less uniform) samplings where t_i is $\frac{2\pi i}{m}$ or $\frac{2\pi(i-1)}{m} + \frac{2\pi}{m^{3/2}}$ according as i is even or odd, and $m = 50, 52, \ldots, 200$, the order of convergence for length with cumulative piecewise-quadratics is estimated as 3.91. Theorem*

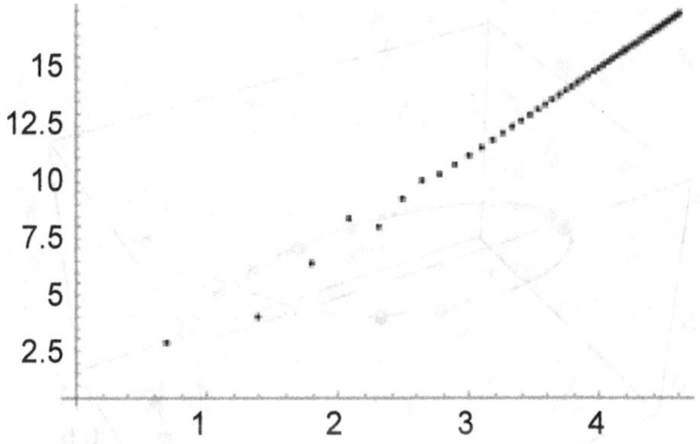

Figure 4.6. Plot of -log $|d(\hat{\gamma}_c) - d(\gamma_c)|$ against $\log m$ for a cubic with inflection point, sampled irregularly, as in Example 4.5.

4.2 asserts at least 3. *Note that here* $m\delta = O(1)$ *also holds. Figure 4.8 shows a cumulative chord piecewise-cubic interpolant* $\hat{\gamma}_h$ *of* 10 *points on the helix. Approximation is visually very good, and* $d(\hat{\gamma}_h) = 8.179$. *Using* $m = 75, 78, \ldots, 300$ *the numerical estimate of order of convergence for length is* 4.013 *confirming sharpness of Theorem 4.2 in respect of piecewise-cubics without strong conditions on sampling.*

Next we verify sharpness of Theorem 4.2 in respect of piecewise-quadratics with sampling conditions of the form (1) and $m\delta = O(1)$.

EXAMPLE 4.7. *Let* $\gamma_c : [0,1] \to \mathbb{R}^2$ *be the cubic* $\gamma_c(t) = (\pi t, (\frac{\pi t+1}{\pi+1})^3)$. *Given* m, *define* $t_i = \frac{i}{m} + \frac{(-1)^{i+1}}{m^{1+\varepsilon}}$ *for* $0 \leq i \leq m$. *Then sampling is* ε-*uniform, and cumulative chord piecewise-quadratic interpolation for* $m = 40, 42, \ldots, 200$, *with* $\varepsilon = 0.0, 0.1, 0.25, 0.5, 0.75, 1, 3$, *yields convergence orders rates for length estimations as* $\alpha_0 = 2.98, 3.09, 3.25, 3.50, 3.76, 4.01, 3.97$, *respectively for length estimates. We found no additional increase in convergence order for* $\varepsilon > 1$.

A key ingredient in the proof of Theorem 4.2 is to show that all coefficients of quadratic and cubic ψ^i are $O(1)$. This need not be the case for the higher degree ψ^i needed to extend the proof to show higher order approximations by cumulative chord piecewise-quartics (see (Kozera, 2003) or (Kozera, 2004)).

EXAMPLE 4.8. *For* m *divisible by* 4 *and* $i = 0, 4, 8, \ldots, m-4$, *consider (very nearly uniform) samplings of the form*

$$t_i = \frac{i}{m}, \quad t_{i+1} = \frac{i+1}{m}, \quad t_{i+2} = \frac{i+2}{m} + \frac{1}{2m}, \quad t_i = \frac{i+3}{m}, \quad t_{i+4} = \frac{i+4}{m}.$$

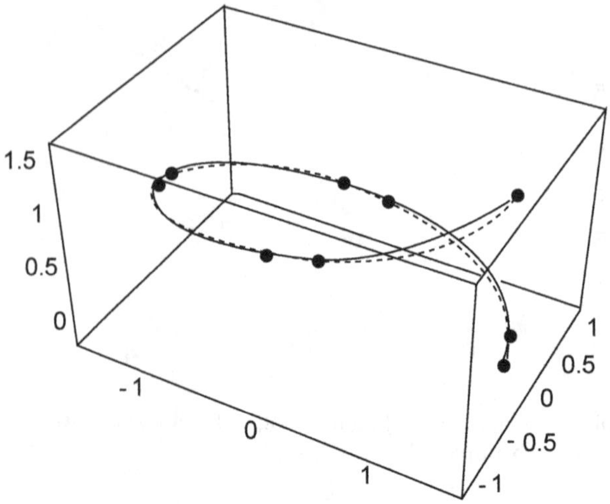

Figure 4.7. 9 points on the elliptical helix γ_h (dashed) in Example 4.6, interpolated by a cumulative chord piecewise-quadratic $\hat{\gamma}_h$ (solid), with $d(\hat{\gamma}_h) = d(\gamma_h) - 7.029 \times 10^{-2}$.

Given C^4 $\gamma : [0, T] \to \mathbb{R}^n$, let $\psi^i : [t_i, t_{i+4}] \to [\hat{t}_i, \hat{t}_{i+4}]$ be the polynomial of degree at most 4 satisfying $\psi^i(t_{i+j}) = \hat{t}_{i+j}$ for $j = 0, 1, 2, 3, 4$. The proof of Lemma 4.2 shows that the $\psi^i[t_{i+j}, t_{i+j+1}, t_{i+j+2}, t_{i+j+3}] = O(1)$ for $j = 0, 1,$ and consequently $\alpha \equiv \psi^i[t_i, t_{i+1}, t_{i+2}, t_{i+3}, t_{i+4}] = O(m)$. Writing $d_j = \|q_{i+j} - q_i\|$ for $j = 1, 2, 3, 4$, a calculation gives

$$\alpha = \frac{m^4}{90}(-3d_1 + 7d_2 - 25d_3 + 5d_4) . \tag{13}$$

As in the proof of Lemma 4.2,

$$d_1 = \frac{1}{m}(1 - \frac{\kappa}{24m^2} + O(\frac{1}{m^3})) , \qquad d_2 = \frac{3}{2m}(1 - \frac{3\kappa}{32m^2} + O(\frac{1}{m^3})) ,$$
$$d_3 = \frac{1}{2m}(1 - \frac{\kappa}{96m^2} + O(\frac{1}{m^3})) , \qquad d_4 = \frac{1}{m}(1 - \frac{\kappa}{24m^2} + O(\frac{1}{m^3})) ,$$

where κ is evaluated at t_i. Substituting into Equation (13),

$$\alpha = -\frac{\kappa}{96}m + O(1) .$$

So, except when γ is affine, α is unbounded.

EXAMPLE 4.9. *Notice that the condition $m\delta = O(1)$ (satisfied implicitly by more-or-less and ε-uniform samplings but not by general admissible*

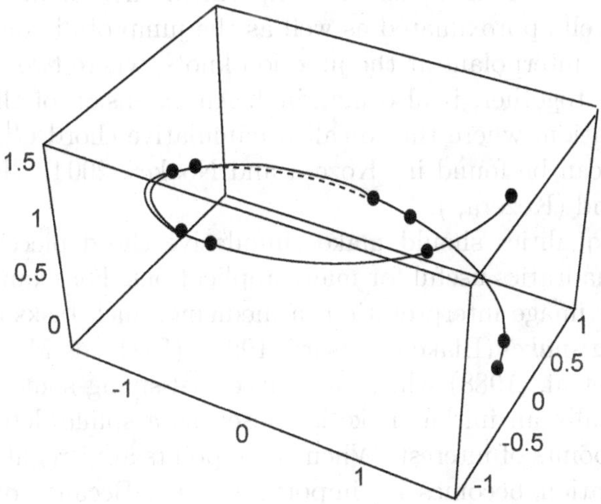

Figure 4.8. 10 points on the elliptical helix γ_h (dashed) in Example 4.6, interpolated by a cumulative chord piecewise-cubic $\hat{\gamma}_h$ (solid), with $d(\hat{\gamma}_h) = d(\gamma_h) + 8.952 \times 10^{-2}$.

samplings, where $\delta \to 0$) is a necessary condition for Theorem 4.2 to hold. Indeed, for γ_c defined as in Example 4.7 and sampled according to

$$t_0 = 0 \quad \text{and} \quad t_i = \frac{1}{\sqrt{m}} + \frac{(i-1)(\sqrt{m}-1)}{(m-1)\sqrt{m}}, \quad 1 \leq i \leq m$$

(for which $m\delta \neq O(1)$) cumulative chord piecewise-quadratic interpolation gives and estimate for $d(\gamma)$ approximation (with $2 \leq m \leq 200$) equal to 2.09 < 3, *claimed by Theorem 4.2 and* $k = 2$. *A similar effect appears for* $k = 3$.

4.5. Conclusions

Quartic and *cubic orders of convergence* for length and trajectory estimates established in Theorem 4.2 for cumulative chord piecewise-cubics and piecewise-quadratics are *sharp*. At least the latter was verified for $n = 2, 3$ in case of computing $d(\gamma)$. The above orders match the corresponding orders of approximations for piecewise-cubics and piecewise-quadratics used with tabular points t_i known (assumed unavailable in our discussion). Curves need not be planar nor convex and sampling need not be more-or-less uniform. For length estimation an extra weak but necessary condition on sampling $m\delta = O(1)$ is needed. Our scheme also performs well on *sporadic data* (i.e. $m << \infty$). As demonstrated experimentally, for m small, both γ and $d(\gamma)$ are well approximated as well as the jump of discontinuities in derivatives of the interpolant at the junction knots, where two consecutive chords are glued together, is also marginal. An extension of this work to smooth interpolation, where the so-called cumulative chord C^1 piecewise-cubics are used can be found in (Kozera and Noakes, 2004), (Kozera and Noakes, 2005) and (Kozera,).

These good qualities should make cumulative chord piecewise-cubics and piecewise-quadratics useful for many applications. For example, image segmentation for image interpretation in medicine. Such tasks are usually approached using *snakes* (Blake and Isard, 1998), (Desbleds-Mansard et al., 2001) or (Kass et al., 1988) which are curves satisfying some variational condition. Typically an initial snake is chosen as a spline determined by specifying data points of interest. When these points are irregularly spaced the parameterization becomes an important issue. Because of the good behavior noted already, cumulative chord piecewise-cubics and piecewise-quadratics seem an excellent choice.

Acknowledgements: This work was supported by the Alexander von Humboldt Foundation Fellowship.

References

Blake, A. and M. Isard: *Active Contours*. Springer, Berlin, 1998.

de Boor, C.: *A Practical Guide to Splines*. Springer, New York, 2001.

de Boor, C., K. Höllig, and M. Sabin: High accuracy geometric Hermite interpolation. *Computer Aided Geom. Design*, **4**:269–278, 1987.

Desbleds-Mansard, C., A. Anwander, L. Chaabane, M. Orkisz, B. Neyran, P. C. Douek, and I. E. Magnin: Dynamic active contour model for size independent blood vessel lumen segmentation and quantification in high-resolution magnetic resonance images. In Proc. *Computer Analysis Images Patterns* (W. Skarbek, editor), pages 264–273, LNCS 2124, Springer, Berlin, 2001.

Epstein, M. P.: On the influence of parameterization in parametric interpolation. *SIAM J. Numer. Anal.*, **13**:261–268, 1976.

Kass, M., A. Witkin, and D. Terzopoulos: Active contour models. *Int. J. Comp. Vision*, **1**:321–331, 1988.

Klingenberg, W.: *A Course in Differential Geometry*. Springer, New York, 1978.

Kozera, R.: Cumulative chord piecewise-quartics for length and trajectory estimation. In Proc. *Computer Analysis Images Patterns* (N. Petkov and M. A. Westenberg, editors), pages 697–705, LNCS 2756, Springer, Berlin, 2003.

Kozera, R.: Asymptotics for length and trajectory from cumulative chord piecewise-quartics. *Fundamenta Informaticae*, **61**:267–283, 2004.

Kozera, R.: Curve modeling via interpolation based on multidimensional reduced data. *Studia Informatica*, **25**, No. 4B(61) :1–140, 2004.

Kozera, R. and L. Noakes: C^1 interpolation with cumulative chord cubics. *Fundamenta Informaticae*, **61**:285–301, 2004.

Kozera, R. and L. Noakes: Smooth interpolation with cumulative chord cubics. In Proc. *Computer Vision Graphics* (R. Kozera, L. Noakes, W. Skarbek, and K. Wojciechowski, editors), Kluwer Academic Publishers. In press (2005).

Kozera, R., L. Noakes, and R. Klette: External versus internal parameterization for lengths of curves with nonuniform samplings. In *Geometry Morphology Computat. Imaging* (T. Asano, R. Klette, and C. Ronse, editors), pages 403–418, LNCS 2616, Springer, Berlin, 2003.

Kvasov, B. I.: *Methods of Shape-Preserving Spline Approximation.* World Scientific, Singapore, 2000.

Lachance, A. and A. J. Schwartz: Four point parabolic interpolation. *Computer Aided Geom. Design*, **8**:143–149, 1991.

Lee, E. T. Y.: Corners, cusps and parameterization: Variations on a theorem of Epstein. *SIAM J. of Numer. Anal.*, **29**:553–565, 1992.

Milnor, J. W.:*Morse Theory.* Annals of Math. Studies **51**, Princeton University Press, 1963.

Mørken, K. and K. Scherer: A general framework for high-accuracy parametric interpolation. *Math. Computat.*, **66**:237–260, 1997.

Noakes, L. and R. Kozera: Interpolating sporadic data. In Proc. *European Conf. Computer Vision* (A. Heyden, G. Sparr, M. Nielsen, and P. Johansen, editors), pages 613–625, LNCS 2351/II, Springer, Berlin, 2002.

Noakes, L. and R. Kozera: More-or-less uniform sampling and lengths of curves. *Quar. Appl. Math.*, **61**:475–484, 2003.

Noakes, L., R. Kozera, and R. Klette: Length estimation for curves with different samplings. In *Digital and Image Geometry* (G. Bertrand, A. Imiya, and R. Klette, editors), pages 339–351, LNCS 2243, Springer, Berlin, 2001.

Noakes, L., R. Kozera, and R. Klette: Length estimation for curves with ε-uniform samplings. In Proc. *Computer Analysis Images Patterns* (W. Skarbek, editor), pages 339–351, LNCS 2124, Springer, Berlin, 2001.

Rababah, A.: High order approximation methods for curves. *Computer Aided Geom. Design*, **12**:89–102, 1995.

Schaback, R.: Interpolation in $\mathbb{R}^2$ by piecewise quadratic visually C^2 Bézier polynomials. *Computer Aided Geom. Design*, **6**:219–233, 1989.

SPHERICAL SPLINES

LYLE NOAKES
School of Mathematics and Statistics
The University of Western Australia

Abstract. The idea of replacing line segments by geodesics in Riemannian manifolds to generalize classical constructions of Bezier polynomials has been around at least since 1985 (Shoemake, 1985), (Duff, 1985). However rather little is known about generalized Bezier curves in spheres. The practical use of generalized Bezier curves for interpolating spherical data is limited by a lack of systematic methods for constructing spherical control polygons. This is a serious defect, because spherical Bezier splines are sensitive to the choice of control polygon, and the control polygons are difficult to visualize even in the practically significant case of S^3. After reviewing some of the difficulties with the standard alternative of normalized quadratic polynomial splines, we focus on elementary properties of generalized Bezier quadratics in spheres. These C^∞ spherical curves remain within the spherical convex hull of the control polygon, resemble the control polygon, and their derivatives at endpoints can be written down explicitly. Using these results, we show how to blend generalized Bezier quadratic segments into generalized C^1 Bezier quadratic *splines*. We then introduce two methods for the automatic construction of control polygons for generalized Bezier quadratic splines, resulting in spherical splines that are optimal according to two different criteria. The J_E-optimal curves are very easy to construct, and much better behaved than normalized quadratic splines. The J_S-optima are more costly to compute, with correspondingly better properties.

Key words: interpolation, spline, Bezier, control polygon, sphere, geodesic

5.1. Introduction

For $m \geq 1$ denote the unit sphere

$$\{x \in E^{m+1} : \|x\| = 1\} \subset E^{m+1}$$

by S^m, where E^{m+1} is Euclidean $m+1$-space, namely $\mathbb{R}^{m+1}$ equipped with the Euclidean norm,

$$\|x\| \equiv \sqrt{x_1^2 + x_2^2 + \ldots + x_{m+1}^2}.$$

The problem of interpolating and approximating spherical data in the m-dimensional manifold S^m is much more widespread than might at first be

R. Klette et al. (eds.), Geometric Properties for Incomplete Data, 77-101.

thought (Angeles and Akras, 1989), (Brady et al., 1982), (Paul, 1979), (Tan and Potts, 1989), (Taylor, 1979), (Watson, 1983), (Fisher, 1993), (Fisher et al., 1987), (Buss and Fillmore, 2001). As explained in Section 5.8, interpolation in S^3 is closely related to interpolation in the group $SO(3)$ of rotations of E^3, and thereby to planning of rigid body trajectories.

Section 5.2 reviews some of significant approaches to problems of this kind. The literature has many tradeoffs between mathematical difficulty, quality of performance, computational speed, and ease of implementation. From the point of view of quality (and mathematical interest) there is a lot to be said for the variational approach, but such methods are complicated to implement and computationally costly. The present chapter focuses on methods that are easy to code and computationally quick, while avoiding the performance problems associated with familiar methods. Interpolants based on classical methods in E^3. like the normalized polynomial splines in Section 5.3, sometimes perform very badly. Readers most interested in engineering implementations may choose to go directly to Section 5.8 for pseudocode for spherical splines adapted to quadratic interpolation in $SO(3)$.

As well as these practical issues, we take the opportunity in Section 5.4 to explore the mathematics of the building blocks for our spherical Bezier splines, namely the spherical Bezier *quadratics*. Although such generalized Bezier curves in spheres have been used since the mid 1980s ((Shoemake, 1985), (Duff, 1985)), little has been said about their mathematical properties. Theorem 5.1 contributes some properties of the derivatives of these nonpolynomial curves. Then, in Section 5.5 Theorem 5.1 is applied to prove properties of spherical Bezier splines (Theorem 5.2). If a spherical Bezier spline is *degenerate* there may be isolated points where the curves are not C^1: such occurrences are rare.

At the end of 5.5 we turn to the practical problem of choosing an initial velocity for a spherical Bezier quadratic spline when this is not given by the data. For ordinary quadratic polynomial splines in Euclidean space this amounts to choosing a control polygon for the interpolant. Spherical Bezier quadratic splines are sensitive to such choices and, unlike the Euclidean situation, their control polygons are difficult to visualise. So an automatic method is needed for choosing spherical control polygons or, equivalently, initial velocities y_1. Section 5.6 introduces a performance measurement J_E for y_1, whose optimum is usually unique, given in closed form by Theorem 5.3. This method is very easy to implement, extremely quick, and much better behaved than normalized polynomial splines (Example 5.4). Of course J_E is only one of many plausible performance measurements.

A more geometrical criterion is given by J_S in Section 5.7. There may be more than one J_S-optimum, and because there is no closed-form formula,

J_S-optima are found by an iterative numerical scheme. The optimization is simple to implement and performs quickly (the pseudocode in Section 5.7 has 7 easy steps). The additional computational effort sometimes gives substantial improvements over the J_E-optimum (Example 5.5). Spherical Bezier quadratic splines using either J_E or J_S-optima offer good compromises between the excellent performance of variational methods, and the speed and ease of implementation of schemes based on polynomial splines.

5.2. Background

There are many methods for interpolation and approximation in manifolds, with a rich variety of interesting features, and no single method has all the advantages of classical polynomial spline interpolation in E^m.

- *Chart-based interpolants* are generally low-quality, but still enjoy a degree of popularity, because they are easily implemented and closely related to classical methods. The idea is to describe smooth m-manifolds locally in terms of subsets of $\mathbb{R}^m$. For instance, using stereographic projection, the whole of S^m is described using only two charts. Then within each chart we have available the whole apparatus of classical approximation theory in Euclidean m-space E^m. There are book-keeping problems in keeping track of local interpolants and piecing them together. More immediate practical diffficulties arise through the geometrical distortions inherent in modelling S^m locally by subsets of E^m. In extreme cases these distortions manifest as near-singularities in charts, with serious consequences for quality and stability of interpolants, and these problems usually occur to some extent unless many charts are used (Noakes, 2003). With more numerous charts book-keeping problems come to the fore, causing different kinds of instabilities.
- A different approach, also building on classical approximation theory, is to normalize interpolants in the ambient space E^{m+1}. Normalized polynomial splines, for instance, are even easier to implement than chart-based interpolants, and falls down in similar ways when consecutive spherical data points are widely spaced, as seen in Section 5.3. Again the most immediate difficulties are caused by geometrical distortions which can cause large variations in speeds of interpolants. Because of these variations, normalized polynomial splines are generally unsuitable for interpolating sparse or censored data, especially where velocity estimates are needed.
- *Variational interpolants* in manifolds (Gabriel and Kajiya, 1985), (Noakes et al., 1989) minimize integrals of Lagrangians, usually of order 2 or higher, which have something to do with the performance of the inter-

polant and usually generalize some accepted performance measurement in classical approximation theory. The variational theory has received quite a lot of attention (Gabriel and Kajiya, 1985), (Noakes et al., 1989), (Chapman and Noakes, 1991), (Brunnett, 1994), (Camarinha et al., 2001), (Camarinha, 1996), (Silva-Leite et al., 2000), (Camarinha et al., 1995), (Chapman and Noakes, 1991), (Crouch and Silva-Leite., 1995), (Krakowski, 2003), (Noakes, 2003), (Noakes (a), 2004), (Noakes (b), 2004), (Giambo et al., 2002), (Giambo et al., 2003), and the interpolants are typically of very high quality: indeed *optimal* in some well-defined sense. The theory has links to classical mechanics, and is mathematically significant, but hard to implement, and not yet ready for real-time applications.

– By replacing line segments in Euclidean space with , some constructions of classical approximation theory can be extended to define in manifolds (Shoemake, 1985), (Duff, 1985). This attractive idea uses geometrical methods to avoid the worst features of chart-based and normalized spline interpolants. On the other hand, although generalized Bezier curves do not quite match the quality of variational interpolants (Noakes, 2003), they are reasonably straightforward to implement. From the beginning, generalized Bezier curves in manifolds were intended to be used as segments of splines. For more recent work see (Crouch et al., 1999), (Altafini, 2001). However there are some major obstacles to the practical use of generalized Bezier curves for interpolation, especially are

  - still some degree of mystery about their mathematical properties, and
  - lack of systematic methods for constructing spherical control polygons to use generalized Bezier curves as segments of generalized *splines*. Comparing Examples 5.4, 5.5 of the present paper, the spherical control polygon is seen to be critical for the quality of the interpolant.

The present paper contributes in both respects. In Section 5.4 some mathematical results are proved about generalized Bezier curves. Using these results, Section 5.6 and Section 5.7 give methods for constructing spherical control polygons that are *optimal* in two different senses, leading to automatic constructions of control polygons. Automatic methods are valuable, because control polygons in S^m are difficult to visualize for $m > 2$, and the case $m = 3$ is significant in applications to rigid body motion. Our methods yield pleasing results in examples where normalized polynomial splines are unsuccessful.

- For still more methods of interpolation in manifolds we refer to (Buss and Fillmore, 2001), (Barr et al., 1992), (Noakes, 1994), (Noakes, 1997), (Noakes, 1998), (Kang and Park, 1999), (Park and Ravani, 2001), (Zefran et al., 1996), (Zefran and Kumar (a), 1996), (Zefran and Kumar (b), 1996), (Zefran and Kumar, 1998), (Zefran et al., 1998), (Belta and Kumar, 2002), (Altafini, 2001), (Altafini, 2001a).

In Section 5.4 we review in S^m, and establish some of their mathematical properties, including relationships between generalized Bezier curves, , and the spherical convex hull of the control vertices. Estimates are given for speeds of generalized Bezier curves. We also prove results on derivatives at endpoints of generalized Bezier curves, and use these in Section 5.5 to construct . The generalized splines depend in an essential way on a choice of parameter $y_1 \in S^m$, which determines the control polygon of the generalized Bezier quadratic spline. Section 5.6 gives a straightforward method for choosing y_1 in an optimal way. The resulting J_E-optimal spherical splines are easy to implement, and have much better properties than normalized quadratic splines, as seen by comparing Examples 5.1, 5.4 where the same data is interpolated by a normalized quadratic polynomial spline and the J_E-optimal spherical spline respectively. In Section 5.7 a second method is introduced, where y_1 optimizes the more geometrical quantity $J_S(y_1)$. This requires more numerical computation than the J_E-optimum, but yields beautiful interpolants in tricky cases like Example 5.5, compared with in Example 5.1 and the J_E-optimum in Example 5.4. First, for background, we review normalized quadratic splines.

5.3. Normalized Quadratic Polynomial Splines

A C^1 quadratic polynomial spline $x : [0, n] \to E^{m+1}$ interpolating $x_0, x_1, \ldots, x_n \in E^{m+1}$ at $0, 1, 2, \ldots, n$ may be given as follows.

- Choose $v_0 \in E^{m+1}$ and let $x|[0,1]$ be the quadratic polynomial satisfying $x(0) = x_0$, $\dot{x}_+(0) = v_0$, $x(1) = x_1$.
- For $1 < i \leq n$ let $x|[i-1, i]$ be the quadratic polynomial satisfying $x(i-1) = x_{i-1}$, $\dot{x}_+(i-1) = \dot{x}_-(i-1)$, $x(i) = x_i$.

As discussed in (de Boor, 2001) Chapter VI, the choice of v_0 may have a significant global effect on x. In the present paper the x_i are given in S^m and x is required to map into S^m, where $m \geq 1$. Unless the x_i are all equal a quadratic polynomial spline x is not a curve in S^m, but if x is never $\mathbf{0}$ its normalization $\hat{x}$, given by

$$\hat{x}(t) \equiv \frac{x(t)}{\|x(t)\|},$$

is C^1 and can be used instead. This works best when the x_i are sampled frequently and regularly from a smooth curve. In other cases normalized polynomial splines can be unsatisfactory.

EXAMPLE 5.1. *For* $m = 2$, $n = 7$, v_0 *the midpoint of* x_0, x_1, *and*

$$\begin{aligned}
x_0 &= (\ \ 1.000000, & 0.0000000, & \ \ 0.0000000)\\
x_1 &= (-0.995244, & 0.0497622, & \ \ 0.0837469)\\
x_2 &= (\ \ 0.990989, & 0.0990989, & \ \ 0.0901104)\\
x_3 &= (-0.988840, & 0.1483260, & \ \ 0.0139545)\\
x_4 &= (\ \ 0.977820, & 0.1955780, & -0.0740071)\\
x_5 &= (-0.965972, & 0.2414930, & -0.0926294)\\
x_6 &= (\ \ 0.957483, & 0.2872450, & -0.0267536)\\
x_7 &= (-0.942049, & 0.3297170, & \ \ 0.0618913)
\end{aligned}$$

the normalized quadratic polynomial spline interpolant $\hat{x} : [0, 7] \to S^2$ *with breakpoints* $1, 2, 3, 4, 5, 6$ *is shown in Figure 5.1, together with the* x_i *labelled by* i. *Although* $\hat{x}$ *is* C^1, *changes in direction near breakpoints are relatively abrupt. Points are plotted on a uniform grid of size* 3500 *in* $[0, 7]$, *showing large variations in speed.*

An alternative is to use a C^1 quadratic polynomial spline interpolant $x : [0, n] \to E^{m+1}$ with breakpoints $3/2, 5/2, \ldots, (2n-3)/2$, following

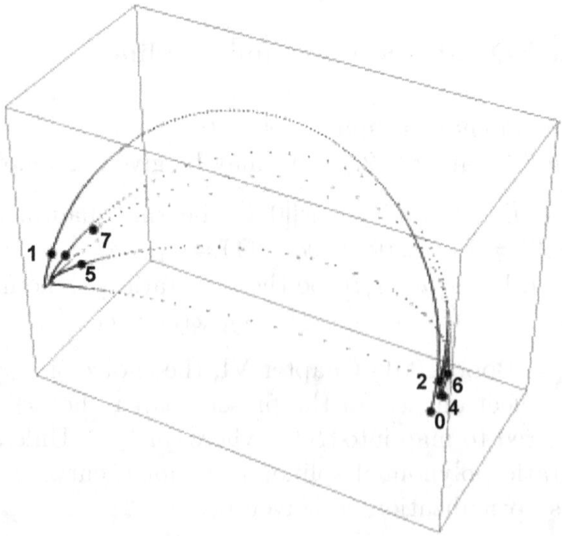

Figure 5.1. $\hat{x} : [0, 7] \to S^2$ in Example 5.1.

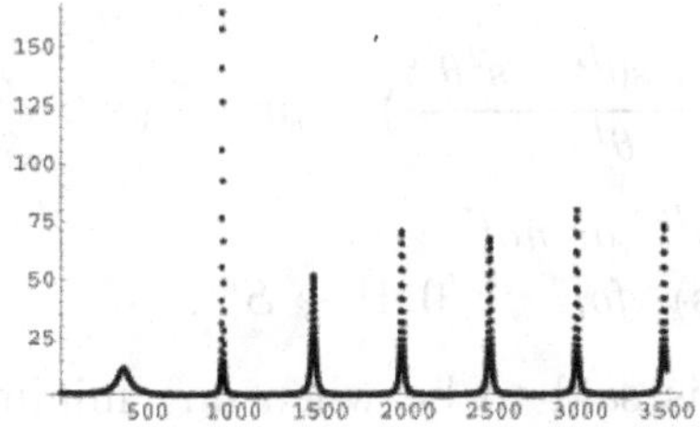

Figure 5.2. Speeds of $\hat{x} : [0, 7] \to S^2$ in Example 5.1.

Chapter VI of (de Boor, 2001) : details are given in an Appendix. If the x_i lie in S^m and x is never $\mathbf{0}$ its normalization is a C^1 interpolant $\hat{x} : [0, n] \to S^m$. However the phenomena observed in Example 5.1 persist. To address these difficulties we first replace normalized quadratic polynomials by *spherical Bezier quadratics*, which are curves in S^m defined using spherical geometry. The task of blending spherical Bezier quadratics into spherical Bezier *splines* is taken up in Section 5.5.

5.4. Spherical Bezier Quadratics

The *spherical distance* $d(y, z)$ for $y, z \in S^m$ is defined as $\arccos < y, z >$, where arccos maps to $[0, \pi]$ and $< \, , \, >$ is the Euclidean inner product. A *great circle arc* is a constant-speed curve $\gamma : [a, b] \to S^m$ whose image is contained in the intersection of S^m with some 2-dimensional vector subspace of E^{m+1}. The arc is *minimal* when $\gamma(s) \neq -\gamma(t)$ for any $a < s < t < b$. In the following definition, and in Theorem 5.1, the most important case is where $\dot{\gamma}^L(1)$ and $\dot{\gamma}^R(0)$ are linearly independent.

DEFINITION 5.1. *Let $\gamma^L, \gamma^R : [0, 1] \to S^m$ be minimal great circle arcs, of lengths θ^L, θ^R, satisfying $\gamma^L(1) = \gamma^R(0)$. Define $\beta \equiv \beta(\gamma^L, \gamma^R) : [0, 1] \to S^m$ as follows.*

- *If $\dot{\gamma}^L(1), \dot{\gamma}^R(0)$ are linearly independent $d(\gamma^L(s), \gamma^R(s)) < \pi$ for $s \in (0, 1)$. Define $\beta(s) = \gamma^{M,s}(s)$, where $\gamma^{M,s} : [0, 1] \to S^m$ is the minimal great circle arc from $\gamma^L(s)$ to $\gamma^R(s)$. Define $\beta(0) = \gamma^L(0)$ and $\beta(1) = \gamma^R(1)$.*
- *If $\theta^R \dot{\gamma}^L(1) = \theta^L \dot{\gamma}^R(0)$, let $\gamma^M : [0, 1] \to S^m$ be the great circle arc from $\gamma^L(0)$ to $\gamma^R(1)$, such that $\gamma^M([0, 1]) = \gamma^L([0, 1]) \cup \gamma^R([0, 1])$. If $\theta^L = \theta^R = 0$ let $\beta(s) = \gamma^L(0)$ for all $s \in [0, 1]$. Otherwise define*

$$\beta(s) \;=\; \gamma^M\left(\frac{(2 - s)s\theta^L + s^2\theta^R}{\theta^L + \theta^R}\right).$$

– *If* $\theta^R\dot{\gamma}^L(1) = -\theta^L\dot{\gamma}^R(0)$, *define*

$$\beta(s) \;=\; \gamma^L(\frac{(2-s)s\theta^L - s^2\theta^R}{\theta^L}) \quad or \quad \beta(\gamma^L,\gamma^R) \;=\; \bar{\beta}(\bar{\gamma}^R,\bar{\gamma}^L),$$

according as $\theta^L \geq \theta^R$ *or not.*
Here $\bar{y}(s) \equiv y(1-s)$ *for* $y : [0,1] \to S^m$.

A subset K of S^m is said to be when any minimal great circle arc whose endpoints are contained in K is entirely contained in K. Therefore S^m is spherically convex, and so is any intersection of its spherically convex subsets. The intersection of all spherically convex subsets of S^m containing a given subset X of S^m is called the (spherical) *convex hull* $< X >$. When X contains a pair of antipodal points $< X >$ is the whole of S^m, but the converse is false, for instance when $X = \{(\cos j\pi/3, \sin j\pi/3) : j = 0,1,2\} \subset S^1$. From Definition 5.1,

$$\beta([0,1]) \subseteq \; < \{\gamma^L(0), \gamma^L(1) = \gamma^R(0), \gamma^R(1)\} > .$$

For $s \in [0,1]$, $d(\beta(s),\gamma^L(0)) \leq s\theta^L + s((1-s)\theta^L + s\theta^R) = s((2-s)\theta^L + s\theta^R)$, and the quantity on the right is maximized when $s = 1$. Therefore, and since

$$\beta(\gamma^L,\gamma^R) \;=\; \bar{\beta}(\bar{\gamma}^R,\bar{\gamma}^L), \tag{1}$$

$$\max\{d(\beta(s),\gamma^L(0)), d(\beta(s),\gamma^R(1))\} \;\leq\; \theta^L + \theta^R.$$

Define $\pi : [0,1] \to S^m$ by $\pi(s) = \gamma^L(2s)$ or $\gamma^R(2s-1)$ for $s \leq 1/2$ or $s \geq 1/2$ respectively. The piecewise great-circular spherical curve π can be thought of as the control polygon of the generalized Bezier curve β. The Bezier curve resembles its control polygon as follows. The following result is proved in an Appendix.

PROPOSITION 5.1. *For* $s \in [0,1/2]$ *and* $s \in [1/2,1]$ *respectively,*

$$\begin{aligned} d(\beta(s),\pi(s)) &\leq (1-s)\theta^L + s\theta^R \;-\; |(1-3s+s^2)\theta^L + (s-s^2)\theta^R| \\ &\leq \frac{3\theta^L + \theta^R}{4} \\ d(\beta(s),\pi(s)) &\leq (1-s)\theta^L + s\theta^R \;-\; |(s-s^2)\theta^L + (-1+s+s^2)\theta^R| \\ &\leq \frac{\theta^L + 3\theta^R}{4}. \end{aligned}$$

COROLLARY 5.1. $d(\beta(1/2),\pi(1/2)) \leq \frac{1}{4}(\theta^L+\theta^R) + \frac{1}{2}\min\{\theta^L,\theta^R\}$ *and, for all* $s \in [0,1]$,

$$d(\beta(s),\pi(s)) \;\leq\; (1-s)\theta^L + s\theta^R.$$

Evidently β is continuous on $[0,1]$, C^∞ on $(0,1)$, $\beta(0) = \gamma^L(0)$, $\beta(1) = \gamma^R(1)$. To study derivatives of β, we first define $D \equiv \{(y,z) \in S^m \times S^m : d(y,z) < \pi\}$ and $\delta : D \times [0,1] \to S^m$ by $\delta(y,z,q) = \gamma_{y,z}(q)$ where $\gamma_{y,z} : [0,1] \to S^m$ is the minimal great circle arc from y to z. Then $\delta(y,z,q) = \delta(z,y,1-q)$, and $\delta(Ry,Rz,q) = R\delta(y,z,q)$ for any $R \in SO(m+1)$. In an Appendix we prove

LEMMA 5.1. *Given* $(y,z) \in D$, $q \in [0,1]$ *and* C^∞ $\omega : [0,1] \to S^m$ *with* $\omega(0) = y$, *write* $\dot\omega(0) = v^{\|} + v^{\perp}$ *where* $v^{\|}$ *is a multiple of* $\dot\gamma_{y,z}(0)$ *and* $< v^{\perp}, \dot\gamma_{y,z}(0) >= 0$. *For* $\theta \equiv d(y,z) \neq 0$,

$$\frac{d}{dt}|_{t=0}\delta(\omega(t),z,q) \;=\; \frac{< v^{\|}, \dot\gamma_{y,z}(0) >}{\theta^2}\,\dot\gamma_{y,z}(q) \;+\; \frac{\sin(1-q)\theta}{\sin\theta}\,v^{\perp}$$

and when $\theta = 0$ *the left hand side is* $\dot\omega(0)$.

From Lemma 5.1 and symmetry of δ,

LEMMA 5.2.

$$\|\frac{\partial\delta(y,z,q)}{\partial y}\| \;\le\; \frac{1}{\sin\theta}, \qquad \|\frac{\partial\delta(y,z,q)}{\partial z}\| \;\le\; \frac{1}{\sin\theta}, \qquad \|\frac{\partial\delta(y,z,q)}{\partial q}\| = \theta.$$

If $\dot\gamma^L(1), \dot\gamma^R(0)$ are linearly dependent then from Definition 5.1

$$\|\dot\beta(s)\| \;\le\; 2\max\{\theta^L, \theta^R\}.$$

PROPOSITION 5.2. *Let* $\dot\gamma^L(1), \dot\gamma^R(0)$ *be linearly independent, and for* $s \in [0,1]$ *let* θ_s *be the length of the minimal great circle arc* $\gamma^{M,s}$. *Then*

$$\|\dot\beta(s)\| \;\le\; \theta_s \;+\; \frac{\theta^L + \theta^R}{\sin\theta_s}.$$

The proof is given in an Appendix. Another Appendix proves much more precise statements about velocities at endpoints:

THEOREM 5.1. *Except when* $\dot\gamma^L(1), \dot\gamma^R(0)$ *are linearly independent with either* $\theta^L = \pi$ *or* $\theta^R = \pi$,

$$\dot\beta(0)_+ \;=\; 2\dot\gamma^L(0) \quad \textit{and} \quad \dot\beta(1)_- \;=\; 2\dot\gamma^R(1). \tag{2}$$

If $\dot\gamma^L(1), \dot\gamma^R(0)$ *are linearly independent with* $\theta^L = \pi$,

$$\dot\beta(0)_+ \;=\; \dot\gamma^L(0) \;+\; \pi\,\frac{\dot\gamma^L(0) + \dot\gamma^R(0)}{\|\dot\gamma^L(0) + \dot\gamma^R(0)\|}. \tag{3}$$

If $\dot\gamma^L(1), \dot\gamma^R(0)$ *are linearly independent with* $\theta^L = \pi$ *and* $\theta^R < \pi$,

$$\dot\beta(1)_- = 2\dot\gamma^R(1). \tag{4}$$

If $\dot\gamma^L(1), \dot\gamma^R(0)$ *are linearly independent with* $\theta^R = \pi$,

$$\dot\beta(1)_- = \dot\gamma^R(1) + \pi \frac{\dot\gamma^R(1) + \dot\gamma^L(1)}{\|\dot\gamma^R(1) + \dot\gamma^L(1)\|}. \tag{5}$$

If $\dot\gamma^L(1), \dot\gamma^R(0)$ *are linearly independent with* $\theta^R = \pi$ *then, according as* $\theta^L < \pi$ *or not,*

$$\dot\beta(0)_+ = 2\dot\gamma^L(0) \quad or \quad \dot\beta(0)_+ = \dot\gamma^L(0) + \pi \frac{\dot\gamma^L(0) + \dot\gamma^R(0)}{\|\dot\gamma^L(0) + \dot\gamma^R(0)\|}. \tag{6}$$

Now we interpolate spherical data by spherical Bezier quadratic *splines*, namely .

5.5. Spherical Bezier Quadratic Splines

Given $x_0, x_1, x_2, \ldots, x_n \in S^m$, define self-adjoint linear endomorphisms ρ_i of E^{m+1} by

$$\rho_i(w) \equiv 2 < w, x_i > x_i - w, \quad \text{where } w \in E^{m+1} \text{ and } 0 < i < n.$$

A (cardinal) *spherical (Bezier quadratic) spline* is a curve $x : [0, n] \to S^m$ constructed as follows.

DEFINITION 5.2. *Choose a minimal great circle arc* $\gamma_1^L : [0,1] \to S^m$ *with* $\gamma_1^L(0) = x_0$. *Set* $y_1 \equiv \gamma_1^L(1)$ *and, for* $0 < i < n$, $y_{i+1} = \rho_i(y_i)$. *Choose minimal great circle arcs* $\gamma_i^R : [0,1] \to S^m$ *from* y_i *to* x_i, *and let* $\gamma_{i+1}^L = \rho_i \circ \bar\gamma_i^R$. *For* $i = 1, 2, \ldots, n$ *and* $t \in [i-1, i]$ *set*

$$x(t) = \beta(\gamma_i^L, \gamma_i^R)(t - i + 1).$$

The spherical spline x is C^∞ away from breakpoints $1, 2, \ldots n-1$, and everywhere-continuous. Also $x(i) = x_i$ for $i = 0, 1, \ldots n$. The arcs γ_i^L, γ_i^R for $i = 1, 2, \ldots n$ constitute the spherical *control polygon*, and with *control vertices* y_i. The spherical spline x is *nondegenerate* when, for all $i = 2, \ldots, n$, $y_i \neq -x_{i-1}$. In an Appendix we prove

THEOREM 5.2. *Let* x *be a spherical spline. Then* $\|\dot x(i)_+\| \le 2\pi$ *for* $i = 0, 1, 2, \ldots, n-1$ *and* $\|\dot x(i)_-\| \le 2\pi$ *for* $i = 1, 2, \ldots, n$. *If* $m = 1$ *or* x *is nondegenerate* x *is* C^1. *If* x *is nondegenerate* $\|\dot x(i)\| < 2\pi$ *for all* $i = 1, 2, \ldots, n$.

So, at least at knots, spherical splines do not achieve very high speeds. Proposition 5.2 gives bounds on speeds between knots of nondegenerate splines. Theorem 5.1 can be used to determine whether a degenerate spherical spline is C^1.

EXAMPLE 5.2. *Let* $x_0 = x_1 = \ldots = x_n$. *If* γ_1^L *and the* γ_i^R *are chosen constant then the spherical spline* x *is constant, and nondegenerate. Alternatively, for* v *a unit vector orthogonal to* x_0, *choose* $\gamma_1^L(s) = x_0 \cos s\pi + v \sin s\pi$ *and* $\gamma_i^R = -\gamma_1^L$ *for* $1 \le i \le n$. *Then* $\gamma_i{}^L = \gamma_1^L$, $y_1 = y_2 = \ldots y_n = -x_0$, $x(t) = x_0 \cos 2\pi t + v \sin 2\pi t$, *and* x *is degenerate if* $n > 1$.

So even when $m > 1$, some degenerate spherical splines are C^1. However most are not.

EXAMPLE 5.3. *Let* $x_0 = x_1 = x_2 = (1,0,0) \in S^2$ *and* $v = (0,1,0)$. *For* $\theta \in (0, \pi]$ *choose* $\gamma_1^L(s) = (\cos s\theta, \sin s\theta, 0)$, $\gamma_1^R(s) = (\cos(1-s)\theta, \sin(1-s)\theta, 0)$, $\gamma_2^R(s) = (\cos(1-s)\theta, -\sin(1-s)\theta, 0)$. *Then* $x(t)$ *is*

$$(\cos 2(1-t)t\theta, \sin 2(1-t)t\theta, 0) \quad \text{or}$$

$$(\cos 2(2-t)(t-1)\theta, -\sin 2(2-t)(t-1)\theta, 0)$$

according as $0 \le t \le 1$ *or* $1 \le t \le 2$, *and* x *is degenerate if and only if* $\theta = \pi$. *Whether* x *is degenerate or not, its image is not smooth at* $x(1/2)$ *or* $x(3/2)$. *Alternatively, choose* $\gamma_1^L(s) = (\cos s\pi, \sin s\pi, 0)$, $\gamma_1^R = -\gamma_1^L$, *and* $\gamma_2^R(s) = -(\cos s\pi, 0, \sin s\pi)$. *Then* x *is degenerate and, by (2), (3) of Theorem 5.1,*

$$\dot{x}(1)_- = 2\pi(0,1,0) \quad \text{and} \quad \dot{x}(1)_+ = \pi(0, 1 + \frac{1}{\sqrt{2}}, -\frac{1}{\sqrt{2}}),$$

as illustrated in Figure 5.3, where the curve is trimmed towards beginning and end, showing left and right tangent directions at $x(1) = (1,0,0)$. *The labels* $1, 2, 3, 4$ *are in increasing order of progression along the curve.*

For quadratic polynomial splines the choice of y_1 affects the interpolant more or less globally, although the effect can be localized, for instance by placing breakpoints at knot averages as in Section 5.3. We adopt an almost opposite strategy, choosing y_1 to moderate the geometry of the spherical control polygon, thereby influencing the spherical spline. In this way the control polygon for an interpolating spherical Bezier quadratic spline is determined automatically, depending on the geometric criterion. First we introduce a performance measurement whose optima can be written down in closed form.

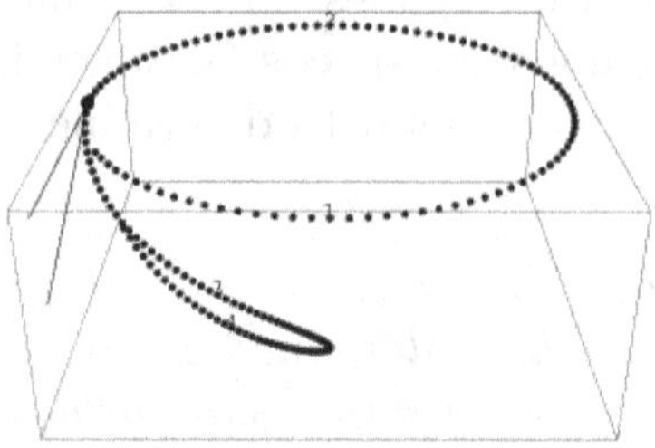

Figure 5.3. The non-C^1 spherical spline in Example 5.3, showing tangent directions $\dot{x}(1)_{\pm}$.

5.6. Moderating Euclidean Accelerations

DEFINITION 5.3. *The* ith Euclidean acceleration *of*

$$x_0, y_1, x_1, y_2, x_2, y_3, \ldots, x_{n-1}, y_n, x_n$$

is

$$\alpha_i \equiv 4(x_i - 2y_i + x_{i-1}) \in E^{m+1}.$$

Define $J_E : S^m \to [0, \infty)$ by $J_E(y_1) \equiv$

$$\sum_{i=1}^n \|\alpha_i\|^2 = 64n + 16\sum_{i=1}^n \|x_{i-1} + x_i\|^2 - 64\sum_{i=1}^n < y_i, x_{i-1} + x_i > =$$

$$64n + 16\sum_{i=1}^n \|x_{i-1} + x_i\|^2 - 64 < y_1, x^* >, \quad \text{where } x^* \equiv \sum_{i=1}^n z_i^* \quad \text{with}$$

$z_1^* \equiv x_1 + x_0$, and, for $2 \leq i \leq n$, $z_i^* \equiv \rho_1 \circ \rho_2 \circ \ldots \circ \rho_{i-1}(x_i + x_{i-1})$.

Writing $x_0^* \equiv x_0$, $x_1^* \equiv x_1$ and $x_i^* \equiv \rho_1 \circ \rho_2 \circ \ldots \circ \rho_{i-1} x_i$ for $i = 2, 3, \ldots, n$,

$$x^* = x_0^* + 2\,(\sum_{i=2}^{n-1} x_i^*) + x_n^*,$$

and we have

THEOREM 5.3. *If* $x^* = \mathbf{0}$, J_E *is constant. For* $x^* \neq \mathbf{0}$, J_E *achieves its minimum at*

$$y_1^* \equiv \frac{x^*}{\|x^*\|},$$

its maximum at $-y_1^*$, *and has no other critical points.*

When y_1 minimizes J_E the spherical spline $x : [0, n] \to S^m$ determined by y_1 for $x_0, x_1, \ldots, x_n$ is called *J_E-optimal.*

EXAMPLE 5.4. *For constant data, as in Examples 5.2, 5.3, the J_E-optimal spherical spline is constant. The J_E-optimum $x : [0, 7] \to S^2$ for the data in Example 5.1 is illustrated in Figure 5.4. Tracing along arcs in order of nodes, the changes in direction are much less abrupt, although the change near x_2 is rather noticeable. Speeds seem much more uniform than in Example 5.1 (as confirmed by Figure 5.6 of Example 5.5).*

5.7. Moderating Variations in Speed

For a nondegenerate spherical Bezier quadratic spline, let θ_0 be the length of the great circle arc γ_1^L, and for $1 \leq i \leq n$ let θ_i be the length of γ_i^R. By Theorem 5.2, $\|\dot{x}(i)\| \leq 2\theta_i$ for $i = 0, 1, 2, \ldots n$, where θ_0 depends on y_1, and θ_j is a function of y_j for $1 \leq j \leq n$. However $y_j = \rho_{j-1} \circ \ldots \circ \rho_2 \circ \rho_1(y_1)$. So all θ_j depend ultimately on y_1. Define $J_S : S^m \to [0, \infty)$ by

$$J_S(y_1) \equiv \sum_{i=1}^{n} (\theta_i - \theta_{i-1})^2.$$

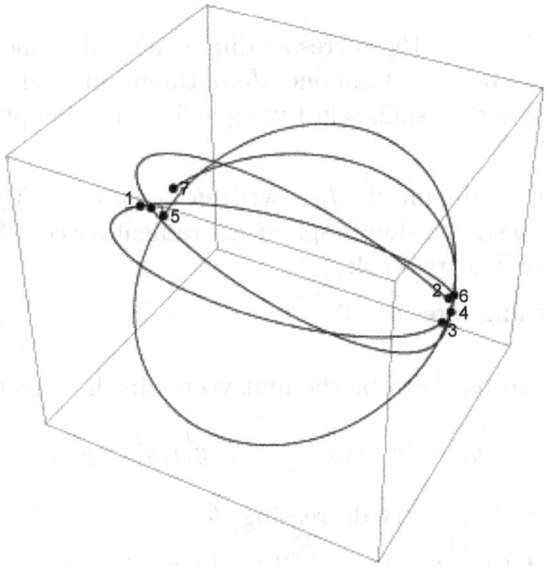

Figure 5.4. The J_E-optimal spherical spline $x : [0, 7] \to S^2$ in Example 5.4.

Then, for $y_1 \in S^m$ and $v_1 \in TS^m_{y_1}$, writing $v_j = \rho_{j-1} \circ \ldots \circ \rho_2 \circ \rho_1(v_1)$ for $1 \leq j \leq n$, $\frac{1}{2}(dJ_S)_{y_1}(v_1) =$

$$(\theta_1 - \theta_0)(d\theta_1(v_1) - d\theta_0(v_1)) + \sum_{i=2}^{n}(\theta_i - \theta_{i-1})(d\theta_i(v_i) - d\theta_{i-1}(v_{i-1}))$$

$$= - < v_1, \tilde{x}(y_1) >$$

where, for $\tilde{x}_i \equiv \frac{x_i}{\sin\theta_i}$,

$$\tilde{x}(y_1) \equiv (\theta_1 - \theta_0)(\tilde{x}_1 - \tilde{x}_0) + (\theta_2 - \theta_1)(\rho_1(\tilde{x}_2) - \tilde{x}_1) +$$

$$\sum_{i=3}^{n} (\theta_i - \theta_{i-1})\, \rho_1 \circ \rho_2 \circ \ldots \circ \rho_{i-2}(\rho_{i-1}(\tilde{x}_i) - \tilde{x}_{i-1}).$$

From this follows

THEOREM 5.4. *The gradient at* $y_1 \in S^m$ *of* $J_S : S^m \to [0, \infty)$ *is*

$$-2(\tilde{x}(y_1) - < \tilde{x}(y_1), y_1 > y_1).$$

If y_1 *is a critical point of* J_S *then either* $\tilde{x}(y_1) = \mathbf{0}$, *or* $\theta_i = 0$ *for some* $i = 0, 1, 2, \ldots, n$, *or*

$$y_1 = \pm \frac{\tilde{x}(y_1)}{\|\tilde{x}(y_1)\|}. \tag{7}$$

When y_1 minimizes J_S the corresponding spherical spline is called J_S-*optimal.* There may be more than one J_S-optimum for a given set of data. For instance the spherical splines in Example 5.2 are J_S-optimal, as is the first in Example 5.3.

The point y_1^* of minimum of J_E, written down explicitly in Theorem 5.3, can be used to start a simple spherical gradient descent for J_S, based on the first part of Theorem 5.4:

1. choose a learning rate $h > 0$,
2. set $y_1^1 = y_1^*$,
3. for $j \geq 1$ take y_1^{j+1} to be the unit vector in the direction of

$$y_1^j + h\,(\tilde{x}(y_1^j) - < \tilde{x}(y_1^j), y_1^j > y_1^j),$$

4. if $J_S(y_1^{j+1}) > J_S(y_1^j)$ try decreasing h,
5. if $J_S(y_1^j) - J_S(y_1^{j+1}) > 0$ is small try increasing h,
6. if $J_S(y_1^{j+1}) \approx J_S(y_1^j)$, after several steps, exit and return y_1^j,
7. otherwise replace j by $j + 1$ and return to 3.

Because J_S measures a more geometrical but related phenomenon, the additional computational effort required to find a J_S-optimum may be well worthwhile.

EXAMPLE 5.5. *For the data in Examples 5.1, 5.4, the value of* J_S *for the* J_E*-optimum* y_1^* *is* 21.77. *Using gradient descent with* $h = 0.01$, J_S *decreases to*

$$18.61,\ 15.34,\ 12.23,\ 9.21,\ 7.25,\ \ldots,\ 0.3658$$

after 70 *iterations. After* 30 *additional iterations* J_S *settles at* 0.365706 *and the corresponding* J_S*-optimum* $x : [0, 7] \to S^2$, *shown in Figure 5.5, is more regular in overall appearance than the* J_E*-optimum and somewhat similar locally.*

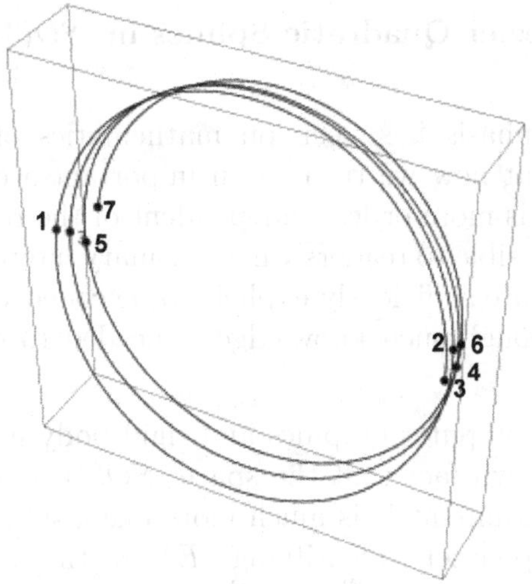

Figure 5.5. The J_S-optimal spherical spline $x : [0, 7] \to S^2$ in Example 5.5.

As noted in Theorem 5.2, speeds at knots for any nondegenerate spherical Bezier quadratic spline are bounded above by 2π. *This bound is approached by the* J_E*-optimum of Example 5.4, whose speeds are shown in Figure 5.6, along with speeds for the* J_S*-optimum. The latter are significantly more uniform, with less abrupt changes of derivatives at breakpoints. Comparing with Figure 5.2, speeds of both spherical splines are far more uniform than those of the normalized quadratic polynomial spline in Example 5.1.*

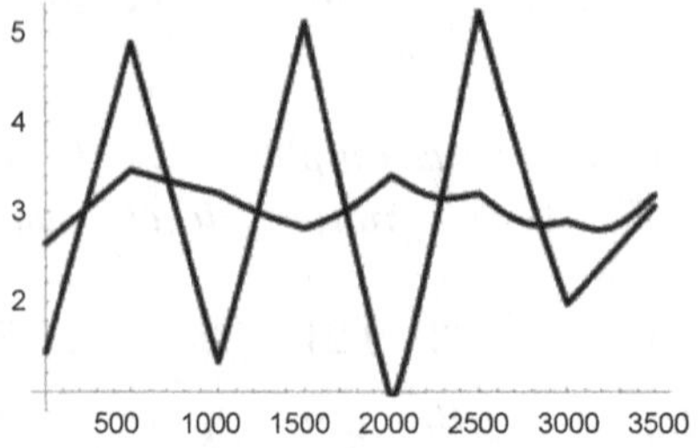

Figure 5.6. Speeds for the J_E (more variable) and J_S optima in Examples 5.4, 5.5.

5.8. Spherical Bezier Quadratic Splines in $SO(3)$

Up to now the emphasis has been on mathematics of spherical Bezier quadratic splines, but now we turn to an important area of applications. The present section is more or less independent of the rest of this chapter, and much more accessible to readers whose primary interests are in applications. The formulae are sufficiently explicit to be coded without knowledge of quaternions, although such knowledge is needed to explain where the formulae come from.

For applications in computer graphics and rigid body motion, it is an important and well-known fact that the space $SO(3)$ of orthogonal 3×3 real matrices of determinant 1 is much more like a sphere than Euclidean 3-space E^3. More precisely, identifying E^4 in the usual way with the space $\mathbb{H}$ of , the unit sphere S^3 in E^4 corresponds to the space of *unit quaternions*. Quaternionic multiplication restricts to a group multiplication on S^3, and the formula

$$\Psi(q)r = qrq^{-1} \tag{8}$$

defines a group homomorphism $\Psi : S^3 \rightarrow SO(3)$ with respect to matrix multiplication on $SO(3)$. In (8) $q \in S^3$ and r is a pure-imaginary quaternion, corresponding to a point in E^3. On the right hand side multiplication and inversion are quaternionic. Then Ψ is C^∞, surjective, and its kernel as a group homomorphism is ± 1. In this way a unit vector

$$q = \begin{bmatrix} q_1 & q_2 & q_3 & q_4 \end{bmatrix}^{\mathbf{T}} \in S^3 \subset E^4$$

gives a rotation

$$\Psi(\pm q) = \mathbf{1} - 2 \begin{bmatrix} q_3^2 + q_4^2 & -q_2q_3 + q_1q_4 & -q_1q_3 - q_2q_4 \\ -q_2q_3 - q_1q_4 & q_2^2 + q_4^2 & q_1q_2 - q_3q_4 \\ q_1q_3 - q_2q_4 & -q_1q_2 - q_3q_4 & q_2^2 + q_3^2 \end{bmatrix}$$

where $\mathbf{T}$ means "transpose". Conversely any $R \in SO(3)$ is $\Psi(q)$ where

$$q_1 = \pm\frac{1}{2}\sqrt{1 + \text{Trace}(R)}. \qquad \text{For } q_1 \neq 0:$$

$$\begin{aligned} q_2 &= \frac{1}{4q_1}(R_{3,2} - R_{2,3}) \\ q_3 &= \frac{1}{4q_1}(R_{1,3} - R_{3,1}) \\ q_4 &= \frac{1}{4q_1}(R_{2,1} - R_{1,2}). \end{aligned}$$

If $q_1 = 0$ set $q_2 = \pm\sqrt{\frac{1+R_{1,1}}{2}}$.

For $q_1 = 0,\ q_2 \neq 0: \ q_3 = \dfrac{1}{2q_2} R_{1,2} \quad q_4 = \dfrac{1}{2q_2} R_{1,3}.$

For $q_1 = q_2 = 0: \ q_3 = \pm\sqrt{\dfrac{1 - R_{3,3}}{2}} \quad q_4 = \pm\sqrt{\dfrac{1 - R_{2,2}}{2}}.$

So we have a C^∞ one-to-one correspondence between rotation matrices $R \in SO(3)$ and pairs of points $\pm q \in S^3$, where the standard Riemannian metric on S^3 corresponds to a bi-invariant (left and right-invariant (Milnor, 1963)-Section 21) Riemannian metric with respect to the Lie group structure of $SO(3)$. This, together with the constructions in Sections 5.4, 5.5, 5.6, 5.7 of spherical Bezier quadratic splines, can be used to construct spherical splines in $SO(3)$ interpolating given rotations $R_0, R_1, \ldots, R_n \in SO(3)$ as follows.

- Choose $x_0 \in S^3$ to be either $\pm q$ where $\Psi(q) = R_0$.
- For $1 \leq i \leq n$ now find q so that $\Psi(q) = R_i$, and choose $x_i \in S^3$ to be whichever of $\pm q$ is nearer x_{i-1} (in case of a tie think, or choose either).
- For $1 \leq i \leq n-1$ define $\rho_i : E^4 \to E^4$ by

  $$\rho_i(w) = 2 < w, x_i > x_i - w$$

  where $< \ , \ >$ is the Euclidean inner product (dot-product).

– Set $x_0^* = x_0$, $x_1^* = x_1$ and, for $2 \leq i \leq n$,

$$x_i^* = \rho_1 \circ \rho_2 \circ \ldots \circ \rho_{i-1}(x_i).$$

– Set $x^* = x_0^* + 2\sum_{i=2}^{n-1} x_i^* + x_n^*$. If $x^* = \mathbf{0}$ think, or choose $y_1^* \in S^3$ arbitrarily. If $x^* \neq \mathbf{0}$ set

$$y_1 = \frac{x^*}{\|x^*\|}$$

where $\| \ \|$ is the Euclidean norm on E^4. Alternatively, let y_1 be a minimizer of J_S, calculated according to the pseudocode in Section 5.7.

– For $1 \leq i \leq n-1$ set $y_{i+1} = \rho_i(y_i)$.

– Follow the procedure in Definition 5.2 to construct $x : [0,n] \to S^3$.

– Set $R = \Psi \circ x : [0,n] \to SO(3)$.

Then R is usually C^1, interpolates $R_0, R_1, \ldots, R_n$ at $0, 1, \ldots, n$, and inherits the good behaviour of spherical Bezier quadratics in S^3. More can be said, but this algorithm is straightforward to implement, and the interpolant is excellent compared with current practice.

5.9. Conclusion

Interpolation by C^1 curves in spheres is a basic task for modelling rigid body motion. The problem is not adequately dealt with by normalized polynomial splines. Interpolation by curves satisfying variational conditions has also received attention (Gabriel and Kajiya, 1985), (Noakes et al., 1989), (Chapman and Noakes, 1991), (Camarinha et al., 1995), (Crouch and Silva-Leite., 1995), (Crouch et al., 1999), (Camarinha et al., 2001), (Noakes, 2003), (Noakes (a), 2004), but is not ready for practical applications.

The alternative of replacing line segments by great circles in the deCastlejau algorithms is well-known (Shoemake, 1985), (Duff, 1985), (Noakes, 1994), (Noakes, 1997), (Noakes, 1998), (Noakes, 1999), (Noakes, 2002), (Crouch et al., 1999), but is problematic because of difficulties in constructing control polygons. By addressing this problem the present paper seeks to make the generalized non-recursive deCastlejau algorithm a more attractive tool for spherical interpolation.

Spherical Bezier quadratic splines are C^∞ away from breakpoints, C^1 everywhere when nondegenerate, and much easier to construct than splines made from Riemannian cubics (Noakes et al., 1989). Given $x_0, x_1, \ldots x_n \in S^m$, a spherical Bezier quadratic spline $x : [0,n] \to S^m$ with breakpoints at $1, 2, \ldots, n-1$, and satisfying

$$x(0) = x_0, \quad x(1) = x_1, \quad x(2) = x_2, \quad \ldots, \quad x(n) = x_n,$$

is determined by a single parameter $y_1 \in S^m$, corresponding to a control polygon for the spherical spline. We choose y_1 to optimize relationships between the data $x_0, x_1, x_2, \ldots, x_n$ and the geometry of the control polygon, according to two different criteria.

According to our first criterion, y_1 is chosen as the typically unique minimizer y_1^* of a sum of squares of so-called *Euclidean accelerations* of the control polygon. It turns out y_1^* is straightforward to compute. Excessive variations in speeds, encountered in normalized quadratic polynomial splines, are reduced by a factor of around 35 in one example, and we are aware of no case where normalized quadratic splines are superior.

In terms of spherical geometry, a more natural performance meaurement is the sum J_S of squares of differences in speed between subsequent knots. The derivative of J_S is simple to write down and, with the J_E optimum as a starting point, a simple kind of gradient descent minimizes J_S. The resulting J_S-optimal spherical Bezier spline is even more attractive than the J_E-optimum. Variations in speed are reduced by a further factor of 4 and the curve has a much more regular appearance.

Although both offer advantages over normalized quadratic polynomial splines, there are striking differences between the J_E-optimal spherical spline in Example 5.4 and the J_S-optimum in Example 5.5. Certainly this kind of data is likely to pose special problems for any method of spherical interpolation, and such marked differences are not observed in all examples. Nonetheless sensitivity of spherical splines to the choice of control polygon is well-demonstrated. This makes it all the more important to have automatic methods for choosing spherical control polygons according to geometric criteria that are relevant to particular applications. The present paper is a step in this direction, with applications to interpolation in $SO(3)$ and planning trajectories of rigid bodies.

Appendix A: A Quadratic Polynomial Spline $x : [0, n] \to E^{m+1}$

The discussion following Example 5.1 concerns a C^1 quadratic polynomial spline interpolant $x : [0, n] \to E^{m+1}$ with breakpoints $3/2, 5/2, \ldots, (2n-3)/2$. This is constructed as the track-sum of the following quadratic polynomials:

– $a_1 s^2 + b_1 s + c_1$ for $s = t \in [0, 3/2]$ interpolating x_0, x_1, y_1 at $s = 0, 1, 3/2$ namely

$$a_1 = \frac{4}{3}y_1 + \frac{2}{3}x_0 - 2x_1, \qquad b_1 = -\frac{4}{3}y_1 - \frac{5}{3}x_0 + 3x_1, \qquad c_1 = x_0.$$

– $a_2 s^2 + b_2 s + c_2$ for $s = t - 2 \in [-1/2, 1/2]$, C^1 at $s = -1/2$ and interpolating x_2, y_2 at $s = 0, 1/2$ namely

$$a_2 = 2(y_1 - 2x_2 + y_2), \qquad b_2 = -y_1 + y_2, \qquad c_2 = x_2,$$
$$\frac{17}{3}y_1 + y_2 = -\frac{1}{3}x_0 + 3x_1 + 4x_2.$$

– for $2 < i < n-1$ and $s = t - i \in [-1/2, 1/2]$, $a_i s^2 + b_i s + c_i$, C^1 at $s = -1/2$ and interpolating y_{i-1}, x_i, y_i at $s = -1/2, 0, 1/2$, namely

$$a_i = 2(y_{i-1} - 2x_i + y_i), \qquad b_i = -y_{i-1} + y_i, \qquad c_i = x_i,$$
$$y_{i-2} + 6y_{i-1} + y_i = 4(x_{i-1} + x_i).$$

– $a_{n-1}s^2 + b_{n-1}s + c_{n-1}$ for $s = t - i + 1 \in [-1/2, 1]$, C^1 at $s = -1/2$ and interpolating x_{n-1}, x_n at $s = 0, 1$, namely

$$a_{n-1} = \frac{4}{3}y_{n-2} - 2x_{n-1} + \frac{2}{3}x_n, \qquad b_{n-1} = -\frac{4}{3}y_{n-2} + x_{n-1} + \frac{1}{3}x_n,$$
$$c_{n-1} = x_{n-1},$$
$$y_{n-3} + \frac{17}{3}y_{n-2} = 4x_{n-2} + 3x_{n-1} - \frac{1}{3}x_n.$$

Solving the $n-2$ linear equations for $y_1, y_2, \ldots y_{n-2}$ also determines the a_i, b_i, c_i.

Appendix B: Proof of Proposition 5.1

For $s \in [0, 1/2]$, $d(\beta(s), \gamma^L(2s)) \leq$

$$\min\{sd(\gamma^L(s), \gamma^R(s)) + s\theta^L,\ (1-s)d(\gamma^L(s), \gamma^R(s)) + s\theta^R + (1-2s)\theta^L\}$$
$$\leq \min\{f^L(s), f^R(s)\}$$

where

$$f^L(s) \equiv s(2-s)\theta^L + s^2\theta^R$$

and

$$f^R(s) \equiv (2 - 4s + s^2)\theta^L + s(2-s)\theta^R.$$

So

$$d(\beta(s), \pi(s)) \leq \frac{1}{2}(f^L(s) + f^R(s) - |f^L(s) - f^R(s)|) =$$
$$(1-s)\theta^L + s\theta^R - |(1 - 3s + s^2)\theta^L + s(1-s)\theta^R|$$
$$\leq \frac{3\theta^L + \theta^R}{4}$$

because f^L is maximized on $[0, 1/2]$ by $s = 1/2$. The result now follows from (1).

Appendix C: Proof of Lemma 5.1

By rotation-invariance, we can suppose $y = (\cos\theta, \sin\theta, 0, \ldots, 0)$ and $z = (1, 0, \ldots, 0)$ for some $\theta \in [0, \pi)$. Also, without loss, either $v^{\|}$ or $v^{\perp}$ is $\mathbf{0}$. For $v^{\|} = \mathbf{0}$, after another rotation, we can assume

$$\omega(t) = (\cos\theta,\ \sin\theta\ \cos t\psi,\ \sin\theta\ \sin t\psi,\ 0,\ \ldots,\ 0)$$

where $\psi \in \mathbb{R}$. Then

$$\delta(\omega(t), z, q) \ = \ (\cos r\theta,\ \sin r\theta\ \cos t\psi,\ \sin r\theta\ \sin t\psi,\ 0,\ \ldots,\ 0)$$

where $r \equiv 1 - q$, and

$$\frac{d}{dt}|_{t=0}\delta(\omega(t), z, q) \ = \ (0,\ 0,\ \psi\ \sin r\theta,\ 0,\ \ldots,\ 0) \ = \ \frac{\sin r\theta}{\sin\theta}\ v^{\perp}.$$

For $v^{\perp} = \mathbf{0}$ we can suppose $\omega(t) \ = \ (\cos(\theta + t\psi),\ \sin(\theta + t\psi),\ 0,\ \ldots,\ 0)$. Then

$$\delta(\omega(t), z, q) \ = \ (\cos(r\theta + t\psi),\ \sin(r\theta + t\psi),\ 0,\ \ldots,\ 0) \quad \text{and}$$

$$\frac{d}{dt}|_{t=0}\delta(\omega(t), z, q) \ = \ \psi\ (-\sin r\theta,\ \cos r\theta,\ 0,\ \ldots,\ 0) \ = \ \psi\ \dot{\gamma}_{y,z}(q).$$

Appendix D: Proof of Proposition 5.2

Since $\dot{\gamma}^L(1), \dot{\gamma}^R(0)$ are linearly independent, $0 \ < \ \min\{\theta^L, \theta^R\} \ < \ \pi$ and, for any $t \in [0, 1]$,

$$0 \ < \ \theta_t \ \le \ (1 - t)\theta^L + s\theta^R \ < \ \pi.$$

So for some $\epsilon > 0$ independent of t, and any

$$q, r \in I_{t,\epsilon} \equiv (t - \epsilon/2, t + \epsilon/2) \cap [0, 1],$$

$$(\delta(\gamma^L(0), \gamma^L(1), q),\ \delta(\gamma^R(0), \gamma^R(1), r) \ \in \ D.$$

Since δ is C^{∞}, so is $\mu : I_{t,\epsilon} \times I_{t,\epsilon} \times [0, 1] \to S^m$ given by

$$\mu(q, r, s) \ = \ \delta(\delta(\gamma^L(0), \gamma^L(1), q),\ \delta(\gamma^R(0), \gamma^R(1), r),\ s),$$

and, by Lemma 5.2,

$$\|\frac{\partial\mu}{\partial q}|_{r=s=t}\| \ \le \ \frac{\theta^L}{\sin\theta_t}, \qquad \|\frac{\partial\mu}{\partial r}|_{q=s=t}\| \ \le \ \frac{\theta^R}{\sin\theta_t}, \qquad \|\frac{\partial\mu}{\partial s}|_{q=r=t}\| \ = \ \theta_t.$$

Since $\beta(t) = \mu(t, t, t)$,

$$\|\dot{\beta}(t)\| \ \le \ \frac{\theta^L + \theta^R}{\sin\theta_t} \ + \theta_t.$$

Appendix E: Proof of Theorem 5.1

If $\dot\gamma^L(1), \dot\gamma^R(0)$ are linearly dependent (2) follows on differentiating the formulae in Definition 5.1 for $\beta(s)$. For $\dot\gamma^L(1), \dot\gamma^R(0)$ linearly independent with $0 \leq \theta^L, \theta^R < \pi$, for some $\epsilon > 0$,

$$(\delta(\gamma^L(0), \gamma^L(1), q), \delta(\gamma^R(0), \gamma^R(1), r)) \in D$$

whenever $q, r \in [0, \epsilon)$. Since δ is C^∞, so is

$$\mu : [0, \epsilon) \times [0, \epsilon) \times [0, 1] \to S^m$$

given by

$$\begin{aligned} \mu(q, r, s) &= \delta(\delta(\gamma^L(0), \gamma^L(1), q), \delta(\gamma^R(0), \gamma^R(1), r), s), \\ \text{and} \quad \frac{\partial\mu}{\partial q}|_{r=s=0} &= \dot\gamma^L(0) = \frac{\partial\mu}{\partial s}|_{q=r=0} \quad \text{while} \quad \frac{\partial\mu}{\partial r}|_{q=s=0} = \mathbf{0}. \end{aligned}$$

For small $s > 0$, $\beta(s) = \mu(s, s, s)$. So

$$\dot\beta(0)_+ = \frac{d}{ds}|_{s=0}\mu(s, s, s) = 2\dot\gamma^L(0),$$

and (1) completes the proof of (2) in this case. If $\dot\gamma^L(1), \dot\gamma^R(0)$ are linearly independent with $\theta^L = \pi$, for small $s > 0$

$$\begin{aligned} \gamma^L(s) &= \gamma^L(0) + s\dot\gamma^L(0) + O(s^2), \\ \gamma^R(s) &= -\gamma^L(0) + s\dot\gamma^R(0) + O(s^2), \\ \gamma^R(s) - \gamma^L(s) &= -2\gamma^L(0) + s\dot\gamma^R(0) - s\dot\gamma^L(0) + O(s^2), \\ < \gamma^R(s) - \gamma^L(s), \gamma^L(s) > &= -2 + O(s^2), \end{aligned}$$

and

$$\begin{aligned} \gamma^R(s) - \gamma^L(s) - < \gamma^R(s) - \gamma^L(s), \gamma^L(s) > \gamma^L(s) = \\ s\dot\gamma^R(0) + s\dot\gamma_i^L(0) + O(s^2). \end{aligned}$$

Because $< \gamma^L(s), \gamma^R(s) > \; = \; -1 + O(s^2)$,

$$d(\gamma^L(s), \gamma^R(s)) = \pi + O(s)$$

and so

$$\beta(s) = \gamma^L(0) + s\dot\gamma^L(0) + s\pi \, \frac{\dot\gamma^R(0) + \dot\gamma^L(0)}{\|\dot\gamma^R(0) + \dot\gamma^L(0)\|} + O(s^2),$$

where the denominator is nonzero by linear independence, and since $\dot\gamma^L(0) = -\dot\gamma^L(1)$. This proves (3). If $\dot\gamma^L(1), \dot\gamma^R(0)$ are linearly independent with

$\theta^L = \pi$ and $\theta^R < \pi$, for small $r \equiv 1 - s$, $\gamma^L(s) = \gamma^R(0) + O(r)$,

$$\begin{aligned} \gamma^R(s) &= \gamma^R(1) - r\dot{\gamma}^R(1) + O(r^2), \\ \gamma^L(s) - \gamma^R(s) &= -\gamma^R(1) + \gamma^R(0) + O(r), \\ < \gamma^L(s) - \gamma^R(s), \gamma^R(s) > &= -1 + < \gamma^R(0), \gamma^R(1) > + O(r), \end{aligned}$$

$$\gamma^L(s) - \gamma^R(s) - < \gamma^L(s) - \gamma^R(s), \gamma^R(s) > \gamma^R(s) =$$
$$\gamma^R(0) - < \gamma^R(0), \gamma^R(1) > \gamma^R(1) + O(r).$$

Also $< \gamma^L(s), \gamma^R(s) > = < \gamma^R(0), \gamma^R(1) > + O(r) = \cos\theta^R + O(r)$. Since $\theta^R \neq 0, \pi$, $d(\gamma^L(s), \gamma^R(s)) = \theta^R + O(r)$. So $\beta(1-r) =$

$$\gamma^R(1) - r\dot{\gamma}^R(1) + r\theta^R \frac{\gamma^R(0) - < \gamma^R(0), \gamma^R(1) > \gamma^R(1)}{\| \gamma^R(0) - < \gamma^R(0), \gamma^R(1) > \gamma^R(1) \|} + O(r^2)$$

$$= \beta(1) - 2r\dot{\gamma}^R(1) + O(r^2).$$

This proves (4). Applying (1) to (3), (4) proves (5), (6).

Appendix F: Proof of Theorem 5.2

Since $\|\dot{\gamma}(i)\| \leq \pi$, $\|\dot{x}(i)_\pm\| \leq 2\pi$ by Theorem 5.1. If, for some $i = 2, \ldots, n$, $y_i = -x_{i-1}$ then $y_{i-1} = -\rho_{i-1}x_{i-1} = -x_{i-1}$, and the minimal great circle arcs $\gamma_{i-1}^R, \gamma_i^L$ are not determined solely by their endpoints, but also by $\dot{\gamma}_{i-1}^R(1)$ which can be any vector w_{i-1} of length π orthogonal to x_{i-1}. Then $\dot{\gamma}_i^L(0) = w_{i-1}$ and, by (5), (2) of Theorem 5.1,

$$\begin{aligned} \dot{x}(i-1)_- &= w_{i-1} + \pi \frac{w_{i-1} + \dot{\gamma}_{i-1}^L(1)}{\|w_{i-1} + \dot{\gamma}_{i-1}^L(1)\|} \quad \text{and} \\ \dot{x}(i-1)_+ &= w_{i-1} + \pi \frac{w_{i-1} + \dot{\gamma}_i^R(0)}{\|w_{i-1} + \dot{\gamma}_i^R(0)\|}, \end{aligned}$$

unless either $\dot{\gamma}_{i-1}^L$ or $\dot{\gamma}_i^R(0)$ happens to be a multiple of w_{i-1}. When $m = 1$ this is precisely what happens, and then $\dot{x}(i-1)_- = 2w_{i-1} = \dot{x}(i-1)_+$ by (3) of Theorem 5.1. Alternatively, for any m, if $y_i \neq -x_{i-1}$ then $\dot{x}(i-1)_- = 2\dot{\gamma}_i^L(0) = \dot{x}(i-1)_+$, again by (3) of Theorem 5.1. So if either $m = 1$ or x is nondegenerate then x is C^1 at $x_1, \ldots, x_{n-1}$. So x is C^1. Furthermore $\|\dot{x}(i)\| = 2\|\dot{\gamma}_i^L(0)\| < 2\pi$ when x is nondegenerate.

References

Altafini, C.: The De Castlejau algorithm on $SE(3)$. In *Nonlinear Control in the Year 2000* (A. Isidori, F. Lamnabhi-Lagarrigue, and W. Respondek, editors), Lecture Notes Control Information Sciences 258, pages 23–34, Springer, London, 2001.

Altafini, C.: Geometric control methods for nonlinear systems and robotic applications. PhD Thesis, Dept. of Mathematics, Royal Inst. of Technology, Stockholm, 2001.

Angeles, J., and R. Akras: Cartesian trajectory planning for 3-DOF spherical wrists. In Proc. *IEEE Conf. Robotics Automation* (G. A. Bekey, J. M. Hollerbach, and A. L. Pai, editors), pages 68–74, 1989.

Barr, A. H., B. Currin, S. Gabriel, and J. F. Hughes: Smooth interpolation of orientations with angular velocity constraints using quaternions. *Computer Graphics*, **26**: 313–320, 1992.

Belta, C. and V. Kumar: Euclidean metrics for motion generation in $SE(3)$. *J. Mech Engineering Science*, Part C 216, **C1**:47–61, 2002.

Brady, J. M., J. M. Hollerbach, T. L. Johnson, T. Lozano-Perez, and M. T. Masson: *Robot Motion: Planning and Control.* MIT Press, Cambridge MA, 1982.

Brunnett, G.: Elastic curves on the sphere. *Adv. Computational Mathematics*, **2**:23–40, 1994.

Buss, S. R. and J. Fillmore: Spherical averages and applications to spherical splines and interpolation. *ACM Trans. Graphics*, **20**:95–126, 2001.

Camarinha, M.: The Geometry of Cubic Polynomials on Riemannian Manifolds. PhD Thesis, Univ Coimbra, Portugal 1996.

Camarinha, M., F. Silva-Leite, and P. Crouch: On the geometry of Riemannian cubic polynomials. *Differential Geom. Appl.*, **15**:107–135, 2001.

Silva-Leite, F., M. Camarinha, and P. Crouch: Elastic curves as solutions of Riemannian and sub-Riemannian control problems. *Math. Control Signals Systems,* **13**:140–155, 2000.

Camarinha, M., F. Silva-Leite, and P. Crouch: Splines of class C^k on non-Euclidean spaces. *IMA J. Math. Control & Information,* **12**:399–410, 1995.

Chapman, P. B. and L. Noakes: Singular perturbations and interpolation - a problem in robotics. *Nonlinear Analysis TMA,* **16**:849–859, 1991.

Crouch, P. and F. Silva-Leite: The dynamic interpolation problem: on Riemannian manifolds, Lie groups, and symmetric spaces. *J. Dynam. Control Systems,* **1**:177–202, 1995.

Giambo, R., F. Giannoni, and P. Piccione: An analytical theory for Rienmannian cubic polynomials. *IMA J. of Math Control and Information,* **19**:445–460, 2002.

Giambo, R., F. Giannoni, and P. Piccione: Higher order interpolation in Riemannian manifolds. preprint, 2003

Crouch, P., G. Kun, and F. Silva-Leite: The De Castlejau algorithm on Lie groups and spheres. *J. Dynam. Control Systems,* **5**:397–429, 1999.

de Boor, C: *A Practical Guide to Splines* (revised edition). Applied Mathematical Sciences 27, Springer-Verlag, 2001.

Duff, T.: Quaternion splines for animating rotations. In Proc. *Montrey Computer Graphics Workshop,* (Usenix Association), pages 54–62, 1985.

Fisher, N. I.: *Statistical Analysis of Circular Data.* Cambridge Univ Press, Cambridge, 1993.

Fisher, N. I., T. Lewis, and B. J. Embleton: *Statistical Analysis of Spherical Data.* Cambridge Univ Press, NY, 1987

Gabriel, S.A. and J.T. Kajiya: Spline interpolation in curved manifolds. Unpublished manuscript, 1985

Kang, I. G. and F. C. Park: Cubic spline algorithms for orientation interpolation. *Int. J. Numerical Methods Engineering,* **46**:45–64, 1999.

Krakowski, K.: *Geometrical Methods of Inference,* PhD Thesis, University of Western Australia, 2003.

Milnor, J.: *Morse Theory.* Annals of Math. Studies 51, Princeton UP, 1963.

Noakes, L.: Asymptotically smooth splines. *World Scientific Series in Approximations and Decompositions,* **4**:131–137, 1994.

Noakes, L.: Riemannian quadratics. In *Curves and Surfaces with Applications in CAGD* (A. Le Mehaute, R. Christopher, and L. L. Schumaker, editors), Vanderbilt University Press, 1, pages 319–328, 1997.

Noakes, L.: Nonlinear corner-cutting. *Advances in Computational Math.,* **8**:165–177, 1998.

Noakes, L.: Accelerations of Riemannian quadratics . *Proc. Amer. Math. Soc.,* **127**: 1827–1836, 1999.

Noakes, L.: Quadratic interpolation on spheres. *Advances Computational Math.,* **17**: 385–395, 2002.

Noakes, L., G. Heinzinger, and B. Paden: Cubic splines on curved spaces. *J. Math. Control & Information,* **6**:465–473, 1989.

Noakes, L.: Null cubics and Lie quadratics. *J. Math. Physics,* **44**:1436–1448, 2003.

Noakes, L.: Duality and Lie quadratics. *Advances in Computational Math.,* in-press, 2004.

Noakes, L.: Non-null Lie quadratics in E^3. *J. Math. Physics,* in-press, 2004.

Noakes, L.: Interpolating camera configurations. In Proc. *Intern. Conf. on Computer Analysis of Images and Patterns,* (N. Petkov and M. Westenberg editors), LNCS 2756, Springer Verlag, 2003.

Park, F. C. and B. Ravani: Smooth invariant interpolation of rotations. *Trans. Graphics,* **16**:277–295, 1997.

Paul, R. P.: Manipulator path control. *IEEE Trans. Syst. Man. Cybern.,* **SMC-9**: 702–711, 1979.

Shoemake, K.: Animating rotation with quaternion curves. *SIGGRAPH,* **19**:245–254, 1985.

Tan, H. H. and R. B. Potts: A discrete path/ trajectory planner for robotic arms. *J.Austral. Math. Soc. Series B,* **31**:1–28, 1989.

Taylor, R. H.: Planning and execution of straight-line manipulator trajectories. *IBM J.Res. Develop.,* **23**:424–436, 1979.

Watson, G. S.: Statistics on Spheres. *Univ. of Arkansas Lecture Notes in Mathematical Sciences*, Wiley, NY, 1983.

Zefran, M., W. Kumar, and C. Croke: Choice of Riemannian metrics for rigid body dynamics. In Proc. *ASME Design Engineering Technical Conference and Computers in Engineering Conference*, Irvine CA, pages 1–11, 1996.

Zefran, M. and V. Kumar: Planning of smooth motions on $SE(3)$. In *IEEE Intern Conf on Robotics and Automation*, pages 121–126, 1996.

Zefran, M. and V. Kumar: Two methods for interpolating rigid body motion. In *IEEE Intern. Conf. on Robotics and Automation*, pages 2922–2927, 1998.

Zefran, M. and V. Kumar: Interpolation schemes for rigid body motions. *Computer-Aided Design*, **30**:179–189, 1989.

Zefran, M., V. Kumar, and C. Croke: On the generation of smooth three-dimensional rigid-body motions. *IEEE Trans. Robotics Automation,* **14**:576–589, 1998.

GRAPH-SPECTRAL METHODS FOR SURFACE HEIGHT RECOVERY FROM GAUSS MAPS

ANTONIO ROBLES-KELLY
National ICT Australia
Canberra Laboratory, Canberra ACT 2601, Australia

EDWIN R. HANCOCK
Department of Computer Science
University of York, York YO1 5DD, United Kingdom

Abstract. In this chapter, we describe the use of graph-spectral methods for purposes of surface height recovery. Here, our input is a 2D field of surface normal estimates, delivered, for instance, by shape-from-shading, shape-from-texture or photometric stereo. We commence by showing how to use the surface normals to obtain a graph whose edge-weight matrix is related to the surface curvature. With this weight matrix at hand, we proceed to recover a path across the field of surface normals whose curvature is minimum. To do this, we present two alternatives. The first of these concerns a combinatorial optimisation over the nodes in the graph. The second one, consists in recovering the random walk making use of a probability matrix, equivalent, by row-normalisation, to the matrix of edge-weights. For both methods, the solution is equivalent, up to scaling, to the leading eigenvector of the edge-weight matrix. Once the minimum curvature path has been recovered, surface integration can be performed by threading the surface normals together along the path. We perform experiments on synthetic and real-wold imagery whose fields of surface normals are delivered by a shape-from-shading algorithm.

Key words: surface reconstruction, Markov chain, curvature, graph seriation

6.1. Introduction

Surface integration is a process that provides the means of converting a field of surface normals (i.e. the projection of the Gauss map of a surface from the unit sphere onto the image plane) into an explicit 3D surface representation. This problem of reconstructing the surface from its Gauss map arises when attempting to infer explicit surface structure from the output of low-level vision modules such as shape-from-shading and shape-from-texture (Rockwood and Winget, 1997). The process involves selecting

R. Klette et al. (eds.), Geometric Properties for Incomplete Data, 103-122.

a path through the surface normal locations. This may be done using either a curvature minimising path or by advancing a wavefront from the occluding boundary or singular points. By traversing the path, the surface may be reconstructed by incrementing the height function using the known distance travelled and the local slope of the surface tangent plane. This is a matter of straightforward trigonometry. It must be stressed that the choice of the path can be critical. If the surface does not satisfy the integrability constraint, (i.e. the x derivative of the y-component of the surface normal must be equal to the y derivative of the x-component of the surface normal, and hence the Hessian must be symmetric), then the shape of the reconstructed surface is affected by the choice of integration path.

6.1.1. RELATED LITERATURE

The most direct method of height recovery is path-based. In the original work on shape-from-shading by Horn and Brooks (Horn and Brooks, 1986; Horn, 1990) the surface height recovery process is applied as a post-processing step. The process proceeds from the occluding boundary and involves incrementing the surface height by an amount determined by the distance traversed and the slope angle of the local tangent plane. Unfortunately, this process is highly sensitive to noise and errors in surface normal direction accumulate rapidly as the front propagates inwards. In an attempt to regulate the errors, Wu and Li (Wu and Li, 1988) average the surface normal directions to obtain height estimates.

More sophisticated path-based methods may be developed if differential geometry is used. For instance, Dupuis and Oliensis (Dupuis and Oliensis, 1992) have described a method which involves propagation in the direction of the steepest gradient from singular points. A fast variant of this algorithm is described by Bichsel and Pentland (Bichsel and Pentland, 1992) who compute the relative height of the surface with respect to the highest intensity point.

An alternative approach to surface reconstruction is to pose the problem of height recovery as one of energy minimisation. Leclerc and Bobick (Leclerc and Bobick, 1991) have developed a direct numerical method for height recovery from shading information which uses curvature consistency constraints. In related work, Tsai and Shah (Tsai and Shah, 1994) describe a fast surface height recovery method, which works well except at the locations of self-shadows and numerically singular points. Recent work by Karaçali and Snyder (Karaçali and Snyder, 2003; Karaçali and Snyder, 2004) has shown how surfaces can be reconstructed under partial integrability constraints. This analysis is applicable when surface height discontinuities are present.

The recovery of the surface height function can also be posed as the solution of a differential equation. In the level-set method of Kimmel, Bruckstein, Kimmia and Siddiqi (Kimmel et al., 1995) surface reconstruction is incorporated as an integral component into the shape-from-shading process. They use the apparatus of level-set theory to simultaneously solve the image irradiance equation and recover the associated surface height-function under constraints provided by surface integrability.

An elegant and ingenious solution is proposed by Frankot and Chellappa (Frankot and Chellappa, 1988) who project the surface normals into the Fourier domain to impose integrability constraints and hence recover surface height. There have been a number of extensions to this work. For instance, a so-called "lawn mower" algorithm has been proposed to impose consistency and surface integrability on the field of surface normals by aligning their directions (Noakes et al., 1999). In a related development, Wei and Klette (Wei and Klette, 2003) have shown how the quality of the recovered surface may be improved by incorporating more complex regularisers into the Fourier domain analysis.

Several of these methods are described in more detail and compared in the comprehensive review paper of Zhang, Tsai, Cryer and Shah (Zhang et al., 1999).

6.1.2. MOTIVATION

The recovery of the integration path is clearly one of optimisation, which can be solved using a number of techniques including direct numerical methods and level set techniques. However, one of the methods that has not received attention is that of posing the problem in a graph-spectral setting and using eigenvector methods to recover the solution. The idea underpinning graph-spectral methods is to abstract the problem in hand using a weighted graph. Here the nodes represent the basic image entities and the weighted edges represent affinity relations between them. By computing the eigenvalues and eigenvectors of the weight matrix, it is possible to find groups or clusters of entities. The graph-spectral method is in fact one of energy minimisation since the eigenvectors can be shown to be minimisers of a quadratic form. In fact, graph-spectral methods have recently proved highly effective in image processing and computer vision. Perhaps the best known method is that of Shi and Malik (Shi and Malik, 2000) which has shown how to locate image regions by recursively bisecting a weighted graph that represents the affinity of pairs of pixels. The method is based on the normalised cut. This is a measure of the relative weight of the edges connecting the two parts of the partition (the cut) to the weight assigned to the edges within the two parts of the bisection (the association). A relaxed solution to the bisection problem is found by locating the eigenvector associated with the second smallest eigenvalue of the Laplacian matrix (the degree matrix minus the affinity weight matrix). Although it is convenient to work with the Laplacian, since it is positive semi-definite, grouping and segmentation can also be performed using an edge-weight or affinity matrix. For instance, both Sarkar and Boyer (Sarkar and Boyer, 1998) and Perona and Freeman (Perona and Freeman, 1998) have developed matrix factorisation methods for line-segment grouping that use eigenvectors of an affinity matrix rather than the associated Laplacian. The Sarkar and Boyer (Sarkar and Boyer, 1998) method can be understood as maximising the association (i.e. the total edge weight) of the clusters.

The methods described above all share the feature of using the eigenvectors of a Laplacian or an affinity matrix to define groups or clusters or objects. However, graph-spectral methods can also be used for path analysis tasks on graphs. For instance, it is well known that the path length distribution can be computed from the spectrum of eigenvalues of the adjacency matrix (Biggs, 1993). Ideas from spectral-graph theory have also been used to analyse the behaviour of random walks on graphs (Lovász, 1993; Chung, 1997; Cvetković et al., 1980). In addition, there are important relationships between the eigenvectors of the edge weight matrix and other quantities related to random walks. These include the access time for a node (i.e. the expected number of time steps that must have elapsed before the node is visited) and the mixing rate (i.e. the rate at which the random walk converges to its steady state). The relationship between the leading eigenvector of the edge weight matrix and the steady state random walk has been exploited in a number of areas including routeing theory and information retrieval (Azar et al., 2000; Kleinberg, 1998).

The advantage of graph-spectral methods is that they can be used to find approximate or relaxed solutions without the need for parallel iterative updates at the pixel-site level. These methods also obviate the need for complex search algorithms. However, although they have been applied to region segmentation and grouping problems, graph-spectral methods have not been applied to curve detection problems of the sort that arise in the determination of the optimal integration path.

Hence, in this paper we pose the recovery of the surface integration path in a graph-spectral setting. According to our graph-based abstraction of the surface reconstruction process, each pixel-site is represented by a node, the edge structure is determined by connectivity on the pixel lattice, and the edge-weights are determined by the change in surface normals directions. By using this graph theoretical setting, we can make use of the edge-connected path on the pixel lattice to perform surface height recovery.

As the path is traversed, the surface height function is incremented by an amount determined by the slope of the local tangent plane.

The integration algorithms presented here have both local and global features. The graph-spectral approach allows to locate an ordering of the sites on the pixel lattice that minimise a quadratic energy function. However, this ordering is not guaranteed to be connected, and a local search procedure must be used to ensure connectivity. Rather than commencing from a specification of the recovery of the integration path in terms of a random walk, we couch the process in an energy minimisation setting. The minimisation problem is formulated so as to encourage path connectivity.

6.2. Affinity Matrix

Here, our goal is to recover the surface height information from the field of surface normals by pursuing a graph-spectral analysis. To this end, we require a transition weight matrix which reflects both, the connectivity of the pixel lattice and the fact that the field is obtained by translating the surface normals from each point on a curved surface to a projection plane. From a computational standpoint, the aim is to find a path on the projection plane along which simple trigonometry may be applied to increment the estimated height function. To be more formal, suppose that the surface under study is $\mathcal{S}$ and that the field of surface normals is constructed on the plane Π. Our aim here is to find a curve Γ_Π across the plane Π that can be used as an integration path to reconstruct the height-function of the surface $\mathcal{S}$. The projection of the curve Γ_Π onto the surface $\mathcal{S}$ is denoted by $\Gamma_\mathcal{S}$. Further, suppose that $\kappa(l)$ is the sectional curvature of the curve $\Gamma_\mathcal{S}$ at the point $q \in \mathcal{S}$ with parametric coordinate l. We seek the path $\Gamma_\mathcal{S}$ that minimises the total squared curvature

$$\mathcal{E}(\Gamma_\mathcal{S}) = \int_{\Gamma_\mathcal{S}} \kappa(l)^2 dl \tag{1}$$

For the surface $\mathcal{S}$ sampled on the plane Π the field of unit surface normals consists of the set $\vec{N}_q$ where $q \in \Pi$. Accordingly, and following do Carmo (Do Carmo, 1976), we let $\Pi_q(\mathcal{S})$ represent the tangent plane to the surface $\mathcal{S}$ at the point q which belongs to the curve $\Gamma_\mathcal{S}$. To compute the sectional curvature $\kappa(l)$ we require the differential of the surface or Hessian matrix $d\vec{N}_q : \Pi_q(\mathcal{S}) \to \Pi_q(\mathcal{S})$. The maximum and minimum eigenvectors λ_1 and λ_2 of $d\vec{N}_q$ are the principal curvatures at the point q. The corresponding eigenvectors $\vec{e}_1 \in \Pi_q(\mathcal{S})$ and $\vec{e}_2 \in \Pi_q(\mathcal{S})$ form an orthogonal basis on the tangent plane $\Pi_q(\mathcal{S})$. At the point q the unit normal vector to the curve $\Gamma_\mathcal{S}$ is $\vec{n}$ and the unit tangent vector is $t_q \in \Pi_q(\mathcal{S})$. The sectional curvature of Γ at q is given by $\kappa(l) = \frac{(\vec{t}_q \cdot \vec{e}_1)^2(\lambda_1 - \lambda_2) + \lambda_2}{\vec{n} \cdot \vec{N}_q}$, where $(\vec{t}_q \cdot \vec{e}_1)^2(\lambda_1 - \lambda_2) + \lambda_2$ is the normal curvature and $\alpha = \arccos \vec{n} \cdot \vec{N}_q$ is the angle between the curve normal and the surface normal.

In practice, we will be dealing with points which are located at discrete positions on the pixel lattice. Suppose that i and j are the pixel indices of neighbouring points sampled on the pixel lattice along the path $\Gamma_{\mathcal{S}}$. With this discrete notation, the cost associated with the path is given by $E(\Gamma_{\mathcal{S}}) = \sum_{(i,j)\in\Gamma_{\mathcal{S}}} E_{i,j} = \sum_{(i,j)\in\Gamma_{\mathcal{S}}} \kappa_{i,j}^2 l_{i,j}$, where $\kappa_{i,j}$ is an estimate of the curvature based on the surface normal directions at the pixel locations i and j, and $l_{i,j}$ is the path distance between these points. The energy associated with the transition between sites i and j is $E_{i,j} = \kappa_{i,j}^2 l_{i,j}$.

In order to compute the path curvature appearing in the expression for the transition energy, we make use of the surface normal directions. To commence, we note that $|\kappa_{i,j}| = \frac{1}{\mathcal{R}_{i,j}}$ where $\mathcal{R}_{i,j}$ is the radius of the local circular approximation to the integration curve on the surface. Suppose that the surface normal directions at the pixel locations i and j are respectively $\vec{N}_i$ and $\vec{N}_j$. The approximating circle connects the points i and j, and has the path segment $l_{i,j}$ as the connecting chord. The change in direction of the radius vector of the circle is $\theta_{i,j} = \arccos \vec{N}_i \cdot \vec{N}_j$, and hence $\cos\theta_{i,j} = \vec{N}_i \cdot \vec{N}_j$. If the angle $\theta_{i,j}$ is small, then we can make the Maclaurin approximation $\cos\theta_{i,j} \simeq 1 - \frac{\theta_{i,j}^2}{2} = \vec{N}_i \cdot \vec{N}_j$. Moreover, the small angle approximation to the radius of curvature of the circle is $\mathcal{R}_{i,j} = \frac{l_{i,j}}{\theta_{i,j}}$. Hence, $\kappa_{i,j}^2 = \frac{2(1-\vec{N}_i\cdot\vec{N}_j)}{l_{i,j}^2}$. The geometry outlined above is illustrated in Figure 6.3.1a.

As a result, we find that the cost associated with the step from the site indexed i to that indexed j is $E_{i,j} = \frac{2}{l_{i,j}}(1 - \vec{N}_i \cdot \vec{N}_j)$. With the energy function at hand, we attempt to find the minimum energy integration path $\Gamma_{\mathcal{S}}$ so as to satisfy the condition $\Gamma_{\mathcal{S}} = \arg\min_{\hat{\Gamma}} E(\hat{\Gamma})$. As mentioned earlier, to do this, we require a suitable matrix representation of the step costs between pairs of sites on the plane Π. Hence, for the pixel-sites indexed i and j, we define the transition weight matrix to have elements

$$W(i,j) = \begin{cases} \exp[-\beta E_{i,j}] & \text{if } j \in N_i \\ 0 & \text{otherwise} \end{cases} \tag{2}$$

where N_i is the set of pixel-neighbours of the pixel-site indexed i and β is a constant. As a result, the curvature weight is only non-zero if sites abut one-another.

6.3. Combinatorics and Random Walks

There are clearly a number of ways in which the energy can be minimised. These might include expectation-maximisation (Leite and Hancock, 1997), relaxation labelling (Zucker et al., 1988) or stochastic methods (Williams and Jacobs, 1997). However, here we choose to present two akin approaches that yield a result which is equivalent up to scaling. These are a combinatorial optimisation method which solves the problem by performing seriation on the edge-weight matrix and a random walk approach that makes use of the apparatus of Markov chains to recover the path whose curvature is minimum.

Thus, the section is organised as follows. We commence by presenting a graph seriation solution that draws on combinatorial optimisation. This is a method that hinges in optimising a penalty function by means of seriating the nodes in the graph subject to connectivity constraints. Having presented a combinatorial approach to the problem, we proceed to show how random walks may be employed to recover the surface height from the field of normals. By casting the problem in a probabilistic setting, we show that the steady state probability of the Markovian process associated to the weight matrix is equivalent, up to scaling, to the solution yielded by graph seriation.

6.3.1. GRAPH SERIATION

In Section 2, we showed how the change in surface normal directions could be used to compute the elements of the transition weight matrix. In this section, we pose the problem of locating an integration path that maximises the total curvature weight as a process of graph-spectral seriation.

To commence, we pose the problem in a graph-based setting. The set of pixel sites can be viewed as a graph $G = (V, E)$ $G = (V, E, W)$ with index-set V and edge-set $E = \{(i, j)|(i, j) \in V \times V, i \neq j\}$. Let the curvature minimising path commence at the node j_1 and proceed via the sequence of edge-connected nodes $\Gamma = \{j_1, j_2, j_3, ...\}$ where $(j_i, j_{i-1}) \in E$. Further, we suppose that the element $W(j_i, j_{i+1})$ of the transition weight matrix

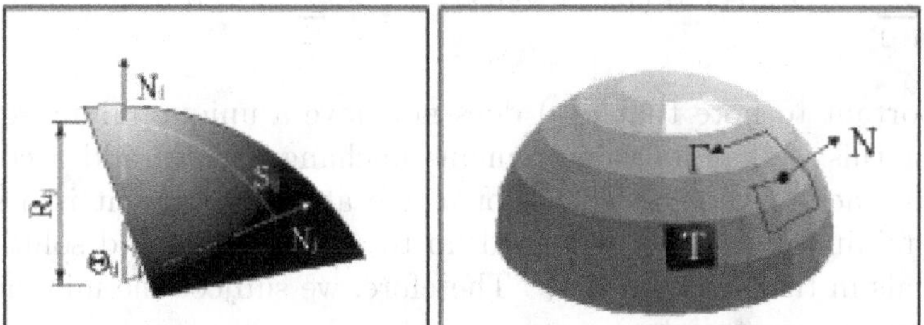

(a) (b)

Figure 6.1. Geometry of the path recovery process.

associated with the move between the nodes j_i and j_{i+1} can be regarded as a pairwise similarity measure. With these ingredients, the problem of finding the path that minimises the curvature between adjacent pixel-sites can be viewed as one of seriation, subject to edge connectivity constraints.

As noted by Atkins, Boman and Hendrikson (Atkins et al., 1998), many applied computational problems, such as sparse matrix envelope reduction, graph partitioning and genomic sequencing, involve ordering a set according to a permutation $\varrho = \{\varrho(j_1), \varrho(j_2), \ldots, \varrho(j_{|V|})\}$ so that strongly related tokens are placed next to one another. The seriation problem is that of finding the permutation ϱ that satisfies the condition

$$\varrho(j_i) < \varrho(j_k) < \varrho(j_l) \Rightarrow \{W(j_i, j_k) \geq W(j_i, j_l) \wedge W(j_k, j_l) \geq W(j_i, j_l)\} \quad (3)$$

This task has been posed as a combinatorial optimisation problem which involves minimising the penalty function

$$g(\varrho) = \sum_{i=1}^{|V|} \sum_{k=1}^{|V|} W(j_i, j_k)(\varrho(j_i) - \varrho(j_k))^2$$

for a set of N elements and a symmetric, real transition weight matrix W.

Unfortunately, the penalty function $g(\varrho)$, as given above, does not impose edge connectivity constraints on the ordering computed during the minimisation process. Furthermore, it implies no directionality in the transition from the node indexed j_i to the one indexed j_{i+1}. To overcome these shortcomings, we turn our attention instead to the penalty function

$$g(\varrho) = \sum_{i=1}^{|V|-1} W(j_i, j_{i+1})(\varrho(j_i) - \varrho(j_{i+1}))^2 \quad (4)$$

where the nodes indexed j_i and j_{i+1} are edge connected. After some algebra, it is straightforward to show that

$$g(\varrho) = \sum_{i=1}^{|V|-1} W(j_i, j_{i+1})(\varrho(j_i)^2 + \varrho(j_{i+1})^2) - 2\sum_{i=1}^{|V|-1} W(j_i, j_{i+1})\varrho(j_i)\varrho(j_{i+1}) \quad (5)$$

It is important to note that $g(\varrho)$ does not have a unique minimiser. The reason for this is that its value remains unchanged if we add a constant amount to each of the coefficients of ϱ. We also note that it is desirable that the minimiser of $g(\varrho)$ is defined up to a constant λ and solutions to polynomials in the elements of W. Therefore, we subject the minimisation problem to the constraints

$$\lambda\varrho(j_i)^2 = \sum_{k=1}^{|V|} W(j_k, j_i)\varrho(j_k)^2 \text{ and } \sum_{k=1}^{|V|} \varrho(j_k)^2 \neq 0 \quad (6)$$

Since the coefficients $\varrho(j_{i+1})$ are inversely proportional to $\lambda - W(j_{i+1}, j_i)$, the coefficient $\varrho(j_{i+1})^2$ increase with decreasing sectional curvature (i.e. the similarity tends to one). The effect of this is to enforce edge connectivity while favouring paths of small local curvature, and also to minimise the overall cost of the path.

Combining the constraint conditions given in Equation 6 with the definition of the penalty function given in Equation 5, it is straightforward to show that the permutation ϱ satisfies the condition

$$\sum_{k=1}^{|V|} \sum_{i=1}^{|V|-1} (W(j_k, j_i) + W(j_k, j_{i+1}))\varrho(j_k)^2 = \lambda \sum_{i=1}^{|V|-1} (\varrho(j_i)^2 + \varrho(j_{i+1})^2) \quad (7)$$

Using matrix notation, we can write the above equation in the more compact form

$$JW\phi = \lambda J\phi \quad (8)$$

where $\phi = \{\varrho(j_1)^2, \varrho(j_2)^2, \ldots, \varrho(j_{|V|})^2\}^T$ and J is the $N \times N$ matrix

$$J = \begin{vmatrix} 1 & 0 & 0 & \ldots & 0 & 0 \\ 0 & 2 & 0 & \ldots & 0 & 0 \\ 0 & 0 & 2 & \ldots & 0 & 0 \\ \vdots & \vdots & \vdots & \ddots & \vdots & \vdots \\ 0 & 0 & 0 & \ldots & 2 & 0 \\ 0 & 0 & 0 & \ldots & 0 & 1 \end{vmatrix} \quad (9)$$

Hence it is clear that locating the permutation ϱ that minimises $g(\varrho)$ can be posed as an eigenvalue problem, and that ϕ is an eigenvector of W. This follows from the fact that Equation 8 can be obtained by multiplying both sides of the eigenvector equation $W\phi = \lambda\phi$ by J. Furthermore, due to the norm condition of the eigenvector, the constraint $\sum_{k=1}^{|V|} \varrho(j_k)^2 \neq 0$ is always satisfied. Taking this analysis one step further, we can premultiply both sides of Equation 8 by ϕ^T to get the matrix equation $\phi^T JW\phi = \lambda\phi^T J\phi$. As a result, it follows that

$$\lambda = \frac{\phi^T JW\phi}{\phi^T J\phi} \quad (10)$$

This expression is reminiscent of the Rayleigh quotient. It also suggests the plausibility of using the mathematical techniques commonly employed to study the asymptotic behaviour of non-homogeneous Markov chains (Bremaud, 2001) to take our analysis further.

We note that the elements of the permutation ϱ are required to be real. Consequently, the coefficients of the eigenvector ϕ are always non-negative.

Since the elements of the matrices J and W are also positive, it follows that the quantities $\phi^T JW\phi$ and $\phi^T J\phi$ are positive. Hence, the set of solutions reduces itself to those that correspond to a constant $\lambda > 0$. We also require the coefficients of the eigenvector ϕ to be linearly independent of the all-ones vector $\mathbf{e} = [1, 1, \ldots, 1]^T$.

With these observations in mind, we focus on proving the existence of a permutation that minimises $g(\varrho)$ subject to the constraints in Equation 6, and demonstrating that this permutation is unique. To this end, we use the Perron-Frobenius theorem (Varga, 2000). This concerns the proof of existence regarding the eigenvalue $\lambda_* = \max_{i=1,2,\ldots,|V|}\{\lambda_i\}$ of a primitive, real, non-negative, symmetric matrix W, and the uniqueness of the corresponding eigenvector ϕ_*. The Perron-Frobenius theorem states that the eigenvalue $\lambda_* > 0$ has multiplicity one. Moreover, the coefficients of the corresponding eigenvector ϕ_* are all positive and the eigenvector is unique. As a result, the remaining eigenvectors of W have at least one negative coefficient and one positive coefficient. If W is substochastic, ϕ_* is also known to be linearly independent of the all-ones vector $\mathbf{e}$. As a result, the leading eigenvector of W is the unique solution of $g(\varrho)$.

6.3.2. RANDOM WALKS

To take our analysis further and make the relationship to the field of surface normals more explicit, we cast the problem of surface height recovery into a random walk setting. In order to profit from a Markov chain approach to the problem, we commence by row-normalising the weight matrix W so its rows sum to unity. To do this, we compute the degree of each node $deg(i) = \sum_{j=1}^{|V|} W(i,j)$. With the diagonal degree matrix $D = diag(deg(1), deg(2), \ldots, deg(|V|))$ at hand, the transition probability matrix is given by $P = D^{-1}W$. The elements of the transition matrix are hence given by $P_{i,j} = \frac{1}{deg(i)} W_{i,j}$. It is interesting to note that the transition matrix P is a row stochastic matrix. Moreover, it is related to the normalised symmetric positive definite matrix $\hat{W} = D^{-\frac{1}{2}} W D^{-\frac{1}{2}} = D^{\frac{1}{2}} P D^{-\frac{1}{2}}$. As a result, we can write $P = D^{-\frac{1}{2}} \hat{W} D^{\frac{1}{2}}$. It is worth noting in passing that the matrix $\hat{W}$ is related to the normalised Laplacian $L = D^{-\frac{1}{2}}(D - W)D^{-\frac{1}{2}} = I - D^{-\frac{1}{2}} W D^{-\frac{1}{2}} = I - \hat{W}$.

Our aim is to use the steady state random walk on the graph G as an integration path for surface height recovery. The walk commences at the pixel j_1 and proceeds via the sequence of pixel sites $\Gamma = \{j_1, j_2, j_3, \ldots\}$. If the random walk can be represented by a Markov chain with transition matrix P, then the probability of visiting the pixel sites in the sequence above is

$$P_{\Gamma_S} = P(j_1) \prod_{l \in \Gamma_S} P_{j_{l+1}, j_l} = \prod_{l \in \Gamma_S} \frac{W_{j_{l+1}, j_l}}{deg(l)} \tag{11}$$

Substituting for the path energy, we have that

$$P_{\Gamma_S} = \frac{\exp\left[-2\beta \sum_{l \in \Gamma_S} \left\{1 - \vec{N}_l.\vec{N}_{l+1}\right\}\right]}{\prod_{l \in \Gamma_S} deg(l)} = \frac{1}{Z_{\Gamma_S}} \exp[-E_{\Gamma_S}] \tag{12}$$

where

$$E_{\Gamma_S} = 2\beta \sum_{l \in \Gamma_S} \left\{1 - \vec{N}_l.\vec{N}_{l+1}\right\} \tag{13}$$

and

$$Z_{\Gamma_S} = \prod_{l \in \Gamma_S} deg(l) \tag{14}$$

Hence, the integration path is a Markov chain with energy function E_{Γ_S} and partition function Z_{Γ_S}. Further, let $Q_t(i)$ be the probability of visiting the pixel site indexed i after t-steps of the random walk and let $Q_t = [Q_t(1), Q_t(2), \ldots]^T$ be the vector whose components are the probabilities of visiting the sites at time t. After t time steps, we have that $Q_t = P^t Q_0$. If $\hat{W}^t$ is the result of multiplying the symmetric positive definite matrix $\hat{W}$ by itself t times, then $P^t = D^{-\frac{1}{2}} \hat{W}^t D^{\frac{1}{2}}$.

To develop a spectral method for locating the steady state random walk, we turn to the spectral decomposition of the normalised affinity matrix $\hat{W}$

$$\hat{W} = D^{-\frac{1}{2}} W D^{-\frac{1}{2}} = \sum_{i=1}^{|V|} \lambda_i \phi_i \phi_i^T \tag{15}$$

where the λ_i are the eigenvalues of $\hat{W}$ and the ϕ_i are the corresponding eigenvectors. By constructing the matrix $\Phi = (\phi_1 | \phi_2 | \ldots | \phi_{|V|})$ with the eigenvectors of $\hat{W}$ as columns and the matrix $\Lambda = diag(\lambda_1, \lambda_2, ..., \lambda_{|V|})$ with the eigenvalues as diagonal elements. We can write the spectral decomposition in the more compact form $\hat{W} = \Phi \Lambda \Phi^T$. Since, the eigenvectors of $\hat{W}$ are orthonormal, i.e. $\Phi \Phi^T = I$, we can have that $\hat{W}^t = \Phi \Lambda^t \Phi^T$. Substituting the spectral expansion of the matrix $\hat{W}$ into the expression for the state-vector of the random walk at time step t, we find

$$Q_t = D^{-\frac{1}{2}} \Phi \Lambda^t \Phi^T D^{\frac{1}{2}} Q_0 = \left\{ \sum_{i=1}^{|V|} \lambda_i^t D^{-\frac{1}{2}} \phi_i \phi_i^T D^{\frac{1}{2}} \right\} Q_o \tag{16}$$

Recall that the leading eigenvalue of $\hat{W}$ is unity. Furthermore, from spectral graph theory (Chung, 1997), provided that the graph G is not a bipartite graph, then the smallest eigenvalue $\lambda_{|V|} > -1$. As a result, when the Markov chain approaches its steady state, i.e. $t \to \infty$, then all but the first term in the above series become negligible. Hence, the steady state random walk is given by

$$Q_s = \lim_{t \to \infty} Q_t = D^{\frac{1}{2}} \phi_* \phi_*^T D^{-\frac{1}{2}} Q_0 \tag{17}$$

This establishes that the leading eigenvector of the normalised affinity matrix $\hat{W}$ determines the steady state of the random walk. It is also important to note that the equilibrium equation for the Markov process is $Q_s = PQ_s$, where Q_s is the vector of steady-state site visitation probabilities. Hence, since the leading eigenvalue of P is unity, then it follows that Q_s is the leading eigenvector of P. For a more complete proof of this result see the book by Varga (Varga, 2000) or the review of Lovasz (Lovász, 1993).

We aim to visit the pixel sites on the lattice in the order of their steady-state state probabilities. Suppose that the initial state vector for the sites is uniform, i.e. $Q_0 = (\frac{1}{|V|}, \ldots, \frac{1}{|V|})^T$. As a result, the steady-state probability of visiting the pixel site i is

$$Q_s(i) = \frac{1}{|V|} \sum_{j=1}^{|V|} \sqrt{\frac{deg(j)}{deg(i)}} \phi_*(i) \phi_*(j) = \frac{1}{|V|} \frac{\phi_*(i)}{\sqrt{deg(i)}} \sum_{j=1}^{|V|} \sqrt{deg(j)} \phi_*(j) \tag{18}$$

Since the summation appearing above is the same for all pixel sites, the probability rank order is determined by the quantity $\psi_*(i) = \frac{\phi_*(i)}{\sqrt{deg(i)}}$. Hence, it is the scaled leading eigenvector

$$\psi_* = D^{-\frac{1}{2}} \phi_* = \left[\frac{\phi_*(1)}{\sqrt{deg(1)}}, \ldots, \frac{\phi_*(|V|)}{\sqrt{deg(|V|)}} \right]^T \tag{19}$$

that determines the probability rank order of the sites in the steady state random walk. Hence, the leading eigenvector of the matrix $\hat{W}$ satisfies the condition

$$\phi_* = \arg\max_{\Phi} \phi^T \hat{W} \phi = \arg\max_{\Phi} \phi^T D^{-\frac{1}{2}} W D^{-\frac{1}{2}} \phi \tag{20}$$

We can make the relationship to the raw field of surface normals more explicit by introducing the matrix $F = (\vec{N}_1 | \vec{N}_2 | \ldots | \vec{N}_{|V|})$ with the surface normals as columns. When the constant β is small, then making use of the Maclaurin expansion of the exponential weighting function we can write $W = \mathbf{e}\mathbf{e}^T - \beta(\mathbf{e}\mathbf{e}^T - F^T F)$, where $\mathbf{e}$ is the all-ones vector of length $|V|$. In practice, we use a 4-pixel neighbourhood to compute the weight matrix, and hence $D \simeq 4I$. As a result, when β is small we can write $D^{-\frac{1}{2}}WD^{-\frac{1}{2}} \simeq \frac{1}{4}(\mathbf{e}\mathbf{e}^T - \beta[\mathbf{e}\mathbf{e}^T - F^TF])$. The path is the one that satisfies the condition

$$\phi_* = \arg\max_{\Phi} \phi^T F^T F \phi = \arg\min_{\Phi} \sum_{i=1}^{|V|}\sum_{j=1}^{|V|} \vec{N}_i \cdot \vec{N}_j \phi(i)\phi(j) \qquad (21)$$

Hence, the integration path will minimise the change in surface normal direction.

Our aim is to use the sequence of pixel sites given by the probability rank order to define a serial ordering for the sites on the pixel lattice. If we visit the sites of the pixel lattice in the order defined by the magnitudes of the coefficients of the leading eigenvector of the normalised affinity matrix, then the path is the steady state of the Markov chain. Unfortunately, the path followed by the steady state random walk is not edge-connected. Hence, we need a means of placing the pixel sites in an order in which edge-connectivity constraints are preserved using the elements of the leading eigenvector ϕ_*.

6.4. Algorithm Description

In the previous section, we presented two alternatives to the problem of recovering the surface height from the field of normals projected onto the image plane. In this section, we elaborate on the description of the first of these two approaches. The reason for preferring the seriation approach over the Markovian one is one of numerical stability. Despite of the fact that both approaches are equivalent, up to scaling, the seriation method is less prone to numerical errors due to the lack of the scaling step for both, the weight matrix and its leading eigenvector.

To recover the minimum curvature path we commence from the pixel site associated with the largest component of ϕ_*, i.e. the largest site probability. We then sort the elements of the leading eigenvector such that they are both in the decreasing magnitude order of the coefficients of the eigenvector, and satisfy edge connectivity constraints on the graph. The procedure is a recursive one that proceeds as follows. At each iteration, we maintain a list of pixel sites visited. At iteration k let the list of pixel sites be denoted by L_k. Initially, $L_0 = j_o$ where $j_0 = \arg\max_j \phi_*(j)$, i.e. j_0 is the component of ϕ_* with the largest magnitude. Next, we search through the set of first neighbours $N_{j_0} = \{k|(j_0, k) \in E\}$ of j_o to find the pixel site associated with the largest remaining component of ϕ_*. The second element in the list is $j_1 = \arg\max_{l \in N_{j_0}} \phi_*(l)$. The pixel site index j_1 is appended to the list of pixel sites visited and the result is L_1. In the kth (general) step of the algorithm we are at the pixel site indexed j_k and the list of pixel sites visited by the path so far is L_k. We search through those first-neighbours of j_k that have not already been traversed by the path. The set of pixel sites is $C_k = \{l | l \in N_{j_k} \wedge l \notin L_k\}$. The next site to be appended to the path list is therefore $j_{k+1} = \arg\max_{l \in C_k} \phi_*(l)$. This process is repeated until no further moves can be made. This occurs when $C_k = \emptyset$ and we denote the index of the termination of the path by T. The serial ordering of the pixel sites that results from this edge-based sorting is the integration path $\Gamma = L_T$.

6.5. Height Recovery

Our surface height recovery algorithm proceeds along the sequence of pixel sites defined by the order of the coefficients of the leading eigenvector associated with the edge-weight matrix. If the path is $\Gamma = (j_1, j_2, j_3, \ldots)$, where the order is established using the method outlined in Section 6.4, we can increment the surface height-function as we move from site to site in this path. In this section, we describe the trigonometry of the height incrementation process.

At step n of the algorithm, we make a transition from the pixel with path-index j_{n-1} to the pixel with path-index j_n. The distance between the pixel-centres associated with this transition is

$$r_n = \sqrt{(x_{j_n}^2 - x_{j_{n-1}})^2 + (y_{j_n} - y_{j_{n-1}})^2} \tag{22}$$

This distance, together with the surface normals $\vec{N}_{j_n} = [N_{j_n}(x), N_{j_n}(y), N_{j_n}(z)]^T$ and $\vec{N}_{j_{n-1}} = [N_{j_{n-1}}(x), N_{j_{n-1}}(y), N_{j_{n-1}}(z)]^T$ at the two pixel-sites may be used to compute the change in surface height associated with the transition. The height increment is given by

$$h_n = \frac{r_n}{2} \left\{ \frac{N_{j_n}(x)}{N_{j_n}(y)} + \frac{N_{j_{n-1}}(x)}{N_{j_{n-1}}(y)} \right\} \tag{23}$$

If the height-function is initialised by setting $z_{j_0} = 0$, then the centre-height for the pixel with path-index j_{n+1} is $z_{j_{n+1}} = z_{j_n} + h_n$. The geometry of this procedure is illustrated in Figure 6.3.1b.

6.6. Experiments

In this section, we present results for our surface height recovery method. In all our experiments, for purposes of recovering the integration path, we have used the graph-spectral seriation method. We have done this in order to be consistent with the algorithm description presented in Section 4. We

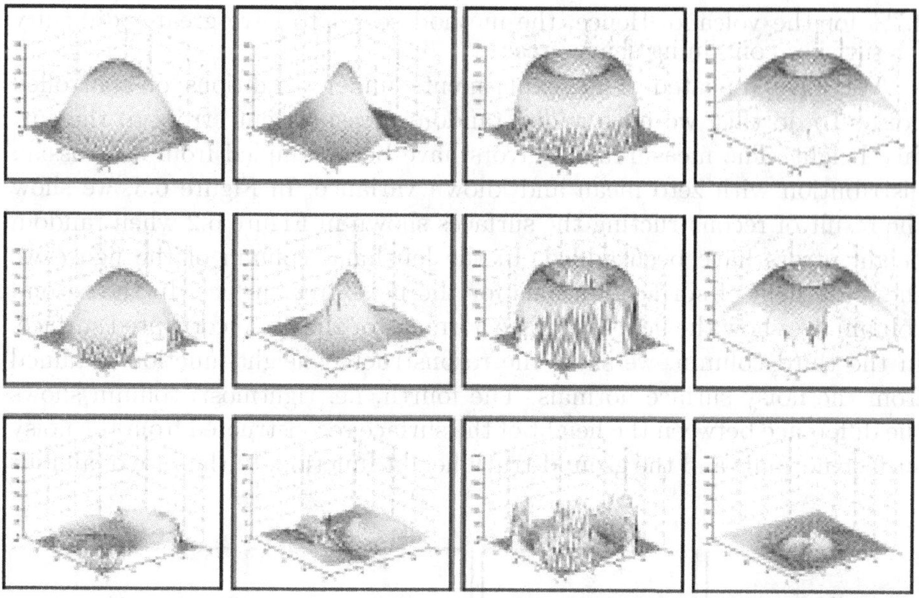

Figure 6.2. Top row: Artificially generated data; Middle row: Reconstructed surface; Bottom row: Error plot.

have divided our experimental evaluation into two parts. We commence with a sensitivity study aimed at evaluating the method on synthetic data. In the second part of the section, we focus on real-world data.

6.6.1. SYNTHETIC DATA

In this section, we provide some experiments on synthetic data. The aim here is to determine the accuracy of the surface reconstruction method. To this end, we have generated synthetic surfaces. From the surfaces, we have computed the field of surface normal directions. We have then applied the graph-spectral method to the field of surface normals to recover an estimate of the surface height.

In Figure 6.2, we show the results obtained for a series of different surfaces. In the top row, we show the original synthetic surface. The middle row shows the surface reconstructed from the field of surface normals. The bottom row shows the absolute error between the ground-truth and the reconstructed surface height. The surfaces studied are a dome, a sharp ridge, a torus and a volcano. In all four cases the surface reconstructions are qualitatively good. For the reconstructed surfaces, the mean-squared errors are 5.6% for the dome, 10.8% for the ridge, 7.8% for the torus and

4.7% for the volcano. Hence, the method seems to have greater difficulty for surfaces containing sharp creases.

We have repeated these experiments under conditions of controlled noise. To do this, we have added random measurement errors to the surface height. The measurement errors have been sampled from a Gaussian distribution with zero mean and known variance. In Figure 6.3, we show the result of reconstructing the surfaces shown in Figure 6.2 when random height errors have been added. In the left-hand column of the figure we show the field of surface normals for the noise-free surface. In the second column, we show the field of surface normals for the noise-corrupted surface. In the third column, we show the reconstructed height-function obtained from the noisy surface normals. The fourth, i.e. rightmost, column shows the difference between the height of the surface reconstructed from the noisy surface normals and the ground-truth height function. In the case of all four

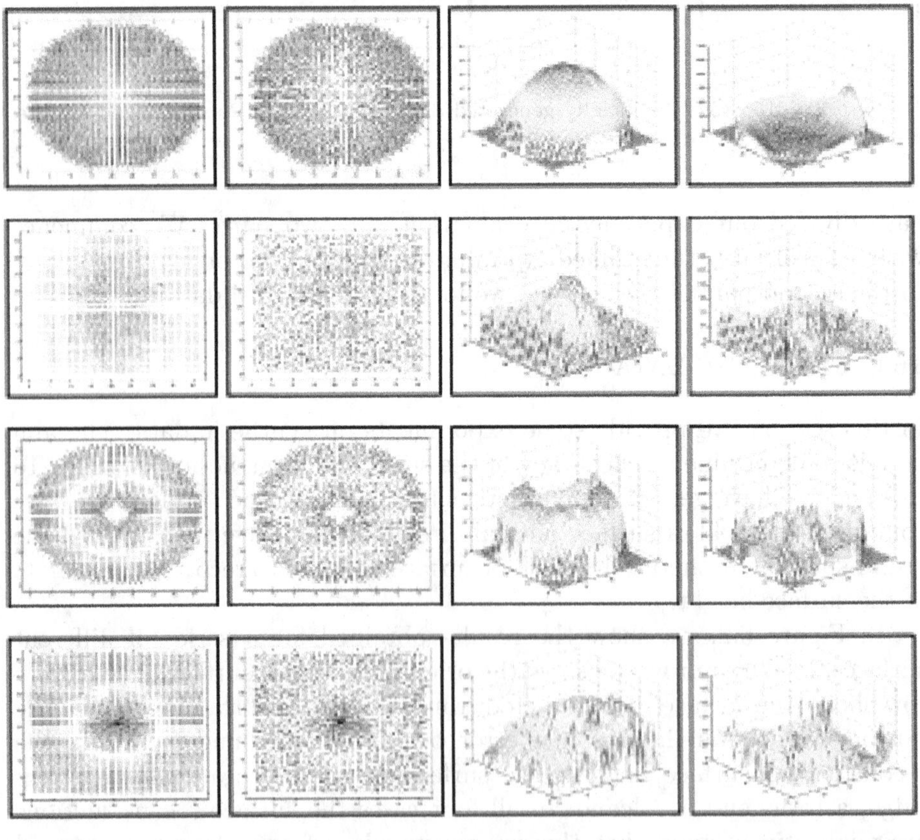

Figure 6.3. Left-hand column: Needle-map without added noise; Second Column: Needle-map with Gaussian noise added (worst case with variance set to unity); Third column: Reconstructed surface; Right-hand column: Error plot.

surfaces, the gross structure is maintained. However, the recovered height is clearly noisy. The height difference plots are relatively unstructured. These are important observations. They mean that our graph-spectral method simply transfers errors in surface normal direction into errors in height, without producing structural noise artefacts.

To investigate the effect of noise further, in Figure 6.4, we plot the mean-squared error for the reconstructed surface height as a function of the standard deviation of the added Gaussian noise. From the plots for the different surfaces shown in Figure 6.2, it is clear that the mean-squared error grows slowly with increasing noise standard deviation.

6.6.2. REAL-WORLD DATA

The second part of our experimental evaluation focusses on real-world imagery. Here, we have applied our surface recovery method to needle-maps extracted making use of the shape-from-shading algorithm of Worthington and Hancock (Worthington and Hancock, 1999). It should be stressed, however, that the method can be used in any situation where surface-data is presented in the form of a field of surface normals sampled on a plane. Hence, it can also be used in conjunction with shape-from-texture and motion analysis.

In the top row of Figure 6.5, we show our first example. The image used is that of a detail of a porcelain urn. The four panels in the top row of the figure convey the following information. The panel in the left-hand side, shows the raw image. The second and third panels show two views of the reconstructed surface. The main features to note here are that both, the convexities and concavities on the surface are well reconstructed. In the far-right panel we show the integration path, i.e. the order of the components of the leading eigenvector, for each site in the path. The path appears to follow the main height contours on the surface patches.

The second and third rows repeat this sequence of images for two images from the University of Central Florida data-base which is used in the comparative study of Zhang *et al.* (Zhang et al., 1999). The images are those of a vase and a bust of Beethoven. The overall shape of the vase is well reconstructed and the fine detail of the face of Beethoven is well reproduced.

6.7. Conclusions

In this chapter, we have described how the surface height may be recovered making use of graph-spectral methods. The work described here can be further extended and improved in a number of different ways. Firstly, there

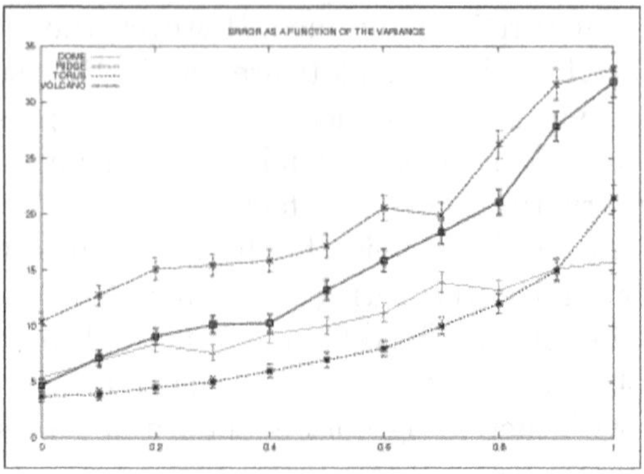

Figure 6.4. Surface reconstruction error versus noise variance.

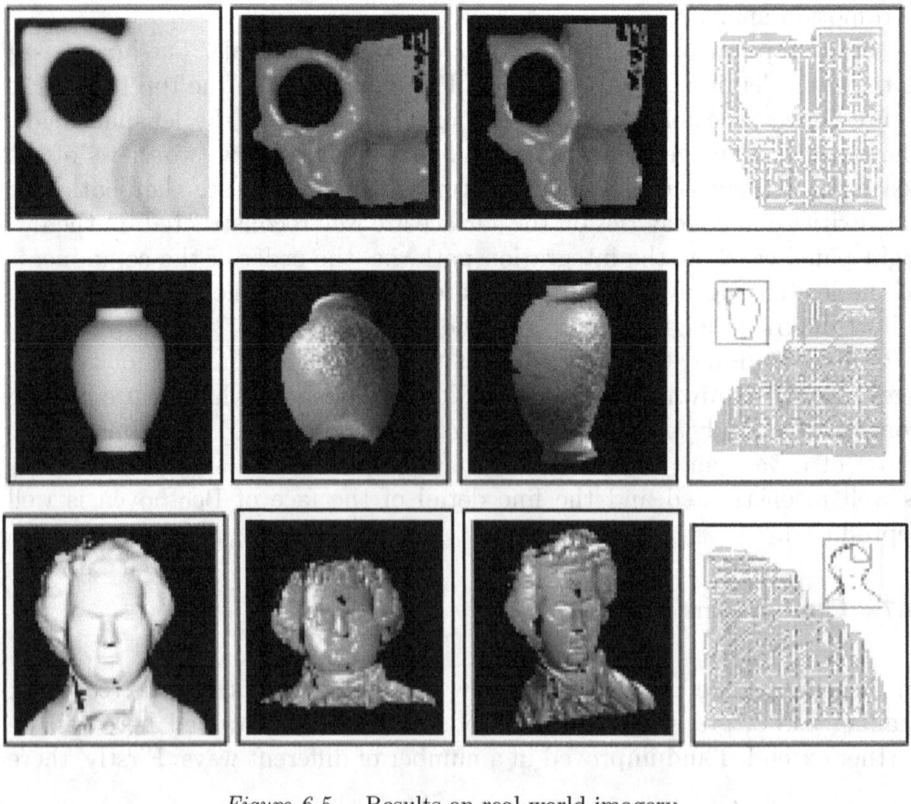

Figure 6.5. Results on real-world imagery.

is clearly scope for using more elaborate differential geometry to construct the affinity matrix. One possibility here is to attempt to ensure consistency of the integration path direction, and that of the local minimum curvature direction. Another possible direction is to develop a more sophisticated model for the integration path. Here, we have sought the path that is the steady state random walk of a Markov chain on a graph. This is a type of diffusion process. A more principled approach may be to pose the recovery of the integration path as the solution of a stochastic differential equation. In other words, the integration path may be posed as the solution of a Fokker-Plank equation. Finally, it may be interesting to investigate whether the idea of recovering a path using graph-spectral methods can be applied to 2D curve enhancement problems. In particular, the behaviour of curves in noisy imagery is frequently posed as a diffusion process. Hence, the behaviour of the transition probabilities is governed by a stochastic differential equation, and the steady state random walk is a natural way to characterise such systems.

References

Atkins, J. E., E. G. Bowman, and B. Hendrickson: A spectral algorithm for seriation and the consecutive ones problem. *SIAM J. Computing*, **28**:297–310, 1998.

Azar, Y., A. Fiat, A. R. Karlin, F. McSherry, and J. Saia: Spectral analysis of data. In Proc. *ACM Symp. Theory Computing*, pages 619–626, 2000.

Bichsel, M. and A. P. Pentland: A simple algorithm for shape from shading. In Proc. *IEEE Conf. Computer Vision Pattern Recognition*, pages 459–465, 1992.

Biggs, N. L.: *Algebraic Graph Theory*. Cambridge University Press, 1993.

Bremaud, P.: *Markov Chains, Gibbs Fields, Monte Carlo Simulation and Queues*. Springer, 2001.

Chung, F. R. K.: *Spectral Graph Theory*. American Mathematical Society, 1997.

Cvetković, D., M. Doob, and H. Sachs: *Spectra of Graphs: Theory and Application*. Academic Press, 1980.

Do Carmo, M. P.: *Differential Geometry of Curves and Surfaces*. Prentice Hall, 1976.

Dupuis, P. and J. Oliensis: Direct method for reconstructing shape from shading. In Proc. *IEEE Conf. Computer Vision Pattern Recognition*, pages 453–458, 1992.

Frankot, R. T. and R. Chellappa: A method of enforcing integrability in shape from shading algorithms. *IEEE Trans. Pattern Analysis Machine Intelligence*, **4**:439–451, 1988.

Horn, B. K. P.: Height and gradient from shading. *Int. J. Computer Vision*, **5**:37–75, 1990.

Horn, B. K. P. and M. J. Brooks: The variational approach to shape from shading. *Computer Vision Graphics Image Processing*, **33**:174–208, 1986.

Karaçali, B. and W. Snyder: Reconstructing discontinuous surfaces from a given gradient field using partial integrability. *Computer Vision Image Understanding*, **92**:78–111, 2003.

Karaçali, B. and W. Snyder: Noise reduction in surface reconstruction from a given gradient field. *Int. J. Computer Vision*, **60**:25–44, 2004.

Kimmel, R., K. Siddiqqi, B. B. Kimia, and A. M. Bruckstein: Shape from shading: Level set propagation and viscosity solutions. *Int. J. Computer Vision*, **16**:107–133, 1995.

Kleinberg, J.: Authoritative sources in a hyperlinked environment. In Proc. *ACM-SIAM Symp. Discrete Algorithms,* pages 668–677, 1998.

Leclerc, Y. G. and A. F. Bobick: The direct computation of height from shading. In Proc. *Computer Vision Pattern Recognition*, pages 552–558, 1991.

Leite, J. A. F. and E. R. Hancock: Iterative curve organisation with the em algorithm. *Pattern Recognition Letters,* **18**:143–155, 1997.

Lovász, L.: Random walks on graphs: a survey. *Bolyai Society Mathematical Studies,* **2**:1–46, 1993.

Noakes, L., R. Kozera and R. Klette: The lawn-mowing algorithm for noisy gradient vector fields. In Proc. *SPIE Conf. Vision Geometry VIII,* (R. A. Melter, A. Y. Wu, and L. J. Latecki, editors), pages 305–316, 1999.

Perona, P. and W. T. Freeman: Factorization approach to grouping. In Proc. *European Conf. Computer Visison,* (H. Burkhardt, B. Neumann, editors), pages 655–670, LNCS 1406, Springer, Berlin, 1998.

Rockwood, A. P. and J. Winget: Three-dimensional object reconstruction from two dimensional images. *Computer-Aided Design,* **29**:279–285, 1997.

Sarkar, S. and K. L. Boyer: Quantitative measures of change based on feature organization: Eigenvalues and eigenvectors. *Computer Vision Image Understanding,* **71**:110–136, 1998.

Shi, J. and J. Malik: Normalized cuts and image segmentation. *IEEE Trans. Pattern Analysis Machine Intelligence,* **22**:888–905, 2000.

Tsai, P. S. and M. Shah: Shape from shading using linear approximation. *Image Vision Computing,* **12**:487–498, 1994.

Varga, R. S.: *Matrix Iterative Analysis,* second edition. Springer, 2000.

Wei, T. and R. Klette: Depth recovery from noisy gradient vector fields using regularization. In Proc. *Computer Analysis Images Patterns,* pages 116–123, 2003.

Williams, L. R. and D. W. Jacobs: Stochastic completion fields: A neural model of illusory contour shape and salience. *Neural Computation,* **9**:837–858, 1997.

Worthington, P. L. and E. R. Hancock: New constraints on data-closeness and needle map consistency for shape-from-shading. *IEEE Trans. Pattern Analysis Machine Intelligence,* **21**:1250–1267, 1999.

Wu, Z. and L. Li: A line-integration based method for depth recovery from surface normals. *Computer Vision Graphics Image Processing,* **43**:53–66, 1988.

Zhang, R., P. S. Tsai, J. E. Cryer and M. Shah: Shape from shading: A survery. *IEEE Trans. Pattern Analysis Machine Intelligence,* **21**:690–706, 1999.

Zucker, S. W., C. David, A. Dobbins and L. Iverson: The organization of curve detection: Coarse tangent fields and fine spline coverings. In Proc. *IEEE Int. Conf. Computer Vision,* pages 568–577, 1988.

Part II

Discrete Geometry

SEGMENTATION OF BOUNDARIES INTO CONVEX AND CONCAVE PARTS

ULRICH ECKHARDT
University of Hamburg, Faculty of Mathematics
Bundesstrasse 55, D–20 146 Hamburg

Abstract. Representing a set by its boundary means a considerable losslesss data reduction by decrease of dimensionality. The price for this reduction is that usually boundaries are topologically much more complex than the sets described by them.

The aim of this paper is to present an approach for handling the boundary of a set. This approach does not make use of differential geometry. It is shown that it is indeed possible to derive important structural properties of a set by inspecting only its boundary.

Key words: planar sets, boundaries, convex and concave parts

7.1. Introduction

A set in the plane (or in higher dimensional space) can be described efficiently by means of its boundary (or surface) whenever the boundary has the Jordan property which means that it separates the plane into two connected components, the interior and the exterior. In this paper the following question is investigated: What can be said about a set if its boundary is probed in a finite number of points?

When investigating the boundary of a set, it is appropriate to impose 'tameness' assumptions on it. In the context of image processing, polygonal or even discrete sets are considered, so the usual tools of differential geometry are not adequate. Consequently, a 'differentialless' geometry in the sense defined in (Latecki and Rosenfeld, 1998) should be adapted. A very efficient description of a set can be found by attributing its boundary with predicates as convex or concave parts (Latecki and Lakämper, 1999). This representation is closely related to the 'Curvature Scale Space' (see e.g. (Mokhtarian, 1997)). In the latter approach the curvature zero crossing points were used as signatures in scale space. These curvature crossing points correspond to points where convex and concave parts of the boundary curve overlap. This

R. Klette et al. (eds.), Geometric Properties for Incomplete Data, 125-144.

means that the description by means of convex and concave parts is more general than the curvature zero crossings since the former does not make use of any concept from differential geometry. Moreover, by labeling parts of the boundary between curvature crossing points as convex or concave parts, ambiguities can be avoided. This is of major importance for shape coding algorithms (Heuer et al., 1999).

The mapping associating to each linear functional the set of its local maximizers on a nonempty compact convex set is upper semi–continuous (this concept will be defined later, see Definition 5.1). Moreover, an upper semi–continuous inverse for this map can be found. The study of this map yields insights into the structure of the boundary of a given set. The mapping can be deformed to yield a homeomorphism from the unit sphere onto the boundary of the set. Moreover, by a finite number of boundary points and tangent directions one can find a directionally convex set containing the given set.

In the general case, things become very complicated. It becomes necessary to rule out 'wild' boundary parts. A minimal requirement is that the boundary of the set under consideration should have the Jordan property. However, some more structure has to be imposed in order to get practical results. A certain regularity condition is stated which is theoretically tractable and practically acceptable.

In ((Scherl, 1987), see also (Eckhardt et al., 1987)) constructed a document processing system. In this system, a set (e.g. a letter, a word, a text line ...) is represented by a rather small subset of boundary points together with tangent information. By means of this representation the amount of data can be reduced very efficiently while retaining sufficient information to perform typical pattern recognition tasks like segmentation of letters, words, text lines, or a classification of different document components (text, structuring elements, pictures), or classification of specific letter styles in the text (serifs, slanted letters).

The aim of this paper is to give a theoretical framework for the concepts mentioned above which is able to cover also the discrete case. First, some known theoretical results about properties of boundaries of sets are given. Under a certain regularity condition the boundary consists of finitely many convex and concave parts which can be used for describing the boundary. In the second part it is shown that the boundary of a convex set in $\mathbb{R}^d$ can be mapped "almost homeomorphic" onto the sphere $\mathfrak{S}_{d-1}$. The generalization to the nonconvex case is indicated. It was not intended here to provide algorithmic details, this is partially done in Helene (Dörksen, 2004).

7.2. Sets and Surfaces in $\mathbb{R}^d$

We consider sets in $\mathbb{R}^d$. Denote by $\langle \cdot, \cdot \rangle$ the ordinary scalar product and by $\| \cdot \|$ the Euclidean norm in $\mathbb{R}^d$. The natural topology of $\mathbb{R}^d$ is generated by declaring the sets (open balls)

$$B_\varepsilon(x) = \left\{ y \in \mathbb{R}^d \mid \|y - x\| < \varepsilon \right\}$$

(for $\varepsilon > 0$) to be open sets. If the specific value of $\varepsilon > 0$ does not matter we write $B(x)$ instead of $B_\varepsilon(x)$.

Let $S \subseteq \mathbb{R}^d$ be a bounded set. Denote by cl S its topological closure, by int S its interior and by $\Gamma = \text{bd } S$ the boundary of S.

In $\mathbb{R}^d$ the sphere $\mathfrak{S}_{d-1}$ is defined by

$$\mathfrak{S}_{d-1} = \left\{ x^* \in \mathbb{R}^d \mid \|x^*\| = 1 \right\}.$$

The elements x^* in $\mathfrak{S}_{d-1}$ are also termed *directions*.

DEFINITION 7.1. *A (closed)* surface *in $\mathbb{R}^d$ is a set Γ which is homeomorphic to the sphere $\mathfrak{S}_{d-1}$. A surface in $\mathbb{R}^2$ is also termed a (closed)* curve.

One important tool of our investigations will be convexity theory (see the books (Eggleston, 1958), (Valentine, 1964) or (Rockafellar, 1970)).

DEFINITION 7.2. *A set $S \subseteq \mathbb{R}^d$ is said to be* convex *if $x, y \in S$ and $0 \leq \lambda \leq 1$ together imply $(1 - \lambda)x + \lambda y \in S$.*

DEFINITION 7.3. *Given any set $S \subseteq \mathbb{R}^d$. The* convex hull *of S is the smallest convex set which contains S. It is denoted by* conv S.

Since the intersection of any system of convex sets is always convex, the concept of the convex hull is well defined.

DEFINITION 7.4. *Let $\mathfrak{D} \subseteq \mathfrak{S}_{d-1}$ be a set of directions.*

The set S is $\mathfrak{D}$–convex if for all $x^* \in \mathfrak{D}$ and for all $x \in \mathbb{R}^d$ the set $S \cap \{x + tx^* \mid t \in \mathbb{R}\}$ is convex (i.e. an interval).

By this Definition, a $\mathfrak{D}$–convex set needs not be connected, in contrast to convex sets. The intersection of any system of $\mathfrak{D}$–convex sets is always $\mathfrak{D}$–convex. However, the intersection of connected $\mathfrak{D}$–convex sets is not necessarily connected (see Example 7.1). By taking the intersection of all $\mathfrak{D}$–convex sets containing a given set S, we obtain the *$\mathfrak{D}$–convex hull* of S which is the smallest $\mathfrak{D}$–convex set containig S. For properties of $\mathfrak{D}$–convex sets see (Fink and Wood, 2004).

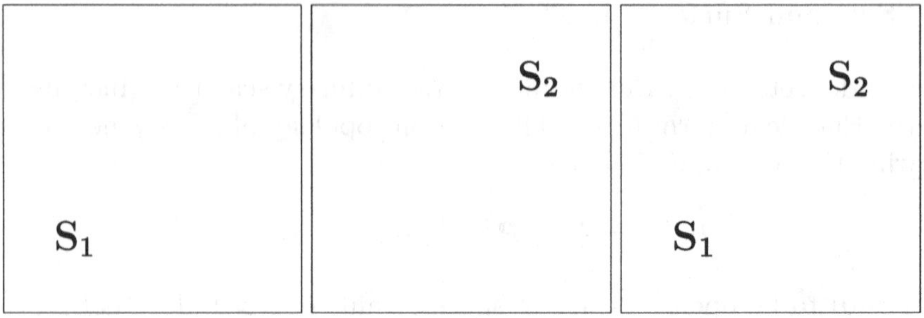

The sets S_1 and S_2 are both connected and $\mathfrak{D}$–convex with respect to $\mathfrak{D} = \{(0,1),(1,0)\}$. Their intersection (shaded area) is $\mathfrak{D}$–convex but not connected.

Figure 7.1. Intersection of two $\mathfrak{D}$–convex sets.

EXAMPLE 7.1. *In Figure 7.1 an example of two sets is given which are convex with respect to two directions in the plane and connected whose intersection, however, is $\mathfrak{D}$–convex but not connected.*

7.3. Convex and Concave Points

Given a set $S \subseteq \mathbb{R}^d$ with boundary Γ. Generally we assume that S is a compact set.

DEFINITION 7.5. $x \in \Gamma$ *is a* convex point *of* S *if there is a neighborhood* $B(x)$ *such that* $B(x) \cap S$ *is convex. Denote by* $\mathcal{T}_0 \subseteq \Gamma$ *the set of all convex points of* S.

$x \in \Gamma$ *is a* concave point *of* S *if there is a neighborhood* $B(x)$ *such that* $B(x) \cap \complement S$ *is convex* ($\complement S = \mathbb{R}^d \setminus S$ *is the complement of* S). *Denote by* $\mathcal{S}_0 \subseteq \Gamma$ *the set of all concave points of* S.

In Figure 7.5 below one can find examples for convex (pictures T, L) and concave points (S, L) as well as a point which is neither convex nor concave (picture I). Generally it is possible that a dense subset of the boundary consists of points of the latter type (see the discussion at the begin of Section 7.5 below). We have to impose regularity conditions in order to rule out such situations.

From the Separation Theorem for Convex Sets (Valentine, 1964, Part II) we conclude:

If x is a convex point of S then

C1 there exists a neighborhood $B(x)$ and a direction $x^* \in \mathfrak{S}_{d-1}$ such that $z \in B(x)$ and $\langle z, x^* \rangle > \langle x, x^* \rangle$ imply $z \notin S$.

If x is a concave point of S then

C2 there exists a neighborhood $B(x)$ and a direction $x^* \in \mathfrak{S}_{d-1}$ such that $z \in B(x)$ and $\langle z, x^* \rangle < \langle x, x^* \rangle$ imply $z \in S$.

This observation leads to the Definition

DEFINITION 7.6. *$x \in \Gamma$ is a* T–point *('Top point') of S if there is a neighborhood $B(x)$ and a direction $x^* \in \mathfrak{S}_{d-1}$ such that $z \in B(x)$ and $\langle z, x^* \rangle > \langle x, x^* \rangle$ imply $z \notin S$. Denote by $\mathcal{T} \subseteq \Gamma$ the set of all* T*–points of S.*

$x \in \Gamma$ is an S–point *('Saddle point') of S if there is a neighborhood $B(x)$ and a direction $x^* \in \mathfrak{S}_{d-1}$ such that $z \in B(x)$ and $\langle z, x^* \rangle < \langle x, x^* \rangle$ imply $z \in S$. Denote by $\mathcal{S} \subseteq \Gamma$ the set of all* S*–points of S.*

REMARK 7.1. *The notations '*T*–point' and '*S*–point' are due to Scherl (*Scherl, 1987*) who introduced these concepts in the context of document analysis.*

For deciding whether a given boundary point is a T*– or* S*–point only information from the boundary is needed together with an indication on which side of the boundary the set is situated. In contrast, for deciding whether a boundary point is convex or not, information has to be gathered from its neighborhood which contains points not on the boundary. Therefore, the concept of* T*– or* S*–points is more well–suited for practical applications than the concept of convex and concave points.*

REMARK 7.2. *There exist examples of nonconvex sets whose boundaries consist only of* T*–points. (*Tietze, 1929*) gave a condition guaranteeing that this situation cannot happen (see condition* **A** *in Section 7.4 below and (*Valentine, 1964, *Theorem 4.4)).*

We can associate to each T– or S–point of Γ a direction $x^* \in \mathfrak{S}_{d-1}$ such that **C1** or **C2**, respectively, holds for this point. This lead (Scherl, 1987) to the following Definition:

DEFINITION 7.7. *The pair (x, x^*) is a* T *or a x^*–*T *(*S *or x^*–*S*)* descriptor *of S if $x \in \Gamma$ is a* T*– (*S*–) point and if* **C1** *(***C2***) with direction x^* holds in x.*

There are points on the boundary which are T–points as well as S–points. For convenience we give them an extra name:

DEFINITION 7.8. *$x \in \Gamma$ is an* L–point *('Line point') of S if x is as well a* T*– as an* S*–point. Denote by $\mathcal{L} \subseteq \Gamma$ the set of all* L*–points of S.*

We define further:

DEFINITION 7.9. *A point $x \in \Gamma$ is an* extreme T–point *if there exists a direction $x^* \in \mathfrak{S}_{d-1}$ such that $\langle z, x^* \rangle > \langle x, x^* \rangle$ implies $z \notin S$. Denote by $\mathcal{ET} \subseteq \Gamma$ the set of all extreme* T*–points of S.*

A point $x \in \Gamma$ is an extreme S–point *if there exists a direction $x^* \in \mathfrak{S}_{d-1}$ such that $\langle z, x^* \rangle > \langle x, x^* \rangle$ implies $z \notin S$. Denote by $\mathcal{ES} \subseteq \Gamma$ the set of all extreme* S*–points of S.*

We state some rather simple topological properties which follow directly from the definitions.

LEMMA 7.1. *The sets $\mathcal{T}_0$ and $\mathcal{S}_0$ are open subsets of Γ.*
$\mathcal{L} = \mathcal{T}_0 \cap \mathcal{S}_0 = \mathcal{T} \cap \mathcal{S}$ is open.
The sets $\mathcal{ET}$ and $\mathcal{ES}$ are closed subsets of Γ.

DEFINITION 7.10. *Given a set $S \subseteq \mathbb{R}^d$ with boundary Γ. The subset Γ_0 of Γ is termed a* convex patch of the boundary *if* conv $\Gamma_0 \subseteq S \cup \Gamma$.

Γ_0 *is termed a* concave patch of the boundary *if* conv $\Gamma_0 \subseteq \mathbb{C}S \cup \Gamma$, *where $\mathbb{C}S = \mathbb{R}^d \setminus S$ is the complement of S.*

Clearly, Γ_0 is a concave patch of the boundary of S if and only if it is a convex patch of the (dlosure of) boundary of $\mathbb{C}S$.

In the following we will investigate questions like these:

- Are boundary patches consisting entirely of convex (concave) points convex (concave) patches?
- If Γ = bd S is the union of finitely many convex patches, does this imply that S is convex?
- What can be said about points of Γ which are neither convex nor concave points?
- How many convex or concave points do exist on the boundary of a — say closed, connected, bounded — set? More mathematically: is the set $\mathcal{T} \cup \mathcal{S}$ dense on Γ?

The first two questions were answered by a couple of Theorems due to Henrich Tietze (see (Tietze, 1929; Tietze, 1928; Tietze, 1927a; Tietze, 1927b) and (Valentine, 1964, Part IV)). The last two questions are not easy to answer. We need very deep results from topology or else very strong assumptions (e.g. the requirement that all boundary points are 'tame' (Latecki and Rosenfeld, 1998)).

7.4. S–Points and $\mathfrak{D}$–Convexity

We now are going to answer the first two questions above concerning convex and concave points or patches, respectively. It should be remarked here, however, that all results proved in this section hold only in the plane $\mathbb{R}^2$.

We introduce the following assumption:

A The interior int S of the set S is connected and S is *regular closed*, i.e. $S = \text{cl int } S$.

LEMMA 7.2. *Let $S \subseteq \mathbb{R}^2$ be a set fulfilling condition* **A**.

Assume that there exists a direction $x^* \in \mathfrak{S}_1$ and a number α such that the set $\{x \in S \mid \langle x, x^* \rangle = \alpha\}$ is not connected.

Then there exists an S*–point $x \in \Gamma$. More precisely, either the pair (x, x^*) or the pair $(x, -x^*)$ is an* S*–descriptor pair of S.*

The assertion of the Lemma is essentially the assertion of the Theorem Léja and Wilkosz (Léja and Wilkosz, 1924) (see (Valentine, 1964, Theorem 4.8)). The assumption involving direction $x^* \in \mathfrak{S}_1$ is equivalent to the assumption that S is not convex.

Lemma 7.2 may be sharpened as follows:

COROLLARY 7.1. *Let $\mathfrak{D} \subseteq \mathfrak{S}_1$.*

If a set $S \subseteq \mathbb{R}^2$ fulfilling condition **A** *does not contain any $\pm x^*$–*S*–descriptor points with $x^* \in \mathfrak{D}$ then it is $\mathfrak{D}$–convex.*

The contrary of the Corollary is not necessarily true as it is illustrated in Figure 7.4. In order to prove the converse of the Corollary we need a nondegeneracy assumption.

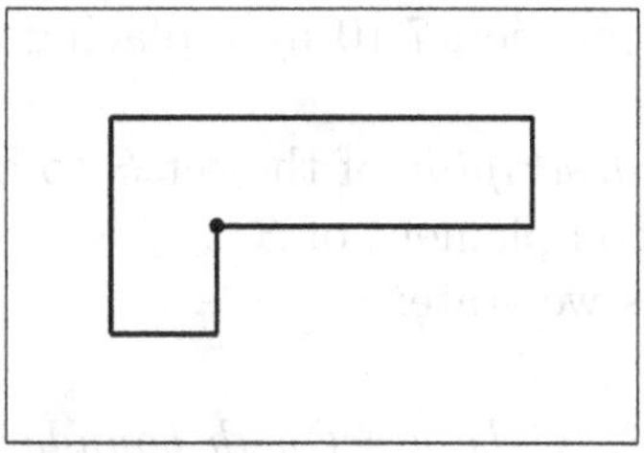

The set S in the picture is $\mathfrak{D}$–convex with respect to the set $\mathfrak{D}$ containing the horizontal and the vertical directions. However, there is a vertex point (marked •) which is an S–point of S.

Figure 7.2. Example of a $\mathfrak{D}$–convex set.

DEFINITION 7.11. *Let (x, x^*) be an* S*–descriptor of a set $S \subseteq \mathbb{R}^d$. x is termed a* strict S–point *with respect to $x^* \in \mathfrak{S}_{d-1}$ if for any line ℓ through x perpendicular to x^*, the component of $\ell \cap S$ containing x is closed.*

REMARK 7.3. *The requirement that a point is a strict* S*–point is not a 'local' one since one has to check in each case when an* S*–point is encountered which is adjacent to a component of $\mathcal{L}$, whether the point on the other end of the component is also an* S*–point. This is the reason why the concept of $\mathfrak{D}$–convexity, which is a very natural concept in the framework of digital geometry (see (*Fink and Wood, 2004*)), is 'harder' to handle than ordinary convexity (see (*Debled–Rennesson et al., 2000*;* Debled–Rennesson and Reveillès, 1995*;* Eckhardt and Reiter, 2003*;* Eckhardt, 2001*)).*

LEMMA 7.3. *Given a set $S \subseteq \mathbb{R}^2$ fulfilling condition* **A**. *Let $\mathfrak{D} \subseteq \mathfrak{S}_1$.*
If Γ contains a strict S*–descriptor (x, x^*) with $x^* \in \mathfrak{D}$ then S is not $\mathfrak{D}$–convex.*

The proof of this Lemma is an extension of the proof of Léja and Wilkosz' Theorem (Valentine, 1964, Theorem 4.8).

We now are able to state a Characterization Theorem for $\mathfrak{D}$–convexity.

THEOREM 7.1. *Given a set $S \subseteq \mathbb{R}^2$ with property* **A**. *Let $\mathfrak{D} \subseteq \mathfrak{S}_1$.*
S is $\mathfrak{D}$–convex if and only if its boundary Γ contains no strict S*–descriptors (x, x^*) with $x^* \in \mathfrak{D}$.*

The proof of this Theorem follows immediately from Lemma 7.3 and Lemma 7.2.

For a subset $\mathfrak{D}$ of directions we define $\mathfrak{D}$*–convex* and $\mathfrak{D}$*–concave patches* of the boundary as in Definition 7.10 by replacing the convex hull by the $\mathfrak{D}$–convex hull.

We define a *strict* T*–descriptor* of the set S to be a strict S–descriptor of the (closure of the) +complement of S.

With these definitions we state:

THEOREM 7.2. *Let $S \subseteq \mathbb{R}^2$ be a set with boundary Γ fulfilling condition* **A**.

1. *A connected subset Γ_0 of Γ is a convex (concave) patch of Γ if and only if it consists entirely of* T*– (*S*–) points.*
2. *A connected subset Γ_0 of Γ is a $\mathfrak{D}$–convex ($\mathfrak{D}$–concave) patch of Γ if and only if does not contain any strict* S*–descriptors (strict* T*–descriptors) (x, x^*) with $x \in \Gamma_0$ and $x^* \in \mathfrak{D}$.*

7.5. Regular Boundaries

In order to answer the questions concerning $\mathcal{T}$ and $\mathcal{S}$ from the end of Section 7.3 we need concepts from topology. The last one of these questions is indeed very deep and it cannot be treated here. There are indeed sets having nontrivial parts of the boundary consisting entirely of points which are neither T– nor S–points. An example of such a set (the 'Warsaw circle') is given in (Giraldo et al., 1999). In this example, however, the boundary is not a Jordan curve. Based on a characterization of Jordan curves in the plane by ((Schönflies, 1900; Schönflies, 1900), see (Rinow, 1975, §40)), Kaufmann (Kaufmann, 1931) was able to prove that for a Jordan boundary the set $\mathcal{T} \cup \mathcal{S}$ is dense on the boundary. The famous von Koch curve (Koch, 1906) is an example of a Jordan curve with the property that all three sets $\mathcal{T}$, $\mathcal{S}$ and also the complement of these both sets are dense on the curve. We need a strong regularity condition to rule out such situations. Generally we assume that the boundary Γ of a set $S \subseteq \mathbb{R}^d$ is a surface, i.e. that it is a homeomorphic image of $\mathfrak{S}_{d-1}$.

DEFINITION 7.12. *$x \in \Gamma$ is a* regular point *of S if there is a neighborhood $B(x)$ such that both $B(x) \cap \mathcal{T}$ and $B(x) \cap \mathcal{S}$ consist of at most finitely many connected components.*

The boundary of a set is called a regular boundary *if it consists only of regular boundary points.*

REMARK 7.4. *In $\mathbb{R}^2$ a point x is regular if and only if there is a neighborhood $B(x)$ such that either*

1. *$B(x) \cap (\mathcal{T} \cup \mathcal{S})$ consists of exactly one connected component, or else*
2. *$(B(x) \cap \Gamma) \setminus \{x\}$ consists of two connected components. Each such component is completely contained in $\mathcal{T}$ or $\mathcal{S}$.*

There remains one class of boundary points on regular boundaries:

DEFINITION 7.13. *$x \in \Gamma$ is an* I–point *('Indifferent point') of S if x is regular and neither a* T*–point nor an* S*–point. Denote by $\mathcal{I} \subseteq \Gamma$ the set of all* I*–points of S.*

By definition, I–points are always isolated points on the (regular) boundary. They separate components of $\mathcal{T}$ and $\mathcal{S}$. In Figure 7.5 examples for all types of regular points in the plane $\mathbb{R}^2$ are shown.

7.6. Upper Semi–Continuous Mappings

For analyzing the boundary of a set wee need some results from the theory of set–valued mappings (see (Aubin and Frankowska, 1990)).

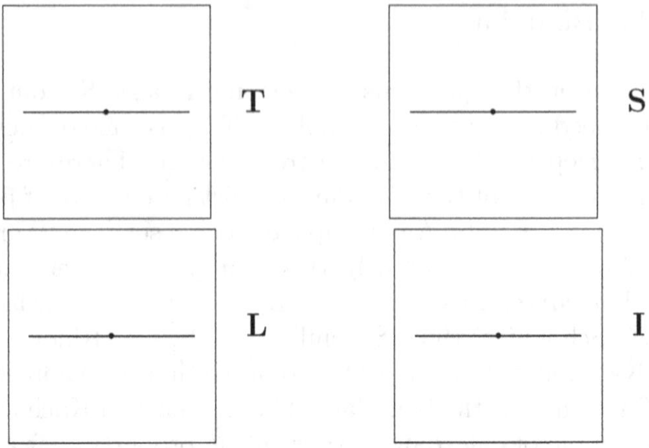

The set S is shaded, the boundary point x under consideration is marked •. The line through the point indicates the line perpendicular to x^*.

Figure 7.3. Types of regular points in the plane.

DEFINITION 7.14. *Let E and F be two topological spaces and denote by $\mathcal{P}(F)$ the collection of all non–empty subsets of F. A point–to–set mapping $f : E \longrightarrow \mathcal{P}(F)$ is said to be* upper semi–continuous *if, for any point $x_0 \in E$ and any open set $U \subseteq F$ with $f(x_0) \subseteq U$, there exists a neighborhood V of x_0 such that $f(x) \subseteq U$ for all $x \in V$.*

Consequences

- Every continuous function is upper semi–continuous.
- Every upper semi–continuous point–to point mapping is continuous
- The composition of upper semi–continuous mappings is upper semi–continuous.

DEFINITION 7.15. *For a set–valued mapping $f : E \longrightarrow \mathcal{P}(F)$ we define a set–valued inverse mapping. For $y \in F$ let*

$$f^{-1}(y) = \{x \in E \mid y \in f(x)\}.$$

The assertions of the following two Theorems follow directly from the definition of upper semi–continuity:

THEOREM 7.3. *Assume that the set–valued mapping $f : E \longrightarrow \mathcal{P}(F)$ is upper semi–continuous and in addition assume that $f(x)$ is connected for all $x \in E$.*

Then for each connected subset $S \subseteq E$ the image $f(S)$ is also connected.

THEOREM 7.4. *The set–valued mapping $f : E \longrightarrow \mathcal{P}(F)$ is upper semi–continuous if and only if for each closed set $S \subseteq F$ the set $f^{-1}(S)$ is closed.*

REMARK 7.5. *As a consequence of the Theorem, if f maps E into the system of all closed subsets of F then f^{-1} maps F into the system of all closed subsets of E.*

7.7. Tangent Mappings

DEFINITION 7.16. *Let S be a nonempty compact subset of $\mathbb{R}^d$. The function $\mu : \mathfrak{S}_{d-1} \longrightarrow \mathbb{R}$ which is defined by*

$$\mu(x^*) = \max_{x \in S} \langle x, x^* \rangle$$

is called the support functional *of S (see (*Kindratenko, 2003, *Section 2.3.)).*

The set–valued mapping $\Psi : \mathfrak{S}_{d-1} \longrightarrow \mathcal{P}(S)$ with

$$\Psi(x^*) = \arg \max_{x \in S} \langle x, x^* \rangle$$

is called the tangent mapping *of S.*

It is well–known (see e.g. (Aubin and Frankowska, 1990)) that μ is continuous and that Ψ is upper semi–continuous whenever S is a nonempty compact subset of $\mathbb{R}^d$.

REMARK 7.6. *For nonempty compact $S \subseteq \mathbb{R}^d$ all sets $\Psi(x^*)$ are closed since*

$$\Psi(x^*) = S \cap \{x \in \mathbb{R}^d \mid \langle x, x^* \rangle = \mu(x^*)\}.$$

THEOREM 7.5. *Let S be a nonempty closed convex subset of $\mathbb{R}^d$.*

The set–valued mapping $\Psi^{-1} : S \longrightarrow \mathcal{P}(\mathfrak{S}_{d-1})$ with

$$\Psi^{-1}(x) = \{x^* \in \mathfrak{S}_{d-1} \mid x \in \Psi(x^*)\}$$

is upper semi–continuous.

LEMMA 7.4. *Assume that the conditions of Theorem 7.5 hold. Let $y \in$ bd S and x_1^*, x_2^* be two directions in $\Psi^{-1}(y)$.*

Then all directions from the set

$$\left\{x \in \mathbb{R}^d \mid x = \lambda_1 x_1^* + \lambda_2 x_2^*, \lambda_1, \lambda_2 \geq 0\right\} \cap \mathfrak{S}_{d-1}$$

belong to $\Psi^{-1}(y)$.

REMARK 7.7. *Lemma 7.4 states that all directions in* $\Psi^{-1}(y)$ *can be obtained as the intersection of a convex cone with vertex* Θ_d (= *origin of* $\mathbb{R}^d$) *and* $\mathfrak{S}_{d-1}$. *This implies that* $\Psi^{-1}(y)$ *is always a connected subset of* $\mathfrak{S}_{d-1}$.

7.8. Structure of Regular Boundaries — The Convex Case

Ψ^{-1}, the inverse of the tangent mapping Ψ, is only a well–defined mapping on Γ if S is a convex set. Whenever S is convex, Ψ^{-1} is even an upper semi–continuous mapping. Hence, the pair (Ψ, Ψ^{-1}) is in the convex case 'nearly' a homeomorphism. We sketch here, how this can be shown rigourosly. First we need a concept from convexity theory:

DEFINITION 7.17. *Let* S *be a closed convex set and* $x \in bd\ S$. *The direction* $x^* \in \mathfrak{S}_{d-1}$ *is termed the* normal vector *of a* supporting hyperplane *at* S *in* x *whenever* $\langle z, x^*\rangle > \langle x, x^*\rangle$ *implies* $z \notin S$ *(see* **C1***).*

The Separation Theorem for Convex Sets (Valentine, 1964, Part II) guarantees that a convex set has at least one supporting hyperplane in each of its boundary points.

DEFINITION 7.18. *Let* S *be a nonempty bounded closed convex set in* $\mathbb{R}^d$.

S *is* smooth *if for each boundary point of* S *the supporting hyperplane is uniquely determined.*

S *is* strictly convex *if all supporting hyperplanes meet the boundary of* S *in exactly one point.*

If a set S is strictly convex then the mapping Ψ is a point–to–point map from $\mathfrak{S}_{d-1}$ to the boundary Γ of S. If S is smooth then the inverse map Ψ^{-1} is a point–to–point map. Hence, if S is smooth and strictly convex, then both Ψ and Ψ^{-1} are continuous and inverse to each other, consequently, $\Psi : \mathfrak{S}_{d-1} \longrightarrow \Gamma$ is a homeomorphism.

THEOREM 7.6.

Given a closed bounded convex set S *with nonempty interior and a real number* $\varepsilon > 0$.

Then there exists a closed bounded convex set S_ε *which is smooth and strictly convex such that* $S \subseteq S_\varepsilon$ *and* $d_H(S, S_\varepsilon) < \varepsilon$.

Here, d_H *denotes the* Hausdorff distance *for sets, this means in this context* ($S \subseteq S_\varepsilon$) *that for each* $x_\varepsilon \in S_\varepsilon$ *there is an* $x \in S$ *such that* $\|x - x_\varepsilon\| < \varepsilon$.

For a proof of this Theorem see (Eggleston, 1958, Theorem 34).

For any closed bounded convex set S with boundary Γ which has non-empty interior we can construct an ε–homeomorphism in the following way:

- Construct S_ε as in Theorem 7.6.
- The mapping $\Psi_\varepsilon : \mathfrak{S}_{d-1} \longrightarrow \text{bd } S_\varepsilon$ is a homeomorphism.
- Select any point $x_0 \in \text{int } S$. The central projection $\Pi_{x_0} : \text{bd } S_\varepsilon \longrightarrow \Gamma$ with center x_0 is

$$\Pi_{x_0}(x) = \{x_0 + \lambda(x - x_0) \mid \lambda > 0 \text{ maximal such that } x_0 + \lambda(x - x_0) \in S\}.$$

  $\Pi_{x_0}(x)$ exists for any $x \neq x_0$ by compactness of S and is a homeomorphism.
- Let $\Psi : \mathfrak{S}_{d-1} \longrightarrow \mathcal{P}(\Gamma)$ be the — generally set–valued — map as defined above. The composite mapping $\Pi_{x_0} \circ \Psi_\varepsilon : \mathfrak{S}_{d-1} \longrightarrow \Gamma$ is a homeomorphism with the property that for given $\varepsilon > 0$ there exists a number C such that for each $x^* \in \mathfrak{S}_{d-1}$ and each $x \in \Psi(x^*)$ there exists an $x_\varepsilon = \Pi_{x_0}\Psi_\varepsilon(x^*)$ such that $\|x - x_\varepsilon\| < C \cdot \varepsilon$ and for each each $x_\varepsilon = \Pi_{x_0}\Psi_\varepsilon(x^*)$ there exists an $x \in \Psi(x^*)$ such that $\|x - x_\varepsilon\| < C \cdot \varepsilon$.

7.9. General Plane Sets with Regular Boundaries

If a set $S \subseteq \mathbb{R}^2$ has a regular boundary then its boundary consists of a finite number of components of $\mathcal{T}$ (convex parts of the boundary) and $\mathcal{S}$ (concave parts).

Generally, for the convex hull conv S of a (compact) set, we can apply the results derived above. Specifically, for each $\varepsilon > 0$ there exists an ε–homeomeorphism bd conv $S \longrightarrow \mathfrak{S}_1$ as shown above.

The *convex defect* $S \setminus \text{conv } S$ consists by regularity of the boundary of a finite number of components. Each of these components consists of finitely many convex or concave parts. We can construct for each of these parts an ε–homeomeorphism on a new copy of $\mathfrak{S}_1$ so that we finally get an ε–homeomeorphism mapping the boundary of S onto a finite number of $\mathfrak{S}_1$'s. Instead of elaborating this procedure formally, we illustrate it by means of a simple example (see Figure 7.10).

Along convex or concave parts of a set in $\mathbb{R}^2$ the succession of descriptors is not arbitrary. Since $\mathfrak{S}_1$ is oriented and since there is an ε–homeomeorphism mapping each convex or concave part of the boundary onto $\mathfrak{S}_1$, also the boundary is oriented. Note, however, that the succession of descriptors is reverted by transition from a convex to a concave part and vice versa. Scherl illustrated this effect by means of so–called legal descriptor cycles (Scherl, 1987, Figure 5.2.4.), see Figure 7.9.

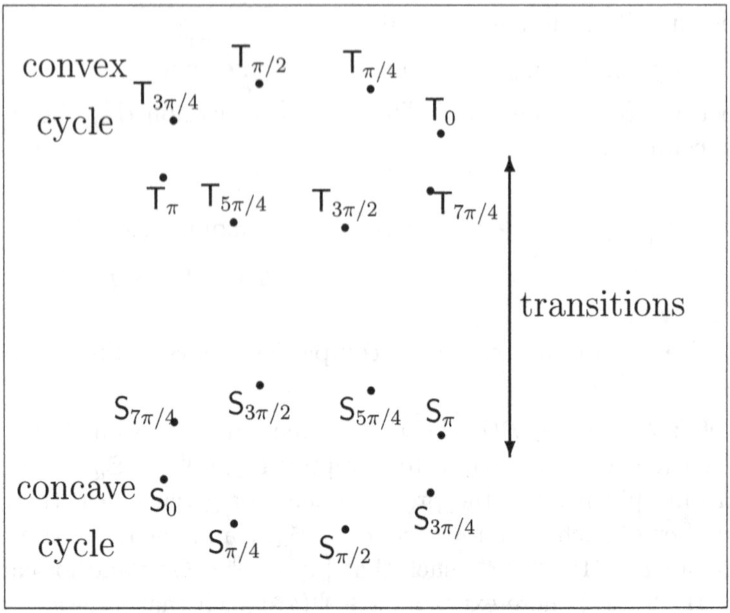

Descriptor directions corresponding to multiples of $\frac{\pi}{4}$ are indicated. These directions are the main directions in the digital plane $\mathbb{Z}^2$.

Figure 7.4. Scheme of descriptor cycles and legal transitions.

7.10. Applications

It has been shown by (Scherl, 1987) by means of numerous examples and also by a prototype implementation that it is possible to describe shape by a relatively small number of descriptor points. As in Scherl's experiments we consider the case of plane sets and the set of descriptors belonging to directions $k \cdot \frac{\pi}{4}$ which are well adapted to the digital plane $\mathbb{Z}^2$ which is the set of all points of the plane having integer coordinates.

The extraction of the descriptors can be done very efficiently as a by–product of boundary extraction at virtually no additional cost. The oriented set of all descriptors (points with a label indicating the corresponding tangent direction and also pointers indicating the succession relation of the descriptor points on the oriented boundary) yields a data reduction while retaining the rough shape of the set under consideration.

The boundary and the descriptors may be viewed as a pyramid structure:

– The bottom of the pyramid is the ordered sequence of boundary points of the set which is coded in some appropriate manner (e.g. by means of a chain code).

- Digital sets are subsets of $\mathbb{Z}^d$. Under certain known conditions (which means that some discrete Jordan Theorem holds, i.e. a set is uniquely determined by its discrete boundary) a digital set can be represented by means of its boundary. Specifically for plane digital sets it is possible to select a subset of the boundary points which can be joined according to the orientation of the boundary as to to yield a *faithful* representation of the set, i.e. the reduced boundary is a polygonal Jordan curve which contains exactly all points of the given set in its interior (Eckhardt and Reiter, 2003).
- A subset of the boundary of a set – or else of any faithful representation of the boundary of a digital set – is given by the oriented set of all descriptors. The tangents corresponding to the T–descriptors belonging to a set $\mathfrak{D}$ of directions yield the $\mathfrak{D}$–convex hull of the set. By joining any two descriptor points which are immediate successors on the oriented boundary, a closed polygonal curve is obtained which, however, in general needs not to be a simple curve. Nevertheless, these curves can be efficiently used for a rough representation of shape.
- The set of all extreme T–descriptors belonging to a set $\mathfrak{D}$ of directions (here all directions $k \cdot \frac{\pi}{4}$) provide the smallest convex polytope whose sides have directions from $\mathfrak{D}$ (a so–called $\mathfrak{D}$*–polytope*) which contains the set under investigation.

The data structure provided by this pyramid can be used for different pattern recognition tasks. For example, the linear time convexity detection algorithm from (Debled–Rennesson et al., 2000) starts at the top of the pyramid with direction set $\mathfrak{D} = \{k \cdot \frac{\pi}{2}\}$. First, the authors verify $\mathfrak{D}$–convexity of the given set. Then the boundary of this set is segmented by means of all descriptor points and descriptor tangents having directions from $\mathfrak{D}$. This results in boundary parts having a very favourable structure considering convexity detection.

An interesting subject is the investigation of this 'Scherl–pyramid' under discrete boundary evolution (Latecki and Lakämper, 1999). Specifically, the information obtained from a faithful representation of a digital set can be used to control the evolution process. This, however, is far beyond the topic of this paper (see (Eckhardt and Reiter, 2003)).

If only information from a finite number of 'probes' of boundary points is available (together with tangent directions) then the observation that the sequence of descriptors along the boundary is oriented can be used to find inclusions for the missing parts of the boundary if it is assumed that the directions between two successive probes lie within a certain interval. Such 'interpolation' assertions can be easily derived. We give one simple example:

THEOREM 7.7. *Given in the plane a set S with oriented boundary Γ. Let (x_1, x_1^*) and (x_2, x_2^*) be two* T*–descriptors of S. Assume that on the boundary part Γ_{12} which is between (in the sense of the orientation of Γ) x_1 and x_2 there are only descriptors (x, x^*) which are between (in the sense of orientation of $\mathfrak{S}_1$) x_1^* and x_2^*. Then Γ_{12} is completely contained in the parallelogram of all $x \in \mathbb{R}^2$ satisfying the inequalities*

$$\langle x_2, x_1^* \rangle \leq \langle x, x_1^* \rangle \leq \langle x_1, x_1^* \rangle ,$$
$$\langle x_1, x_2^* \rangle \leq \langle x, x_2^* \rangle \leq \langle x_2, x_2^* \rangle .$$

The proof of this assertion follows from investigating the convex hull of Γ_{12}. In Figure 7.10 an illustrative example is shown. It is possible to derive inclusions for other situations where also S–descriptors are taken into account.

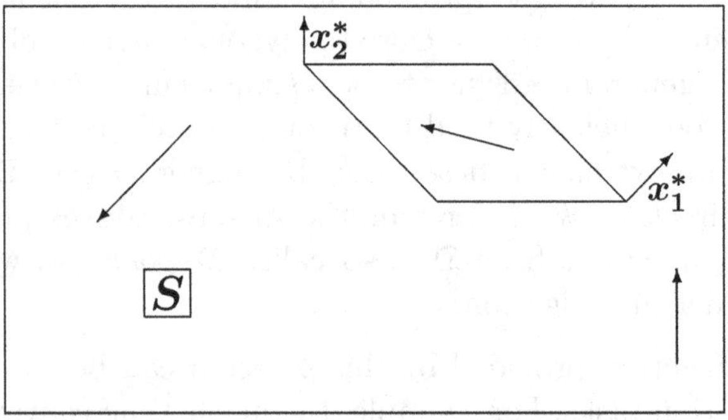

Long arrows indicate the orientation of the boundary.

Figure 7.5. Example for Theorem 7.7.

We conclude this discussion with a simple example. In Figure 7.10 a digital set and its boundary is given. The process of finding a stack of $\mathfrak{S}_1$'s such that the convex and concave parts of the (outer) boundary can be homeomorphically mapped on this stack is illustrated in Figure 7.10. In Figure 7.10 the smallest convex $\mathfrak{D}$–polygon containing the set as obtained from the extreme T–descriptors is shown as well as the polygonal approximation which is found by joining the descriptor points by line segments.

7.11. Conclusions

Under suitable conditions it is possible to derive properties of boundaries of sets using only tools from convexity theory without making any differentiability assumptions. It was shown that the boundary of a set can be mapped 'almost' homeomorphically to a stack of spheres.

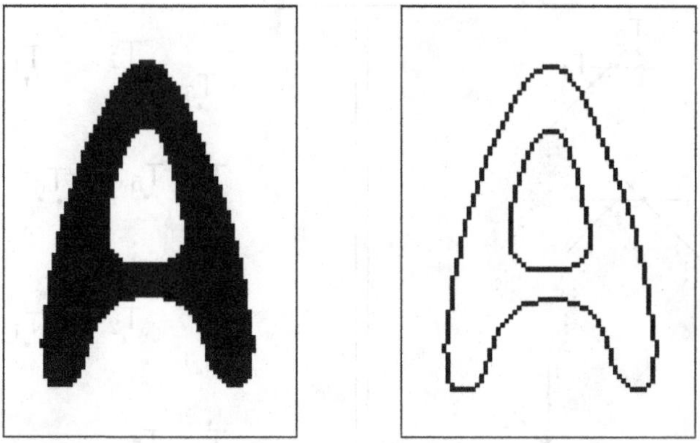

In the left picture a digital set is given. The right picture shows the boundary of the set. The boundary can be understood to consist of two closed polygonal curves in $\mathbb{R}^2$.

Figure 7.6. Digitization of letter 'A'.

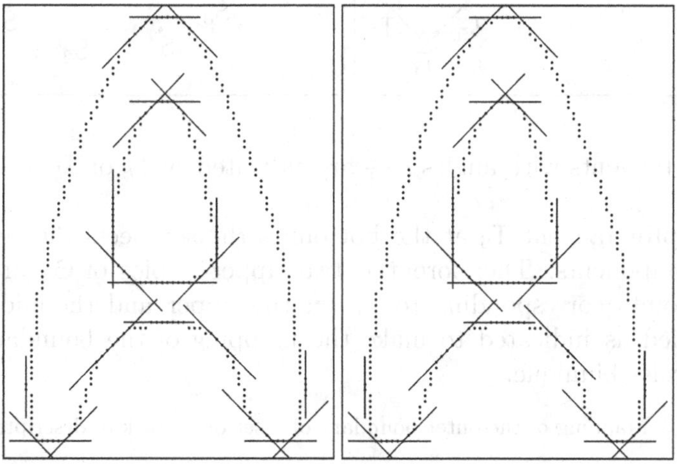

The set $\mathfrak{D}$ consists of all directions $k \cdot \frac{\pi}{4}$.

Figure 7.7. $\mathfrak{D}$–convex hull (left) and descriptor approximation (right).

By gathering informations from a finite number of points along the boundary one can extract properties which are relevant for the shape of a set. These properties can be arranged in a hierarchical manner as a pyramid structure. The informations obtained in this way can be used for defining convex and concave parts of the boundary and for investigating and controlling discrete evolution of boundary curves.

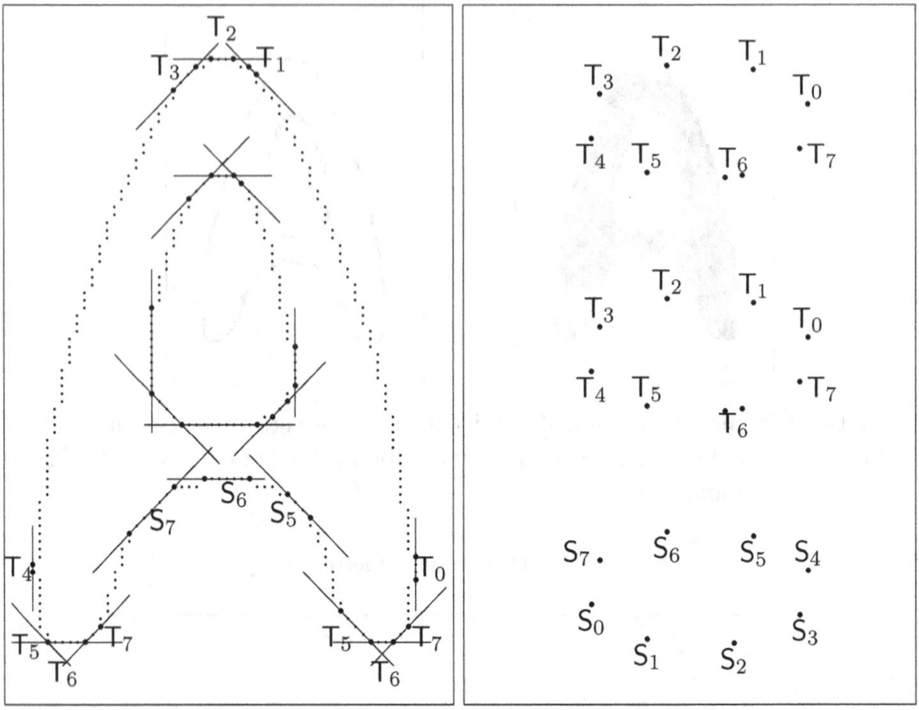

Descriptor tangents with angles $k \cdot \frac{\pi}{4}$ are indicated by T_k or S_k, respectively.

The descriptor tangent T_6 at the bottom of the set meets the set in two disjoint components. Therefore the two upper copies of $\mathfrak{S}_1$ are cut up and the points corresponding to T_6 on the upper and the middle copy are identified as indicated to make the mapping of the boundary to the stack of circles biunique.

Figure 7.8. Mapping of the outer boundary of a set on a stack of descriptor cycles.

There are two important topics which are not treated here. One of them is the extension to higher dimensions. It is possible to derive properties of higher dimensional sets by investigating two–dimensional sections of them. The second problem not treated here is much more difficult. Usually in applications sets are given in a discrete manner as 'digital sets'. Therefore it is desirable to have a completely discrete theory. However, it turns out that the discrete case is much more complicated than the continuous one (Debled–Rennesson et al., 2000; Debled–Rennesson and Reveillès, 1995; Dörksen, 2004; Eckhardt and Reiter, 2003; Eckhardt, 2001).

References

Aubin, J.–P. and H. Frankowska: *Set–Valued Analysis*. Birkhäuser-Verlag, Boston, Mass., 1990.

Debled–Rennesson, I., J.–L. Rémy, and J. Rouyer–Degli: Detection of the discrete convexity of polyominoes. In Proc. *Discrete Geometry for Computer Imagery,* (G. Borgefors, I. Nyström and G. Sanniti di Baja), pages 491–504, LNCS 1953, Springer, Berlin, 2000.

Debled–Rennesson, I. and J.-P. Reveillès: A linear algorithm for segmentation of digital curves. *Int. J. Pattern Recognition Artificial Intelligence*, **9**:635–662, 1995.

Dörksen, H: Shape Representations of Digital Sets based on Convexity Properties. PhD Thesis, Universität Hamburg, Fachbereich Mathematik, Juni 2004.

Eckhardt, U. and H. Reiter: Polygonal representations of digital sets. *Algorithmica*, **38**: 5–23, 1902.

Eckhardt, U.: Digital lines and digital convexity. In *Digital and Image Geometry: Advanced Lectures Editors,* (G. Bertrand, A. Imiya, R. Klette, editors), pages 207–226, LNCS 2243, Springer, Berlin, 2001.

Eckhardt, U., W. Scherl, and Z. Yu: Representation of plane curves by means of descriptors in Hough space. I. Continuous theory. TR, *Rechenzentrum der Universität Hamburg*, 1987.

Eggleston, H.G.: *Convexity*. Cambridge University Press, Cambridge, 1958.

Fink, E. and D. Wood: *Restricted–Orientation Convexity*. Springer, Berlin, 2004.

Giraldo, A., A. Gross, and L. J. Latecki: Digitizations preserving shape. *Pattern Recognition*, **32**:365–376, 1999.

Heuer, J., A. Kaup, U. Eckhardt, L. J. Latecki, and R. Lakämper: Region localization in MPEG–7 with good coding and retrieval performance. ISO/IEC JTC1/SC29/WG11/M5417, MPEG 99, Maui, USA, 1999.

Kaufmann, B.: Über die Konvexitäts– und Konkavitätsstellen auf Jordankurven. *J. reine und angewandte Mathematik*, **164**:112–127, 1931.

Kindratenko, V.V.: On using functions to describe the shape. *J. Mathematical Imaging Vision*, **18**:225–245, 2003.

von Koch, H.: Sur une courbe continue sans tangente obtenue par une construction géométrique élémentaire. *Acta Mathematica*, **30**:145–174, 1906.

Latecki, L.J. and R. Lakämper: Convexity rule for shape decomposition based on discrete contour evolution. *Computer Vision Image Understanding*, **73**:441–454, 1999.

Latecki, L.J. and A. Rosenfeld: Supportedness and tameness. Differentialless geometry of plane curves. *Pattern Recognition*, **31**:607–622, 1998.

Latecki, L.J.: *Discrete Representation of Spatial Objects in Computer Vision*. Kluwer, Dordrecht, 1998.

Léja, F. et W. Wilkosz: Sur une propriété des domaines concaves. *Annales de la Société Polonaise de Mathématique*, **2**:222–224, 1924.

Mokhtarian, F.: Silhouette–based occluded object recognition through curvature scale space. *Machine Vision Applications*, **10**:87–97, 1997.

Lehrbuch der Topologie. Deutscher Verlag der Wissenschaften, Berlin, 1975.

Rockafellar, R.T.: *Convex Analysis*. Princeton University Press, Priceton, NJ, 1970.

Scherl, W.: *Bildanalyse allgemeiner Dokumente*. Springer, Berlin, 1987.

Schönflies, A.: Die Entwicklung der Lehre von den Punktmannigfaltigkeiten. *Jber. Deutsch. Math.-Verein.*, **VIII**:1–250, 1900.

Schönflies, A.: Die Entwicklung der Lehre von den Punktmannigfaltigkeiten. Bericht erstattet der Deutschen Mathematiker-Vereinigung Teil II. *Jber. Deutsch. Math.-Verein.*, Ergänzungsband II, 1900.

Tietze, H: Bemerkungen über konvexe und nichtkonvexe Figuren. *J. reine und angewandte Mathematik*, **160**:67–69, 1929.

Tietze, H.: Über Konvexheit im kleinen und im großen und über gewisse den Punkten einer Menge zugeordnete Dimensionszahlen. *Mathematische Zeitschrift*, **28**:697–707, 1928.

Tietze, H.: Eine charakteristische Eigenschaft der abgeschlossenen konvexen Punktmengen. *Math. Annalen*, **99**:394–398, 1927.

Tietze, H.: Über konvexe Figuren. *J. reine und angewandte Mathematik*, **158**:168–172, 1927.

Valentine, F.A.: *Convex Sets*, McGraw–Hill, New York, 1964.

CONVEX AND CONCAVE PARTS OF DIGITAL CURVES

HELENE DÖRKSEN-REITER
Hamburg University
Faculty of Mathematics, Hamburg, Germany

ISABELLE DEBLED-RENNESSON
Equipe Adage
LORIA Nancy, France

Abstract. Decomposition digital curves into convex and concave parts is of relevance in several scopes of image processing.

In digital plane, convexity cannot be observed locally. It becomes an interesting question, how far one can decide whether a part of a digital curve is convex or concave by a method which is "as local as possible". In (Eckhardt and Reiter, 2004), it was proposed to define meaningful parts of a digital curve as meaningful parts of the corresponding polygonal representation. This technique has an approximative character.

In our considerations, we use geometry of arithmetical discrete line segments (Reveillès, 1991; Debled-Rennesson and Reveillès, 1995). We will introduce an exact method to define convex and concave parts of a digital curve.

Key words: digital geometry, digital convexity, discrete line, convex and concave curves, polygonal representation

8.1. Introduction

Roots of digital geometry can be found in practical applications of digital image processing and computer graphics. A considerable part of books on digital geometry is devoted to convexity (see e.g. (Voss, 1993, Chapter 4.3)). It is a simple observation that convex parts of objects determine their visual parts. They are of importance, for example, for recognition objects by comparing with given shapes from a database. However, the problem is that many significant parts are not convex, since a visual part may have concavities. So, one is interested in decomposition the boundary of a digital set into convex and concave parts.

In an earlier paper (Latecki and Lakämper, 1999), such a partition was performed by segmenting the boundary into digital line segments. In

R. Klette et al. (eds.), Geometric Properties for Incomplete Data, 145-159.

an other paper (Eckhardt and Reiter, 2004), it was proposed to define meaningful parts of the boundary by meaningful parts of the corresponding polygonal representation. The first method is much rougher, however, both techniques have an approximative character. Also, recent publications, whose discussions are related to the considered problem, e.g. (Coeurjolly et al., 2004) is about digital arc segmentation, (Brimkov and Klette, 2004) elucidates new aspects of digital curves and surfaces, shall attract attention.

Decomposition into meaningful parts allows simplification shapes by discrete evolution (Latecki and Lakämper, 2002). The aim of this paper is to investigate the discrete evolution process and its properties specifically with respect to convex and concave parts of the boundary from a low-level point of view. The principal idea is to use geometrical properties of arithmetical discrete lines (Reveillès, 1991; Debled-Rennesson and Reveillès, 1995).

In 1987, a method based on sets of descriptors was proposed (Scherl, 1987). Descriptors are points of local support with respect to a finite number of directions. In a certain sense, Scherl's descriptors segment the boundary of a set into components which are "suspicious candidates" for being convex or concave parts.

In Section 1 of this paper, we recall definitions about discrete curves. Then, in Section 2, we present definition and geometrical properties of arithmetical discrete lines which are used to define fundamental segments of 8-curves introduced in Section 3. Thanks to this notion, an adaptation of the concept of convex and concave curves is proposed in Section 4. In Section 5, the decomposition curves into maximal convex and concave parts is studied. Finelly, in Section 6, an application to fundamental polygonal representations of digital curves is presented.

8.2. Digital Curves on $\mathbb{Z}^2$

The *digital space* $\mathbb{Z}^d$ is the set of all points in Euclidean space $\mathbb{R}^d$ having integer coordinates. Digital space $\mathbb{Z}^2$ is also called *digital plane*. Subsets of digital plane are termed *digital sets*, often they are also called *digital objects* or *digital images*. Single elements of $\mathbb{Z}^2$ are termed *grid points*. A digital set consisting of all grid points which are lying on a horizontal, vertical or diagonal real line in $\mathbb{R}^2$ is called a *horizontal, vertical* or *diagonal grid line*.

The neighborhood structure is a significant concept in the study of digital objects. In our considerations, we concentrate on 8-neighborhood structure for sets and 4-neighborhood structure for their complements. Generally, the choice of two different notions, one for the object and another for its complement is related to avoiding certain paradoxes (Kong, 1989).

DEFINITION 8.1. *Given an 8-connected digital set $K \subseteq \mathbb{Z}^2$. K is called a* digital 8-curve *whenever each point $x \in K$ has exactly two 8-neighbors in*

K with the possible exception of at most two points, the end points *of the curve, having exactly one neighbor in K.*

A curve without end points is termed a closed curve.

Each digital 8-curve can be ordered (or oriented) in a natural manner. Let us consider an ordered finite 8-curve $K = (\kappa_1, \cdots, \kappa_n)$, where κ_i is an 8-neighbor of κ_{i+1} for $i = 1, 2, \cdots, n-1$. Then, the curve K can be described by means of a simple compact ordered data structure. It contains coordinates of κ_1 and a sequence of code numbers $\{0, 1, \cdots, 7\}$. Codes indicate for each point of K which of its neighbors will be the next point on the curve. This data structure was proposed in (Freeman, 1961) and is known as *chain code.*

DEFINITION 8.2. *Given an ordered digital* 8-curve $K = (\kappa_1, \cdots, \kappa_n)$. *For* $k \in \{0, 1, \cdots, 7\}$ *the curve K is called a* $(k, k+1(\text{mod } 8))$-curve *whenever the chain code representation of K consists exclusively of two chain codes k and $k+1(mod\, 8)$.*

In 1987, so-called *shape descriptors* (Scherl, 1987) were introduced. Descriptors are boundary points of an 8-connected digital set belonging to local extrema of linear functionals with main directions in digital plane$\frac{k\pi}{4}$, $k = 0, 1, \cdots, 7$. Hence, Scherl's descriptors are segments of horizontal, vertical and diagonal grid lines. They belong to locally convex or concave parts of the set and termed T- or S-*descriptors*, respectively. In Figure 8.1, descriptor points of a digital image are demonstrated.

The succession of descriptor points on the oriented boundary is not arbitrary (Eckhardt and Reiter, 2004; Scherl, 1987). It can be shown that the boundary of an 8-connected digital set can be decomposed into $(k, k+1 \,(\text{mod } 8))$-curves. Moreover, the part between two such successive curves consists only of descriptor points (Eckhardt and Reiter, 2004).

Since each $(k, k+1(\text{mod } 8))$-curve is an image by a rotation of some $(0,1)$-curve, in the later sections we may concentrate, without loss of generality, on $(0,1)$-curves.

8.3. Discrete Lines and Convexity

The arithmetical definition of discrete lines was introduced in (Reveillès, 1991).

DEFINITION 8.3. *A* discrete line *with a* slope a/b, $b \neq 0$ *and* $pgcd(a,b) = 1$, lower bound μ, arithmetical thickness ω *is the set of grid points which satisfies the double diophantine inequality with all integer parameters.*

$$\mu \leq ax - by < \mu + \omega$$

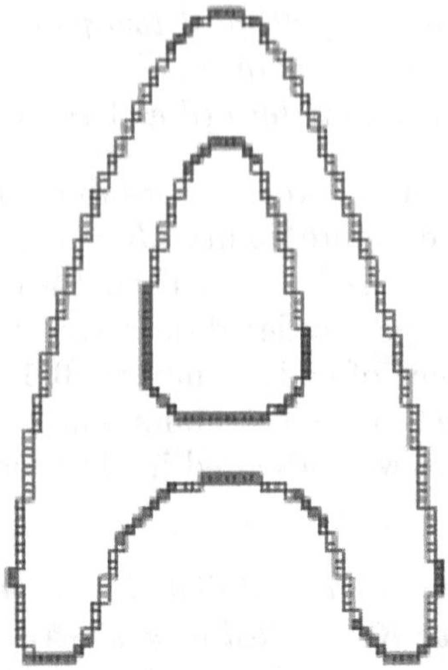

Figure 8.1. Scherl's descriptors of 'Letter A' are indicated by dark voxels.

A (finite or infinite) subsequence of a discrete line is called a discrete line segment.

We denote the preceding discrete line by $D(a, b, \mu, \omega)$. We are mostly interested in *naïve* lines which verify $\omega = \sup(|a|, |b|)$, we shall denote them by $D(a, b, \mu)$. Without loss of generality, we may consider discrete lines under restrictions $a, b > 0$ and $a < b$, therefore $\omega = \max(a, b) = b$.

Real straight lines $ax - by = \mu$ and $ax - by = \mu + b - 1$ are called *upper leaning line* and *lower leaning line* of $D(a, b, \mu)$, respectively. Grid points which satisfy the mentioned equalities are called *upper* or *lower leaning points*. We remark that the distinction between lower and upper leaning points depends on the equation, there is here no geometrical invariancy. Considering a discrete line segment of $D(a, b, \mu)$ with minimal a and b, we may denote by U_F (L_F) the upper (lower) leaning point of the segment whose x-coordinate is minimal. In the same way, we denote by U_L (L_L) the upper (lower) leaning point whose x-coordinate is maximal. In Figure 8.2, an example of a discrete line segment is demonstrated.

It can be shown (Reveillès, 1991) that a discrete line $D(a, b, \mu)$ with slope $a/b \leq 1$ has exactly one grid point on each vertical line. If $a/b < 1$, then the intersection between $D(a, b, \mu)$ and any horizontal line is composed by $[b/a]$ or $[b/a] + 1$ successive grid points, where [ ] means the integer part.

A discrete line $D(a, b, \mu)$, where $0 < a < b$, is 8-connected and satisfies *chord property* (Debled-Rennesson and Reveillès, 1995; Reveillès, 1991), i.e.

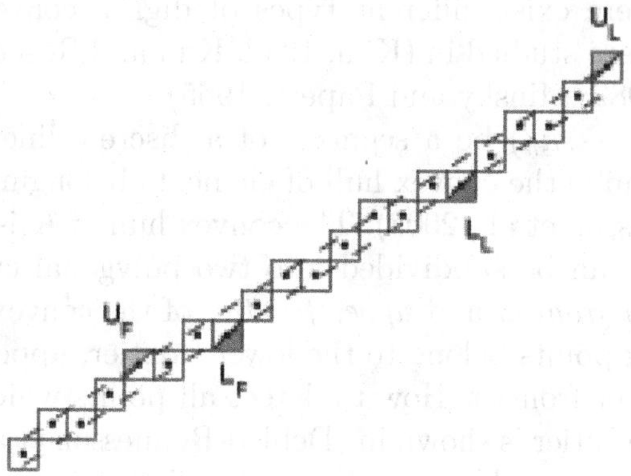

Figure 8.2. Segment of a discrete line $D(5, 8, -4)$. Dashed lines represent upper and lower leaning lines of the segment, leaning points are indicated by pale and dark triangles.

for $P, Q \in D(a, b, \mu)$ and for $U = (x_U, y_U)$ lying on the straight line segment joining P and Q there exists $R = (x_R, y_R) \in D(a, b, \mu)$ such that $\max(|x_R - x_U|, |y_R - y_U|) < 1$. It follows that a discrete line segment is a digital straight line segment (Hübler et al., 1981). Finite digital curves satisfying chord property are discrete line segments. There are infinite digital curves which possess chord property and, however, are not discrete lines (Hübler, 1989).

LEMMA 8.1. **(Reveillès)** *A discrete line* $D(a, b, \mu)$, *whose parameters satisfy* $0 < a < b$, *is an 8-curve.*

The movement from left to right along a discrete line with $0 < a < b$ occurs by using two translations, either $(x, y) \mapsto (x + 1, y)$ or $(x, y) \mapsto (x + 1, y + 1)$. In our denotation, it is a $(0, 1)$-curve.

LEMMA 8.2. *Each discrete line* $D(a, b, \mu)$, *whose parameters satisfy* $0 < a < b$, *is a* $(0, 1)$*-curve.*

The term of convexity is a central subject of many geometrical investigations. Particularly, in application oriented disciplines of geometry, it plays an important role. Basic constructions of digital geometry are discrete lines, discrete line segments and digitally convex sets. They belong since beginning of the research in digital geometry to frequently examined objects.

In Euclidean geometry, a set in $\mathbb{R}^d$ is said to be convex if whenever it contains two points, it also contains the line segment joining them. Already in the two-dimensional case, there were observed difficulties by direct transfer of this definition into digital circumstances (see e.g. (Hübler, 1989)). In the literature, there exist different types of digital convexity. The most common of them are studied in (Kim, 1982; Kim and Rosenfeld, 1980; Kim and Rosenfeld, 1982; Minsky and Papert, 1965).

Let $K = (\kappa_1, \cdots, \kappa_n)$ be a segment of a discrete line $D(a, b, \mu)$. The problem to determine the convex hull of elements belonging to K is solved in (Debled-Rennesson et al., 2003). The convex hull of K is a closed polygonal curve which can be subdivided into two polygonal curves joining κ_1 and κ_n : the *lower frontier* and *upper frontier* of the convex hull. It is clear that lower leaning points belong to the lower frontier, upper leaning points belong to the upper frontier. How to detect all points which belong to the lower and upper frontier is shown in (Debled-Rennesson et al., 2003, Proposition 3). Since the curve K is a segment of a discrete line, the intersection of K and its convex hull consists only of elements of K. This fact justifies the following concept of a *digitally convex curve*:

DEFINITION 8.4. *A digital curve K is said to be* lower digitally convex (upper digitally convex) *if there is no grid point between K and the lower (upper) frontier of the convex hull of K.*

Algorithm SegConv for testing convexity of digital curves is proposed in (Debled-Rennesson et al., 2003). This algorithm has linear time complexity.

8.4. Fundamental Segments of 8-curves

DEFINITION 8.5. *Let $K = (\kappa_1, \cdots, \kappa_n)$ be a $(0,1)$-curve. Parameters a and b in discrete line segments considered below are assumed to be minimal. A part $(\kappa_i, \cdots, \kappa_j)$ is called a* fundamental segment *of K whenever one of the following conditions is true:*

- *$i = 1$, $j = n$ and $(\kappa_1, \cdots, \kappa_n)$ is a segment of $D(a, b, \mu)$. Then K consists of one single fundamental segment.*
- *$i = 1$, $j < n$ and $(\kappa_1, \cdots, \kappa_j)$ is a segment of $D(a, b, \mu)$ such that $(\kappa_1, \cdots, \kappa_{j+1})$ is not a segment of any discrete line. Here, $(\kappa_1, \cdots, \kappa_j)$ is the first fundamental segment of K.*
- *$i > 1$, $j = n$ and $(\kappa_i, \cdots, \kappa_n)$ is a segment of $D(a, b, \mu)$ such that $(\kappa_{i-1}, \cdots, \kappa_n)$ is not a segment of any discrete line. Here, $(\kappa_i, \cdots, \kappa_n)$ is the last fundamental segment of K.*
- *$i > 1$, $j < n$ and $(\kappa_i, \cdots, \kappa_j)$ is a segment of $D(a, b, \mu)$ such that $(\kappa_{i-1}, \cdots, \kappa_j)$ and $(\kappa_i, \cdots, \kappa_{j+1})$ are not segments of any discrete line.*

The fundamental segment $(\kappa_i, \cdots, \kappa_j)$ will be denoted by $F(a, b, \mu)$.

This definition means that the convex hull of a fundamental segment of K and the left or right added point consists at least of one grid point of the complement of K (Debled-Rennesson et al., 2003, Remark 6). Hence, fundamental segments are maximal possible subsets of K belonging to discrete lines.

By definition, fundamental segments do not depend on the orientation of K. All fundamental segments can be ordered in the sense of the oriented curve, we mark these by $F_i(a_i, b_i, \mu_i)$, $i = 1, \cdots, m$. It is clear that two successive fundamental segments have always different slopes, their common part is not empty, and it is always a segment of a discrete line. In addition, more than two fundamental segments can possess common parts of K.

Clearly, the decomposition of a $(0,1)$-curve into fundamental segments is unique. The problem to find this decomposition is equivalent to the problem to determine subsets of the curve having constant tangents. A linear algorithm for computing such subsets is proposed in (Feschet and Tougne, 1999).

In Figure 8.3, fundamental segments of a digital $(0,1)$-curve are indicated.

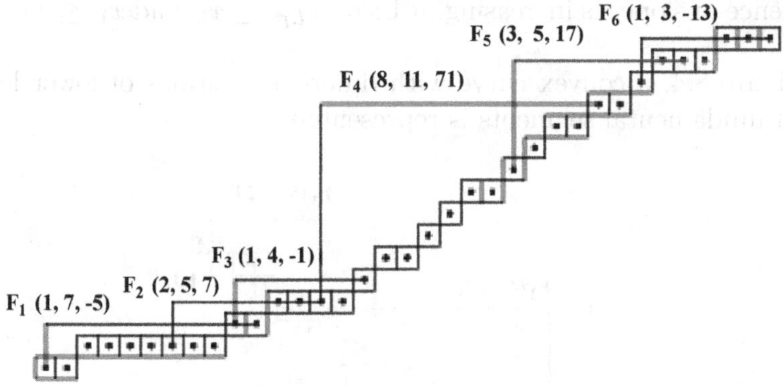

Figure 8.3. Fundamental segments of a digital curve. The first point of the curve is $(0,0)$. Lower bounds μ_i, $i = 1, \cdots, 6$ are computed with respect to $(0,0)$.

8.5. Convex and Concave Curves

Fundamental segments allow an adaption of the concept of *convex* and *concave curves* from continuous theory.

DEFINITION 8.6. *Let K be a $(0,1)$-curve and $F_i(a_i, b_i, \mu_i)$, $i = 1, \cdots, m$ be successive fundamental segments of K. The curve K is called* convex (concave) *whenever the sequence of slopes of fundamental segments is increasing (decreasing), i.e.* $\frac{a_j}{b_j} < \frac{a_{j+1}}{b_{j+1}}$ $\left(\frac{a_j}{b_j} > \frac{a_{j+1}}{b_{j+1}}\right)$, $1 \leq j \leq m-1$.

If K is a discrete line segment, then it is convex as well as concave.

Since the sequence of fundamental segments does not depend of the orientation of K, the concave curve is a convex one if the orientation of K is turned back. In following considerations, we will prove only convex case, the concave case can be formulated analogously and shown by duality.

Leaning points of fundamental segments of convex curves are located not arbitrarily. Namely, they appear in the successive order on the curve. This statement is proved in the next lemma.

We mark the x- and y-coordinate of a point P as x_P and y_P, respectively.

LEMMA 8.3. *Let K be a convex $(0,1)$-curve and $F_i(a_i, b_i, \mu_i)$, $i = 1, \cdots, m$ be fundamental segments of K. Then $x_{L_{L_j}} \leq x_{L_{F_{j+1}}}$ for all $1 \leq j \leq m-1$.*

Proof Given fundamental segments $F_j(a_j, b_j, \mu_j)$ and $F_{j+1}(a_{j+1}, b_{j+1}, \mu_{j+1})$ of K. Let Π be the polygonal set consisting of edges e_1: $a_j x - b_j y = \mu_j + b_j - 1$ and e_2: $a_{j+1}x - b_{j+1}y = \mu_{j+1} + b_{j+1} - 1$ which are lower leaning lines of $F_j(a_j, b_j, \mu_j)$ and $F_{j+1}(a_{j+1}, b_{j+1}, \mu_{j+1})$. One single vertex $V = (x_V, y_V)$ of Π is the intersection point of e_1 and e_2. Obviously, K is above Π. Since the sequence of slopes is increasing, it holds $x_{L_{L_j}} \leq x_V$ and $x_V \leq x_{L_{F_{j+1}}}$. □

In Figure 8.4, a convex curve with different locations of lower leaning points of fundamental segments is represented.

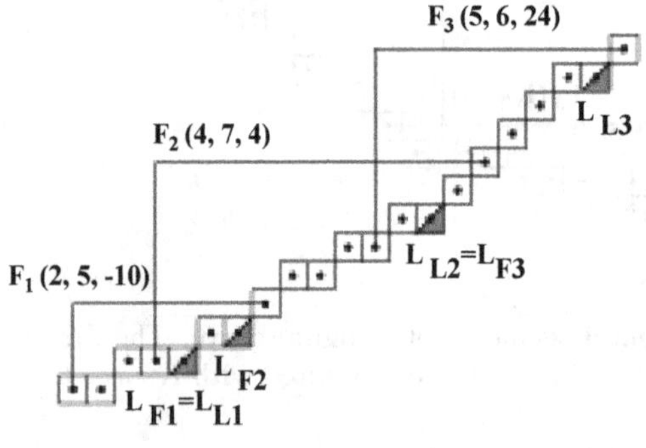

Figure 8.4. Convex curve and its lower leaning points of fundamental segments $F_i(a_i, b_i, \mu_i)$, $i = 1, 2, 3$. For lower leaning points holds $x_{L_{L_1}} < x_{L_{F_2}}$ and $x_{L_{L_2}} = x_{L_{F_3}}$.

Let us assume that for leaning points L_{L_j} and $L_{F_{j+1}}$ of fundamental segments $F_j(a_j, b_j, \mu_j)$ and $F_{j+1}(a_{j+1}, b_{j+1}, \mu_{j+1})$ of a convex curve K holds $x_{L_{L_j}} = x_{L_{F_{j+1}}}$. Then $L_{L_j} = L_{F_{j+1}}$ is a vertex of the lower frontier of the convex hull of K. Before the case $x_{L_{L_j}} < x_{L_{F_{j+1}}}$ will be examined, we introduce concept of *supporting lines*.

A real line L is called a *lower supporting line in $P \in K$* (briefly, LSL) if $P \in L$ and there exists a (continuous) neighborhood $N(P)$ of P such that all elements in $K \cap N(P)$ are lying on or above L. A convex curve K with a fundamental segment $F(a, b, \mu)$, whose leaning points are L_F and L_L, is lying on or above lower leaning line of $F(a, b, \mu)$. Hence, $ax - by = \mu + b - 1$ is a LSL in L_F, L_L and all grid points which belong to K on the real line segment $[L_L, L_F]$. Moreover, there exists no grid point between the segment $(L_F, \cdots, L_L)$ of K and $[L_F, L_L]$. If the whole curve K is on or above a LSL, then the LSL is also called a *global lower supporting line* (briefly, GLSL).

If an arbitrary $(0,1)$-curve K with m fundamental segments $F_i(a_i, b_i, \mu_i)$ and a GLSL in $P \in K$ such that $P \in (L_{F_1}, \cdots, L_{L_m})$ are given, then P belongs to one of lower leaning lines of $F_i(a_i, b_i, \mu_i)$. Obviously, there can exist a GLSL in P which is before L_{F_1} or after L_{L_m}, however, this case does not play any role for our further considerations.

LEMMA 8.4. *Let K be a convex $(0,1)$-curve and $F_i(a_i, b_i, \mu_i)$, $i = 1, \cdots, m$ be fundamental segments of K. Let us assume $x_{L_{L_j}} < x_{L_{F_{j+1}}}$ for some $1 \le j \le m-1$. Then there is no grid point between fundamental segments $F_j(a_j, b_j, \mu_j)$ and $F_{j+1}(a_{j+1}, b_{j+1}, \mu_{j+1})$ and the polygonal set with successive edges e_1: $a_j x - b_j y = \mu_j + b_j - 1$, e_2: the real line through L_{L_j} and $L_{F_{j+1}}$, and e_3: $a_{j+1}x - b_{j+1}y = \mu_{j+1} + b_{j+1} - 1$.*

Proof Real lines e_1 and e_3 are lower leaning lines of fundamental segments $F_j(a_j, b_j, \mu_j)$ and $F_{j+1}(a_{j+1}, b_{j+1}, \mu_{j+1})$. It follows that there is no grid point between the polygonal set with both edges, whose intersection point is the vertex V, and these fundamental segments. Thus, we only must show that elements of $(L_{L_j}, \cdots, L_{F_{j+1}})$ are lying above e_2.

Let us assume $P \in K$ is one single point inside the triangle with vertices L_{L_j}, $L_{F_{j+1}}$, V. Illustration is given in Figure 8.5. We deduce that there exists

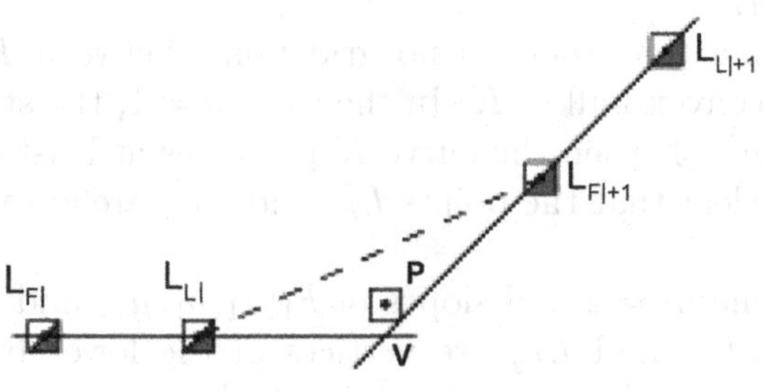

Figure 8.5. Illustration to Lemma 8.4.

a GLSL in P. Hence, P is on one of lower leaning lines of fundamental segments, but not on e_1 or e_3, i.e. there must exists a fundamental segment between $F_j(a_j, b_j, \mu_j)$ and $F_{j+1}(a_{j+1}, b_{j+1}, \mu_{j+1})$. It leads to a contradiction that both fundamental segments are successive.

Analogously, the case, where more than one points are inside the triangle, leads to this contradiction. □

Now we are able to show the equivalence between convex and lower digitally convex curves.

THEOREM 8.1. *Let $K = (\kappa_1, \cdots, \kappa_n)$ be a $(0,1)$-curve. The curve K is convex if and only if K is lower digitally convex.*

Proof Let $F_i(a_i, b_i, \mu_i)$, $i = 1, \cdots, m$ be fundamental segments of K.

1. Let us assume K is convex. We consider a polygonal curve Π consisting of vertices of the lower frontier of the convex hull of $F_1(a_1, b_1, \mu_1)$ before L_{F_1}, intersection points of lower leaning lines of $F_j(a_j, b_j, \mu_j)$ and $F_{j+1}(a_{j+1}, b_{j+1}, \mu_{j+1})$, $j = 1, \cdots, m-1$ and vertices of the lower frontier of the convex hull of $F_m(a_m, b_m, \mu_m)$ after L_{L_m}. Since K is convex, Π possesses increasing slopes and there is no grid point between Π and the curve K. By Lemma 8.3, for two successive fundamental segments $F_j(a_j, b_j, \mu_j)$ and $F_{j+1}(a_{j+1}, b_{j+1}, \mu_{j+1})$ holds $x_{L_{L_j}} \leq x_{L_{F_{j+1}}}$. If $x_{L_{L_j}} = x_{L_{F_{j+1}}}$, then $L_{L_j} = L_{F_{j+1}}$ is a vertex of Π. In the case $x_{L_{L_j}} < x_{L_{F_{j+1}}}$ for the slope s of the real line segment $[L_{L_j}, L_{F_{j+1}}]$ holds $\frac{a_j}{b_j} < s < \frac{a_{j+1}}{b_{j+1}}$. According to Lemma 8.4, there is no grid point between $(L_{L_j}, \cdots, L_{F_{j+1}})$ and $[L_{L_j}, L_{F_{j+1}}]$. In this case, we modify the polygonal curve Π by replacing the vertex which is intersection point of lower leaning lines of $F_j(a_j, b_j, \mu_j)$ and $F_j(a_{j+1}, b_{j+1}, \mu_{j+1})$ by vertices L_{L_j} and $L_{F_{j+1}}$. Hence, modified Π has all successive vertices of the lower frontier of the convex hull of K and there is no grid point between K and the frontier.

2. We assume that there is no grid point between K and the lower frontier of the convex hull of K. In the case $m = 1$, the statement is, obviously, true. If $m > 1$, then the curve K possesses at least two fundamental segments. It is clear that the points L_{F_1} and L_{L_m} are always vertices of the lower frontier.

Let us assume $m = 2$ and slopes of $F_1(a_1, b_1, \mu_1)$ and $F_2(a_2, b_2, \mu_2)$ are decreasing, i.e. L_{F_1} and L_{L_2} are vertices of the lower frontier and there exists no other vertex between them. Then, there must be at least one grid point between $(L_{F_1}, \cdots, L_{L_2})$ and $[L_{F_1}, L_{L_2}]$. Otherwise, $(L_{F_1}, \cdots, L_{L_2})$ is a discrete line segment belonging to the curve contradicting the consecutivity of $F_1(a_1, b_1, \mu_1)$ and $F_2(a_2, b_2, \mu_2)$.

For $m > 2$ the similar arguments lead to a contradiction when we assume that slopes of $F_j(a_j, b_j, \mu_j)$ and $F_{j+1}(a_{j+1}, b_{j+1}, \mu_{j+1})$, $1 \leq j \leq m-1$ are decreasing. □

8.6. Decomposition of Curves into Meaningful Parts

Let us consider a digital $(0,1)$-curve K. By means of fundamental segments of K and their slopes, we are able to define convex and concave parts of K which are maximal.

DEFINITION 8.7. *Let K be a finite $(0,1)$-curve and $F_i(a_i, b_i, \mu_i)$, $i = 1, \cdots, m$ be fundamental segments of K. A part consisting of successive fundamental segments $F_u(a_u, b_u, \mu_u), \cdots, F_v(a_v, b_v, \mu_v)$, $1 \le u \le v \le m$ is called a* maximal convex part *of K whenever one of the following conditions is true:*

- $u = 1$, $v = m$ *and* $\frac{a_j}{b_j} < \frac{a_{j+1}}{b_{j+1}}$, $1 \le j \le m-1$.
- $u \ne 1$, $v \ne m$, $\frac{a_{u-1}}{b_{u-1}} > \frac{a_u}{b_u}$, $\frac{a_v}{b_v} > \frac{a_{v+1}}{b_{v+1}}$ *and* $\frac{a_j}{b_j} < \frac{a_{j+1}}{b_{j+1}}$ *for all* $u \le j \le v-1$.
- $u = 1$, $v \ne m$, $\frac{a_v}{b_v} > \frac{a_{v+1}}{b_{v+1}}$ *and* $\frac{a_j}{b_j} < \frac{a_{j+1}}{b_{j+1}}$ *for all* $1 \le j \le v-1$.
- $u \ne 1$, $v = m$, $\frac{a_{u-1}}{b_{u-1}} > \frac{a_u}{b_u}$ *and* $\frac{a_j}{b_j} < \frac{a_{j+1}}{b_{j+1}}$ *for all* $u \le j \le m-1$.

A maximal concave part of K is defined in the same manner by replacing the signs '$<$' and '$>$' in the above definition.

It is clear that a convex (concave) curve consists exactly of one single maximal convex (concave) part. Maximal parts overlap each other. If the curve is neither convex nor concave, then each of its maximal convex and concave part has at least two fundamental segments. The common component of two successive meaningful parts consists exactly of one fundamental segment.

Let us concentrate on Figure 8.6, where the curve from Figure 8.3 is represented again. Slopes of 6 fundamental segments are $\frac{a_i}{b_i} = 0.1429$, 0.4, 0.25, 0.7273, 0.6, 0.3333. We deduce that the curve possesses four maximal parts: convex consisting of $F_1(a_1, b_1, \mu_1)$ and $F_2(a_2, b_2, \mu_2)$; concave with $F_2(a_2, b_2, \mu_2)$ and $F_3(a_3, b_3, \mu_3)$; convex with $F_3(a_3, b_3, \mu_2)$ and $F_4(a_4, b_4, \mu_4)$; concave consisting of $F_4(a_4, b_4, \mu_4)$, $F_5(a_5, b_5, \mu_5)$ and $F_6(a_6, b_6, \mu_6)$. The point P belonging to each maximal part is indicated.

We are interested in partitionning an arbitrary curve K into maximal possible meaningful parts. First, we are able to decompose K into $(0,1)$-curves using Scherl's descriptors (see Section 8.2). Next, the pre-decomposition into convex and concave parts can be defined in the following manner: the segment of K between descriptor points is convex or concave whenever it is convex or concave part of the corresponding $(0,1)$-curve. Obviously, the pre-decomposition possesses no maximal possible parts. Finely, using the fact that T-descriptors belong to a convex part and S-descriptors to a concave part we can determine the maximal possible meaningful parts of the curve.

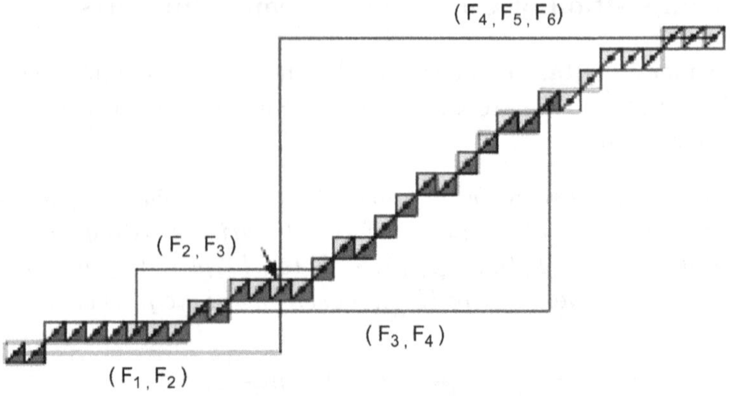

Figure 8.6. Maximal convex and concave parts of the curve from Figure 8.3.

8.7. Fundamental Polygonal Representations of Digital Curves

In this section, we shortly discuss an important application of decomposition curves into fundamental segments. Let K be a finite $(0,1)$-curve and $F(a,b,\mu)$ a fundamental segment of K. The (whole) segment $F(a,b,\mu)$ is located above the lower leaning line $ax - by = \mu + b - 1$ and under the upper leaning line $ax - by = \mu$. Moreover, there is no grid point between the segment and these leaning lines. We consider two successive fundamental segments $F_j(a_j,b_j,\mu_j)$ and $F_{j+1}(a_{j+1},b_{j+1},\mu_{j+1})$. The common part of segments is located above (under) both lower (upper) leaning lines $a_jx - b_jy = \mu_j + b_j - 1$ and $a_{j+1}x - b_{j+1}y = \mu_{j+1} + b_{j+1} - 1$ ($a_jx - b_jy = \mu_j$ and $a_{j+1}x - b_{j+1}y = \mu_{j+1}$). Hence, segments $F_j(a_j,b_j,\mu_j)$ and $F_{j+1}(a_{j+1},b_{j+1},\mu_{j+1})$ are above (under) the polygonal curve Π with edges given by mentioned real lines and the vertex given by their intersection point. There exists no grid point between these polygonal curves and fundamental segments.

Considerations above allow to introduce a concept of *fundamental polygonal representations*:

DEFINITION 8.8. *Let K be a finite $(0,1)$-curve and $F_i(a_i,b_i,\mu_i)$, $i = 1,\cdots,m$ fundamental segments of K. A polygonal curve Π with edges given by lower (upper) leaning lines of $F_i(a_i,b_i,\mu_i)$ and vertices given by their intersection points in successive order is called lower (upper) fundamental polygonal representation of K.*

Figure 8.7 shows fundamental polygonal representations of the curve from Figure 8.3.

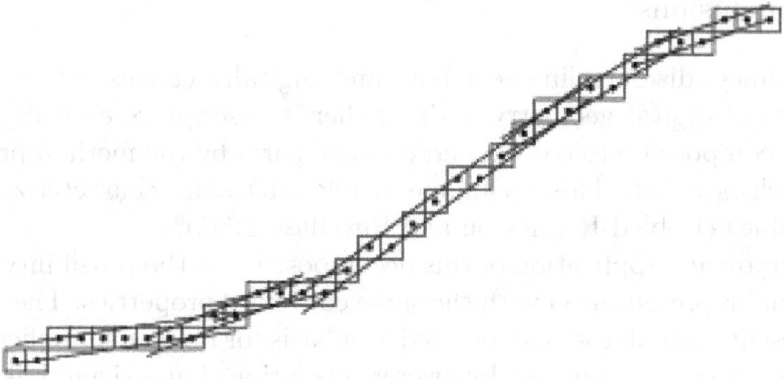

Figure 8.7. Fundamental polygonal representations of the curve from Figure 8.3.

We collect some simple properties of fundamental polygonal representations:

1. There is no grid point between fundamental polygonal representations and the digital curve.
2. Vertices of fundamental polygonal representations are, generally, not grid points.
3. The lower (upper) fundamental polygonal representation of a convex (concave) digital curve possesses only convex (concave) vertices.
4. Fundamental polygonal representations have the same convexity properties as the digital curve.
5. Representations are translations of each other and have the same Euclidean lengths.
6. If for lower (upper) leaning points of fundamental segments $F_i(a_i, b_i, \mu_i)$, $i = 1, \cdots, m$ of a convex (concave) curve holds

$$L_{L_j} = L_{F_{j+1}} \quad (U_{L_j} = U_{F_{j+1}}) \quad 1 \leq j \leq m-1$$

then vertices of the lower (upper) fundamental polygonal representation between L_{F_1} and L_{L_m} (U_{F_1} and U_{F_m}) are vertices of the lower (upper) frontier of the convex hull of K.

In the end, we will in short discuss on the computational complexity of presented procedures.

Partitionning the boundary of a digital set into $(0, 1)$-curves corresponds to detection descriptor points on the boundary. It can be shown (Scherl, 1987) that descriptor points can be obtained within linear time. On the basis of the algorithm (Feschet and Tougne, 1999), partitions of $(0, 1)$-curves into maximal convex and concave parts lead to a linear algorithm. In the same manner, it allows to construct a linear algorithm for determining fundamental polygonal representations.

8.8. Conclusions

Discrete lines, discrete line segments and digitally convex sets are basic constructs of digital geometry. Using Scherl's descriptors each digital set can be decomposed into convex and concave parts by the method proposed here which is exact. This technique is related to the characterization of discrete lines (Debled-Rennesson and Reveillès, 1995).

An important application of this decomposition is the possibility to find a polygonal representation with the same convexity properties. The polygonal representation of a set can be used as a basis for further simplification of the representing polygonal set by discrete evolution (Latecki and Lakämper, 2002). However, in spite of the precision, there is a disadvantage of the presented decomposition. It is the fact that the corresponding polygonal representation can possess vertices whose coordinates are not integers.

As alternative polygonal representations would be representations having "only few" uncorresponding convex or concave parts, but whose vertices are elements of $\mathbb{Z}^2$. Our group is preparing a further paper about linear algorithm for such polygonal representations of digital sets.

References

Brimkov, V. E., and R. Klette: Curves, hypersurfaces, and good pairs of adjacency relations. CITR-TR-144, Computer Science Dept., The University of Auckland, 2004.

Coeurjolly, D., Y. Gerard, L. Tougne, and J.-P. Reveillès: An elementary algorithm for digital arc segmentation. *Discrete Applied Mathematics*, **139**:31–50, 2004.

Debled-Rennesson, I. and J.-P. Reveillès: A linear algorithm for segmentation of digital curves. *Int. J. Pattern Recognition Artificial Intelligence*, **9**:635–662, 1995.

Debled-Rennesson, I., J.-L. Rémy, and J. Rouyer-Degli: Detection of discrete convexity of polyominoes. *Discrete Applied Mathematics*, **125**:115–133, 2003.

Eckhardt, U. and H. Reiter: Polygonal representations of digital sets. *Algorithmica*, **38**: 5–23, 2004.

Feschet, F. and L. Tougne: Optimal time computation of the tangent of a discrete curve: application to the curvature. In *Discrete Geometry Computer Imagery*, (G. Bertrand, M. Couprie, and L. Perroton, editors), pages 31–40, LNCS 1568, Springer, Berlin, 1999.

Freeman, H: On the encoding of arbitrary geometry configurations. *IRE Trans. EC*, **10**:260–268, 1961.

Hübler, A: *Diskrete Geometrie fuer die digitale Bildverarbeitung.* Dissertation B, Jena University, 1989.

Hübler, A, R. Klette, and K. Voss: Determination of the convex hull of a finite set of planar points within linear time. *Elektronische Informationsverarbeitung Kybernetik*, **17**:121–139, 1981.

Kim, C. E.: Digital convexity, straightness and convex polygons. *IEEE Trans. Pattern Analysis Machine Intelligence*, **4**:618–626, 1982.

Kim, C. E. and A. Rosenfeld: On the convexity of digital regions. *Pattern Recognition*, **5**:1010–1015, 1980.

Kim, C. E. and A. Rosenfeld: Digital straight lines and convexity of digital regions. *IEEE Trans. Pattern Analysis Machine Intelligence*, **4**:149–153, 1982.

Kong, T. Y. and A. Rosenfeld: Digital topology: introduction and survey. *Computer Vision Graphics Image Processing*, **48**:357–393, 1989.

Latecki, L. and R. Lakämper: Convexity rule for shape decomposition based on discrete contour evolution. *Computer Vision Image Understanding*, **73**:441–454, 1999.

Latecki, L. and R. Lakämper: Application of planar shape comparison to object retrieval in image databases. *Pattern Recognition*, **35**:15–29, 2002.

Minsky, M. and S. Papert: *Perceptrons. An Introduction to Computational Geometry.* MIT Press, Cambridge, London, 1965.

Reveillès, J. P.: *Géométrie discréte, calcul en nombres entiers et algorithmique.* Thése d'État, Strasbourg, 1991.

Scherl, W.: *Bildanalyse allgemeiner Dokumente.* Informatik-Fachberichte 131, Springer, Berlin, 1987.

Voss, K.: *Discrete Images, Objects and Functions in* $\mathbb{Z}^n$. Springer, Berlin, 1993.

POLYGONALISATION AND POLYHEDRALISATION BY OPTIMISATION

TRUONG KIEU LINH
School of Science and Technology, Chiba University, Japan
ATSUSHI IMIYA
IMIT, Chiba University
Yayoi-cho 1-33, Inage-ku, Chiba 263-8522, Japan
AKIHIKO TORII
School of Science and Technology, Chiba University, Japan

Abstract. In this paper, we first derive a set of inequalities for the parameters of a Euclidean linear object from sample pixels on a plane (voxel in a space), and an optimisation criterion with respect to this set of constraints for the recognition of the Euclidean object. Second, using this optimisation problem, we prove uniqueness and ambiguity theorems for the reconstruction of a Euclidean object. Finally, we develop a polygonalisation (polyhedralisation) algorithm for the boundary of a discrete shape on a plane (in a space).

Key words: linear discrete object, integer linear programming, polygonalisation, polyhedralisation

9.1. Introduction

Combinatorial geometry provides a methodology on the construction of efficient algorithms for geometric data processing, such as CAD-modelling, geographical information systems, robotics and so on. On the other hand, since geometric data are expressed as digital data in the computers which achieve the data manipulation on geometric data, we are required to design appropriate data processing expression.

Digital geometry provides a traditional data expression in the computers. As interface between digital geometry and combinatorial geometry, in this paper, we aim to develop a class of algorithms to transform digital data, which are dealt in digital geometry, to discrete data such as polygons

R. Klette et al. (eds.), Geometric Properties for Incomplete Data, 161-182.

and polyhedra, which are dealt in computational geometry. The transforms which yields polygons and polyhedra from digital data are called polygonalisation and polyhedralisation, respectively. In this paper, we develop algorithms for these transformations employing integer linear programming (Korte and Vygen, 2000). In this paper, we deal with the recognition stage as a combinatorial optimisation problem (Korte and Vygen, 2000). The algorithm achieves invertible data compression of digital objects, since the algorithm transforms collections of pixels and voxels to collections of line and plane parameters, respectively, the number of elements of which are usually smaller than the number of pixels and voxels of digital objects.

Recently, a linear-programming based method for the recognition of linear manifolds has been proposed (Francon et al., 1996; Buzer, 2003). This method is based on the mathematical property that a point set determines a system of linear inequalities for the parameters of a line and a plane, and the recognition process for a plane is converted to the computation of the feasible region for this system of inequalities.

The other class for recognition of linear manifold is based on the binary relation among local configurations of voxels in 3×3 and $3\times3\times3$ regions, on a plane and in a space, respectively, which characterise local properties of discrete planes (Barneva et al., 2000; Schramm, 1997; Vittone and Chassery, 1999; Reveilles, 1995; Sivigion et al., 2004).

Our method proposed in this paper is based on the former method for the derivation of constraints on parameters of the Euclidean plane that passes through sample voxels. Furthermore, we derive a minimisation criterion for the parameters of the Euclidean plane with respect to the constraints yielded from a set of pixels and voxels on a line and a plane. Linear programming framework solves the l_1-norm criterion (Ben-Tel and Nemirovski, 2001) for the plane fitting, which is robust against outer layers and noise. Therefore, the method is suitable for the reconstruction of planes from samples with intervals (Neumaier, 2001). Treating samples of voxels as samples with intervals, we derive an algorithm for the reconstruction of the Euclidean planes from collection of sample voxels.

9.2. Optimisation Problem for Recognition

9.2.1. RECOGNITION OF PLANE IN A SPACE

Setting $\boldsymbol{p} = (p, q, r)^\top$ to be a point in three-dimensional discrete space $\mathbf{Z}^3$, a cube

$$v(\boldsymbol{p}) = \{\boldsymbol{x} |\, |x - p| \leq \frac{1}{2}, |y - q| \leq \frac{1}{2}, |z - r| \leq \frac{1}{2}, \}, \tag{1}$$

for $\boldsymbol{x} = (x, y, z)^\top$ in three-dimensional Euclidean space $\mathbf{R}^3$, is called a voxel of $\mathbf{R}^3$. Hereafter, $P = \{\boldsymbol{x}_i\}_{i=1}^k$ stands for the set of the centres of the voxels $V = \{v(\boldsymbol{x}_i)\}_{i=1}^N$. We call an element of P a grid.

For vector $\boldsymbol{a} = (a, b, c)^\top$ in $\mathbf{Z}^3$ such that $\gcd(a, b, c) = 1$ and integer μ, the supercover of the plane $\boldsymbol{a}^\top \boldsymbol{x} + \mu = 0$ is the collection of voxels which satisfy the inequality

$$|\boldsymbol{a}^\top \boldsymbol{x} + \mu| \leq \frac{1}{2}|\boldsymbol{a}|_1. \tag{2}$$

where $|\boldsymbol{x}|_1$ is the l_1 norm of vector $\boldsymbol{x}$.

We can consider that $\boldsymbol{a} \geq 0$, $(a \geq 0, b \geq 0, c \geq 0)$, substituting $(x, -y, z)^\top$, $(x, y, -z)^\top$, and $(x, -y, -z)^\top$ to (x, y, z) in Equation (2). for the cases $ab < 0$, $ac < 0$, and $ab, ac < 0$, respectively. Therefore, assuming that $\boldsymbol{a} \geq 0$, we develop an algorithm for the reconstruction of a Euclidean plane from sample voxels.

DEFINITION 9.1. *If all elements in a collection of girds P are elements of the supercover of a plane, we call that the elements of P are coplanar.*

This definition leads to the definition of recognition and reconstruction of a supercover.

DEFINITION 9.2. *For a collection of grids P, the process to examine coplanarity of elements of a collection of grids is recognition of a supercover. The computation of the parameters of the plane from coplanar grids is the reconstruction of plane.*

These definitions of recognition and reconstruction imply that the computation of parameters of a plane from sample voxels achieves both recognition and reconstruction. Therefore, we develop an algorithm for the computation of parameters of a plane from a supercover.

Computation of parameters of a plane from sample voxels is stated as the following problem.

PROBLEM 9.1. *For a collection of sample grids $\{\boldsymbol{x}_i\}_{i=1}^k$, if there exists a Euclidean plane whose supercover contains all voxels $\{v(\boldsymbol{x}_i)\}_{i=1}^k$, compute parameters $\boldsymbol{a}$ and μ.*

This problem is mathematically equivalent to finding parameters $\boldsymbol{a}$ and μ which satisfy the system of inequalities

$$|\boldsymbol{a}^\top \boldsymbol{x}_i + \mu| \leq \frac{1}{2}|\boldsymbol{a}|_1, \ i = 1, 2, \cdots, k. \tag{3}$$

Setting $\boldsymbol{s}_{\boldsymbol{x}} = (\mathrm{sgn}x, \mathrm{sgn}y, \mathrm{sgn}z)^\top$ for $\boldsymbol{x} = (x, y, z)^\top$, Equation (3) derives the system of inequalities

$$\boldsymbol{a}^\top(\boldsymbol{x}_{ij} + \boldsymbol{s}_{\boldsymbol{a}}) \geq 0,\ \boldsymbol{x}_{ij} = \boldsymbol{x}_i - \boldsymbol{x}_j \tag{4}$$

$$-\boldsymbol{a}^\top\boldsymbol{x}_i - \frac{1}{2}|\boldsymbol{a}|_1 \leq \mu \leq -\boldsymbol{a}^\top\boldsymbol{x}_i + \frac{1}{2}|\boldsymbol{a}|_1, \tag{5}$$

for $i, j = 1, 2, \cdots, k$, $i \neq j$. Since the vector $(\boldsymbol{x}_{ij} + \boldsymbol{s}_{\boldsymbol{a}})$, such that $i \neq j$, is a constant for $i, j = 1, 2, \cdots, N$, Equation (4) determines a cone in $\mathbf{Z}^3$ as shown in Figure 9.1 (a). We call this integer cone defined by Equation (5) the feasible region of the parameters for the recognition of a plane from sample voxels.

Using the inequality

$$\frac{1}{2}|\boldsymbol{a}|_1 \leq \frac{1}{2}(|\boldsymbol{a}|_1 + |\mu|) \leq |\boldsymbol{a}|_1 + |\mu|, \tag{6}$$

we solve the following problem for the computation of the parameter of a plane from sample voxels.

PROBLEM 9.2. *Find $\boldsymbol{a} \in \mathbf{Z}^3$ and $\mu \in \mathbf{Z}$ which minimises $J = I + |\mu|$, for $J \in \mathbf{Z}$ with respect to*

$$-\boldsymbol{a}^\top\boldsymbol{x}_i + \frac{1}{2}y \leq \mu \leq \boldsymbol{a}^\top\boldsymbol{x}_i + \frac{1}{2}y, \tag{7}$$

and for I which minimises $I = |\boldsymbol{a}|_1$, for $\boldsymbol{a} \in \mathbf{Z}^3$ with respect to

$$\boldsymbol{a}^\top\boldsymbol{y}_{ij} \geq 0, i, j = 1, 2, \cdots, k,\ i \neq j, \tag{8}$$

where

$$\boldsymbol{y}_{ij} = \boldsymbol{x}_i - \boldsymbol{x}_j + \boldsymbol{s}_{\boldsymbol{a}}. \tag{9}$$

In the following, we prove the uniqueness theorem for the reconstructed plane by minimisation of J in Problem 9.2.

THEOREM 9.1. *If and only if $|\boldsymbol{a}|_1 = 2n$, the supercover $|\boldsymbol{a}^\top\boldsymbol{x} + \mu| \leq \frac{1}{2}|\boldsymbol{a}|_1$ contains $2 \times 2 \times 2$ cubes.*

Proof If $|\boldsymbol{a}|_1 = 2n$, the Euclidean line $\boldsymbol{a}^\top\boldsymbol{x} + \mu = 0$ passes through the point $\boldsymbol{x} + \frac{1}{2}\boldsymbol{e}$ for $\boldsymbol{x} \in \boldsymbol{Z}^3$. Moreover, for voxels $p(\boldsymbol{x})$ and $p(\boldsymbol{x} + \boldsymbol{e}_i)$, if $|\boldsymbol{a}^\top\boldsymbol{x} + \mu| \leq \frac{1}{2}|\boldsymbol{a}|_1$, the Euclidean plane $\boldsymbol{a}^\top\boldsymbol{x} + \mu = 0$ passes through the point $\boldsymbol{x} + \frac{1}{2}\boldsymbol{e}$, and $|\boldsymbol{a}|_1 = 2n$. □

This theorem implies the following corollary.

COROLLARY 9.1. *For grids in P which satisfy the system of double inequality $|\boldsymbol{a}^\top \boldsymbol{x}_i + \mu| \leq \frac{1}{2}|\boldsymbol{a}|_1$, if $|\boldsymbol{a}|_1 = 2n$, then only $\boldsymbol{a}^\top \boldsymbol{x} + \mu = 0$ passes through all voxels in V.*

This corollary leads to the conclusion that for a supercover, $2 \times 2 \times 2$ cubes guarantee the uniqueness of the Euclidean reconstruction of a plane. We state another theorem:

THEOREM 9.2. *For grids P which satisfy the system of double inequality $|\boldsymbol{a}^\top \boldsymbol{x}_i + \mu| \leq \frac{1}{2}|\boldsymbol{a}|_1$, if $|\boldsymbol{a}|_1 = 2n+1$, for $|\gamma| \leq \frac{1}{2}$, a line $\boldsymbol{a}^\top \boldsymbol{x} + \mu + \gamma = 0$ passes through all voxels in V.*

Proof If $|\boldsymbol{a}|_1 = 2n+1$, we have $0 \leq \boldsymbol{a}^\top \boldsymbol{x}_i + \mu + \gamma + \frac{|\boldsymbol{a}|_1}{2} \leq |\boldsymbol{a}|_1$ for $|\gamma| \leq \frac{1}{2}$. Therefore, $\boldsymbol{a}^\top \boldsymbol{x}_i + \mu + \gamma = 0$ and $\boldsymbol{a}^\top \boldsymbol{x}_i + \mu = 0$ determine the same set of voxels. □

For $|\boldsymbol{a}|_1 = 2n+1$, planes exist in the strip whose centre is $\boldsymbol{a}^\top \boldsymbol{x} + \mu = 0$ and width is $1/2^n|\boldsymbol{a}^\top \boldsymbol{a}|$ for $n = 3$. For an integer λ and $|\gamma| < \frac{1}{2}$, if $\lambda\mu + \lambda\gamma$ is an integer, both $\boldsymbol{a}^\top \boldsymbol{x} + \mu + \gamma = 0$ and $\lambda \boldsymbol{a}^\top \boldsymbol{x} + \lambda\mu + \lambda\gamma = 0$ are the same plane. Therefore, $|\lambda|$ is larger than 1, since

$$|\boldsymbol{a}|_1 + |\mu| < |\lambda|(\boldsymbol{a}_1 + |\mu|) < |\boldsymbol{a}'|_1 \tag{10}$$

for $\boldsymbol{a}' = (\lambda \boldsymbol{a}^\top, \lambda(\mu + \gamma))^\top$. This inequality geometrically means that the plane which minimises $(|\boldsymbol{a}|_1 + |\mu|)$ is the central plane in the strip region. From these theorems, we have the following property on the solution of the Problem 9.2.

THEOREM 9.3. *The solution which minimises both J and I uniquely determines a plane, if the feasible region is not the empty set.*

Hereafter, we call a $2 \times 2 \times 2$ cube a bubble. For $\gcd(a, b, c, \mu) = 1$, let $\boldsymbol{b} = (a', b', c')$, for $\gcd(a, b, c) = g$, $a' = \frac{a}{g}$, $b' = \frac{b}{g}$, $c' = \frac{c}{g}$.

For the supercover of the plane L, namely $ax + by + cz + \mu = 0$, elementary number theory derives relations in Tables 9.1 and 9.2, on the uniqueness of the Euclidean reconstruction of planes with bubbles and with out bubbles, respectively, from the geometrical and algebraic properties of bubbles. In the tables, Q is the set of all quotient numbers. In Tables 9.1 and 9.2, the centre plane of the supercover L is the plane which minimises the optimisation criterion.

The equivalent planes of L are planes which define the same supercover with L. The universal planes of L are the planes which contain all voxels of the supercover of L. The supercover of the universal plane of L always contains bubbles. These relations imply that if $\gcd(a, b, c) = 1$, the plane which minimises the criterion is uniquely computed. Furthermore, if $a + b + c =$ odd, the plane reconstructed from the supercover does not pass through the corners of voxels.

TABLE 9.1. Reconstruction of a Plane From the Supercover with Bubbles.

$\gcd(a,b,c)$	$a+b+c$ $a'+b'+c'$	Equivalent Plane	Universal Plane
1	$a+b+c$: even	L	$\emptyset$
2	$a'+b'+c'$: odd	L	$\emptyset$
>2	×	×	×

TABLE 9.2. Reconstruction of a Plane from the Supercover without Bubbles.

$\lvert\boldsymbol{a}\rvert_1$ $\lvert\boldsymbol{b}\rvert_1$	Centre plane	Equivalent plane	Universal plane
		$\gcd(a,b,c)=1$	
$\lvert\boldsymbol{a}\rvert_1$ odd	L	$k\boldsymbol{a}^\top\boldsymbol{x}+k\mu+k\varepsilon=0$ where $k\in Z$, $\lvert\varepsilon\rvert<\frac{1}{2}$, $k\varepsilon\in Z$	$2\boldsymbol{a}^\top\boldsymbol{x}+2\mu\pm1=0$
		$\gcd(a,b,c)=2$	
$\lvert\boldsymbol{b}\rvert_1$ even	L	$k\boldsymbol{a}^\top\boldsymbol{x}+k\mu+k\varepsilon=0$ where $k\in Z$, $\lvert\varepsilon\rvert<1$ $k\varepsilon\in Z$	$\boldsymbol{a}^\top\boldsymbol{x}+\mu\pm1=0$
		$\gcd(a,b,c)>2$	
$\lvert\boldsymbol{b}\rvert_1$ odd	$\boldsymbol{a}'^\top\boldsymbol{x}+\mu'=0$ where $\mu'\in Z$, $\frac{\mu}{g}-\frac{1}{2}<\mu'<\frac{\mu}{g}+\frac{1}{2}$	$k\boldsymbol{a}^\top\boldsymbol{x}+k\mu'+k\varepsilon=0$ where $k\in Z$, $\lvert\varepsilon\rvert<\frac{1}{2}$, $k\varepsilon\in Z$	$2\boldsymbol{a}'^\top\boldsymbol{x}+2\mu'\pm1=0$
$\lvert\boldsymbol{b}\rvert$ even	$2\boldsymbol{a}'^\top\boldsymbol{x}+2\mu'+1=0$ where $\mu'\in Z$, $\frac{\mu}{g}-1<\mu'<\frac{\mu}{g}$	$k\boldsymbol{a}'^\top\boldsymbol{x}+k\mu'+k\varepsilon=0$ where $k\in Z$, $0<\varepsilon<1$, $k\varepsilon\in Z$	$\boldsymbol{a}'^\top\boldsymbol{x}+\mu'=0$ or $\boldsymbol{a}^\top\boldsymbol{x}+\mu'+1=0$

The minimisation criterion determines a series of parallel planes $\boldsymbol{a}^\top \boldsymbol{e} = k$ for $\boldsymbol{e} = (1,1,1)^\top$ and integer k. We set $S_k = \{\boldsymbol{a}|\boldsymbol{a} > 0,\ \boldsymbol{a}^\top \boldsymbol{e} = k,\ k > 3\}$. Here, S_k is a convex polygon on the plane $\boldsymbol{a}^\top \boldsymbol{e} = k$. $\boldsymbol{a}^\top \boldsymbol{y}_{ij} \geq 0$, $\boldsymbol{a} \geq 0$, and $\boldsymbol{e}^\top \boldsymbol{P}_a \boldsymbol{a} = 0$, where $\boldsymbol{P}_a$ is the orthogonal projector to plane $a = 0$, determine a convex polygon in on a plane $a = 0$. The projection of S_k to the plane $a = 0$ is a convex polygon. Figure 9.1 (b) shows the projection of the common set of a plane and a cone. We set this projection of S_k to the plane $a = 0$, as $S_k(a)'$.

If we can determine integer points $\boldsymbol{a}' = (0,b,c)^\top$ in $S_k(a)'$, $a_n = k - \boldsymbol{a}'^\top \boldsymbol{e}$. This geometric property leads to the conclusion that it is possible to reduce the dimensions of the search space. Therefore, we have the following algorithm:

step 1:	Set $\boldsymbol{a}^\top \boldsymbol{e} = k$ for $k \geq 3$
step 2:	Compute the vertices of S_k
step 3:	Project S_k to the plane $a_k = 0$ and set it as $S_k(i)'$.
step 4:	If there exist an integer points in $S_k(i)'$, then set them $\boldsymbol{a}'$, else go to $k := k+1$ and go to **step1**.
step 5:	For the integer point $S_k(i)'$, set $a_i = k - \boldsymbol{a}'^\top \boldsymbol{e}$.
step 6:	Compute μ.

9.2.2. RECOGNITION OF LINE ON A PLANE

Setting $\boldsymbol{p} = (p,q)^\top$ to be a point in two-dimensional discrete space $\mathbf{Z}^2$, a square,

$$p(\boldsymbol{p}) = \{\boldsymbol{x}|\, |x-p| \leq \frac{1}{2}, |y-q| \leq \frac{1}{2}\}, \tag{11}$$

for $\boldsymbol{x} = (x,y)^\top$ is called a pixel of $\mathbf{R}^2$.

For two-dimensional vector $\boldsymbol{a} = (a,b)^\top$ of $\mathbf{Z}^2$ and integer μ, the supercover of line $\boldsymbol{a}^\top \boldsymbol{x} + \mu = 0$, for $\gcd(a,b) = 1$, is the collection of pixels which satisfy the inequality

$$|\boldsymbol{a}^\top \boldsymbol{x} + \mu| \leq \frac{1}{2}|\boldsymbol{a}|_1. \tag{12}$$

Setting $c = 0$ the relations in Tables 9.1 and 9.2 are satisfied for lines on a plane. Therefore, the algorithm developed in the previous subsection also reconstruct a line from sample pixels on a plane setting $c = 0$. The fan of the feasible region of parameters is shown in Figure 9.1 (c).

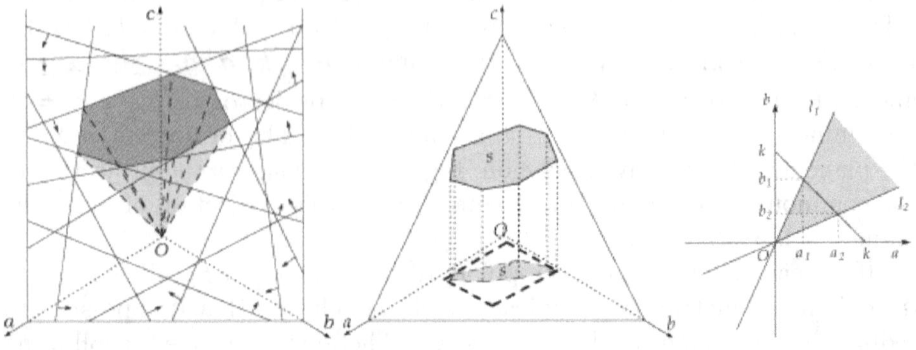

(a) (b) (c)

Figure 9.1. Feasible Regions of a Plane and a Line: (a) The feasible region of a plane in a space. (b) The projection of a common region of this feasible region and a plane $a+b+c=k$ for $a,b,c \geq 0$. These projections reduce the sizes of the search space. (c) The feasible region of a line in a plane. The algorithm searches solution which minimises $|a|+|b|$ in a two-dimensional cone.

9.2.3. RECOGNITION OF A LINE IN 3D SPACE

The reconstruction of a line in 3D space is decomposed into the reconstruction of supercovers on planes $x = 0$, $y = 0$, and $z = 0$. In this case, setting $\boldsymbol{P}_\alpha$, $\alpha = 1,2,3$ to be the orthogonal projection matrix to planes perpendicular to the vector $\boldsymbol{e}_\alpha$, $\boldsymbol{e}_1 = (1,0,0)^\top$, $\boldsymbol{e}_1 = (0,1,0)^\top$, and $\boldsymbol{e}_1 = (0,0,1)^\top$, respectively, a spatial line is expressed as

$$|\boldsymbol{a}_\alpha^\top \boldsymbol{P}_\alpha \boldsymbol{x}_\alpha + \mu_\alpha| \leq \frac{1}{2}|\boldsymbol{a}_\alpha|_1,\ \alpha = 1,2,3. \tag{13}$$

Therefore, the recognition of spatial line is expressed as the following problem.

PROBLEM 9.3. *For a collection of sample grids $\{\boldsymbol{x}_i\}_{i=1}^k$ in $\mathbf{Z}^3$, if there exists a Euclidean line whose supercover contains all voxels $\{v(\boldsymbol{x}_i)\}_{i=1}^N$, compute parameters $\boldsymbol{a}_\alpha$ and μ_α for $\alpha = 1,2,3$.*

Figures 9.2 (a), (b), and (c) show the supercovers of a plane, a line in a space, and a line on a plane, respectively.

The supercover of a line in a space yields superovers of lines in mutually orthogonal three planes which are perpendicular to $\boldsymbol{e}_1$, $\boldsymbol{e}_2$, and $\boldsymbol{e}_3$. Therefore, bubbles on planes are the projections of bubbles in a space. In a space there exists four class of bubbles, $(2\times2\times1)$- $(1\times2\times2)$- $(2\times1\times2)$-parallelepipeds, whose projections are bubbles on x-y, y-z, and z-x planes,

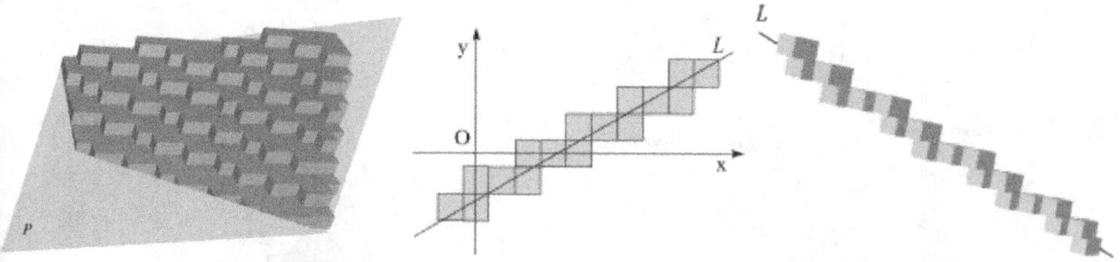

(a) (b) (c)

'igure 9.2. Supercover of Linear Object in a Space and on a Plane: (a), (b), and (c) re supercover of a plane, aline in a space, and a line on a plane, respectively.

espectively and $(2\times2\times2)$-cubes, whose projections are bubbles on all x-y, -z, and z-x planes, respectively. These properties lead to the conclusion hat geometrical property of the supercover of a line in a space is described s the combination of geometrical properties of three projected supercovers n three planes.

We have the next theorem for the bubbles of the supercover of a line in space.

'HEOREM 9.4. *For the supercover of a line in a space*

$$\begin{aligned} 0 &\leq ax + bz + \mu_1 + \tfrac{|a|+|b|}{2} \leq |a| + |b| \\ 0 &\leq ay + cz + \mu_2 + \tfrac{|a|+|c|}{2} \leq |a| + |c| \\ 0 &\leq cx - by + \mu_3 + \tfrac{|b|+|c|}{2} \leq |b| + |c| \\ & a, b, c, \mu_1, \mu_2 \in \mathbf{Z}, \mu_3 = \tfrac{c\mu_1 - b\mu_2}{a}, \end{aligned} \tag{14}$$

iff

$$\frac{\mu_1 + \frac{a+b}{2}}{\gcd(a,b)}, \frac{\mu_2 + \frac{a+c}{2}}{\gcd(a,c)}, \frac{\frac{c\mu_1 - b\mu_2}{a} + \frac{c-b}{2}}{\gcd(b,c)} \in \mathbf{Z}, \tag{15}$$

the supercover contains bubbles.

If a line in a space is described as,

$$\begin{aligned} a_1x + b_1y + c_1z + d_1 = 0 \\ a_2x + b_2y + c_2z + d_2 = 0, \end{aligned} \tag{16}$$

'e have the equations

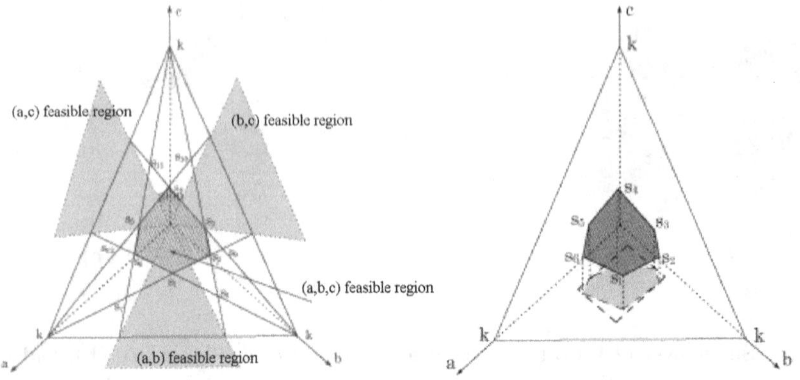

(a) (b)

Figure 9.3. Feasible Region of a Line in a Space: (a) The feasible region of a line in a space is a polyhedral cone determined as the common cone of three cones, which are feasible regions of projections of a line to planes $x = 0$, $y = 0$, and $z = 0$. (b) The projection of the feasible region to the plane $c = 0$ is used to reduce the dimension of the search space.

$$\left\{ \begin{array}{l} (a_1b_2 - a_2b_1)x + (c_1b_2 - c_2b_1)z + (d_1b_2 - d_2b_1) = 0 \\ (a_1b_2 - a_2b_1)y + (a_1c_2 - a_2c_1)z + (a_1d_2 - a_2d_1) = 0, \end{array} \right. \tag{17}$$

This property implies that the system of inequalities

$$\left\{ \begin{array}{l} 0 \leq (a_1b_2 - a_2b_1)x + (c_1b_2 - c_2b_1)z + (d_1b_2 - d_2b_1) + \frac{\omega_1}{2} \leq \omega_1 \\ 0 \leq (a_1b_2 - a_2b_1)y + (a_1c_2 - a_2c_1)z + (a_1d_2 - a_2d_1) + \frac{\omega_2}{2} \leq \omega_2 \\ 0 \leq (a_1c_2 - a_2c_1)x - (c_1b_2 - c_2b_1)y + (d_1c_2 - d_2c_1) + \frac{\omega_3}{2} \leq \omega_3, \end{array} \right. \tag{18}$$

where

$$\begin{aligned} \omega_1 &= |a_1b_2 - a_2b_1| + |c_1b_2 - c_2b_1|, \\ \omega_2 &= |a_1b_2 - a_2b_1| + |a_1c_2 - a_2c_1|, \\ \omega_3 &= |a_1c_2 - a_2c_1| + |c_1b_2 - c_2b_1|, \end{aligned} \tag{19}$$

determines the supercover of the line defined as Equation (16), that is, the supercover of a line in a space is expressed as the supercover of lines on planes $x = 0$, $y = 0$, and $z = 0$. This geometrical property derives the following algorithm for the reconstruction of a Euclidean line in a space from a set of voxels.

step 1:	Set $k := 3$.
step 2:	Compute feasible regions for $(a,b),(a,c),(c,b)$.
step 3:	Set $a+b+c=k$.
step 4:	Compute vertices of the feasible region s on the plane $a+b+c=k$.
step 5:	Compute the projection of the feasible region s onto the plane $c=0$, and set it as s'.
step 6:	Compute integer solutions in s'.
step 7:	Compute μ_1, μ_2, and μ_3.
step 8:	If all of μ_1, μ_2, and μ_3 are not integers then $k := k+1$ and go to step 3, else output solutions .

In the algorithm above, we used the method for the computing feasible region of a Euclidean line in a plane. Figures 9.3 (a) and (b) show the feasible region and its projection to the plane $c = 0$, respectively for the reconstruction of a line in a space.

9.3. Polygonalisation and Polyhedralisation

9.3.1. POLYGONALISATION ON PLANE

Setting **P** to be a digital curve which is a sequence of 4-connected pixels, the polygonalisation on a plane is described as follows.

PROBLEM 9.4. *For a digital boundary curve* $\mathbf{P}$, *setting* $\{p_{ij}\}_{j=1}^{n(i)} = \mathbf{P}_i$, *derive a partition of* $\mathbf{P}$, $\mathbf{P} = \cup_{i=1}^{n}\mathbf{P}_i$, *such that* $|\mathbf{P}_i \cap \mathbf{P}_{i+1}| = \varepsilon$, *where* ε *is an appropriate integer, which minimises*

$$J = \sum_{i=1}^{n}(|\boldsymbol{a}_i|_1 + \mu_i) \tag{20}$$

for the system of inequalities,

$$|\boldsymbol{a}_i^\top \boldsymbol{x}_{ij} + \mu_i| \leq \frac{1}{2}|\boldsymbol{a}_i|_1, \tag{21}$$

for $i = 1, 2, \cdots, n$ *and* $j = 1, 2, \cdots, n(i)$.

Using geometrical properties of the supercover, we introduce the following algorithm.

---

step1 :	Input a pixel sequence $\mathbf{P} = \{\boldsymbol{p}_i\}_{i=0}^{n}$.
step2 :	Set $head = 0, j = 0$.
step3 :	Set$tail = head + 3$.
step4 :	$\mathbf{L}_j = \{\mathbf{p}_i\}_{tail}^{head}$.
step5 :	If there exists a line l_j which passes through $\mathbf{L}_j = \{\mathbf{p}_i\}_{tail}^{head}$ with the condition that $\lvert a\rvert + \lvert b\rvert$ is odd, then set $tail = tail + 1$ and go to step 3
step6 :	If $j = 0$, then set $j = j + 1$, $head = tail$ and go to step 2.
step7 :	if $j > 0$, then compute the common point A_{j-1} of $l_{j-1}$and l_j.
step8 :	If A_{j-1} exists and lies in $\mathbf{L}_j$ or $\mathbf{L}_{j-1}$, then go to step 10.
step9 :	Set $head = head - 1$, and go to step3.
step10 :	Output $\mathbf{L}_{j-1}, l_{j-1}$
step11 :	If $tail < n$, then set $head = tail$, $j = j + 1$ and go to step3.
step12 :	If $tail = n$, then stop.

According to the greedy property of the algorithm, this algorithm stops and fulfils the uniqueness of the solution for the starting point.

9.3.2. POLYGONALISATION IN SPACE

Setting $\mathbf{P}$ to be a digital curve which is a sequence of 6-connected voxels, the polygonalisation in a space is described as follows.

PROBLEM 9.5. *For a digital curve* $\mathbf{P}$, *setting* $\{\boldsymbol{p}_{ij}\}_{j=1}^{n(i)} = \mathbf{P}_i$, *derive a partition of* $\mathbf{C}$

$$\mathbf{P} = \cup_{i=1}^{n}\mathbf{P}_i, \ \lvert\mathbf{P}_i \cap \mathbf{P}_{i+1}\rvert = \varepsilon, \tag{22}$$

where ε *is an appropriate integer, which minimises*

$$J = \sum_{j=1}^{n}(\lvert\boldsymbol{a}_j\rvert_1 + \lvert\boldsymbol{\mu}_j\rvert_1) \tag{23}$$

where $\boldsymbol{a}_j = (a_j, b_j, c_j)^\top$ *and* $\boldsymbol{\mu} = (\mu_{1j}, \mu_{2j}, \mu_{3j})$, *for the system of inequalities,*

$$\begin{cases} 0 \leq a_j x_i + b_j z_i + \mu_{1j} + \frac{\omega_1}{2} \leq \omega_{1j} \\ 0 \leq a_j y_i + c_j z_i + \mu_{2j} + \frac{\omega_2}{2} \leq \omega_{2j} \\ 0 \leq c_j x_i - b_j y_i + \mu_{3j} + \frac{\omega_3}{2} \leq \omega_{3j} \end{cases} \tag{24}$$

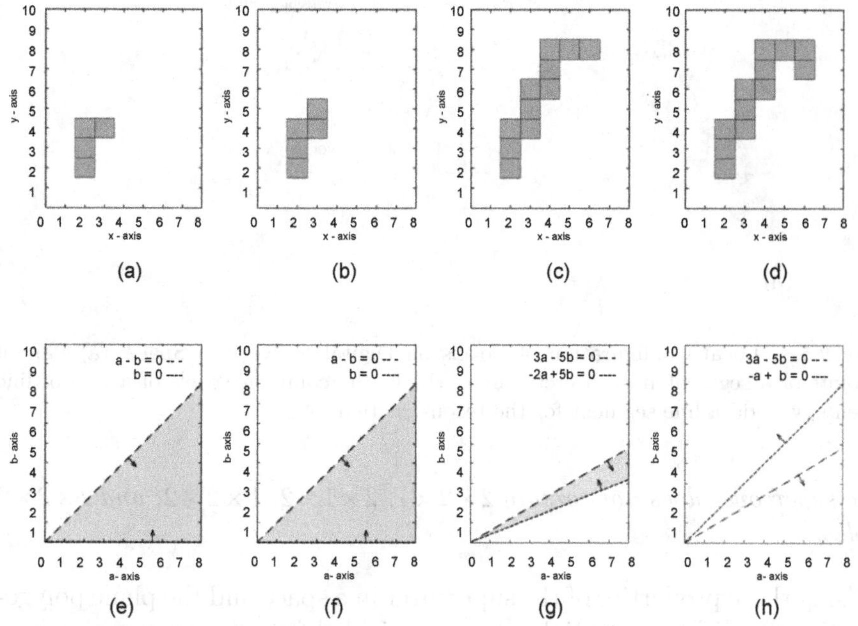

Figure 9.4. An Example of Parts of a Pixel Sequence and the Feasible Regions for parts: (a), (b), (c), and (d) are 4, 5, 11, and 12 successive pixels from a sequence. (e), (f), (g), and (h) are feasible regions for these parts of the sequence. For the configuration of (a), Euclidean lines are exist since the feasible regions not empty. For the configuration of (d), there exists no answer, since the feasible region is the empty set.

determines the same supercover, where $\omega_{1j} = |a_j| + |b_j|$, $\omega_{2j} = |a_j| + |c_j|$, *and* $\omega_{3j} = |c_j| + |b_j|$ *and* $\mu_{3j} = \frac{c_j\mu_{1j} - b_j\mu 2j}{a_j}$.

To solve this minimisation problem, we prepare the following theorem.

THEOREM 9.5. *Setting* $\boldsymbol{p}_1 = (x_1, y_1, z_1)^\top$ *and* $\boldsymbol{p}_2 = (x_2, y_2, z_2)^\top$ *to be a pair of points on supercover*

$$\begin{cases} 0 \le ax + bz + \mu_1 + \frac{\omega_1}{2} \le \omega_1 \\ 0 \le ay + cz + \mu_2 + \frac{\omega_2}{2} \le \omega_2 \\ 0 \le cx - by + \mu_3 + \frac{\omega_3}{2} \le \omega_3 \end{cases} \tag{25}$$

determines the same supercover, where $\omega_1 = |a| + |b|$, $\omega_2 = |a| + |c|$, *and* $\omega_3 = |c| + |b|$, *the number of voxels between* $\boldsymbol{p}_1$ *and* $\boldsymbol{p}_2$ *along this supercover is*

$$N(\boldsymbol{p}_1, \boldsymbol{p}_2) = |\boldsymbol{p}_1 - \boldsymbol{p}_2| + 1, \tag{26}$$

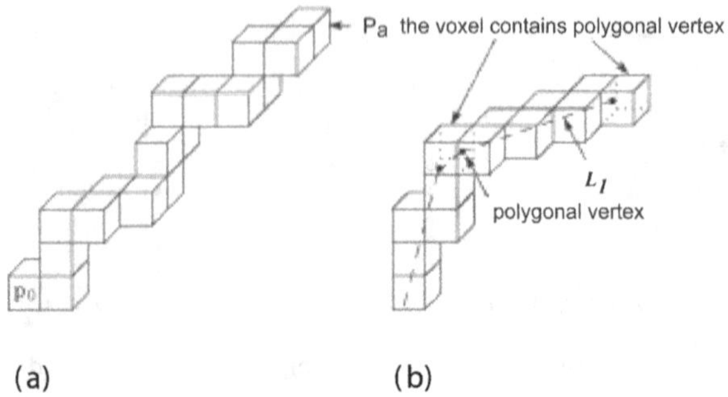

Figure 9.5. Local Configuration of Voxels on Digital Curve in a Space: (a) Pair of endpoints of a segment in a discrete curve. (b) In the common voxels of a pair of line segments, we add a line segment for the reconstruction.

if the supercover does not contain $2\times2\times1$, $2\times1\times2$, $1\times2\times2$, *and* $2\times2\times2$ *bubbles.*

Using these properties of the supercover in a space and the plane polygonalisation algorithm, a spatial polygon is yielded from a sequence of voxels.

---

step 1: Input $\mathbf{P} = \{\mathbf{p}_i\}_0^n$.
step 2: Set $head = 0$, $j = 0$, $L_j = \{\mathbf{p}_i\}_{head}^{tail}$.
step 3: Set $tail = head + 3$
step 4: If there exist a supercover of a 3D line L_j contains L_j, then go to step 6, else go to step6.
step 5: $tail = tail + 1$ and go to step 4.
step 6: Compute the polygonal vertices and output L_{j-1} and L_{j-1}
step 7: If $tail < n$, then set $head = tail$, $j = j + 1$ and go to step 3, else stop.

If a pair of discrete line segments L_i and L_{i+1} share a unique voxel $\boldsymbol{v}_{ij}$, that is $L_i \cap L_{i+1} = \{v(\boldsymbol{x}_i)\}$ as shown in Figure 9.5, we call L_i and L_j is crossing. Step 5 detects these discrete line segments and Euclidean lines whose supercover coincides discrete lines. However, the detected Euclidean lines usually do not cross. Therefore, in step 6, the algorithm generates a line segment which connects L_i and L_{i+1} in the voxel $\boldsymbol{v}_{ij}$. This line segment $L_{i\,i+1}$ is also described by integer parameters, since this line connects points whose elements are rational numbers.

9.3.3. POLYHDERALISATION

Let $C(\boldsymbol{x})$ to be the union of the 6-neighbourhood of voxels $v(\boldsymbol{x})$ and point $v(\boldsymbol{x})$. Furthermore, for a connected voxel-set $\mathbf{A}$, $C(\mathbf{A})$ expresses the union of voxel-set $\mathbf{A}$ and voxels which are 6-connected to the boundary of $\mathbf{A}$ and $S(P_j)$ expresses the supercover of plane P_j. Moreover, $\mathrm{V}_{1,2,4}$ means one of V_1, V_2 and V_4. Then, setting $|\mathbf{A}|$ to be the cardinality of a set $\mathbf{A}$, for 6-connected, the relation $1 \leq |C(\boldsymbol{x})| \leq 7$ is satisfied. Furthermore, a point $\boldsymbol{x}$ which satisfies the relation $2 \leq |C(\boldsymbol{x})| \leq 3$ lies on an edge or a vertex. According to these geometric properties, we select a collection of connected points which satisfy the relation $3 \leq |C(\boldsymbol{v}_i)| \leq 4$. These geometrical conditions derive the following algorithm.

---

step 1: Input 6-connected voxels $\mathbf{V} = \{\mathbf{v_i}\}_\mathbf{0}^\mathbf{n}$

step 2: Select start voxel $\mathbf{v}_{head}$ and set $j = 0$,
where $N_6(\mathbf{v}_{head}) \geq 4$.

step 3: Compute $CCCC(\mathbf{v}_i)$ and set $V = \{\mathbf{v}_i\}_{head}^{tail}$.

step 4: For $V = \{\mathbf{v}_i\}_{head}^{tail}$ compute supercover,
if a supercover exists, then set it as S_j and go to step 5,
else go to step 2.

step 5: Compute $C(V)$ and set $V = C(V)$.

step 6: For $V = \{\mathbf{v}_i\}_{head}^{tail}$ compute supercover,
if a supercover exists, then set it as S_j and go to step 5,
else go to step 7.

step 7: In $\mathbf{V} \setminus V$, if a portion of S_j exists,
add them to V, with the condition that
the new pint-set $V = \{\mathbf{v}_i\}_{head}^{tail}$ is connected

step 8: Select a voxel from $C(V) \setminus V$ and add it to V.

step 9: For V compute supercover, if a supercover exists,
then set it as S_j and go to step 5,
else output S_j and $j = j + 1$.

step 10: If $tail < n$, then set $head = N_6 C^{-1}(V)$
and go to step2, else stop.

The algorithm assumes that, the 6-connected boundary voxels are extracted from discrete object $\mathbf{D}$. For example, setting $\mathbf{N}_{26}$ to be the 26-neighbourhood of the origin in $\mathbf{Z}^3$, the boundary voxels $\partial\mathbf{D}$ is extracted as

$$\partial \mathbf{D} = \mathbf{D} \setminus (\mathbf{D} \ominus \mathbf{N}_{26}), \tag{27}$$

where $\mathbf{A} \ominus \mathbf{B}$ is the Minkowski subtraction of $\mathbf{B}$ from $\mathbf{A}$.

Parameters of faces allow us to classify voxels of discrete objects using the relation,

$$\boldsymbol{x}_i \begin{cases} \in L(\boldsymbol{a}_j, \mu_j), & \text{if } |\boldsymbol{a}_j \boldsymbol{x}_i^\top + \mu_j| \leq \frac{1}{2}|\boldsymbol{a}_j|_1, \\ \notin L(\boldsymbol{a}_i, \mu_j), & \text{otherwise,} \end{cases} \tag{28}$$

where

$$L(\boldsymbol{a}, \mu) = \left\{ \boldsymbol{x} \mid |\boldsymbol{a}^\top \boldsymbol{x} + \mu| \leq \frac{1}{2}|\boldsymbol{a}|_1 \right\}, \tag{29}$$

and voxels in $L(\boldsymbol{a}, \mu)$ are 6-connected. Since voxels on the plane is uniquely determined using the relation

$$L(\boldsymbol{a}_i, \mu_i) = \left\{ \boldsymbol{x} \mid |\boldsymbol{a}_i \boldsymbol{x}^\top + \mu_i| \leq \frac{1}{2}|\boldsymbol{a}_i|_1 \right\}, \tag{30}$$

we can reconstruction a collection of voxels $\mathbf{D}$ from the collection of parameters $\mathbf{P}$ through the boundary voxels $\partial\mathbf{D}$. Therefore, our algorithm achieves data compression by transforming the collection of voxels $\mathbf{D}$ to a collection of parameters of planes $\mathbf{P}$ through the collection of boundary voxels $\partial\mathbf{D}$.

Since the vector $\boldsymbol{a}$ for $(\boldsymbol{a}^\top, \mu)^\top$, which determines a plane segment on the discrete boundary, is the normal vector of the plane $\boldsymbol{a}^\top \boldsymbol{x} + \mu = 0$, $\boldsymbol{a}$ is the normal vector of the plane segment $L(\boldsymbol{a}, \mu)$. This geometric property implies that the algorithm also estimates the normal of the discrete boundary.

9.4. Numerical Examples

In this section, we first show numerical examples which suggest asymptotical uniqueness of the reconstructed results for 2D polygonalisation. For a sampled circle of fixed radius, we have evaluated the length of the perimeter and the area encircled by the reconstructed polygon, selecting each point as the starting point of the polygonalisation. The result in Figure 9.6 shows that the length and area of the reconstructed polygon from a digitised circle is numerically independent to the starting points. The second and third examples show the error analysis for the lengths and areas of the reconstructed circles against the length of radius. For the evaluation, we show the same geometric features of minimum perimeter polygons reconstructed from the same collections of sample points. These examples show that the reconstructed circle by our method asymptotically converges to the original circles by increasing the resolution, since the evolution of the radius is mathematically equivalent to the inclusion of the resolution.

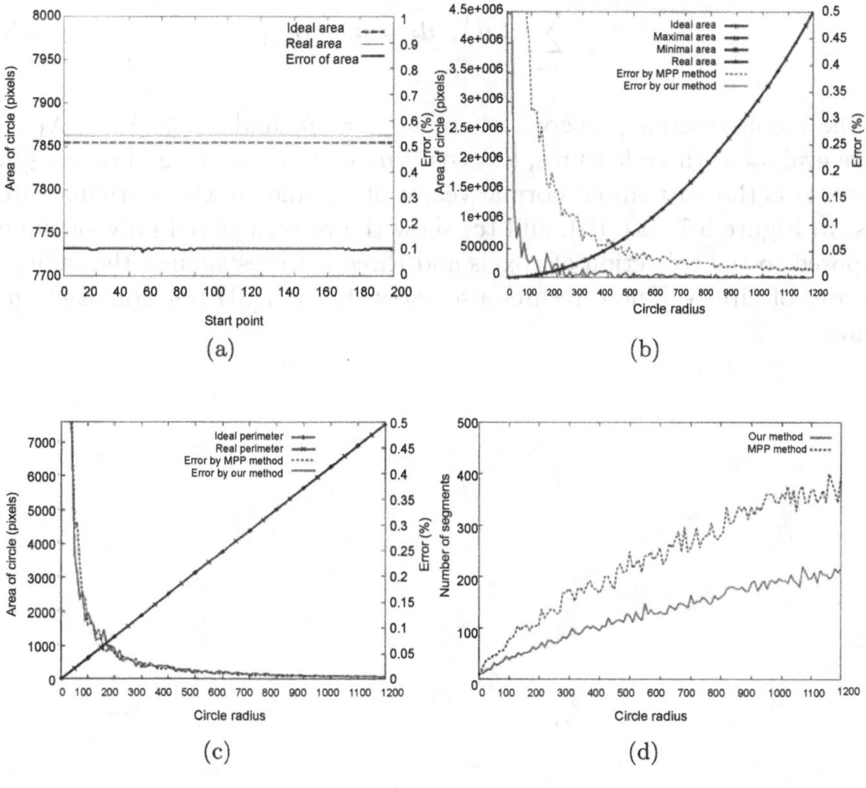

Figure 9.6. Error Analysis in 2D: (a) For a circle with a fixed radius, selecting each pixel as the start point, we evaluated the area encircled by the reconstructed polygon. Numerically, the area is approximately independent to the selection of the start point. (b) For circles with various radii, we evaluated the area encircled by the reconstructed polygon. For the comparison, we evaluated the areas of the outer isotetic polygon, the inner isotetic polygon, the minimum perimeter polygon, and the polygon reconstructed by our method. The error against the ideal area of circle degrees when the radius of the circle increases. (c) For circles with various radii, we evaluated the perimeter length of the reconstructed polygon. For the comparison, we evaluated the areas of the outer isotetic polygon, the inner isotetic polygon, the minimum perimeter polygon, and the polygon reconstructed by our method. The error against the ideal area of circle degrees when the radius of the circle increases. (d)For circles with various radii, we evaluated the the number of edges of the reconstructed polygon. The number of edges of polygon reconstructed by our method is smaller than that of edges of the minimum perimeter polygon.

Next, we show error analysis on flatness of polygonalisation in a space. For the evaluation of the flatness, setting λ_i and $\boldsymbol{u}_i$ for $i = 1, 2, 3$ to be the eigenvalues and eigenvectors of the structure tensor $\boldsymbol{S}$

$$\boldsymbol{S} = \sum_{n-1}^{n} \boldsymbol{d}_i \boldsymbol{d}_i^{\top}, \ \boldsymbol{d}_i = \boldsymbol{p}_i - \boldsymbol{p}_{i-1}, \tag{31}$$

of the reconstructed polygon, where $\boldsymbol{p}_{n+i} = \boldsymbol{p}_i$ and $\lambda_1 \geq \lambda_2 \geq \lambda_3$. We have evaluated three features, $\theta = \cos^{-1} \boldsymbol{u}_1^{\top} \boldsymbol{n}$, $c_1 = \lambda_2/\lambda_1$, and $c_2 = \lambda_3/\lambda_1$, where $\boldsymbol{n}$ is the normalised normal vector of a plane on which original circle lies. In Figure 9.7, (a), (b), and (c) show the reconstructed polygon superimposed to the collection of voxels and three features against the radius of a series of circles. These results also show that our 3D polygonalisation is stable.

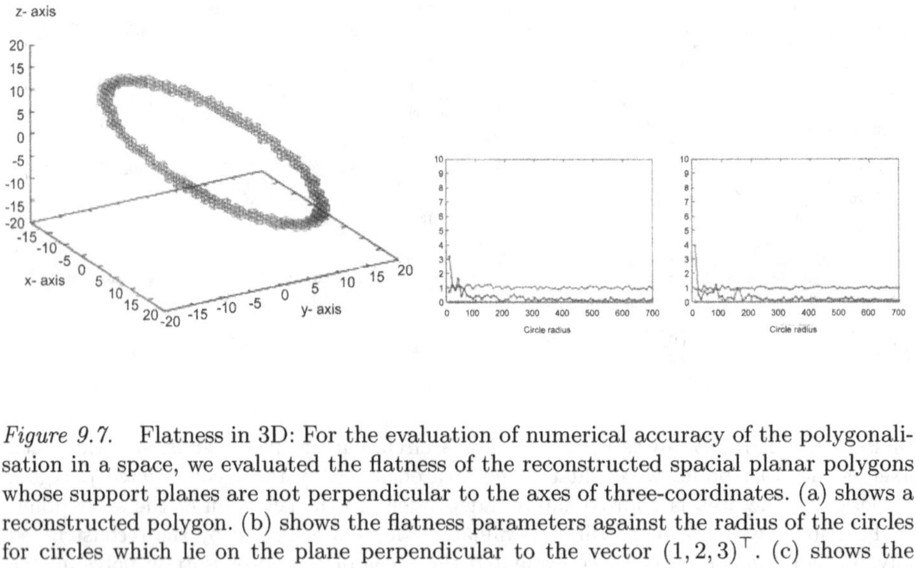

Figure 9.7. Flatness in 3D: For the evaluation of numerical accuracy of the polygonalisation in a space, we evaluated the flatness of the reconstructed spacial planar polygons whose support planes are not perpendicular to the axes of three-coordinates. (a) shows a reconstructed polygon. (b) shows the flatness parameters against the radius of the circles for circles which lie on the plane perpendicular to the vector $(1, 2, 3)^{\top}$. (c) shows the flatness parameters against the radius of the circles for circles which lie on the plane perpendicular to the vector $(1, 1, 1)^{\top}$. For these two numerical examples, graphs show the reconstructed polygons are planar.

In Figure 9.8, we show (a) a sequence of pixels and (b) a reconstructed polygon from pixels of (a). The algorithm extracted 202 edges from 1994 pixels for the planar problem. Figure 9.8 (c) shows the reconstructed polygon superimposed on a sequence of voxels.

For the polyhedralisation, since the qualitative evaluation criteria are still under consideration, we show the extracted parameters from real data. Furthermore, we show a result for the classification of voxels on the boundary.

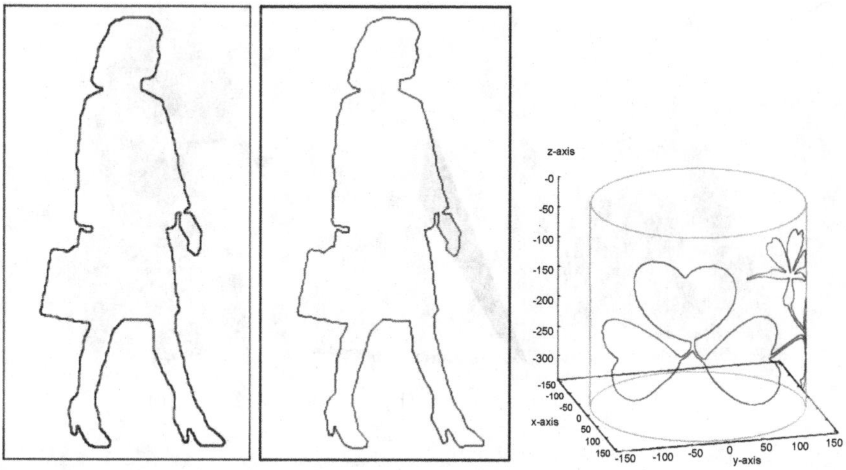

(a) (b) (c)

Figure 9.8. Results of Polygonalisation in 2D and 3D: (a) A sequence of pixels. (b) A reconstructed polygon from pixels of (a). (c) 3D polygon superimposed on a sequence of voxels.

For the complete polyhedralisation, we are required to apply spatial polygonalisation process to these voxels. In Table 9.3, we show parameters extracted from the boundary voxels of Figure 9.9 (c). For example voxels,

$$\begin{array}{llll} (-3,6,-13)^\top, & (-11,9,-11)^\top, & (-9,2,-8)^\top, & (-15,1,-4)^\top \\ (-13,10,-11)^\top & (-12,10,-12)^\top & (-17,10,-9)^\top & (-8,4,-10)^\top \\ (-17,2,-4)^\top & (-8,3,-9)^\top & (-9,5,-10)^\top & (-18,3,-4)^\top \\ (-16,10,-10)^\top & & & \end{array}$$

lie on plane $3x + 4y + 6z + 62 = 0$. Therefore, the normal vector of these voxels is $(3,4,6)^\top$ for the collection of these voxels.

9.5. Concluding Remarks

We have dealt with supercover models on a plane and in a space. We first derived a set of inequalities for the parameters of a Euclidean linear manifold from sample points, and an optimisation criterion with respect to this set of constraints for the recognition of a Euclidean linear manifold. Then using this optimisation problem, we proved uniqueness and ambiguity theorems for the reconstruction of Euclidean linear manifolds. Finally, we developed an algorithm for the computation of the parameters of a Euclidean linear

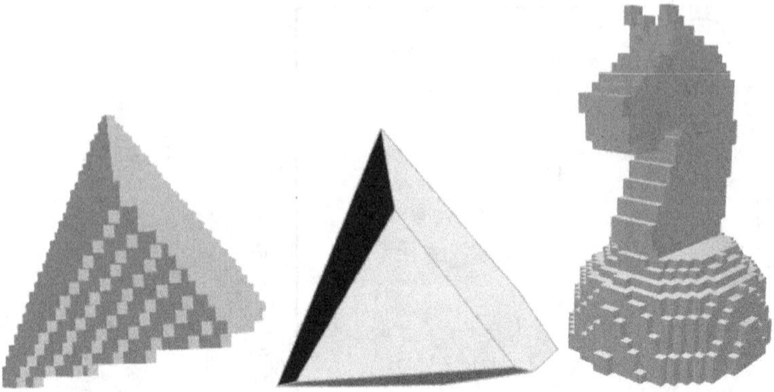

(a) (b) (c)

Figure 9.9. 3D Reconstruction: (a) shows a digital tetrahedron, and (b) shows the reconstructed polyhedron. (b) is digital non-convex polyhedron which we extracted parameters of surface elements.

manifold from pixels and voxels on a plane and in a space, respectively. The theory proposed in this paper is valid for the recognition of naive and standard manifolds.

For the 18-connectivity, Sivignon, et al. (Sivigion et al., 2004) proposed a polyhedralisation algorithms. We proposed a polygonalisation algorithm for 4-connected planar discrete objects (Linh and Imiya, 2003). The paper is extention of our method to 3-dimensional objects. We detect linear objects from pixels and voxels in a plane and a space using parameter space expres-

TABLE 9.3. Parameters of faces: 36 Euclidean faces extracted from Small Knight with 5861 boundary faces.

Parameters of the planes			
$(11,1,1,233)^\top$	$(3,-10,4,-8)^\top$	$(19,-13,13,238)^\top$	$(0,0,1,9)^\top$
$(1,-1,9,104)^\top$	$(1,1,-3,-62)^\top$	$(1,0,-6,-87)^\top$	$(3,2,0,-31)^\top$
$(2,-1,-18,-347)^\top$	$(3,2,0,-52)^\top$	$(1,-4,28,525)^\top$	$(2,-9,0,-89)^\top$
$(1,4,0,29)^\top$	$(1,2,0,17)^\top$	$(2,-2,-7,-155)^\top$	$(9,-1,1,194)^\top$
$(2,14,3,-75)^\top$	$(3,3,1,20)^\top$	$(3,2,2,38)^\top$	$(9,1,-1,158)^\top$
$(3,4,6,62)^\top$	$(5,2,8,110)^\top$	$(1,0,2,25)^\top$	$(8,0,-1,143)^\top$
$(0,1,0,6)^\top$	$(1,-4,-4,-107)^\top$	$(8,-11,-2,-39)^\top$	$(1,0,-6,-103)^\top$
$(11,21,1,-78)^\top$	$(1,0,0,-7)^\top$	$(11,3,-9,-102)^\top$	$(2,2,1,21)^\top$
$(1,0,2,32)^\top$	$(0,0,1,15)^\top$	$(0,1,0,-3)^\top$	$(0,0,1,12)^\top$

sion of linear objects, (Bhattacharya and Rosenfeld, 2003; Bhattacharya and Rosenfeld, 1994; Sivigion et al., 2004). We showed that geometry of feasible regions of parameters in the parameter space for polygonalisation and polyhedralisation.

References

Horng, J.-H.: Improving fitting quality of polygonal approximation by using the dynamic programming technique. *Pattern Recognition Letters,* **23**:1657-1673, 2003.

Andres, E., P. Nehlig, and J. Francon: Supercover of straight lines, planes, and triangles. In Proc. *Discrete Geometry for Computer Imagery,* (E. Ahronovitz and C. Fiorio, editors), pages 243-254, LNCS 1347, Springer, Berlin, 1997.

Couprie, M., G. Bertrand, and Y. Kenmochi: Discretization in 2D and 3D orders. *Graphical Models,* **65**:77-91, 2003.

Ronse, Ch. and M. Tajine: Hausdorff discretization for cellular distances and its relation to cover and supercover siscretizations, *Journal of Visual Communication and Image Representation,* **12**:1690-200, 1999.

Francon, J., J. M. Schramm, and M. Tajine: Recognizing arithmetic straight lines and planes. In *Discrete Geometry for Computer Imagery,* (S. Miguet, A. Montanvert, S. Ub?da, editors), pages 141-150, LNCS 1176, Springer, Berlin, 1996.

Buzer, L.: A linear incremental algorithm for naive and standard digital lines and planes recognition. *Graphical Models,* **65**:61-76, 2003.

Barneva, R. P., V. E. Brimkov, and P. Nehlig: Thin discrete triangular meshes. *Theoretical Computer Science,* **246**:73-105, 2000.

Neumaier, A.: *Introduction to Numerical Analysis,* Cambridge University Press, Cambridge, 2001.

Schramm, J. M.: Coplanar tricubes. In Proc. *Discrete Geometry for Computer Imagery,* (E. Ahronovitz and C. Fiorio, editors), pages 87-98, LNCS 1347, Springer, Berlin, 1997.

Vittone, J. and J. M. Chassery: Digital naive planes understanding. In Proc SPIE 3811 *Vision Geometry VIII* (L. J. Latecki, R. A. Melter, D. M. Mount, A. Y. Wu, editors), pages 22-32, 1999.

Reveilles, J.-P.: Combinatorial pieces in digital lines and planes. In. Proc. SPIE 2573 *Vision Geometry IV,* (R. A. Melter, A. Y. Wu, F. L. Bookstein, W. D. Green, editors), pages 23-34, 1995.

Andres, E.: Discrete linear objects in dimension n: The standard model. *Graphical Models,* **65**:92-111, 2003.

Jonas, A. and N. Kiryati: Digital representation schemes for 3-D Curves. *Pattern Recognition,* **30**:1803-1816, 1997.

Bhattacharya, P. and A. Rosenfeld: Convexity properties of space curves. *Pattern Recognition Letters,* **24**:2509-2517, 2003.

Bhattacharya, P. and A. Rosenfeld: Polygons in three dimensions. *J. Visual Commun. Image Representation,* **5**:139-147, 1994.

Korte, B. and J. Vygen: *Combinatorial Optimization: Theory and Algorithms.* Springer, Berlin, 2000.

Ben-Tel, A. and A. Nemirovski: *Lectures on Modern Convex Optimization.* SIAM, 2001.

Sivigion, I., F. Dupont, and J.-M. Chassery: Decomposition of three-dimensional discrete object surface into discrete plane pieces. *Algorithmica,* **38**:25-43, 2004.

Linh, T.-K. and A. Imiya: Nonlinear Optimisation for Polygonalization. In *Discrete Geometry for Computer Imagery,* I.Nystr?m,G. Sanniti di Baja, S. Svensson, editors), pages 444-453, LNCS 2886, Springer, Berlin, 2003.

Signs & Symbols. The Pepin Press-Agile Rabbit Edition, Amsterdam, 2001.

Graphic Ornaments. The Pepin Press-Agile Rabbit Edition, Amsterdam, 2001.

BINARY TOMOGRAPHY
BY ITERATING LINEAR PROGRAMS

STEFAN WEBER, CHRISTOPH SCHNÖRR,
THOMAS SCHÜLE
University of Mannheim
Dept. of Mathematics and Computer Science, CVGPR-Group, D-68131 Mannheim, Germany

AND JOACHIM HORNEGGER
Friedrich-Alexander University Erlangen-Nürnberg
Dept. of Computer Science, D-91058 Erlangen, Germany

Abstract. A novel approach to the reconstruction problem of binary tomography from a small number of X-ray projections is presented. Based on our previous work, we adopt a linear programming relaxation of this combinatorial problem which includes an objective function for the reconstruction, the approximation of a smoothness prior enforcing spatially homogeneous solutions, and the projection constraints. We supplement this problem with an unbiased concave functional in order to gradually enforce binary minimizers. Application of a primal-dual subgradient iteration for optimizing this enlarged problem amounts to solve a sequence of linear programs, where the objective function changes in each step, yielding a sequence of solutions which provably converges.

Key words: discrete tomography, combinatorial optimization, linear programming, d.c. programming

10.1. Introduction

Discrete Tomography is concerned with the reconstruction of discrete-valued functions from projections. Historically, the field originated from several branches of mathematics like, for example, the combinatorial problem to determine binary matrices from its row and column sums (see the survey (Herman and Kuba, 1999, chapter 1). Meanwhile, however, progress is not only driven by challenging theoretical problems (Gardner and Gritzmann, 1997; Gritzmann et al., 1998) but also by real-world applications where discrete tomography might play an essential role (cf. (Herman and Kuba, 1999, chapters 15–21)).

R. Klette et al. (eds.), Geometric Properties for Incomplete Data, 183-197.

The work presented in this paper is motivated by the reconstruction of volumes from *few* projection directions within a *limited* range of angles. From the viewpoint of established mathematical models (Natterer and Wübbeling, 2001), this is a severely ill-posed problem. The motivation for considering this difficult problem relates to the observation that in some specific medical scenarios (see below), it is reasonable to assume that the function f to be reconstructed is *binary-valued*. This poses one of the essential questions of discrete tomography: how can knowledge of the discrete range of f be exploited in order to regularize and solve the reconstruction problem?

10.1.0.1. *Medical Application*
A potential application of discrete tomography in the field of medical imaging is the 3D reconstruction from Digital Subtraction Angiography (DSA) images. DSA is a common technique for separating contrast-filled vessels from the background. To this end, two images of the same scenery are taken, one with contrast-agent and another one without (see Figure 10.1). This results in low-noise projection images as input data for the reconstruction of a function which is assumed to be *binary*.

Figure 10.1. Illustration Digital Subtraction Angiography (DSA) imaging. A pair of images is taken from each projection direction, one (left) with and another one (center) without contrast agent. Subtraction of both images yields an image (right) that shows the distribution of the contrast agent only. The images at hand were taken from a dough phantom. We simulated the absence of the contrast agent by simply removing the phantom from the scenery.

10.1.0.2. *Problem Statement*
The imaging geometry is represented by a linear system of equations $Ax = b$. Each projection ray corresponds to a row of matrix A, and its projection value is the corresponding component of b. The row entries of A represent the length of the intersection of pixels (voxels in the 3D case) of the (arbitrarily) discretized volume and the corresponding projection ray (see Fig. 10.2). This corresponds to the assumption that the function to

be reconstructed is binary-valued, i.e. x is a binary-valued vector. Each component $x_i \in \{0,1\}$ indicates whether the corresponding pixel (belongs to the reconstructed object, $x_i = 1$, or not, $x_i = 0$ (see Fig. 10.2). The reconstruction problem is to compute the binary indicator vector x from the *under*-determined linear system of projection equations:

$$Ax = b, \quad x = (x_1, ..., x_n)^\top \in \{0,1\}^n \tag{1}$$

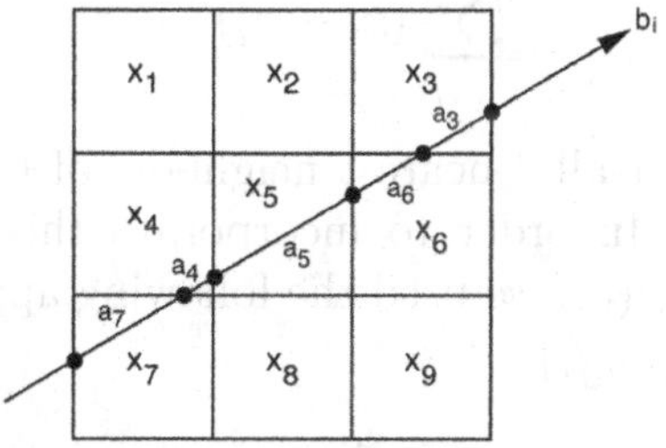

Figure 10.2. Discretization model leading to the algebraic representation of the reconstruction problem: $Ax = b$, $x \in \{0,1\}^n$.

10.2. Previous Work and Contribution

Linear programming in the context of binary tomography was originally suggested in (Aharoni et al., 1997) in order to find invariant elements, i.e. pixels that have the same value for all feasible solutions to a given reconstruction problem, see (Matej et al., 1999) as well.

Due to noise in the measurement vector b when dealing with real data, (1) is likely to have no feasible solution. In order to take advantage of continuous problem formulations and numerical interior point methods, Fishburn et al. (Fishburn et al., 1997) considered the relaxation $x_i \in [0,1]$, $i = 1, \cdots, n$, and investigated the following linear programming approach for computing a feasible point:

$$\min_{x \in [0,1]^n} \langle 0, x \rangle, \quad Ax = b \tag{2}$$

In particular, the information provided by feasible solutions in terms of additivity and uniqueness of subsets $\mathcal{S} \subset \mathbb{Z}^n$ is studied in (Fishburn et al., 1997).

Gritzmann et al. (Gritzmann et al., 2000) introduced the following linear *integer* programming problem for binary tomography

$$\max_{x \in \{0,1\}^n} \langle e, x \rangle, \quad e := (1, \ldots, 1)^\top, \quad Ax \leq b\,, \tag{3}$$

and suggested a range of greedy approaches within a general framework for local search. Compared to (2), the objective function (3), called *best-inner-fit (BIF)* in (Gritzmann et al., 2000), looks for the maximal set compatible with the measurements. Furthermore, the formulation of the projection constraints is better suited to cope with measurement errors and noise.

In (Weber et al., 2003; Weber et al., 2004), we studied the relaxation of (3) $x_i \in [0,1]$, for all i, supplemented with a standard smoothness prior enforcing spatial coherency of solutions

$$\sum_{\langle i,j \rangle} (x_i - x_j)^2 \tag{4}$$

Here, the sum runs over all 4 nearest neighbors of the pixel grid (6 neighbors in the 3D case). In order to incorporate this prior into the linear programming approach (3), we used the following approximation by means of auxiliary variables $\{z_{\langle i,j \rangle}\}$:

$$\min_{x \in [0,1]^n, \{z_{\langle i,j \rangle}\}} \; -\langle e, x \rangle + \frac{\alpha}{2} \sum_{\langle i,j \rangle} z_{\langle i,j \rangle} \tag{5}$$

$$\text{subject to} \quad Ax \leq b \,, \quad z_{\langle i,j \rangle} \geq x_i - x_j \,, \; z_{\langle i,j \rangle} \geq x_j - x_i$$

10.2.0.3. *Contribution*

A global minimizer of the linear program (5) can straightforwardly be computed using an interior point method. In (Weber et al., 2004) we showed that for sparse volume structures, like blood vessels in the brain, in principle, rather accurate 3D-reconstructions may result from solving (5), provided an additional user parameter determining the rounding $[0,1] \ni x_i \to \{0,1\}$, $\forall i$, is set properly in a postprocessing step.

To get rid of this parameter, we supplement (5) with a concave functional enforcing binary solutions $x \in \{0,1\}^n$. Applying a two-step subgradient minimization technique leads to a *sequence* of programs of type (5), whose solutions converge to a local binary-valued minimizer.

Our approach may be regarded as an alternative to (Kleinberg and Tardos, 1999; Censor, 2001) where different techniques have been suggested for rounding solutions of relaxed optimization problems. Rather than rounding in a postprocessing step, we integrate both objective functionals for reconstruction and binary-valued solutions into a single optimization problem, and solve it with a suitable mathematical programming approach.

10.3. Optimization Approach

Our approach reads:

$$\min_{x\in[0,1]^n,\{z_{\langle i,j\rangle}\}} \quad -\langle e,x\rangle + \frac{\alpha}{2}\sum_{\langle i,j\rangle} z_{\langle i,j\rangle} + \frac{\mu}{2}\langle x, e-x\rangle \tag{6}$$

$$\text{subject to} \quad Ax \le b, \quad z_{\langle i,j\rangle} \ge x_i - x_j\,,\; z_{\langle i,j\rangle} \ge x_j - x_i$$

Compared to (5), we supplemented in (6) the concave functional

$$\frac{\mu}{2}\langle x, e-x\rangle = \frac{\mu}{2}\sum_i x_i - x_i^2\,, \tag{7}$$

which is minimal at the vertices of the domain $[0,1]^n$. Furthermore, since it vanishes at $\{0,1\}^n$, it does not alter binary minimizers of the original problem. Our strategy is to choose an increasing sequence of values for μ and to minimize for each of them (6).

Problem (6) is no longer convex, of course. To explain our approach for computing a minimizer, we put

$$z := (x^\top, \cdots, z_{\langle i,j\rangle}, \cdots)^\top \tag{8}$$

and rewrite all constraints of (6)

$$0 \le x_i \le 1\,, \quad Ax \le b\,, \quad z_{\langle i,j\rangle} \ge x_i - x_j\,,\; z_{\langle i,j\rangle} \ge x_j - x_i$$

in the form

$$\tilde{A}z \le \tilde{b}\,, \tag{9}$$

Using the notation

$$\delta_C(z) = \{0\,,\; z\in C + \infty,\; z \notin C$$

for the indicator functions of a convex set C, problem (6) then reads:

$$\min_z f(z)\,,$$

where (cf. definition (8))

$$f(z) \;=\; -\langle e,x\rangle + \frac{\alpha}{2}\sum_{\langle i,j\rangle} z_{\langle i,j\rangle} + \frac{\mu}{2}\langle x, e-x\rangle + \delta_K(\tilde{b} - \tilde{A}z)\,, \tag{10}$$

$$\;=\; g(z) - h(z)\,, \tag{11}$$

$K = \mathbb{R}^n_+$ is the standard cone of nonnegative vectors, and

$$g(z) \;=\; -\langle e,x\rangle + \frac{\alpha}{2}\sum_{\langle i,j\rangle} z_{\langle i,j\rangle} + \delta_K(\tilde{b} - \tilde{A}z)\,, \tag{12}$$

$$h(z) \;=\; \frac{\mu}{2}\langle x, x-e\rangle\,. \tag{13}$$

Note that both functions $g(z)$ and $h(z)$ are convex, and that $g(z)$ is non-smooth due to the linear constraints.

To proceed, we need the following basic concepts (Rockafellar, 1972) defined for a function $f : \mathbb{R}^n \to \overline{\mathbb{R}} := \mathbb{R} \cup \{-\infty, +\infty\}$, and a set $C \subset \mathbb{R}^n$:

$$\operatorname{dom} f := \{x \in \mathbb{R}^n \mid f(x) < +\infty\} \qquad \text{effective domain of } f$$
$$f^*(y) := \sup_{x \in \mathbb{R}^n} \{\langle x, y\rangle - f(x)\} \quad \text{conjugate function}$$
$$\partial f(\overline{x}) := \{v \,|\, f(x) \geq f(\overline{x}) + \langle v, x - \overline{x}\rangle\,,\ \forall x\} \quad \text{subdifferential of } f \text{ at } \overline{x}$$

We adopt from (Pham Dinh and Elbernoussi, 1998; Pham Dinh and Hoai An, 1998) the following two-step subgradient algorithm for minimizing (11):

Subgradient algorithm:
Choose $z^0 \in \operatorname{dom} g$ arbitrary.
For $k = 0, 1, \cdots$, compute:

$$y^k \in \partial h(z^k)\,, \tag{14}$$
$$z^{k+1} \in \partial g^*(y^k)\,. \tag{15}$$

The investigation of this algorithm in (Pham Dinh and Hoai An, 1998) includes the following results:

PROPOSITION 10.1. ((Pham Dinh and Hoai An, 1998)). *Assume* $g,\ h : \mathbb{R}^n \to \overline{\mathbb{R}}$ *be proper*[1], *lower-semicontinuous and convex, and*

$$\operatorname{dom} g \subset \operatorname{dom} h\,, \quad \operatorname{dom} h^* \subset \operatorname{dom} g^*\,. \tag{16}$$

Then

(i) the sequences $\{z^k\}, \{y^k\}$ *according to (14), (15) are well-defined,*
(ii) $\{g(z^k) - h(z^k)\}$ *is decreasing,*
(iii) every limit point[2] z^* *of* $\{z^k\}$ *is a critical point*[3] *of* $g - h$.

[1] A function is called proper if its domain is non-empty.

[2] An limit point is a point which is the limit of a sequence, also called a accumulation point.

[3] A critical point of function is a point were the subgradient of this function includes zero.

10.4. Reconstruction Algorithm

We apply (14), (15) to problem (6). Condition (16) holds, because obviously $\operatorname{dom} g \subset \operatorname{dom} h$, and $g^*(y) = \sup_z \{\langle z, y\rangle - g(z)\} < \infty$ for any finite vector y.

(14) reads

$$\begin{aligned} y^k &= \nabla h(z^k) \\ &= \mu(x^k - \frac{1}{2}e) \end{aligned} \tag{17}$$

since

$$\partial h(\overline{z}) = \{\nabla h(\overline{z})\}$$

if h is differentiable (Rockafellar, 1972). To compute (15), we note that g is proper, lower-semicontinuous, and convex. It follows (Rockafellar, 1972) that

$$\partial g^*(\overline{y}) = \{z \mid g^*(y) \geq g^*(\overline{y}) + \langle z, y - \overline{y}\rangle,\ \forall y\} \tag{18}$$

$$= \operatorname{argmax}_z\{\langle \overline{y}, z\rangle - g(z)\}\ , \tag{19}$$

which is a *convex* optimization problem. Hence, (15) reads:

$$z^{k+1} \in \operatorname{argmin}_z\{g(z) - \langle y^k, z\rangle\}$$

Inserting y^k from (17), we finally obtain by virtue of (12), (9), and (8):

Reconstruction algorithm (μ fixed)
Choose $z^0 \in \operatorname{dom} g$ arbitrary.
For $k = 0, 1, \ldots$, compute z^{k+1} as minimizer of the linear program:

$$\min_{x\in[0,1]^n, \{z_{\langle i,j\rangle}\}} \ -\left\langle e + \mu(x^k - \frac{1}{2}e), x \right\rangle + \frac{\alpha}{2} \sum_{\langle i,j\rangle} z_{\langle i,j\rangle} \tag{20}$$

$$\text{subject to} \quad Ax \leq b\,, \quad z_{\langle i,j\rangle} \geq x_i - x_j\,,\ z_{\langle i,j\rangle} \geq x_j - x_i$$

Here, $Ax \leq b$ are the original constraints from (6).

In practice, we start with $\mu = 0$ and repeat the reconstruction algorithm for increasing values of μ, starting each iteration with the previous reconstruction z^k. This outer iteration loop terminates when $\forall i,\ \min\{x_i, 1 - x_i\} < \varepsilon$.

Note that for $\mu = 0$, we minimize (5), whereas for $\mu > 0$ it pays to shift in (20) the current iterate in the direction of the negative gradient of the "binarization" functional (7). While this is an intuitively clear modification of (5), convergence of the sequence of minimizers of (20) is not obvious. Proposition 10.1, however, proves the convergence.

10.5. Experimental Results

We compare iterative linear programming (20), with the regularized best inner fit approach (5).

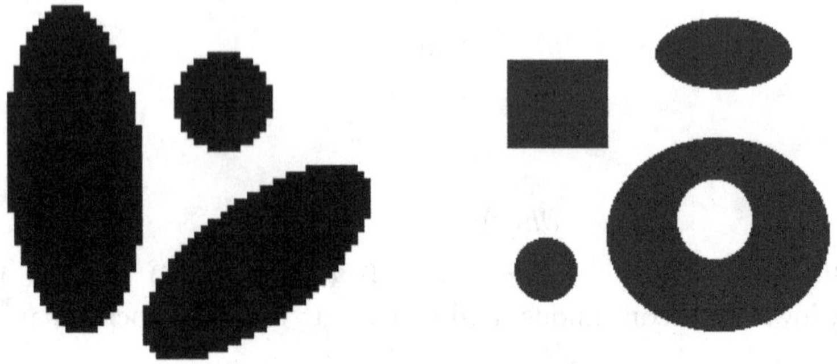

(a) Original, 64×64, exp. 1. (b) Original, 256×256, exp. 2.

Figure 10.3. From each image, 3 projections (noiseless), 0°, 45°, and 90° were taken for setting up the two reconstruction problems used in our evaluation.

For evaluation purposes, we created two reconstruction problems from the images shown in figure 10.3. From each image, three projections (noiseless) were taken, 0°, 45°, and 90°. Figure 10.4 shows the reconstruction results of the regularized best inner fit approach (5). This result illustrates that both reconstruction problems are not easy to solve due to the large area covered by the objects and the corresponding amount of self-occlusions.

Throughout all experiments, the parameter μ was initialized with 0. After each iteration μ was increased by 0.1 in the first experiment and 0.05 in the second one. Further, the regularization parameter α was chosen as 0.5 in the first and 1.0 in the second experiment.

Comparison of the results for (20) and (5) in figures 10.5 and 10.7, respectively, shows the superior performance of the approach (20). The reason is that, through iterating the linear programs, rounding is not done as a separate post-processing step, but during optimization, while taking into account the projection constraints. Figures 10.8 and 10.9 illustrate intermediate results for both reconstruction problems after different numbers of iterations. One can see how the solution converges towards a binary vector because of the increasing influence of the functional (7). Figure 10.6 further illustrates this process.

In further experiments, we tested the behavior of our approach in the presence of noise. Therefore, we added a normal distributed error, $\mu = 0.0$

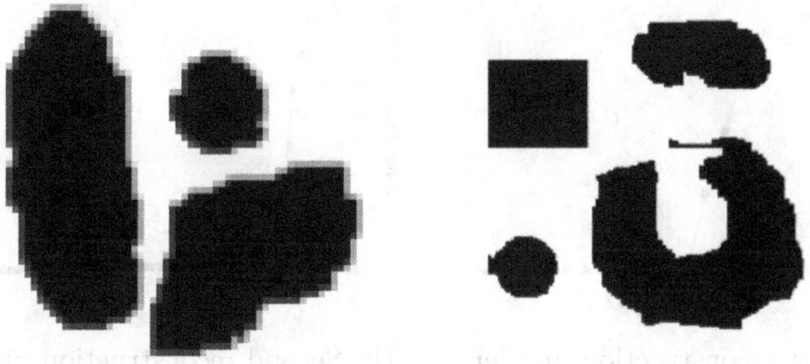

(a) Regularized BIF, exp. 1. (b) Regularized BIF, exp.2.

Figure 10.4. Results obtained by the regularized best inner fit approach (5).

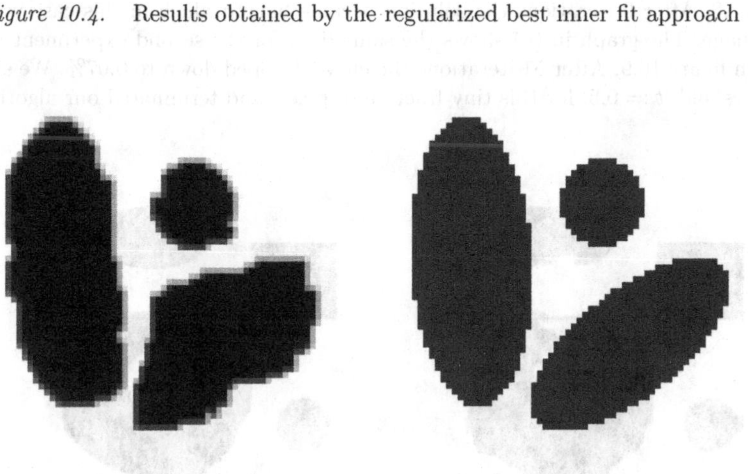

(a) Regularized BIF. (b) Iterated LPs (10 Iterations).

Figure 10.5. Comparison between the regularized best inner fit approach and our approach proposed in this paper.

and $\sigma \in \{1.0, 2.0\}$, to the projection data from the first experiment and computed reconstructions from this data. Results of these experiments are shown in the figures 10.10 and 10.11. In case of noiseless projections $\alpha = 0.25$ is a good choice. However, it turned out that in case of noisy projections it is preferable to choose a higher value of α. We checked different values of $\alpha \in \{0.25, 0.5, 0.75\}$.

Concerning computation time, a single iteration (solving one LP) of the 64×64 image costs about 7 seconds on a 3 GHz Intel Pentium 4, while it was about 4 minutes and 6 seconds for the 256×256 image.

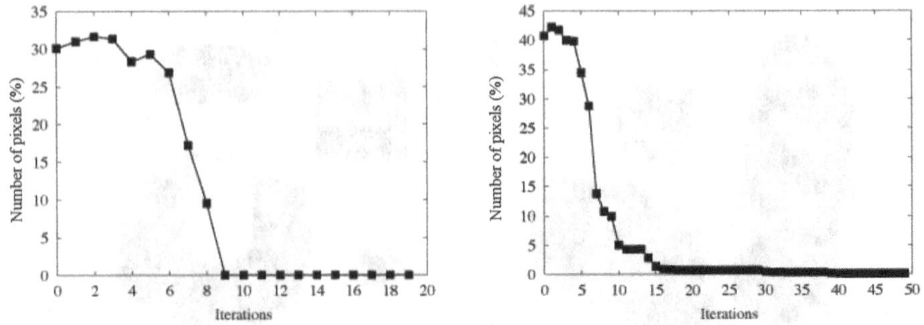

(a) First reconstruction problem. (b) Second reconstruction problem.

Figure 10.6. Both graphs show the percentage of non-binary pixels per iteration. The graph in (a) corresponds to the first reconstruction experiment and to the images shown in figure 10.8. After 9 iterations the solution became binary which in this case was the original image. The graph in (b) shows the same data for the second experiment which is shown in figure 10.9. After 51 iterations the curve dropped down to 0.07%. We simply used a threshold, $t := 0.5$, for this tiny fraction of pixels and terminated our algorithm.

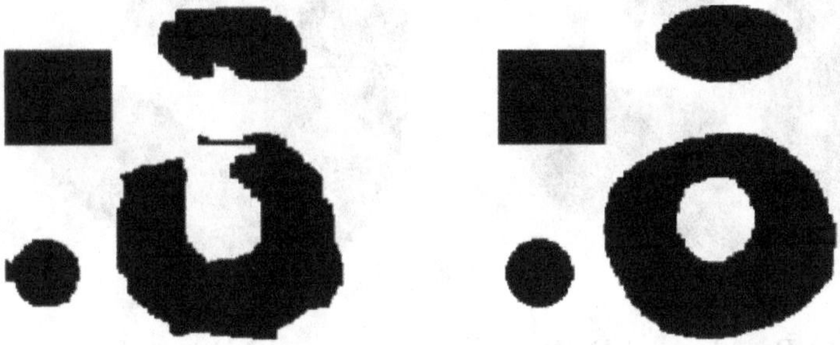

(a) Regularized BIF. (b) Iterated LPs (51 Iterations).

Figure 10.7. Comparison between regularized BIF and iterating LPs for the second experiment. We terminated the iterated LPs after 51 iterations and set the remaining non-binary pixels (0.07%) to zero in order to obtain a binary solution.

10.6. Conclusion and Further Work

In this paper we have shown a new reconstruction approach based on linear programming for the problem of discrete tomography. Unlike other LP methods, the rounding process is now explicitly done within the reconstruction process and not as a postprocessing step after the reconstruction.

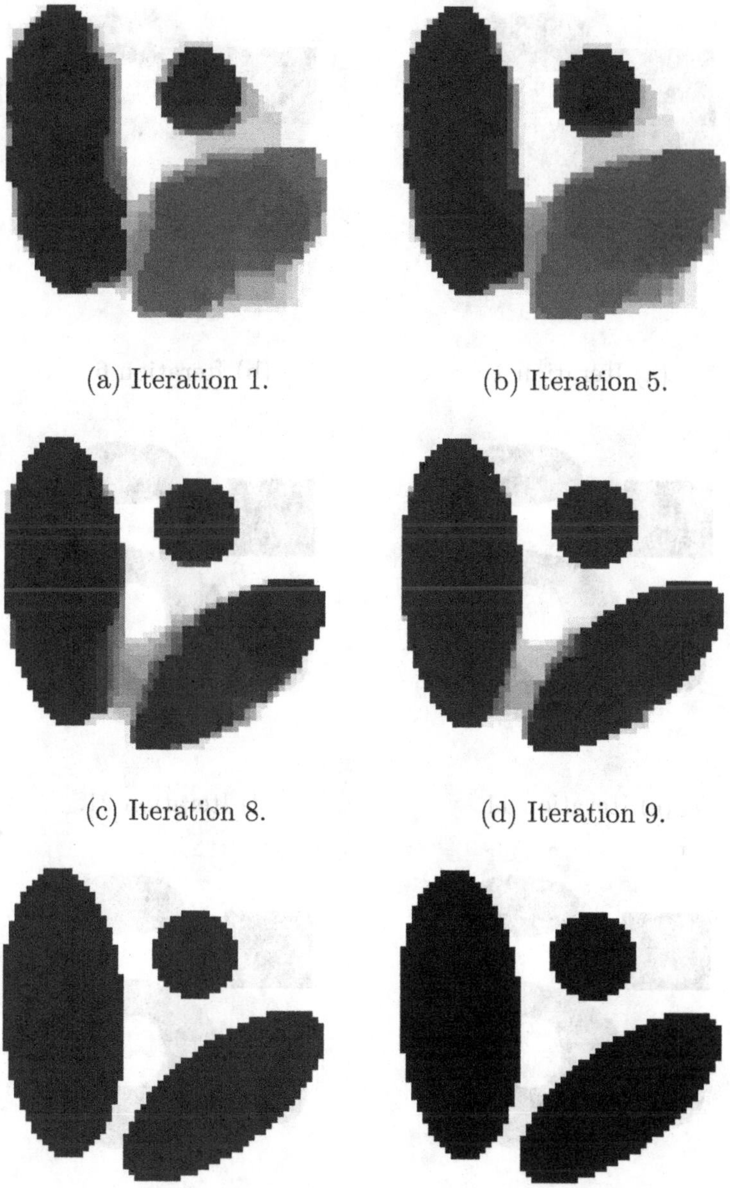

(a) Iteration 1. (b) Iteration 5.

(c) Iteration 8. (d) Iteration 9.

(e) Iteration 10. (f) Original.

Figure 10.8. (a)–(e) Results at different iterations of our proposed reconstruction method. The original image is shown in figure 10.8(f) from which three projections, 0°, 45°, and 90° were taken.

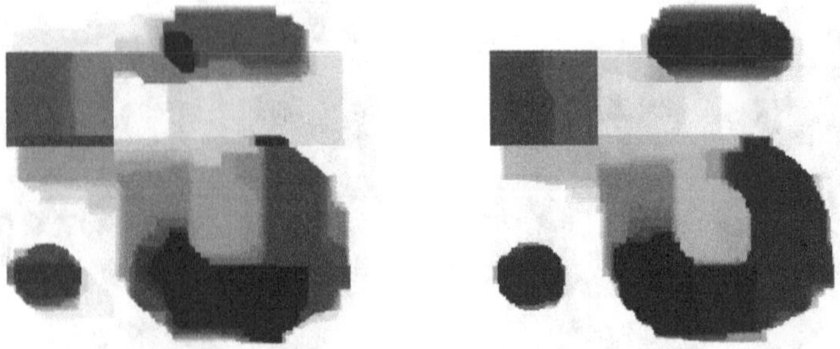

(a) Iteration 1. (b) Iteration 6.

(c) Iteration 11. (d) Iteration 21.

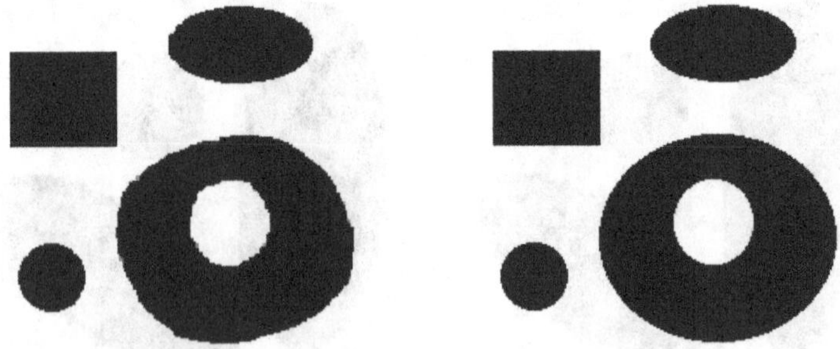

(e) Iteration 51. (f) Original.

Figure 10.9. (a)-(e) show the results at different iteration steps. The original image is shown in (f) from which three projections, 0°, 45°, and 90°, were taken.

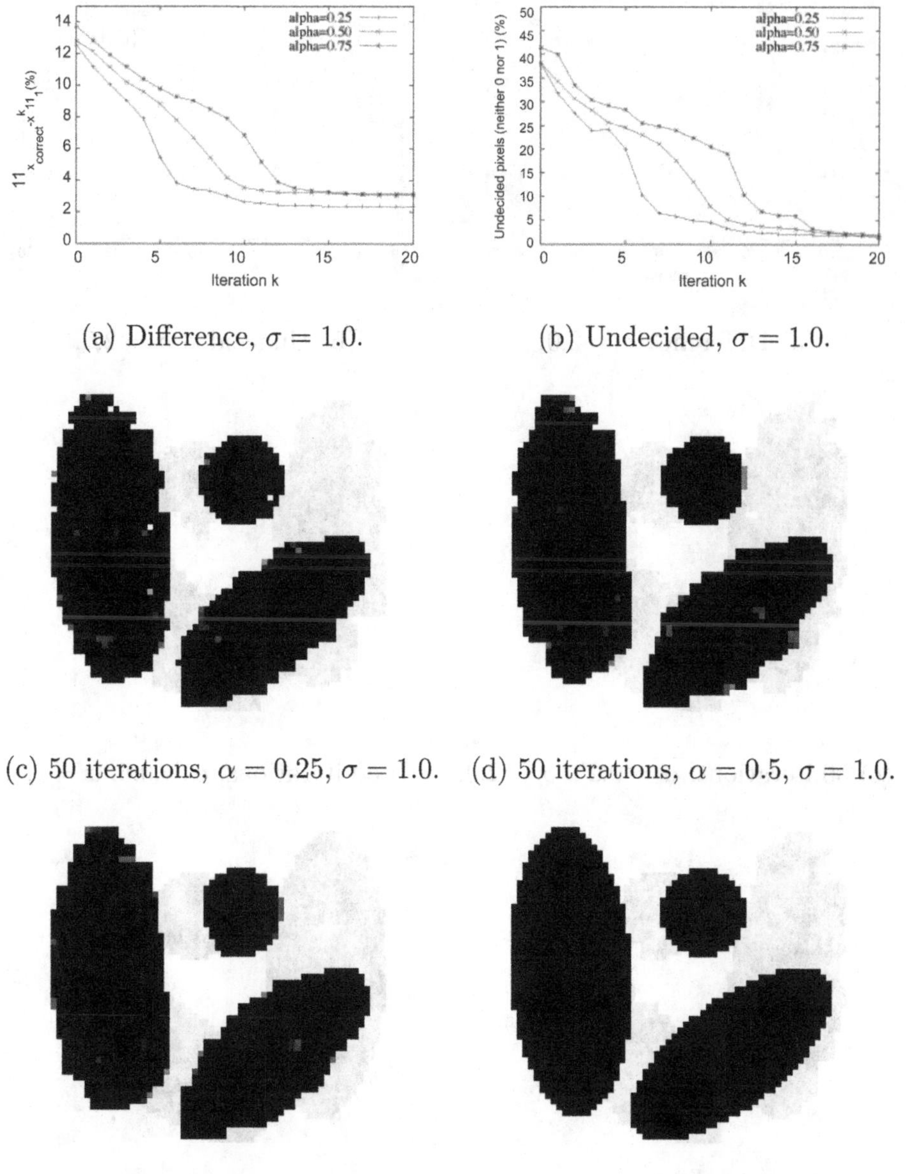

(a) Difference, $\sigma = 1.0$.

(b) Undecided, $\sigma = 1.0$.

(c) 50 iterations, $\alpha = 0.25$, $\sigma = 1.0$.

(d) 50 iterations, $\alpha = 0.5$, $\sigma = 1.0$.

(e) 50 iterations, $\alpha = 0.75$, $\sigma = 1.0$.

(f) Original image.

Figure 10.10. In order to test the behavior of our algorithm in the presence of noise we added a normal distributed ($\mu = 0.0$, $\sigma = 1.0$) error to the projection data (again 3 projections, 0°, 45°, and 90°). (a) Percentage of the absolute difference between the original image and the solutions of the first 20 iterations. (b) Percentage of the undecided pixels (neither 0 nor 1) of the first 20 iterations. (c)-(e) Reconstructions (50 iterations) with different choices of $\alpha \in \{0.25, 0.5, 0.75\}$. (f) Original image.

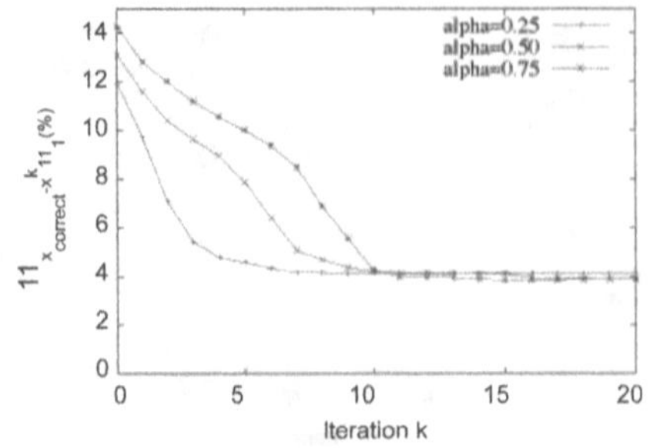

(a) Difference $\sigma = 2.0$.

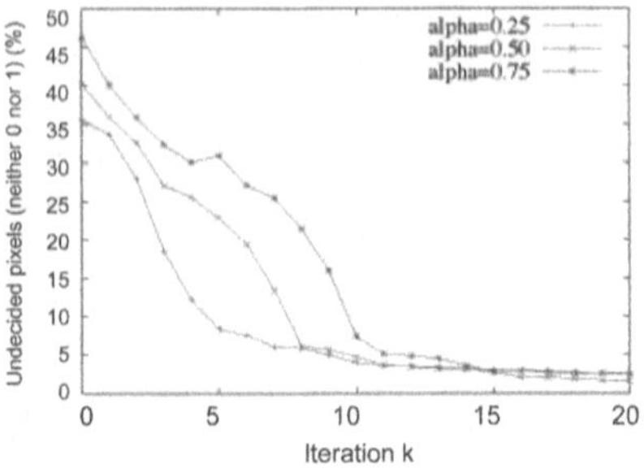

(b) Undecided $\sigma = 2.0$.

(c) 50 iterations, $\alpha = 0.25$, $\sigma = 2.0$.

(d) 50 iterations, $\alpha = 0.5$, $\sigma = 2.0$.

(e) 50 iterations, $\alpha = 0.75$, $\sigma = 2.0$.

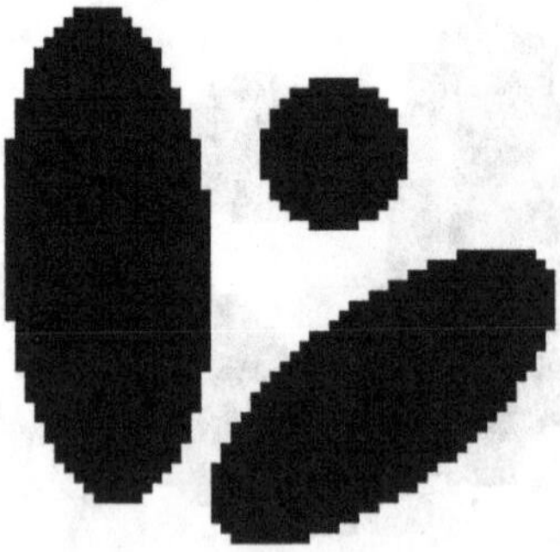

(f) Original.

Figure 10.11. Reconstructions from noisy projection data, normal distributed error ($\mu = 0.0$, $\sigma = 2.0$) error added to the projection data. Three projections, 0°, 45°, and 90°. (a) Percentage of the absolute difference between the original image and the solutions of the first 20 iterations. (b) Percentage of the undecided pixels (neither 0 nor 1) of the first 20 iterations. (c)-(e) Reconstructions (50 iterations) with different choices of $\alpha \in \{0.25, 0.5, 0.75\}$. (f) Original image.

Hence, the problem constraints of the linear program do affect the rounding. On the other hand, one has to solve a sequence of LPs instead of a single one which of course leads to more computationally effort. However, the linear programs do not differ too much from each other, as only the target vector c has to be modified. Therefore, it would be interesting to see if this can be exploited in order to speed-up computations. For instance, the decomposition of linear programs appears to be attractive in this context since the decomposition of matrix A has to be done only once and could then be used in all iterations.

References

Aharoni, R., G. T. Herman, and A. Kuba: Binary vectors partially determined by linear equation systems. *Discrete Mathematics*, **171**:1–16, 1997.

Censor, Y.: Binary steering in discrete tomography reconstruction with sequential and simultaneous iterative algorithms. *Lin. Algebra and its Appl.*, **339**:111–124, 2001.

Pham, D.-T. and L. T. Hoai An: A d.c. optimization algorithm for solving the trust-region subproblem. *SIAM J. Optim.*, **8**:476–505, 1998

Pham, D.-T., and S. Elbernoussi: Duality in d.c. (difference of convex functions) optimization subgradient methods. In *Trends in Mathematical Optimization* (K.-H. Hoffmann, editor), Int. Series of Numer. Math. 84, pages 277–292. Birkäuser, Basel, 1988.

Fishburn, P., P. Schwander, L. Shepp, and R. Vanderbei: The discrete radon transform and its approximate inversion via linear programming. *Discr. Appl. Math.*, **75**:39–61, 1997.

Gardner, R. J., and P. Gritzmann: Discrete tomography: Determination of finite sets by x-rays. *Trans. Amer. Math. Soc.*, **349**:2271–2295, 1997

Gritzmann, P., S. de Vries, and M. Wiegelmann: Approximating binary images from discrete X-rays. *SIAM J. Optimization*, **11**:522–546, 2000.

Gritzmann, P., D. Prangenberg, S. de Vries, and M. Wiegelmann: Success and failure of certain reconstruction and uniqueness algorithms in discrete tomography. *Int. J. Imag. Syst. Technol.*, **9**:101–109, 1998.

Herman, G., and A. Kuba: *Discrete Tomography: Foundations, Algorithms, and Applications.* Birkhäuser, Boston, 1999.

Kleinberg, J. M., and E. Tardos: Approximation algorithms for classification problems with pairwise relationships: Metric labeling and Markov random fields. In *IEEE Symp. Foundations of Comp. Science*, pages 14–23, 1999.

Matej, S., A. Vardi, G. T. Herman, and E. Vardi: Binary tomography using Gibbs priors. In *Discrete Tomography: Foundations, Algorithms, and Applications* (G. T. Herman and A. Kuba, editors), pages 191–212. Birkhäuser, Boston, 1999.

Natterer, F., and F. Wbbeling: *Mathematical Methods in Image Reconstruction. SIAM*, Philadelphia, 2001

Rockafellar, R. T.: *Convex Analysis.* Princeton Univ. Press, Princeton, NJ, 1972.

Weber, S., C. Schnrr, and J. Hornegger: A linear programming relaxation for binary tomography with smoothness priors. In Proc.*Int. Workshop on Combinatorial Image Analysis* (A. Del Lungo, V. Di Gesù, and A. Kuba, editors), 2003.

Weber, S., T. Schle, C. Schnrr, and J. Hornegger: A linear programming approach to limited angle 3D reconstruction from DSA projections. *Methods of Information in Medicine*, **4**:320–326, 2004.

CASCADE OF DUAL LDA OPERATORS FOR FACE RECOGNITION

WŁADYSŁAW SKARBEK, KRZYSZTOF KUCHARSKI
Institute of Radioelectronics
Warsaw University of Technology, Poland

AND MIROSŁAW BOBER
Visual Information Laboratory
Mitsubishi Electric, Guildford, United Kingdom

Abstract. In this paper, we propose a cascade of Dual-LDA (DLDA) operators for Face Recognition. We show that such an approach results in efficient and low-dimensional feature space for face representation with enhanced discriminatory power. Comparative results to classical LDA and cascade of classical LDA algorithms are presented, showing significantly improved performance. A theoretical analysis for Fisher and DLDA is also presented. Experimental evaluation of the proposed FR algorithm, conducted on MPEG test set with over 8000 images of 929 individuals, shows state-of-the-art performance.

Key words: cascade of classifiers, linear discriminant analysis, dual linear discriminant analysis, feature extraction, face recognition, face descriptor

11.1. Introduction

Face recognition (FR) is a very complex problem, mainly due to significant intra-class variations in appearance due to pose, facial expression, aging, illumination and imaging conditions, hair style, etc. Despite constant improvements in the performance and robustness of FR systems, which accelerated recently, even the most advanced systems often fail to meet the requirements of many key applications.

A great number of different approaches to face recognition have been proposed in the last two decades - a good overview of the first decade can be found in (Chellappa *et al.*, 1995). Recently, newly emerging techniques attempt to improve performance by explicitly modelling or compensating for the variations in facial appearance (Bronstein *et al.*, 2004). They use 3D head shape models and head pose estimation, texture models, and illumination models, but still data acquisition requirements limit their use to the range of very specialized applications.

R. Klette et al. (eds.), Geometric Properties for Incomplete Data, 199-219.

So far, appearance-based approaches operating directly on 2D images and processing them as holistic patterns seem to be the most successful group when dealing with standard video cameras. They usually use principal component analysis (PCA) or linear discriminant analysis (LDA) for dimensionality reduction and feature selection. LDA attempts to find a linear transformation for the learning sequence which gives the best separation for classes. The measure of separation is based on data variances in each class and for class means, too. LDA in its certain special forms is known from the 1930s and it is linked to the name of R.A. Fisher – the famous English statistician (Fisher, 1936). Most general cases of LDA were introduced when pattern recognition theory had established its position about thirty years later (compare (Devijver *et al.*, 1982)).

There is a common perception that LDA-based algorithms outperform PCA-based algorithm because they select a low-dimensionality representation by trying to maximize its discriminatory capabilities. The first application of the LDA to face recognition was presented by Belhumeur *et al.* almost 10 years ago (Belhumeur et al., 1997). Recently, it has become apparent that classical LDA has several weaknesses and that its performance depends strongly on the implementation (i.e. how the Fisher criterion is maximized). Several new realizations of the LDA algorithm with improved performance have been proposed: Direct-LDA (Yu *et al.*, 2001), fractional-step LDA (Lu *et al.*, 2003), etc.

In this paper, we introduce a novel approach to Face Recognition based on a cascade of Dual-LDA (DLDA) operators. DLDA has been recently proposed by the authors (Skarbek *et al.*, 2004) and shown to provide superior discriminatory performance compared with classical LDA. Here, we further extend the framework and introduce an algorithm which uses a cascade of DLDA operators applied to the subsets of the facial images data. The paper consists of four sections, describing the theoretical background behind LDA and DLDA, the design of our FR algorithm based on Cascade of DLDA operators, experimental evaluation of the performance using the Equal Error Rates(EER) measure, and comparison with other proposals which took part in the MPEG-7 contest over the years 2001-2002.

11.2. LDA theoretical background

The multidimensional linear discriminant analysis assumes the projection of data from $\mathbb{R}^N$ into $\mathbb{R}^r$ for certain $1 < r < N$. Two separation measures are analyzed: the original Fisher's ratio (between-class variance over within-class variance) and its inverse ratio (within-class variance over between-class variance).

11.2.1. WITHIN AND BETWEEN-CLASS VARIANCES OF VECTORS

Let $Y = [y_1, \ldots, y_L]$ be the projected data in r dimensional space, i.e. $y_i = W^t x_i$, $x_i \in \mathbb{R}^N$, $W \in \mathbb{R}^{N \times r}$, $y_i \in \mathbb{R}^r$. Let the number of elements x_i which represents class $j = 1, \ldots, J$ be L_j, i.e. $L = L_1 + \ldots + L_J$. We can identify elements in Y extracted from $j-$th class by the index set I_j.

The class separation $s(Y)$ is measured by the ratio of *between-class variance* $\mathrm{var}_b(Y)$ and *within-class variance* $\mathrm{var}_w(Y)$ ($\triangleq$ accounts for a definition):

$$s(Y) \triangleq \frac{\mathrm{var}_b(Y)}{\mathrm{var}_w(Y)} \tag{1}$$

The *unbiased vector between-class variance* is defined with use of the Euclidean norm $\|\cdot\|$ as follows :

$$\mathrm{var}_b(Y) \triangleq \frac{1}{J-1} \sum_{j=1}^{J} L_j \|\overline{y}^j - \overline{y}\|^2 \tag{2}$$

where the class vector mean $\overline{y}^j$ and grand vector mean $\overline{y}$ are:

$$\overline{y}^j \triangleq \frac{1}{L_j} \sum_{i \in I_j} y_i \overline{y} \triangleq \frac{1}{L} \sum_{i=1}^{L} y_i \tag{3}$$

The *unbiased vector within-class variance* has the form:

$$\mathrm{var}_w(Y) \triangleq \frac{1}{L-J} \sum_{j=1}^{J} \sum_{i \in I_j} \|y_i - \overline{y}^j\|^2 \tag{4}$$

Since the LDA projection transforms the point x_i onto the point $y_i = W^t x_i$, the above statistics for vector data Y can be expressed by statistics on vector data X. Within-class variance for the projected data Y depends on the input data X through the matrix $S_w(X)$ called the *within-class scatter matrix:*

$$S_w(X) = \frac{1}{L-J} \sum_{j=1}^{J} \sum_{i \in I_j} (x_i - \overline{x}^j)(x_i - \overline{x}^j)^t \tag{5}$$

$$\mathrm{var}_w(Y) = \mathrm{trace}(W^t S_w(X) W) \tag{6}$$

Similarly, between-class variance for the projected data Y depends on the input data X through the matrix $S_b(X)$ called the *between-class scatter matrix:*

$$S_b(X) = \frac{1}{J-1}\sum_{j=1}^{J} L_j(\overline{x}^j - \overline{x})(\overline{x}^j - \overline{x})^t \tag{7}$$

$$\mathrm{var}_b(Y) = \mathrm{trace}(W^t S_b(X) W) \tag{8}$$

Optimization of vector variances can be reduced to the optimization of scalar variances of $Y_k \triangleq [(y_1)_k, \ldots, (y_L)_k], k = 1, \ldots, r$ using the following formulas:

$$\mathrm{var}_w(Y) = \sum_{k=1}^{r} \mathrm{var}_w(Y_k) \tag{9}$$

$$\mathrm{var}_b(Y) = \sum_{k=1}^{r} \mathrm{var}_b(Y_k) \tag{10}$$

Therefore, the class separation measure can be expressed as follows:

$$s(Y) \triangleq \frac{\mathrm{var}_b(Y)}{\mathrm{var}_w(Y)} = \frac{\mathrm{trace}(W^t S_b W)}{\mathrm{trace}(W^t S_w W)}$$

11.2.2. WITHIN AND BETWEEN-CLASS COVARIANCE MATRICES

A natural requirement for $W = [w_1, \ldots, w_r]$ can be stated: find such projection vectors w_k, $k = 1, \ldots, r$ that the within-class variance of each projected component is one, i.e.:

$$\mathrm{var}_w(Y_k) = w_k^t S_w w_k = 1, \;\; w_k \perp \ker(S_w)$$

and each between-class variance $w_k^t S_b w_k$ is maximal for $k = 1, \ldots, r$.

However, the requirements for the best solution of such a problem should take into account mutual relationships between component variables.

We introduce now the concept of covariance to have the measure for relationships between components which are related to class context.

The *within-class covariance matrix* $cov_w(Y)$ is defined as follows:

$$cov_w(Y_k, Y_l) \triangleq \frac{1}{L-J}\sum_{j=1}^{J}\sum_{i \in I_j}((y_i)_k - (\overline{y}^j)_k)((y_i)_l - (\overline{y}^j)_l)$$

$$cov_w(Y) \triangleq [cov_w(Y_k, Y_l)]_{1 \le k,l \le r} \tag{11}$$

The *between-class covariance matrix* $cov_w(Y)$ has the form:

$$
\begin{aligned}
cov_b(Y_k, Y_l) &\triangleq \frac{1}{J-1}\sum_{j=1}^{J} L_j((\overline{y}^j)_k - (\overline{y})_k)((\overline{y}^j)_l - (\overline{y})_l) && (12)\\
cov_b(Y) &\triangleq [cov_b(Y_k, Y_l)]_{1\le k,l\le r} && (13)
\end{aligned}
$$

Basing on above definitions it can be easily proved that covariance matrices are actually the projected scatter matrices:

$$
\begin{aligned}
cov_w(Y) &= W^t S_w(X) W && (14)\\
cov_b(Y) &= W^t S_b(X) W && (15)
\end{aligned}
$$

11.2.3. MULTIDIMENSIONAL LDA PROBLEM

The natural requirement for our multidimensional projection is the decorrelation between the components of the projected data, in both the within and between class contexts

$$cov_w(Y_k, Y_l) = 0,\ cov_b(Y_k, Y_l) = 0,\ k, l = 1, \ldots, r,\ l \neq k$$

The imposed conditions for component variances and covariances can be written in the compact way:

$$W^t S_w W = I_{r\times r},\ W^t S_b W \text{ is diagonal} \qquad (16)$$

Finally, we can formulate the multidimensional LDA problem.

Given data matrix $X = [x_1, \ldots, x_L]$, $x_i \in \mathbb{R}^N$, find a projection matrix $W = [w_1, \ldots, w_r] \in \mathbb{R}^{N\times r}$, rank$(W) = r$, $w_k \perp ker(S_w)$ which performs at the same time the diagonalization of both scatter matrices S_w and S_b, makes component within-class variances equal to one, and maximizes the between-class variance $var_b(W^t X)$:

$$W^* \triangleq \arg \max_{W\perp \ker(S_w), W^t S_w W = I;\ W^t S_b W \text{ is diagonal}} \sum_{k=1}^{r} w_k^t S_b w_k \qquad (17)$$

The solution W of the problem defined by the equation Eq. 17 is found among members of the LDA models family $\mathcal{F}_r(X)$ that can be compactly characterized as follows (Skarbek *et al.*, 2003):

$$\mathcal{F}_r(X) \triangleq \{ \; W \in \mathbb{R}^{N\times r} : W = U_{q0}\Lambda_{q0}^{-1/2}V_r$$
$$S_w(X) \overset{\text{REVD}}{=} U_{q0}\Lambda_{q0}U_{q0}^t,$$
$$A \triangleq U_{q0}\Lambda_{q0}^{-1/2}, \; A^t S_b(X)A \overset{\text{REVD}}{=} V_{r0}\Sigma_{r0}V_{r0}^t \tag{18}$$
$$1 \leq r \leq r_0$$
$$\}$$

11.2.4. DUAL MULTIDIMENSIONAL LDA PROBLEM – DLDA

The dual problem minimizes the reversed goal function under the dual constrains:

$$W^t S_b W = I_{r\times r}, W^t S_w W \text{ is diagonal} \tag{19}$$

The dual LDA problem formulation is as follows.

Given data matrix $X = [x_1, \ldots, x_L]$, $x_i \in \mathbb{R}^N$, *find a projection matrix* $W = [w_1, \ldots, w_r] \in \mathbb{R}^{N\times r}$, *rank*$(W) = r$, $w_k \perp ker(S_b)$ *which performs at the same time the diagonalization of both scatter matrices* S_b *and* S_w, *makes component between-class variances equal to one, and minimizes the within-class variance* $var_w(W^tX)$:

$$W^* \triangleq \arg \min_{W\perp\ker(S_b), W^tS_bW=I;\; W^tS_wW \text{ is diagonal}} \sum_{k=1}^{r} w_k^t S_w w_k \tag{20}$$

The solution $W = AV_r$ of the problem defined by the equation Eq. 20 is found in terms of matrices A, V_r, as shown below.

Let us consider the reduced eigenvalue decomposition (REVD) for $S_b = U_{q0}\Lambda_{q0}U_{q0}^t$, where the first $q_0 = \text{rank}(S_b)$, columns in U and Λ are chosen, $\Lambda = \text{diag}(\lambda_1, \ldots, \lambda_N)$, $\lambda_1 \geq \ldots \geq \lambda_N$. Then the search space has the form:

$$\begin{aligned} \mathcal{B} &= \{a : a^t S_b a = 1, \; a \perp \ker(S_b)\} \\ &= \{A\alpha : A \triangleq U_{q0}\Lambda_{q0}^{-1/2}, \; \alpha \in \mathbb{R}^{q_0}, \|\alpha\| = 1\} \end{aligned}$$

Now, the behavior of the objective function $w^t S_w w$ can be analyzed using REVD for $A^t S_w A \triangleq V_{r0}\Sigma_{r0}V_{r0}^t$, where $r_0 = \text{rank}(A^t S_w A)$: $w^t S_w w = \alpha^t A^t S_w A\alpha = \alpha^t V_{r0}\Sigma_{r0}V_{r0}^t\alpha$.

Minimization of $w^t S_w w$ with the constraint $1 - \alpha^t\alpha = 0$ by Lagrangian multipliers leads to the stationary points $\alpha_k = v_k$ with value σ_k, $k = 1,\ldots,r_0$. Therefore, the optimal point for the goal function $f(W) = f(w_1,\ldots,w_r)$ can be combined from stationary points $w_k = Av_k$ of the quadratic form

$w^t S_w w$ for $r \leq r_0$:

$$f(W) = \operatorname{trace}(W^t S_w W) = \sum_{k=1}^{r} w_k^t S_w w_k \tag{21}$$

$$f(W) \geq \sum_{k=r_0-r+1}^{r_0} v_k^t A^t S_w A v_k = \sum_{k=r_0-r+1}^{r_0} v_k^t V_{r_0} \Sigma_{r_0} V_{r_0}^t v_k \tag{22}$$

$$= \sum_{k=r_0-r+1}^{r_0} \sigma_k \tag{23}$$

Hence, the optimal projection $W = AV_r$, $r \leq r_0$, and DLDA models can be compactly characterized dually to LDA as follows:

$$\begin{aligned} \mathcal{D}_r(X) \triangleq \{ \; & W \in \mathbb{R}^{N \times r} : \; W = U_{q0} \Lambda_{q0}^{-1/2} V_{r_0-r+1..r_0} \\ & S_b(X) \overset{\text{REVD}}{=} U_{q0} \Lambda_{q0} U_{q0}^t, \\ & A \triangleq U_{q0} \Lambda_{q0}^{-1/2}, \; A^t S_w(X) A \overset{\text{REVD}}{=} V_{r_0} \Sigma_{r_0} V_{r_0}^t \\ & 1 \leq r \leq r_0 \\ \} & \end{aligned} \tag{24}$$

where $V_{r..r'}$ denotes submatrix of matrix V consisting of columns with indices from r to r'.

If both scatter matrices S_b and S_w are nonsingular then the domains for both goal functions are equal and the DLDA problem is equivalent to the LDA problem. However, in practice scatter matrices are singular with different kernel subspaces. Therefore, in general, dual LDA will produce different results. This conclusion is clearly supported by experiments presented in Section 4, showing that DLDA finds a more compact feature subspace compared to LDA.

It appears that if W is the dual LDA projection, $Y = W^t X = [y_1, \ldots, y_L]$ and the random variable E_w is the within class error $y_k - y_l$ for randomly chosen vectors $y_k, y_l \in Y$ from the same class, i.e. $k \in I_j$ and $l \in I_j$ for certain $j = 1, \ldots, J$, we can identify variances for these error vectors without any additional computation.
Let *the covariance matrix* for within-class errors $R(E_w(Y))$ be defined as follows:

$$R(E_w(Y)) \triangleq \sum_{j=1}^{J} \frac{1}{L_j^2} \sum_{k,l \in I_j} (y_k - y_l)(y_k - y_l)^t$$

As can be proved this covariance matrix is proportional to the within-class scatter matrix $S_w(Y)$:

$$R(E_w(Y)) = 2S_w(Y) = 2W^tS_w(X)W = 2\Lambda'$$

Λ' is the diagonal matrix in the diagonalization of $S_w(Y)$.

Therefore, the distance $\delta(y_k, y_l)$ between DLDA features $y_k, y_l \in \mathbb{R}^r$, measured by likelihood of feature error $y_k - y_l$, is the weighted Euclidean distance, where component weights are inverses of eigenvalues for the covariance matrix $R(E_w(Y))$:

$$\delta(y_k, y_l) = \sum_{i=1}^{r} \frac{(y_{1,i} - y_{2,i})^2}{\lambda'_i}. \tag{25}$$

11.2.5. ALGORITHM 2SS4DLDA

The proposed algorithm is derived using the observation that REVD on scatter matrix can be replaced by singular value decomposition (SVD) working directly on appropriately normalized original data matrix X. This normalization actually computes within-class errors for original data vectors and between-class errors for group means.

In the LDA case our algorithm finds features which on average produce the maximum variance of between-class error on a good approximation of within-class errors. By the above LDA properties the algorithm 2SS4LDA (*two singular subspaces for LDA*) has been proposed (Skarbek *et al.*, 2004).

On the other hand the DLDA algorithm finds features which on average produce the minimum variance of within-class error on a good approximation of between-class error.

Algorithm 2SS4DLDA

Input:
- Data matrix $X = [x_1, \ldots, x_L], x_i \in \mathbb{R}^N$
- Class membership vector I
- Singular subspace dimensions q, r

Output:
- DLDA model $W \in \mathbb{R}^{N \times r}$
- Corrected q, r

Method: Steps 1-9:

1. Compute the global centroid c and class centroids: $C \leftarrow [c_1, \ldots, c_J]$;
2. Perform centroid shifting and normalization for data matrices X, C :

$$\text{if } i \in I_j \text{ then } y_i \leftarrow (x_i - c_j)/\sqrt{L_j}, i = 1, \ldots, L,$$

$$d_j \leftarrow (c_j - c)\sqrt{L_j/J},\ j = 1, \ldots, J$$

3. Find the singular subspace of $D = [d_1, \ldots, d_J]$ by SVD for Y obtaining $U_{q_0} \leftarrow [u_1, \ldots, u_{q_0}]$ corresponding to singular values

$$\Lambda_{q_0}^{1/2} \leftarrow \left[\sqrt{\lambda_1}, \ldots, \sqrt{\lambda_{q_0}}\right],\ q_0 \triangleq \mathrm{rank}(D)$$

4. If $q > q_0$ then $q \leftarrow q_0$; If $q < q_0$ then $U_q \leftarrow [u_1, \ldots, u_q]$ and

$$\Lambda_q^{1/2} \leftarrow \left[\sqrt{\lambda_1}, \ldots, \sqrt{\lambda_q}\right]$$

5. Compute the whitening projection matrix: $A_q \leftarrow U_q \Lambda_q^{-1/2}$;
6. Make the whitening projection for normalized data vectors:

$$y_i \leftarrow A_w^t y_i,\ i = 1, \ldots, L$$

7. Find the singular subspace of $Y \triangleq [y_1, \ldots, y_L]$ by SVD for Y obtaining $V_{r_0} \leftarrow [v_1, \ldots, v_{r_0}]$ corresponding to all positive singular values:

$$(\Lambda')_{r_0}^{1/2} \leftarrow \left[\sqrt{\lambda'_1}, \ldots, \sqrt{\lambda'_{r_0}}\right],\ r_0 \triangleq \mathrm{rank}(Y)$$

8. If $r > r_0$ then $r \leftarrow r_0$; If $r < r_0$ then $V_{(r_0-r+1)..r_0} \leftarrow [v_{r_0-r+1}, \ldots, v_{r_0}]$ and

$$(\Lambda')_{(r_0-r+1)..r_0}^{1/2} \leftarrow \left[\sqrt{\lambda'_{r_0-r+1}}, \ldots, \sqrt{\lambda'_{r_0}}\right]$$

9. Compute DLDA model, i.e. the projection matrix W: $W \leftarrow A_q V_{(r_0-r+1)..r_0}$.

The 2SS4DLDA algorithm is based on two singular value approximations applied respectively to the normalized class means data matrix and the normalized multi-class input data matrix.

In Step 5 of the algorithm the extreme points of a hyper-ellipsoid. $\mathcal{L}'_1 \triangleq \{a \in \mathbb{R}^N : a \perp \ker(S_b),\ a^t S_b a = 1\}$ are found and inserted as columns into the matrix A. With this operation the domain of the class separation function $f(W) \triangleq s(Y)$ is narrowed to make values of $f(W)$ bounded.

Next, in Step 6 optimization of $f(W)$ on its domain is reduced to the optimization on the unit sphere.

The 2SS4DLDA algorithm is controlled by subspace dimension parameters q and r. The first singular subspace of dimension q is designed for normalized class means and is used to compute new coordinates for the original data. The second singular subspace is built in this new coordinates. In a sense it is nested SVD procedure. The feature vectors are computed using r left singular vectors spanning the second singular subspace.

11.3. Cascade of operators framework

This section presents in detail our framework and illustrates how the algorithm derived in Section 2 is used to ensure convenient optimization of its performance.

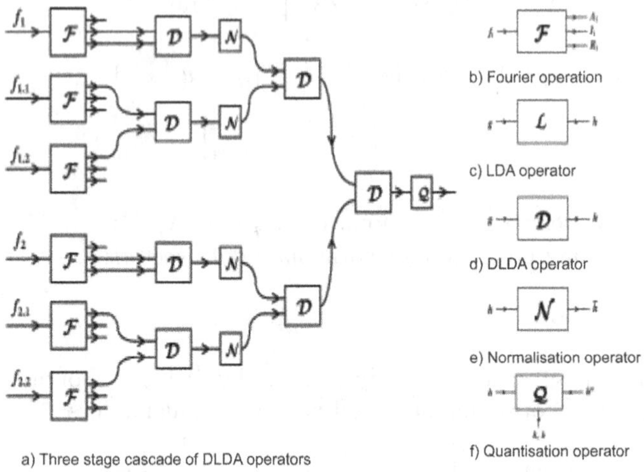

Figure 11.1. Three stage cascade of DLDA operators.

Our Face Recognition framework is shown in Figure 11.1. It forms a three stage cascade of operators with DLDA playing the main role as a tool for providing a compact set of features of great discriminative power. The cascade is a kind of complex feature extractor returning for every input facial image its corresponding descriptor.

The operators may be perceived as black boxes taking a signal on input and producing a signal on output. Each is controlled by several external parameters. Therefore, the behaviour of the whole cascade and its performance in FR applications depends on the values set for each individual operator as well as the manner they are connected in a framework.

Figure 11.2 shows which regions of a typical facial image are used at the input of the cascade. Six types of facial images may be manually distinguished: holistic one (1), upper and lower half of the holistic image (1.1, 1.2), central image (2) and analogously its upper and lower half (2.1, 2.2). This particular choice is quite intuitive and other types of image selection may be utilized as well.

It should be noted however, that the central image, by focusing on the center of the normalized image is less influenced by variations of the background or hair style. Furthermore, the use of the upper and lower component images, in combination with the cascade architecture and DLDA

Figure 11.2. Facial image parts on cascade's input.

algorithm, enables our algorithm to extract local features linked to the eye region and the mouth-chin region.

The Fourier operator presented in Figure 11.1b) transforms the image intensity input into the frequency domain using the 2D Discrete Fourier Transform. Apart from the change of representation, it also makes a selection of the coefficients included in output feature vector. The selection process controlled by two bound spatial frequencies ω_x and ω_y is illustrated in Figure 11.3. Process of finding ω_x and ω_y usually goes independently for every input facial component.

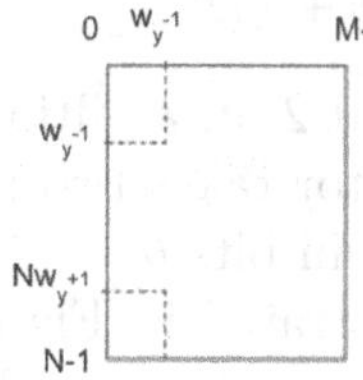

Figure 11.3. The way of selecting coefficients in the Fourier operator block.

The Fourier operator has outputs corresponding to the real parts R_i, imaginary parts I_i and amplitudes A_i of the selected spectrum coefficients. Scanning the blocks of coefficients row-wise and putting them into a vector g forms the input data for DLDA operators that follow next.

The main advantage of applying the Fourier preprocessing is increased robustness to image misalignment (vertical and horizontal shift) and consequently to pose variations. In addition, by selecting only certain frequency coefficients at this step, we can reduce problem dimensionality already at the input stage.

One can effortlessly replace DFT with any other transformation e.g. DCT or fractal operator to find out suitable input feature space for the given sort of problem.

The LDA and DLDA operators (Figure 11.1c,d)) apply respective linear transformation to the input feature vector g, the former performing extraction using 2SS4LDA algorithm and latter using 2SS4DLDA algorithm. Each algorithm is controlled by two parameters q and r setting the appropriate singular subspace dimensions and determining dimensionality of the output feature vector.

The input vector may be composed of multiple outputs of cascade previous stages which are simply merged in that case (compare Figure 11.1a)).

The normalization operator shown in Figure 11.1e) produces on output the vector of unit length $\tilde{h}$ obtained by dividing each component of the input vector h by its length. The normalization is usually needed to ensure the comparability of vectors obtained by processing inputs from different sources of the cascade i.e. different image parts.

Figure 11.1f) presents the quantization operator appearing in the last stage of cascade and converting vector component value x into discrete level using following operations:

1. thresholding: $\tilde{x} \leftarrow \frac{x}{k\sigma}$, where σ is the standard deviation of vector component and k is the threshold factor
2. clamping: if $\tilde{x} < -1$ then $\tilde{x} \leftarrow -1$ else if $\tilde{x} > 1$ then $\tilde{x} \leftarrow 1 - 2^{-2b}$
3. level assignment: $x^q \leftarrow \lfloor(\tilde{x} + 1)2^{b-1}\rfloor$

Instead of operations 1 and 2 an additional normalization may be applied. The quantization operator takes two parameters: threshold factor k and size of the face descriptor in bits b.

Although the cascade illustrated in Figure 11.1a) has exactly defined architecture, a concept of making the cascade from operator blocks allows for flexibility in using many possible configurations depending on the given FR problem properties. As it was mentioned various image partitionings and different transformation operators could be applied to tune the cascade performance. However, the process of choosing DLDA parameters has to be repeated for every change in order to obtain the best recognition quality. In the next section we present the possible solution of this task.

11.4. Results of the FR experiments

The face recognition experiments were set up according to MPEG-7 Visual Core Experiment (VCE) (Bober , 2002). 11845 gray-level photos of 1433 different persons in various poses and illumination conditions were utilized of which 3655 images of 504 persons constituted the training set and the rest i.e. 8190 images of 929 persons formed the testing set. Every facial image had size 46x56 and eyes manually located.

The distance δ between two facial images was computed using formula Eq. 25 where eigenvalues are directly obtained in Step 8 of 2SS4DLDA algorithm.

The enhanced distance δ' between the feature vector q of the query image and the image feature vector f in the database also takes into account the distance from the average feature vector $\bar{f}$ obtained for all images for the given person (excluding the query image):

$$\delta'(q,f) \triangleq \frac{\delta(q,f) + \delta(q,\bar{f})}{2} .$$

As the measure of recognition algorithm performance the Equal Error Rate (EER) was used. This is defined as a point on the Receiver Operator Characteristic (ROC) curve where false acceptance rate equals to false rejection rate and is commonly employed in face recognition experiments, specially in verification task. Other important factors affecting the usefulness of the cascade especially for the indexing applications were extraction complexity and descriptor size.

In the next subsections the results for cascades of increasing complexity will be presented starting from a single stage through the two stage case coming finally to the three stage architecture presented on the Figure 11.1. The results are presented with stress put on comparison between facial feature extractors using LDA and DLDA transformations.

In all experiments the descriptor's components are quantized to 5 bits ($b = 5$) to conform to both versions of the MPEG-7 face descriptor.

11.4.1. SINGLE STAGE CASCADE

The first experiment investigates the behaviour of the LDA and DLDA single operators applied only to the holistic image with or without the presence of the Fourier operator. Figure 11.4 illustrates a relationship between

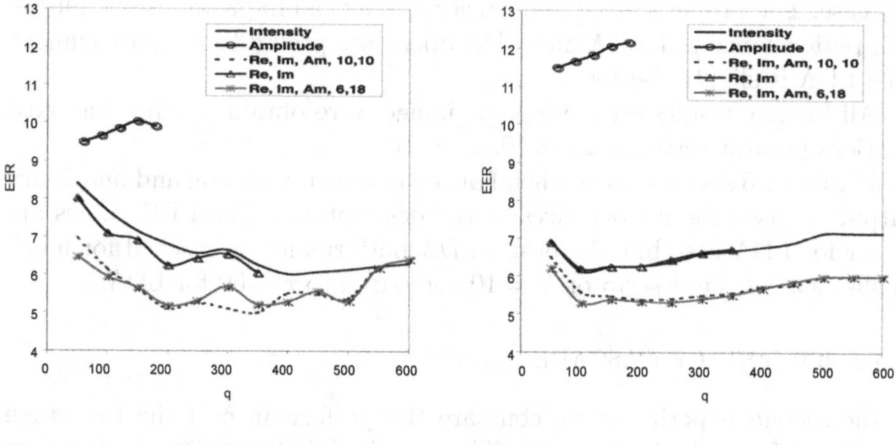

Figure 11.4. EER as function of first singular subspace dimension q for LDA (left) and DLDA (right); r=20; Two bottom lines are for different ω_x and ω_y value pairs.

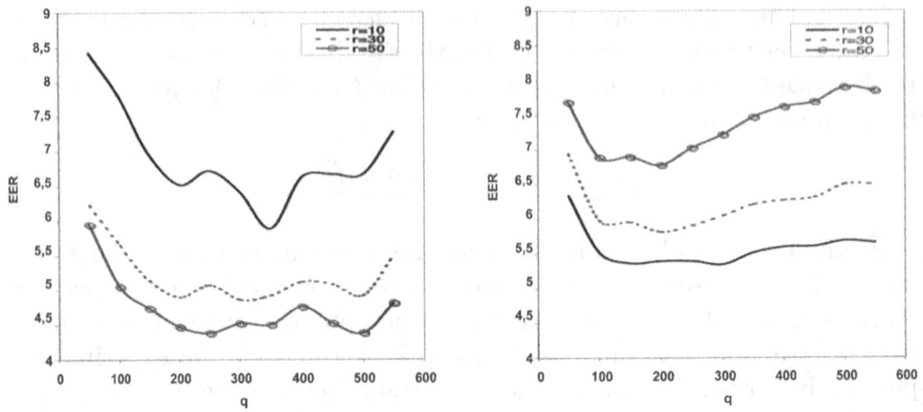

Figure 11.5. EER as function of first singular subspace dimension q for various descriptor sizes r; LDA on the left, DLDA on the right.

parameter q of the LDA/DLDA operator and the EER for the descriptor size set to 20.

The use of various combinations of Fourier outputs is examined: real and imaginary parts, amplitudes alone, all three together as well as the influence of the Fourier operator's parameters ω_x, ω_y (i.e. bound spatial frequencies) on the recognition performance.

It can be seen that adding a Fourier operator improves the performance, but also results in decreasing of the extraction complexity due to using only selected coefficients. In the situation considered, the size of the Fourier features vector is at most 570 while the intensity vector is 2576 components large.

The amplitudes alone show very poor performance, but together with real and imaginary parts of the spectrum outperform noticeably the intensity case. The proper spatial frequencies setting can improve results a little. The performance of DLDA and LDA operators proves to be quite similar, even LDA is slightly better.

All further results on the holistic image were obtained using the fixed Fourier operator settings: $\omega_x = 10$, $\omega_y = 10$.

Figure 11.5 shows results when Fourier operator with real and imaginary outputs is used for various sizes of the descriptor r. The EER values are better for LDA case but the best DLDA performance is reached for much smaller size of the descriptor; $r = 10$ compared to $r = 50$ for LDA.

11.4.2. TWO STAGE CASCADE

In the second experiment we compare the performance of the two stage cascade with the single stage case. The meaningful illustration of advantage such an approach is presented in Figure 11.6b). As a result of processing e real, imaginary parts and amplitudes through separate DLDA operators d then merging them using the DLDA operator on the second stage we noticeably smaller EER values than merging them to form an input of e single DLDA operator (Figure 11.4b)).

Better results can be achieved when the amplitudes from the holistic age are replaced by the merged amplitudes obtained from Fourier opera- s applied separately to the upper and lower halves of the holistic image able 11.1, $\omega_x = 11$, $\omega_y = 9$). EER performance of the LDA/DLDA single ge operating only on these merged amplitudes is presented in Figure 7; it is very low compared to the results shown in Figure 11.4. That t proves that such a partitioning of the facial image provides additional ormation from the discriminatory point of view.

From Table 11.1 we also see that combining optimal DLDA operators as as the single stage is concerned does not usually mean the best result for stage cascade. This may happen due to low-dimensional inputs to the DA operator on the second stage. They simply provide too few features this DLDA operator to find their better combinations. Significantly, the proving properties of the cascade can be noted only for DLDA operators mpare Figure 11.5a) and 11.6a)).

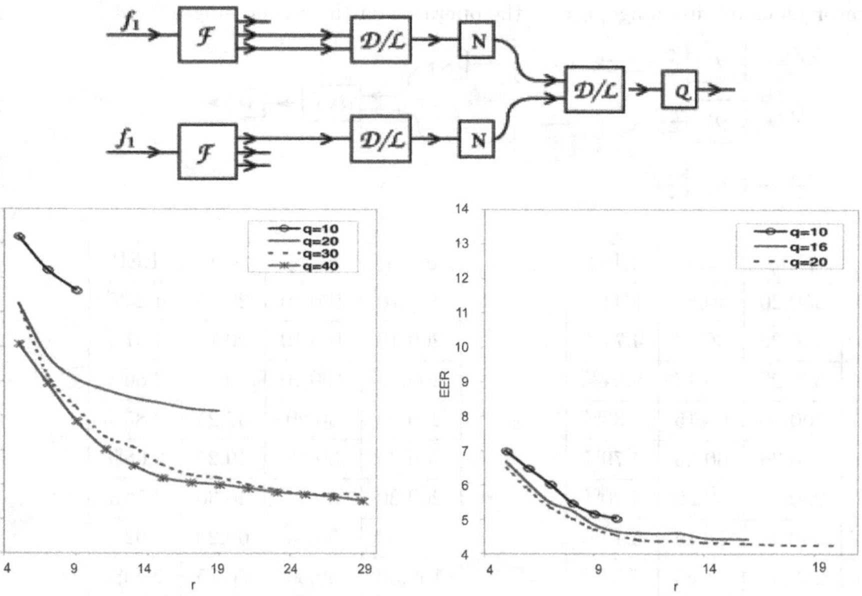

ure 11.6. EER as function of the descriptor size r for various values of first singular space dimension q for the given cascade architecture (top); LDA on the left, DLDA he right.

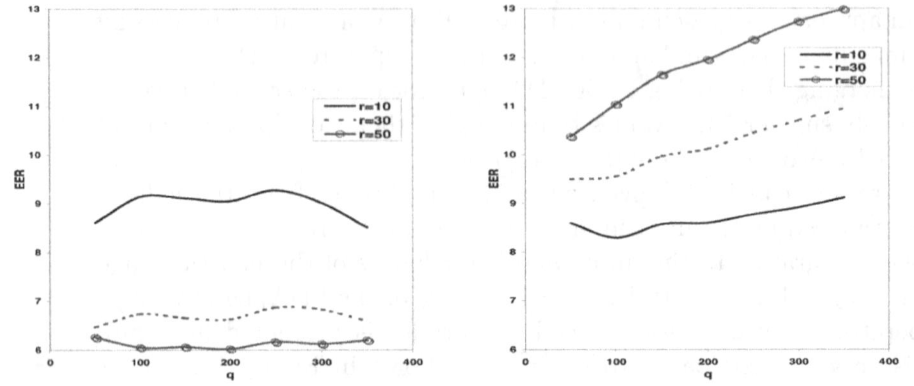

Figure 11.7. EER as function of first singular subspace dimension q for various descriptor sizes r for $f_{1.1}$ and $f_{1.2}$ input images; LDA on the left, DLDA on the right.

Table 11.2 contains analogous results for the two stage cascade acting on the central image and its upper and lower half. These are substantially

TABLE 11.1. Selected best results for two stage cascade (top) of LDA (bottom left) and DLDA (bottom right) operators on holistic image. The pair q_1, r_1 corresponds to upper operator block on first stage; q, r to the operator on the second stage.

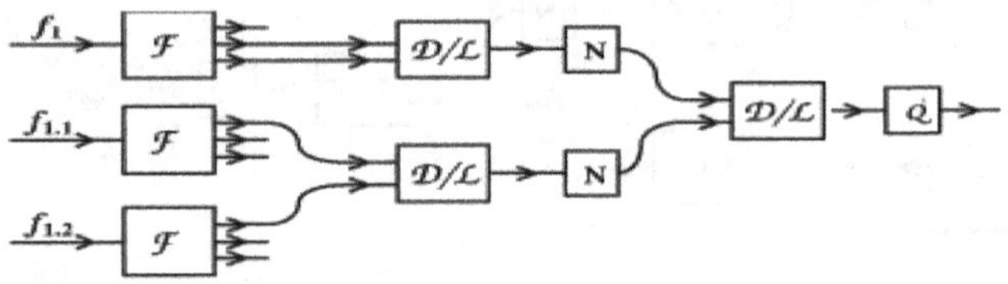

q_1, r_1	q_2, r_2	q, r	EER
350,20	350,20	40,13	5.41%
350,20	350,20	36,19	4.74%
350,20	350,20	40,17	8.04%
350,30	200,30	60,15	5.33%
350,30	200,30	60,25	4.79%
350,30	200,30	60,20	4.90%
350,40	200,40	75,15	5.13%
350,40	200,40	75,25	4.52%
350,40	200,40	75,20	4.71%

q_1, r_1	q_2, r_2	q, r	EER
200,10	100,10	20,10	4.23%
200,10	100,10	20,12	4.07%
200,10	100,10	20,17	3.86%
200,20	50,20	37,25	2.85%
200,20	50,20	40,25	2.68%
200,20	50,20	40,30	2.58%
150,30	50,30	60,20	2.92%
150,30	50,30	60,35	2.46%
150,30	50,30	60,40	2.35%

worse than in the holistic image case due to a smaller amount of information provided on the input of the cascade.

TABLE 11.2. Selected best results for two stage cascade (top) of LDA (bottom left) and DLDA (bottom right) operators on central image. The pair q_1, r_1 corresponds to upper operator block on first stage; q, r to the operator on the second stage.

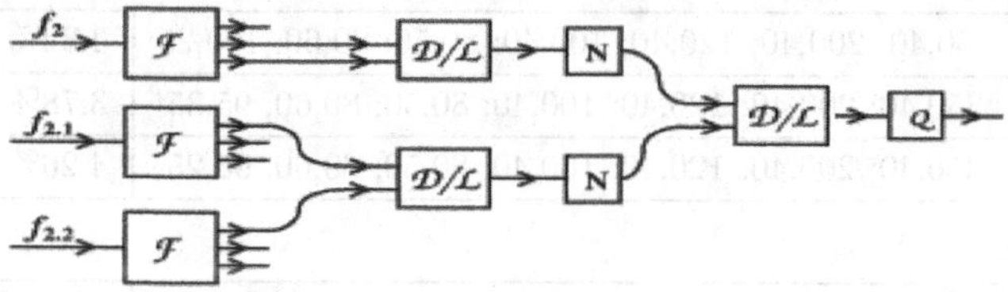

q_1, r_1	q_2, r_2	q, r	EER
120,30	80,30	57,30	4.39%
120,30	80,30	60,30	4.30%
120,30	80,30	60,20	4.66%
120,40	100,40	80,30	4.32%
120,40	100,40	80,20	4.59%
120,40	100,40	80,25	4.39%

q_1, r_1	q_2, r_2	q, r	EER
60,20	40,20	40,20	3.94%
60,20	40,20	40,25	3.81%
60,20	40,20	40,30	3.72%
60,30	40,30	60,25	3.88%
60,30	40,30	60,35	3.57%
60,30	40,30	60,30	3.71%

11.4.3. THREE STAGE CASCADE

In the third experiment we follow with the optimization making use of the best results found among two stage cascades analyzed. In Table 11.3 the best selected results are summarized.

The last entry contains the best result found not by extending the best cascades as we do in this section but by exhaustive search of an arbitrary chosen subset of the available parameters. It is an important observation showing that the full optimization of the cascade can only be done by checking every combination of cascade parameters, for both the DLDA and Fourier operators.

Nonetheless, the improvement in terms of EER with respect to single stage DLDA operator is about 250%.

11.5. Performance comparison

The MPEG-7 Video Experts Group set up in 2000 (WG11, 2000) a experiments framework to encourage researchers to develop new methods of face recognition. In their call for proposals the stress was put on face

TABLE 11.3. Selected best results for three stage cascade of LDA (top) and DLDA (bottom) operators. The parameter sequence corresponds to top-to-bottom and left-to-right layout of the cascade as seen in Figure 11.1.

parameter sequence	EER
350,40; 200,40; 120,40; 100,40; 80,50; 80,60; 95,20	4.56%
350,40; 200,40; 120,40; 100,40; 80,50; 80,60; 100,25	4.32%
350,40; 200,40; 120,40; 100,40; 80,50; 80,60; 95,35	3.78%
350,40; 200,40; 120,40; 100,40; 80,50; 80,60; 95,25	4.26%

parameter sequence	EER
200,20; 50,20; 60,30; 40,30; 40,30; 60,40; 70,45	2.04%
200,20; 50,20; 60,30; 40,30; 40,30; 60,40; 70,30	2.18%
200,20; 50,20; 60,30; 40,30; 40,30; 60,40; 70,40	2.07%
200,20; 50,20; 60,30; 40,30; 40,30; 60,40; 70,35	2.11%
200,20; 100,80; 120,30; 70,70; 100,44; 100,54; 98,48	1.90%

retrieval (FIR) accuracy, both matching and feature extraction complexity, and face descriptor's size. In (Bober , 2002) the person identification (PID) accuracy was added as the next objective as well as strict rules for the core experiments were settled. These rules considering facial data used are described in Section 11.4.

The Average Normalized Modified Retrieval Rank (ANMRR) was used as the FIR accuracy measure (WG11, 2000). In the PID experiment the Average Success Rate (ASR) was used (Bober , 2002).

To fully test proposed solutions against various pose and illumination conditions this experiment was divided into seven parts denoted PID1,...,PID7.

Altough many different proposals were submitted during the contest, including number of various extensions of the Principal Component Analysis (e.g. (Skarbek, 2002)), Independent Component Analysis ((Hyunwoo *et al.*, 2002)) and 2D Hidden Markov Models ((Nefian *et al.*, 2001)) the Linear Discriminant Analysis (LDA) applied to the selected set of Fourier features ((Tae-Kyun *et al.*, 2002), (Kamei *et al.*, 2002), (Kamei *et al.*, 2002)) proved to be the most successful approach.

In the Table 11.4 the best proposals in terms of ANMRR and ASR are summarized where the Advanced Face Recognition (AFR) (Kamei *et al.*, 2002) descriptor is a result of combining the first two proposals. It has two versions, mandatory (mAFR) that became a part of standard, and extended (eAFR) taking additionally into consideration the intensity information as well as applying pose compensation to the input facial image.

The concept of making use of discriminative information from various parts of image inspired us to elaborate the more generic and flexible solution that was presented in this paper. The results placed in the last row of the Table 11.4 were obtained using the best three stage cascade architecture found in process of optimization described in Section 11.4.3 (compare Table 11.3).

TABLE 11.4. FIR and PID accuracy for selected best proposals in MPEG-7 contest compared to best 3 stage cascade of DLDA operators (cDLDA)

	FIR	PID1	PID2	PID3	PID4	PID5	PID6	PID7
Kamai *et al.*	0.39	0.94	0.81	0.61	0.82	0.62	0.94	0.82
Tae-Kyun *et al.*	0.35	0.97	0.81	0.57	0.92	0.58	0.96	0.79
mAFR	0.33	0.97	0.83	0.64	0.94	0.65	0.96	0.81
eAFR	0.27	0.98	0.88	0.74	0.94	0.74	0.98	0.86
cDLDA	0.31	0.95	0.85	0.67	0.89	0.68	0.95	0.8

The computational complexity of all these solutions is determined by the extraction step i.e. the sequence of operations required to obtain a face descriptor from a single intensity image given the pre-computed model. In the MPEG-7 contest the complexity was expressed in terms of multiplications and additions number. The approximate results comparing both versions of AFR descriptor and our best three stage cascade architecture are given in the Table 11.5.

TABLE 11.5. Extraction complexity for AFR descriptor in comparison to best 3 stage cascade of DLDA operators

	mAFR	eAFR	best 3 stage DLDA cascade
multiplications	291K	952K	131K
additions	291K	952K	131K
size in bits	240	640	240

11.6. Conclusions

In the paper we have proposed a new Face Recognition framework which uses a multi-layer cascade of DLDA operators. DLDA is shown to give a very concise and efficient feature representation, compared to classical LDA approaches. By appropriately cascading DLDA operators, further significant gains in performance can be achieved. Clearly, the cascade architecture is superior in extracting much of the discriminant information present in facial images.

We have also shown how to design an FR algorithm using the proposed framework, analyzing its behaviour and showing performance at each stage of the cascade.

The proposed algorithm uses a Fourier-based representation applied to six regions of the facial image, which are fed into a three-layer cascade of DLDA operators. Extensive experiments on a collection of facial databases, performed according to the protocols defined by the MPEG-7 group, show that the proposed algorithm outperforms single stage DLDA and matches the performance of the state-of-the-art FR algorithms.

References

Belhumeur, P. N., J. P. Hespanha, and D. J. Kriegman: Eigenfaces vs. fisherfaces: recognition using class specific linear projection. *IEEE Trans. Pattern Analysis Machine Intelligence*, **19**:711–720, 1997.

Bober, M.: Description of MPEG-7 visual core experiments. ISO/IEC JTC1/SC29/WG11, report N4749, Klagenfurt, 2002.

Call for proposals for face recognition technology. ISO/IEC JTC1/SC29/WG11, report N3676, La Baule, 2000.

Skarbek, W.: Extending face recognition descriptor using hierarchical fuzzy local PCA. ISO/IEC JTC1/SC29/WG11, proposal M9022, Shanghai, 2002.

Nefian, A. and B. Davies: Standard support for automatic face recognition. ISO/IEC JTC1/SC29/WG11, proposal M7251, Sydney, 2001.

Hyunwoo, K., K. Tae-Kyun, L. Jong Ha, H. Wonjun, and K. Seok Cheol: Component-based 2nd order ICA face descriptor. ISO/IEC JTC1/SC29/WG11, proposal M7914, Jeju Island, 2002.

Kamei, T. and A. Yamada: Proposal of a face recognition descriptor based on Fourier spectral. ISO/IEC JTC1/SC29/WG11, proposal M7953, Jeju Island, 2002.

Hyunwoo, K., K. Tae-Kyun, H. Wonjun, and K. Seok Cheol: Component-based LDA face descriptor. ISO/IEC JTC1/SC29/WG11, proposal M8243, Fairfax, 2002.

Kamei, T., A. Yamada, K. Hyunwoo, K. Tae-Kyun, H. Wonjun H, K. Seok Cheol: Advanced face descriptor using Fourier and intensity LDA features. ISO/IEC JTC1/SC29/WG11, proposal M8998, Shanghai, 2002.

Chellappa, R., C. L. Wilson, and S. Sirohey: Human and machine recognition of faces: A survey. *Proc. IEEE*, **83**:705–740, 1995.

Devijver, P. A., and J. Kittler: *Pattern Recognition: A Statistical Approach.* Prentice Hall, Englewood Cliffs, N.J., 1982.

Fisher, R. A.: The use of multiple measurements in taxonomic problems. *Annals of Eugenics*, **7**:179–188, 1936.

Lu, J., K. N. Plataniotis, and A. N. Venetsanopoulos: Face recognition using LDA based algorithms. *IEEE Trans. Neural Networks*, **14**:117–126, 2003.

Skarbek, W., M. Bober, and K. Kucharski: Face recognition by fisher and scatter linear discriminant analysis. In Proc. *Computer Analysis Images Patterns* (N.Petkov and M.A. Westenberg, editors), LNCS 2756, pages 638–645, Springer, Berlin, 2003.

Skarbek, W., K. Kucharski, and M. Bober: Dual linear discriminant analysis for face recognition. *Fundamenta Informaticae*, **61**:303–334, 2004.

Bronstein, A. M., M. M. Bronstein, A. Spiro, and R. Kimmel: Face recognition from facial surface metric. In Proc. *Europ. Conf. Computer Vision* Part II (T. Pajdla and J. Matas, editors), LNCS 3022, pages 225–237, Springer, Berlin, 2004.

Yu, H. and H. Yang: A direct LDA algorithm for high-dimensional data - with application to face recognition. *Pattern Recognition*, **34**:2067–2070, 2001.

PRECISION OF GEOMETRIC MOMENTS IN PICTURE ANALYSIS

MARTIN N. HUXLEY
School of Mathematics, Cardiff University
23 Senghennydd Road, Cardiff CF 24 4YH, United Kingdom

REINHARD KLETTE
CITR, University of Auckland
Tamaki Campus, Building 731, Auckland, New Zealand

JOVIŠA ŽUNIĆ*
Computer Science Department, Exeter University
Harrison Building, Exeter EX4 4QF, United Kingdom

Abstract. Moments play an important role in picture analysis, pattern recognition, or classification. The presentation of Euclidean sets by digital pictures causes an inherent loss of information. There are infinitely many different preimages with identical corresponding digital pictures. The problem we discuss here is how efficiently real geometric moments of planar convex shapes can be reconstructed from their digital pictures in dependency on the applied grid resolution and order of considered moments.

Key words: digital shapes, moments, discrete moments

12.1. Introduction

Moments have been widely used in shape recognition and identification. The moment-concept has been introduced by Hu (Hu, 1962) into picture analysis. Since then a variety of new moment-types and moment-based methods has been developed and used. We mention a few of them: object recognition (Dudani *et al.*, 1977), reconstruction of geometric properties of regions (Jain *et al.*, 1995), and motion analysis (Pei *et al.*, 1994).

These methods and applications are the reason for the ongoing strong interest in moment calculations. In general, the (a, b)-*moment*, denoted by

* J. Žunić is also with the Mathematical Institute, Serbian Academy of Sciences and Arts, Belgrade.

R. Klette et al. (eds.), Geometric Properties for Incomplete Data, 221-235.

$m_{a,b}(S)$, of a planar set S is defined by

$$m_{a,b}(S) = \iint_S x^a y^b \, dx \, dy \, .$$

The moment $m_{a,b}(S)$ has the order $a+b$. The moments $m_{a,b}$ are also called *geometric moments*. In applications of picture analysis and pattern recognition real objects are acquired as binary pictures at a given resolution. In the manifold of different digitization models, we specify that for a set S its digitization is defined to be the union of all grid squares (i.e., pixels in the grid cell model) whose centers belong to S, which is called *Gauss digitization* in (Klette *et al.*, 2004).

Formally speaking, if the centers of pixels are presented as points in the integer grid and if $\mathbf{r}$ denotes the picture resolution (i.e., the number of pixels per measure unit), then the set

$$\mathbf{D}(\mathbf{r} \cdot S) = \{(i,j) \mid (i,j) \in \mathbf{r} \cdot S \cap \mathbf{Z}^2\} \tag{1}$$

is the *binary picture* of set S (also called *shape*) on a regular orthogonal grid having resolution $\mathbf{r}$. As usual, $\mathbf{r} \cdot S = \{(\mathbf{r} \cdot x, \mathbf{r} \cdot y) \mid (x,y) \in S\}$ is the dilation of S by the factor $\mathbf{r}$. For $\mathbf{r} = 1$, $\mathbf{D}(S) = \{(i,j) \mid (i,j) \in S \cap \mathbf{Z}^2\}$ is the digitization on the integer grid (i.e., one pixel per measure unit).

In picture analysis or pattern recognition, the exact values of moments $m_{a,b}(S)$ remain unknown. Therefore it seems appropriate (in accordance with (1)) to approximate $m_{a,b}(S)$ as follows:

$$\begin{aligned} m_{a,b}(S) &= \frac{1}{\mathbf{r}^{a+b+2}} \cdot m_{a,b}(\mathbf{r} \cdot S) \\ &= \frac{1}{\mathbf{r}^{a+b+2}} \cdot \iint_{\mathbf{r} \cdot S} x^a y^b \, dx \, dy \approx \frac{1}{\mathbf{r}^{a+b+2}} \cdot \sum_{(i,j) \in \mathbf{D}(\mathbf{r} \cdot S)} i^a \cdot j^b \end{aligned} \tag{2}$$

Throughout the article, for a planar set S and natural numbers a and b, the corresponding *discrete moment* of S is denoted as $\mu_{a,b}(\mathbf{D}(S))$ and defined by

$$\mu_{a,b}(\mathbf{D}(S)) = \sum_{(i,j) \in S \cap \mathbf{Z}^2} i^a \cdot j^b$$

The order of $\mu_{a,b}(\mathbf{D}(S))$ is $a+b$ and $\mu_{0,0}(\mathbf{D}(S)) = \#\mathbf{D}(S)$.

In this article we assume that grid points have nonnegative coordinates, and that the origin is placed in the lower left corner of the considered array (i.e., a set of pixels carrying a picture). Under these assumptions, if a shape S is given then $\mu_{a,b}(\mathbf{D}(S))$ equals the number of integer points inside of the 3D body $B_{a,b}(S)$ defined as

$$B_{a,b}(S) = \{(x,y,z) \mid (x,y) \in S, \; 0 < z \le x^a \cdot y^b\} \tag{3}$$

In other words,

$$\mu_{a,b}(\mathbf{D}(S)) = \#\left(B_{a,b}(S) \cap \mathbf{Z}^3\right) \tag{4}$$

The article is focused on the error in the approximation $m_{a,b}(S) \approx \mathbf{r}^{-(a+b+2)} \cdot \mu_{a,b}(\mathbf{D}(\mathbf{r} \cdot S))$, where real moments are estimated by the corresponding discrete moments. This problem is equivalent (see (2)) to the study of the order of magnitude of

$$|m_{a,b}(\mathbf{r} \cdot S) - \mu_{a,b}(\mathbf{D}(\mathbf{r} \cdot S))| \tag{5}$$

This article deals with planar convex shapes only, but due to the definition of geometric moments, our results can also be extended to sets which are unions, intersections and set differences of a finite number of convex sets.

In the rest of this article it will always be assumed that the studied shapes are bounded by "piecewise smooth" curves. To be precise, by a piecewise smooth curve C we mean that C is composed of finitely many pieces C_i which can be given by an equation $y = \phi(x)$ or $x = \theta(y)$, and the functions $\phi(x)$ and $\theta(y)$ have at least continuous derivatives up to the order of three.

Throughout the article we assume that the following smoothness conditions are satisfied on a curve C bounding a given shape S:

(1) The radius of curvature ρ and its derivative $\frac{d\rho}{d\psi}$ exist on each piece C_i and are continuous functions of ψ on C_i.
(2) On each piece C_i, the radius of curvature ρ has a maximum value and a non-zero minimum value.
(3) On each piece C_i, the radius of curvature has a bounded number of local maxima and minima.

12.2. Related Results

The number of lattice points inside of convex bodies is intensively studied in number theory ((Davenport, 1951; Krätzel, 1981)). Taking into account (4), a direct application of Davenport's result (Davenport, 1951) (for the case of sets specified here) says that

$$|m_{a,b}(\mathbf{r} \cdot S) - \mu_{a,b}(\mathbf{r} \cdot S)|$$

is upper bounded by 1 plus the total sum of projections of $B_{a,b}(\mathbf{r} \cdot S)$ onto the xy-plane, xz-plane, yz-plane, x-axis, y-axis, and z-axis. In other words, we have

$$|m_{a,b}(\mathbf{r} \cdot S) - \mu_{a,b}(\mathbf{r} \cdot S)| = \left|m_{a,b}(\mathbf{r} \cdot S) - \#\left(B_{a,b}(\mathbf{r} \cdot S) \cap \mathbf{Z}^3\right)\right|$$

$$\leq \int_0^{\mathbf{r}\cdot x_{max}} (\mathbf{r}\cdot y_{max})^b \cdot x^a \cdot dx + \int_0^{\mathbf{r}\cdot y_{max}} (\mathbf{r}\cdot x_{max})^a \cdot y^b \cdot dy$$

$$+ \quad \mathbf{r}^2 \cdot x_{max} \cdot y_{max} + x_{max}^a \cdot y_{max}^b \cdot \mathbf{r}^{a+b} + \mathbf{r}\cdot x_{max} + \mathbf{r}\cdot y_{max} + 1$$

$$= \left(\frac{x_{max}^{a+1} \cdot y_{max}^b}{a+1} + \frac{x_{max}^a \cdot y_{max}^{b+1}}{b+1} \right) \cdot \mathbf{r}^{a+b+1} + \mathbf{r}^2 \cdot x_{max} \cdot y_{max}$$

$$+ x_{max}^a \cdot y_{max}^b \cdot \mathbf{r}^{a+b} + (x_{max} + y_{max}) \cdot \mathbf{r} + 1 \qquad (6)$$

where $x_{min} = \min\{x \mid (x,y) \in S\}$, $x_{max} = \max\{x \mid (x,y) \in S\}$, $y_{min} = \min\{y \mid (x,y) \in S\}$, and $y_{max} = \max\{y \mid (x,y) \in S\}$. If $a+b$ is bounded ($a, b \geq 0$), and the set S is fixed, we have

$$|m_{a,b}(\mathbf{r}\cdot S) - \mu_{a,b}(\mathbf{r}\cdot S)| = O\left(\mathbf{r}^{a+b+1}\right). \qquad (7)$$

Moments having an order of up to two, and limitations in estimating basic geometric features from corresponding digital pictures, are studied in (Klette *et al.*, 1999). The results have been extended in (Dudani *et al.*, 2000) to moments of arbitrary bounded order.

In this article we will demonstrate how the use of Huxley's result (see Theorem 2) can lead to an estimate for (5) which is better than the estimate (7) if applied to convex sets which satisfy some smoothness conditions (which are defined in a way to be not very restrictive) on their boundaries.

12.3. The Number of Grid Points inside a Closed Curve

Let C be a piecewise smooth convex closed curve in the plane. Generally speaking, a lattice (or grid) L in the plane is the set of all points with vectors $m \cdot \tilde{\omega}_1 + n \cdot \tilde{\omega}_2$, where $\tilde{\omega}_1$ and $\tilde{\omega}_2$ are linearly independent vectors, and m and n run through all integers, positive, negative or zero. The points of L are called lattice points. We are mainly interested in the case

$$\tilde{\omega}_1 = \left(\frac{1}{\mathbf{r}}, 0\right) \qquad \text{and} \qquad \tilde{\omega}_2 = \left(0, \frac{1}{\mathbf{r}}\right)$$

where $\mathbf{r}$ is large and positive. Let $f(x,y)$ be a weighting function. We want to evaluate the sum $\sum f(m,n)$ over points $m \cdot \tilde{\omega}_1 + n \cdot \tilde{\omega}_2$ of L which lie inside or on the curve C.

The tangent angle ψ is defined by the ratios

$$\cos\psi \,:\, \sin\psi : 1 \;=\; 1 \,:\, \phi'(x) \,:\, \sqrt{1+\phi'(x)^2}$$

or

$$\cos\psi \; : \; \sin\psi : 1 \; = \; \theta'(y) \; : \; 1 \; : \; \sqrt{1+\theta'(x)^2}$$

The arc length s satisfies

$$\left(\frac{ds}{dx}\right)^2 \; = \; 1 + \phi'(x)^2$$

or

$$\left(\frac{ds}{dy}\right)^2 \; = \; 1 + \theta'(y)^2$$

The radius of curvature is

$$\rho = \left|\frac{ds}{d\psi}\right|$$

Once again, our discussion is restricted to the lattice with

$$\tilde{\omega}_1 \; = \; \left(\frac{1}{\mathbf{r}}, 0\right) \qquad \text{and} \qquad \tilde{\omega}_2 \; = \; \left(0, \frac{1}{\mathbf{r}}\right).$$

By rescaling with factor $\mathbf{r}$, lattice L becomes the integer grid, and the curve C is expanded by factor $\mathbf{r}$ to the homothetic curve $\mathbf{r} \cdot C$. We must consider the contour lines $E(t)$ of the function $f(x, y)$, the locus of points where $f(x, y) = t$. The contour line $E(t)$ may be made up of disjoint components. We define the inside of $E(t)$ to be that side of $E(t)$ on which $f(x, y) > t$. Now let $D(t)$ be the closed region of the (x, y) plane consisting of points which are both inside or on the closed curve $\mathbf{r} \cdot C$, and inside or on the contour line $E(t)$. Using the Riesz interchange

$$f(x, y) = \int_{(x,y)\in D(t)} dt$$

we convert the calculation of the sum $\sum f(m, n)$ into the problem of counting all the lattice points in $D(t)$, possibly with averaging over t. We suppose that the components of $E(t)$ are either straight lines, or curves convex inwards, or curves convex outwards. We can express the region $D(t)$ as a finite sum or difference of convex sets $D_j(t)$.

The boundary of the set $D(t)$ consists of parts of the curve $\mathbf{r} \cdot C$ and parts of the contour line $E(t)$, so the same is true for the boundary $C_j(t)$ of each component $D_j(t)$.

Let

$$x_j(t) \; = \; \max_{D_j(t)} x \; - \; \min_{D_j(t)} x$$

$$y_j(t) = \max_{D_j(t)} y - \min_{D_j(t)} y.$$

Further, let $R_j(t)$ be the maximum radius of curvature of the boundary $C_j(t)$; if a part of $C_j(t)$ is a straight line, then $R_j(t) = \infty$.

Let $A_j(t)$ be the area of $D_j(t)$.

We state two previously known results.

THEOREM 12.1. **(Krätzel 1981).** *The number of lattice points in the convex set $D_j(t)$ is equal to*

$$A_j(t) + O\left(x_j(t) + y_j(t) + 1\right).$$

Note. The theorem in presented form is proved by Krätzel (Krätzel, 1981) but he attributed that result to Gauss. Even that Gauss nowhere states an estimate for the remainder term, his argument in (Gauss, 1876) for ellipses leads to the estimate that the error term is of order no bigger than the length of the major axis.

THEOREM 12.2. **(Huxley 2003).** *Suppose that $C_j(t)$ consists of boundedly many components, each of which satisfies the conditions (1), (2), and (3) above. Then there is a dimensionless constant A, calculated from the shapes of the components of $C_j(t)$ but independent of the scale of measuring the length, such that if the minimum radius of curvature of each $C_j(t)$ is at last A, then the number of points of the integer lattice in $D_j(t)$ is*

$$A_j(t) + O\left(R_j(t)^{\frac{131}{208}} \cdot \left(\log R_j(t)\right)^{\frac{18627}{8320}}\right)$$

The constant implied in the order of magnitude notation is also calculated from the shapes of the components of $C_j(t)$ but independent of the length scale.

Gauss's argument already makes it clear that the error is a boundary effect. The proof of Theorem 2 begins by dividing the boundary into four or more parts, and on each part choosing a coordinate transform of the form

$$X = a \cdot x + b \cdot y, \qquad Y = -b \cdot x + a \cdot y$$

where a and b are integers. The equation of the boundary becomes $Y = g(X)$. If the boundary does not contain a straight line, then we can arrange for the function $g(X)$ to have "nice" properties, such as the derivatives $g''(X)$ and $g^{(3)}(X)$ being bounded away from zero and infinity. This construction is explained in Section 18 of the monograph (Huxley, 1996).

12.4. Moment Calculations

We need to describe the contour lines $E(t)$

$$f(x,y)=t$$

of the function $f(x,y)$. We write f_1, f_2 for $\dfrac{\partial f}{\partial x}, \dfrac{\partial f}{\partial y}$, and so on for further partial derivatives. The gradient vector of $f(x,y)$ is $(f_1, f_2,)$ so the anti-clockwise tangent vector to the curve $E(t)$ is $(-f_2, f_1)$, making an angle ψ with the x-axis, where

$$\tan\psi = \frac{dy}{dx} = \frac{-f_1}{f_2}$$

The arc length s on $E(t)$ satisfies the following equation:

$$\frac{\partial s}{\partial x} = \sec\psi = \frac{\sqrt{f_1^2+f_2^2}}{f_2}$$

To find the radius of curvature we have as follows:

$$\begin{aligned}
\frac{f_1^2+f_2^2}{f_2^2}\cdot\frac{\partial\psi}{\partial x} &= \sec^2\psi\cdot\frac{\partial\psi}{\partial x} = \frac{\partial(\tan\psi)}{\partial x} = \frac{\partial}{\partial x}\left(\frac{-f_1}{f_2}\right)\\
&= \frac{f_1}{f_2^2}\cdot\left(f_{12}+\frac{dy}{dx}\cdot f_{22}\right) - \frac{f_2}{f_2^2}\cdot\left(f_{11}+\frac{dy}{dx}\cdot f_{12}\right)\\
&= \frac{f_1}{f_2^2}\cdot\left(f_{12}-\frac{f_1\cdot f_{22}}{f_2}\right) - \frac{f_2}{f_2^2}\cdot\left(f_{11}-\frac{f_1\cdot f_{12}}{f_2}\right)\\
&= \frac{-(f_1^2\cdot f_{22}-2\cdot f_1\cdot f_2\cdot f_{12}+f_2^2\cdot f_{11})}{f_2^3}
\end{aligned}$$

It follows that the radius ρ of curvature is equal to

$$\frac{ds}{d\psi} = \frac{-\left(f_1^2+f_2^2\right)^{3/2}}{f_1^2\cdot f_{22}-2\cdot f_1\cdot f_2\cdot f_{12}+f_2^2\cdot f_{11}}$$

The following special case shows some inherent difficulties. We consider the monomial weight

$$f(x,y)=x^a\cdot y^b$$

with $a\geq 0$, $b\geq 0$. Since $f(x,y)$ is homogeneous, we write the contour lines as

$$f(x,y)=\lambda\cdot\mathbf{r}^{a+b}$$

where λ is a bounded parameter for points inside the curve $\mathbf{r} \cdot C$. If $a = 0$ or $b = 0$, then contour lines $E(\lambda \cdot \mathbf{r}^{a+b})$ are straight lines, and Theorem 2 is not applicable. However, the equations $Y = g(X)$ of the contour lines are so simple that we can appeal directly to the method underlying Theorem 2, which is to estimate sums

$$\sum_m \rho(g(m))$$

where $\rho(t)$ is the row-of-teeth function

$$\rho(t) = [t] - t + \frac{1}{2}$$

The only coordinate changes that we need are of the form $X = \pm x$, $Y = \pm y$, or $X = \pm y$, $Y = \pm x$, with

$$g(X) = \mathbf{r} \cdot \left(\frac{\lambda \cdot \mathbf{r}^a}{X^a} \right)^{1/b} \qquad \text{or} \qquad g(X) = \mathbf{r} \cdot \left(\frac{\lambda \cdot \mathbf{r}^b}{X^b} \right)^{1/a}$$

If $a = 0$, then the contour lines are ordinates $Y = \lambda^{1/b} \cdot \mathbf{r}$, and if $b = 0$, then the contour lines are $Y = \lambda^{1/a} \cdot \mathbf{r}$. We consider these cases later.

We compute the radius of curvature for $a > 0$ and $b > 0$. We have

$$f_1 = \frac{a \cdot f}{x}, \quad f_2 = \frac{b \cdot f}{y},$$

$$f_{11} = \frac{a \cdot (a-1) \cdot f}{x^2}, \quad f_{12} = \frac{a \cdot b \cdot f}{xy}, \quad f_{22} = \frac{b \cdot (b-1) \cdot f}{y^2}$$

and

$$\rho = \frac{(b^2 \cdot x^2 + a^2 \cdot y^2)^{3/2}}{a \cdot b \cdot (a+b) \cdot x \cdot y} \tag{8}$$

We consider the first quadrant, where x and y are positive. If $x \leq y$, then (8) gives the following:

$$\rho \leq \frac{((a^2 + b^2) \cdot y^2)^{3/2}}{a \cdot b \cdot (a+b) \cdot x \cdot y} \leq \frac{(a+b)^2}{a \cdot b} \cdot \frac{y^2}{x} \leq \frac{(a+b)^2}{a \cdot b} \cdot \left(\frac{y^{2 \cdot a + b}}{\lambda \cdot \mathbf{r}^{a+b}} \right)^{1/a}$$

Similarly, if $y \leq x$ then (8) gives

$$\rho \leq \frac{(a+b)^2}{a \cdot b} \left(\frac{x^{a + 2 \cdot b}}{\lambda \cdot \mathbf{r}^{a+b}} \right)^{1/b}$$

We consider the path of the contour line $E(\lambda \cdot M^{a+b})$ which lies inside the curve $\mathbf{r} \cdot C$, so that x and y have an order of at most $O(\mathbf{r})$ and

$$\rho = O\left(\mathbf{r} \cdot \max \left\{ \frac{1}{\lambda^{1/a}}, \frac{1}{\lambda^{1/b}} \right\} \right)$$

We write the error bound of Theorem 2 as

$$O\left(R^k \cdot (\log R)^\mu\right)$$

where R is the maximum radius of curvature.

We use Theorem 2 if

$$R^k \cdot (\log R)^\mu = O(\mathbf{r}) \tag{9}$$

(i.e., $\log R = O(\mathbf{r})$), and the error bound in Theorem 2 becomes

$$O\left(\left(\lambda^{-k/a} + \lambda^{-k/b}\right) \cdot \mathbf{r}^k \cdot (\log \mathbf{r})^\mu\right).$$

If

$$\lambda \leq \max\left\{\mathbf{r}^{\frac{-a\cdot(1-k)}{k}}, \mathbf{r}^{\frac{-b\cdot(1-k)}{k}}\right\} \tag{10}$$

then (9) is false and we use Gauss's Theorem 1. We write λ_0 for the value on the right of (10), and λ_1 for the maximum of λ on D.

The corresponding continuous moment is

$$W = \iint_D f(x,y)\,dxdy = \iint_D \left(\int_{(x,y)\in D(t)} dt\right) dxdy$$

$$= \iiint_{D(t)} dxdydt = \int \sum_j A_j(t)dt$$

with sign changes if some of the regions $D_j(t)$ count negatively. The discrete moment is

$$V = \sum_{\substack{m \\ (m,n)\in D}} \sum_{n} f(m,n) = \sum_{\substack{m \\ (m,n)\in D(t)}} \sum_{n} \int_{(m,n)\in D} dt = \int \left(\sum_{\substack{m \\ (m,n)\in D(t)}} \sum_{n} 1 \right) dt$$

$$= \mathbf{r}^{a+b} \cdot \int_0^{\lambda_1} \left(\sum_{\substack{m \\ (m,n)\in D(\lambda\cdot\mathbf{r}^{a+b})}} \sum_{n} 1 \right) d\lambda$$

$$= \mathbf{r}^{a+b} \cdot \int_{\lambda_0}^{\lambda} \sum_j \left(A_j(\lambda\cdot\mathbf{r}^{a+b}) + O(\mathbf{r})\right) d\lambda$$

$$+\mathbf{r}^{a+b} \cdot \int_{\lambda_0}^{\lambda} \sum_j \left(A_j(\lambda \cdot \mathbf{r}^{a+b}) + O\left(\mathbf{r}^k \cdot \max\left\{ \frac{1}{\lambda^{k/a}}, \frac{1}{\lambda^{k/b}} \right\} \cdot (\log \mathbf{r})^{\mu} \right) \right) d\lambda$$

$$= \int_{\lambda_0}^{\lambda} \sum_j A_j(\lambda \cdot \mathbf{r}^{a+b}) d\lambda + O\left(\lambda_0 \cdot \mathbf{r}^{a+b+1} \right)$$

$$+O\left(\mathbf{r}^{a+b+k} \cdot (\log \mathbf{r})^{\mu} \cdot \max\left\{ \lambda_1^{1-k/a}, \lambda_1^{1-k/b} \right\} \right)$$

$$= W + \left(\mathbf{r}^{a+b+k} \cdot \left((\log \mathbf{r})^{\mu} \cdot \max\left\{ \mathbf{r}^{-(\frac{a}{k}-1)\cdot(1-k)}, \mathbf{r}^{-(\frac{b}{k}-1)\cdot(1-k)} \right\} \right) \right)$$

$$= W + O\left(\mathbf{r}^{a+b+k} \cdot (\log \mathbf{r})^{\mu} \right)$$

since λ_1 is bounded. We assume that $\min\{a, b\} > k$.

If $b = 0$ and $a > 0$, then the contour lines $E(t)$ are abscissae $x = c$, and Theorem 1 is directly applicable. The method in the proof of Theorem 1 shows that the error term depends on the position of the straight line with respect to the integer lattice. For an integer m, values $t = (m + \frac{1}{2})^a$ specify the contour line $E(t)$ to be defined by $x = m + \frac{1}{2}$ (i.e., a straight line equidistant from the two closest lattice lines), and the contribution of $E(t)$ to the edge error is zero. The rest of the boundary of $D(t)$ is a part of the curve $\mathbf{r} \cdot C$, and the method in the proof of Theorem 2 ((Huxley, 2003)) tells us that for $t = (m + \frac{1}{2})^a$, in the notation of Theorem 2, the number of integer points in $D_j(t)$ is equal to

$$A_j(t) + O\left(R^k \cdot (\log R)^{\mu} \right)$$

where R is the maximum radius of curvature of the curve $\mathbf{r} \cdot C$, with $R = O(M)$. We have, for some integer m_0,

$$V = m_0^a \cdot \sum_{\substack{m \\ (m,n)\in D}} \sum_{n} 1 \; + \sum_j \sum_{\substack{m \geq m_0 \\ (m,n)\in D_j((m+1/2)^a)}} \sum_n ((m+1)^a - m^a)$$

$$= m_0^a \cdot \left(A + O\left(R^k \cdot (\log R)^{\mu} \right) \right)$$

$$+ \sum_{m \geq m_0} ((m+1)^a - m^a) \cdot \sum_j \left(A_j \cdot ((m+1/2)^a) + O\left(R^k \cdot (\log R)^{\mu} \right) \right)$$

where A denotes the area of D. We can write

$$V = U + O\left(\mathbf{r}^{a+k} \cdot (\log \mathbf{r})^{\mu} \right) \tag{11}$$

and

$$U = \iint\limits_{D} [[x]]^a dxdy \tag{12}$$

where U is the discretized moment involving $[[x]]$ (i.e., the nearest integer to x).

It remains to estimate

$$W - U = \iint\limits_{D} (x^a - [[x]]^a)dxdy = \sum_{m \leq m_0} \int\limits_{m-1/2}^{m+1/2} \int\limits_{(x,y)\in D} (x^a - m^a)dydx$$

The upper and lower branches of the curve $\mathbf{r} \cdot C$ have equations of the form $y = g(x)$, with a radius of curvature

$$\rho = \frac{(1 + g'^2)^{3/2}}{|g''|}$$

The radius of curvature has an order of magnitude M, so we have

$$g'' = O\left(\frac{(1 + g'^2)^{3/2}}{M}\right) \tag{13}$$

In general we wish to estimate integrals of the type

$$\int\limits_{m-\frac{1}{2}}^{m+\frac{1}{2}} (x^a - m^a) \cdot g(x)dx$$

$O(\mathbf{r}^a)$ is a trivial estimate. We use this if $g'(x)$ is very large. The trapezium role estimate is

$$\frac{1}{2}\left(\left(\left(m + \frac{1}{2}\right)^a - m^a\right) g\left(m + \frac{1}{2}\right) + \left(\left(m - \frac{1}{2}\right)^a - m^a\right) g\left(m - \frac{1}{2}\right)\right)$$

$$+ O\left(\max\left|\frac{d^2}{dx^2}(x^a - m^a) \cdot g(x)\right)\tag{14}$$

The explicit term in (14) is

$$\frac{1}{2} \cdot \left(\left(\frac{1}{2} \cdot a \cdot m^{a-1} + O\left(\mathbf{r}^{a-2}\right)\right) \cdot g\left(m + \frac{1}{2}\right)\right.$$

$$+\left(-\frac{1}{2}\cdot a\cdot m^{a-1}+O\left(\mathbf{r}^{a-2}\right)\right)\cdot g\left(m-\frac{1}{2}\right)\Big)$$

$$=\frac{1}{4}\cdot a\cdot m^{a-1}\left(g\left(m+\frac{1}{2}\right)-g\left(m-\frac{1}{2}\right)\right)+O\left(\mathbf{r}^{a-1}\right)$$

$$=O\left(\mathbf{r}^{a-1}\cdot\left(1+\max|g'(x)|\right)\right) \tag{15}$$

For the error term in (14) we have

$$\frac{d^2}{dx^2}(x^a-n^a)\cdot g(x)=a\cdot(a-1)\cdot x^{a-2}\cdot g(x)+2\cdot a\cdot x^{a-1}\cdot g'(x)+(x^a-m^a)\cdot g''(x)$$

Following (13), we are able to estimate this expression as

$$O\left(\mathbf{r}^{a-1}\cdot\left(1+\max|g'(x)|+\frac{\max|g'(x)|^3}{\mathbf{r}}\right)\right) \tag{16}$$

The maxima in (15) and (16) are over the interval $m-\frac{1}{2}\le x\le m+\frac{1}{2}$. Since the radius of curvature has $\mathbf{r}$ as an order of magnitude, for $t\ge 1$ the interval of x-values where $|g'(x)|\ge t$ has the length

$$O\left(\frac{\mathbf{r}}{\sqrt{1+t^2}}\right)=O\left(\frac{\mathbf{r}}{t}\right) \tag{17}$$

We use the trivial estimate $O\left(\mathbf{r}^a\right)$ if

$$\max|g'(x)|\ge \mathbf{r}^{2/3}$$

which happens for $O\left(\mathbf{r}^{1/3}\right)$ values of m; see (17). We split the remaining range of m-values into such subintervals that $g'(x)\le 1$ or $2^r\le|g'(x)|\le 2^{r+1}$, for some integer r. The values of m with $|g'(x)|\le 1$ contribute $O\left(\mathbf{r}^a\right)$ to (12), and values with $2^r\le|g'(x)|\le 2^{r+1}$ contribute

$$O\left(\frac{\mathbf{r}}{2^r}\cdot\mathbf{r}^{a-1}\cdot\left(1+2^r+\frac{2^{3r}}{\mathbf{r}}\right)\right)=O\left(\mathbf{r}^a+2^{2r}\cdot\mathbf{r}^{a-1}\right) \tag{18}$$

Finally we sum (18) over $2^r\le\mathbf{r}^{2/3}$, and we see that

$$W-U=O\left(\mathbf{r}^{a+1/3}\right)$$

This is smaller than the error term in (11).

THEOREM 12.3. *Let C be a simple closed curve satisfying the smoothness conditions piecewise on finitely many pieces C_j.*

(1) The radius of curvature ρ and its derivative $\frac{d\rho}{d\psi}$ exists on C_i, and are continuous functions of ψ on C_i.
(2) On each piece C_i, the radius of curvature has a maximum value and a non-zero minimum value.
(3) On each piece C_i, the radius of curvature ρ has a bounded number of local maxima and minima.

Let $\mathbf{r} \cdot C$ denote the curve C expanded by a factor $\mathbf{r}$. Let D be the closed region of points inside or on the curve $\mathbf{r} \cdot C$. Let $a \geq 0$, $b \geq 0$ be integers. We consider the continuous moment

$$W = m_{a,b}(D) = \iint_D x^a \cdot y^b dxdy$$

and the discrete moment

$$V = \mu_{a,b}(D) = \sum_{\substack{m \\ (m,n)\in D}} \sum_{n} m^a \cdot n^b$$

then

$$V = W + O\left(\mathbf{r}^{a+b+k} \cdot (\log \mathbf{r})^{\mu}\right)$$

where $k = \frac{131}{208} = 0.6298...$ and $\mu = \frac{18627}{8320}$. The constant implied in the O sign depends on a and b.

12.5. Conclusions

In this article we considered how efficiently the real moments $m_{a,b}(S)$ of a given planar shape S can be computed from the digital pictures of the shape S. By equations (3) and (4) the problem can be transformed into the estimation of the number of lattice points inside a 3D body (see (3)).

How efficiently the number of integer points $N(R)$ in a closed bounded region R approximates the volume of R in a given d-dimensional space? This question has already been studied in literature. The answer is very simple if R is obtained from a fixed region R_1 by uniform dilation about the origin with linear ratio of dilation $\mathbf{r}$. Then it follows that the number $N(R)$ of integer points in $R = \mathbf{r} \cdot R_1$ is equal to

$$N(R) = \mathbf{r}^d \cdot Volume\_of\_(R_1) + O\left(\mathbf{r}^{d-1}\right) \tag{19}$$

This means that in $3D$ the error term has an order of magnitude bounded by $O\left(\mathbf{r}^2\right)$.

However, this result can not directly be applied to the problem considered here because $B_{a,b}(S)$ from (3) is not of the form $\mathbf{r} \cdot R_1$.

Davenport's result (Davenport, 1951) gives a solution for the studied problem (see (6) and (7)). The error term is estimated as $O\left(\mathbf{r}^{a+b+1}\right)$, if $a+b$ is the order of the considered moment.

In this article we have shown how a better estimate can be derived, assuming that we estimate real moments of real shapes having no straight line segment on their boundaries. The derivation is based on a result in (Huxley, 2003). If straight line sections on boundaries are allowed then the previous estimate is the best possible.

Davenport's result (6) can be applied directly to estimating the error term in the approximation $m_{a,b}(\mathbf{r} \cdot S) \approx \mu_{a,b}(\mathbf{r} \cdot S)$, even for unbounded $a+b$. It seems reasonable that the error term which comes from (6), can be improved by extending the method presented here to cases of a reasonable relation between $\mathbf{r}$ (applied resolution) and $a+b$ (the order of moments). A more careful calculation would elucidate the dependence on the exponents a and b, allowing Theorem 2 to be used as long as a and b do not exceed some small fractional power of $\mathbf{r}$. In pattern recognition or picture analysis, it seems appropriate to assume $a+b \ll \mathbf{r}$; but such an estimation remains as a problem for a future investigation.

References

Davenport, H.: On a principle of Lipschitz. *J. London Math. Soc.*, **26**:179–183, 1951.

Dudani, S. A., K. J. Breeding, and R. B. McGhee: Aircraft identification by moment invariants. *IEEE Trans. Computers*, **26**:39–46, 1977.

Gauss, C. F.: De nexu inter multitudinem classium, in quas formae binariae secundi gradus distribuuntur, earumque det erminantem. (Commentatio prior societati regiae exhibita 1834). In: Werke, Zweiter Band, pages 269–303, Königliche Gesellschaften der Wissenschaften, Göttingen, 1876.

Hu, M.: Visual pattern recognition by moment invariants. *IRE Trans. Information Theory*, **8**:179–187, 1962.

Huxley, M. N.: Exponential Sums and Lattice Points. *Proc. London Math. Soc.*, **60**: 471– 502, 1990.

Huxley, M. N.: Exponential Sums and Lattice Points II. *Proc. London Math. Soc.*, **66**: 279–301, 1993.

Huxley, M. N.: Exponential Sums and Lattice points III. *Proc. London Math. Soc.*, **87**:591–609, 2003.

Huxley, M. N.: *Area, Lattice points, and Exponential Sums.* London Math. Soc. Monographs 13., Oxford University Press, 1996.

Jain, R., R. Kasturi, and B. G. Schunck: *Machine Vision.* McGraw-Hill, New York, 1995.

Jiang, X. Y. and H. Bunke: Simple and fast computation of moments. *Pattern Recognition*, **24**:801–806, 1991.

Klette, R. and A. Rosenfeld: *Digital Geometry - Geometric Methods for Digital Picture Analysis.* Morgan Kaufmann, San Francisco, 2004.

Klette, R. and J. Žunić: Digital Approximation of Moments of Convex Regions. *Graphical Models Image Processing*, **61**:274–298, 1999.

Klette, R. and J. Žunić: Multigrid Convergence of Calculated Features in Image Analysis. *J. Mathematical Imaging Vision*, **13**:173–191, 2000.

Krätzel, R.: *Zahlentheorie*. VEB Deutscher Verlag der Wissenschaften, Berlin, 1981.

Leu, J.-G.: Computing a shape's moments from its boundary. *Pattern Recognition*, **24**:949-957, 1991.

Pei, S.-C. and L.-G. Liou: Using moments to acquire the motion parameters of deformable objects without correspondence. *Image Vision Computing*, **12**:475-485, 1994.

Peter, M.: The local contributions of zeros of curvature to lattice point asymptotics. *Math. Zeitschrift.*, **233**:803-815, 2000.

Singer, M. H.: A general approach to moment calculation for polygons and line segments. *Pattern Recognition*, **26**:1019-1028, 1993.

SHAPE-FROM-SHADING BY ITERATIVE FAST MARCHING FOR VERTICAL AND OBLIQUE LIGHT SOURCES *

ARIEL TANKUS
School of Computer Science
Tel-Aviv University, Tel-Aviv, 69978 Israel

NIR SOCHEN
School of Mathematics
Tel-Aviv University, Tel-Aviv, 69978 Israel

YEHEZKEL YESHURUN
School of Computer Science
Tel-Aviv University, Tel-Aviv, 69978, Israel

Abstract. Shape-from-Shading (SfS) is a fundamental problem in Computer Vision. Its goal is to solve the image irradiance equation. One prominent solution is the Fast Marching Method of Kimmel and Sethian. When the light source is oblique, Kimmel and Sethian proposed to rotate the image to the light source coordinate system and then solve an 'almost' Eikonal equation. This paper presents a new iterative variant of the Fast Marching Method which copes better with images taken under oblique light sources. Robustness is achieved by avoiding the change of coordinate system. The advantages of the proposed method are demonstrated on synthetic and real images.

Key words: Shape-from-Shading, Eikonal equation, Fast Marching, oblique light sources

13.1. Introduction

Shape-from-Shading (SfS) is one of the fundamental problems in Computer Vision. First introduced by Horn in the 1970s (Horn, 1977), its goal is to solve the image irradiance equation, which relates the reflectance map to

* This research has been supported in part by Tel-Aviv University fund, the Adams Super-Center for Brain Studies, the Israeli Ministry of Science, the ISF Center for Excellence in Applied Geometry, the Minerva Center for geometry, and the A.M.N. fund.

R. Klette et al. (eds.), Geometric Properties for Incomplete Data, 237-258.

image intensity. An efficient way to solve this equation numerically is the celebrated Fast Marching Method of Sethian (Kimmel and Sethian, 2001), (Sethian, 1999).

Various methodologies have been proposed since the introduction of the field of Shape-from-Shading by Horn (Horn, 1975), (Horn, 1977), (Horn, 1986) in the 1970s. Horn's book (Horn and Brooks, 1989) reviews the early approaches which include characteristic strips and Calculus of Variations. (Zhang *et al.*, 1999) categorizes Shape-from-Shading techniques by their modus operandi. Namely, minimization approaches: (Zheng and Chellappa, 1991), (Lee and Kuo, 1993); propagation approach: (Bichsel and Pentland, 1992); local approach: (Lee and Rosenfeld, 1985); linear approaches: (Pentland, 1984), (Tsai and Shah, 1994). A newer minimization approach is that of citehancock:model-sfs, which uses the Mumford-Shah functional to derive diffusion kernels. Other researchers put topological properties of the surface to use (e.g., citekimmel:globalsfs) or employ deformable models (e.g., (Samaras and Metaxas, 1999)). These are only examples, as the amount of work in the field of Shape-from-Shading is too large to describe herein.

Of particular relevance to this paper are works which utilize Level-Set and Fast Marching methodologies (see (Sethian, 1999) for a deep insight). These approaches refer to the image irradiance equation as describing the motion of a front (e.g., (Osher and Sethian, 1988), (Kimmel *et al.*, 1995)). The Fast Marching Method re-orders the computation, to make it a one-pass solution of the Eikonal equation, based on the observation that the upwind difference structure of the numerical approximation allows us to propagate information "one way", that is from smaller values to larger values ((Sethian, 1996a), (Sethian, 1996b)). (Sethian, 1996a) proves the Fast Marching Method converges to the viscosity solution (see: (Crandall and Lions, 1983), (Lions, 1982) for the definition and properties of viscosity solutions).

(Kimmel and Sethian, 2001) implemented the Fast Marching Method as an optimal algorithm for surface reconstruction. They referred to the image irradiance equation as an Eikonal equation for vertical light sources. Solution of the equation for oblique light sources is obtained by rotation of the image coordinate system to that of the light source (as inspired by (Lee and Rosenfeld, 1985)).

While the Fast Marching Method is a highly efficient numerical solution to the image irradiance equation for vertical light sources, it is suboptimal for oblique light sources. For non-vertical light sources, the rotation of coordinate system requires an a-priori knowledge of the depth of the surface. As this knowledge is exactly the goal of the algorithm, one must employ an approximation, which reduces the robustness of the algorithm. This paper presents two new ways to employ the Fast Marching Method for oblique light sources as well. The first algorithm iteratively repeats the rotation with improved depth maps, while the second algorithm iteratively applies the complete Fast Marching Method for the Eikonal equation in the case of an oblique light source and avoids any rotation. Comparison with the original algorithm (Kimmel and Sethian, 2001) would demonstrate that the second algorithm overcomes the limitations of the original.

The paper is organized as follows. First, we present the notation and basic assumptions (Section 13.2), and review the Fast Marching Method (Section 13.3). We then propose the two iterative methods for improved accuracy in cases where the light source is oblique (Sections 13.4, 13.5). Section 13.6 compares the original method with the two new ones on both synthetic and real-life images. Finally, Section 13.7 draws the conclusions.

13.2. Notation and Assumptions

Let us first describe the notation and assumptions that hold throughout this paper. Photographed surfaces are assumed representable by functions of real-world coordinates. $z(x, y)$ denotes the depth function in a real-world Cartesian coordinate system whose origin is at camera plane. A real-world coordinate $(x, y, z(x, y))$ is projected orthographically onto image point (x, y).The intensity and surface normal at this image point are denoted: $I(x, y)$ and $\vec{N}(x, y)$, respectively. The intensity function $I(x, y)$ is assumed to be a positive, Lipschitz continuous function and lower than 1 (in order to ensure the existence of the strict viscosity subsolution) (see (Rouy and Tourin, 1992) for details). The scene object is Lambertian, and is illuminated by a point light source at infinity whose direction is: $\vec{L} = (p_s, q_s, -1)$.

13.3. The Fast Marching Method

This section reviews the Fast Marching method of (Kimmel and Sethian, 2001) for vertical and oblique light sources.

13.3.1. MOTIVATION

The Shape-from-Shading problem for a Lambertian surface under directional light (assuming orthographic projection) is not well posed and may have infinitely many solutions (see, for example: (Brooks and Chojnacki, 1994), (Brooks *et al.*, 1992a), (Deift and Sylvester, 1981), (Brooks *et al.*, 1992b), (Oliensis, 1991), (Dupuis and Oliensis, 1992), (Kozera, 1997), (Klette *et al.*, 1998)). Various methodologies were suggested in the literature to deal with the ill-posedness (one such example is Photometric Stereo: (Ikeuchi *et al.*, 1986), (Woodham, 1989), (Onn and Bruckstein, 1990), (Klette *et al.*, 1999)). However, one may enforce uniqueness by adding the Dirichlet boundary conditions. Indeed, in many applications it is unrealistic to assume that one has these data, which are the goal of the algorithm. But as we would see, in the case of Fast Marching, Dirichlet boundary conditions are required merely at critical image points, not on image boundaries. At critical points it is possible to obtain the true depth by global topology solvers (e.g., (Kimmel and Bruckstein, 1995); see (Kimmel and Sethian, 2001) for more details). We are therefore interested in Fast Marching techniques.

The next subsection would present a consistent and monotone ("upwind") numerical scheme which lies at the heart of the Fast Marching Method, and is the key to obtaining a unique viscosity solution. For the Eikonal equation, (Rouy and Tourin, 1992) showed that an iterative algorithm based on this scheme with Dirichlet boundary conditions on image boundaries and at all critical points converges towards the viscosity solution with the same boundary conditions. While the algorithm of (Rouy and Tourin, 1992) requires the Dirichlet boundary conditions on image boundaries and at all critical points, the Fast Marching Method needs the Dirichlet conditions only at critical points. Existence of the viscosity solution was proven in (Lions, 1982) and uniqueness, in (Rouy and Tourin, 1992) and (Ishii, 1987). (Sethian, 1996a) proved that the Fast Marching Method produces a solution that everywhere satisfies the discrete version of the Eikonal equation. We next describe the algorithm.

13.3.2. FAST MARCHING FOR VERTICAL LIGHT SOURCES

The algorithm of (Kimmel and Sethian, 2001) stems from the orthographic image irradiance equation:

$$I(x,y) = \vec{L} \cdot \vec{N}(x,y) = \frac{p_s z_x + q_s z_y + 1}{\|\vec{L}\ Vert\sqrt{z_x^2 + z_y^2 + 1}} \tag{1}$$

For a vertical light source, that is $\vec{L} = (0, 0, -1)$, the equation becomes an Eikonal equation which can be written as:

$$p^2 + q^2 = \tilde{F}^2 \tag{2}$$

where $p \overset{\text{def}}{=} z_x$, $q \overset{\text{def}}{=} z_y$ and $\tilde{F} = \sqrt{(I(x,y))^{-2} - 1}$.

Following (Kimmel and Sethian, 2001), we use the numerical approximation (originally introduced in (Rouy and Tourin, 1992) as a modification of the scheme of (Osher and Sethian, 1988)):

$$\begin{aligned} p_{ij} &\approx \max\{D_{ij}^{-x}z, -D_{ij}^{+x}z, 0\} \\ q_{ij} &\approx \max\{D_{ij}^{-y}z, -D_{ij}^{+y}z, 0\} \end{aligned}$$

where

$$D_{ij}^{-x}z \stackrel{\text{def}}{=} \frac{z_{ij} - z_{i-1,j}}{\Delta x}$$

is the standard backward derivative and

$$D_{ij}^{+x}z \stackrel{\text{def}}{=} \frac{z_{i+1,j} - z_{ij}}{\Delta x},$$

the standard forward derivative in the x-direction ($z_{ij} \stackrel{\text{def}}{=} z(i\Delta x, j\Delta y)$). $D_{ij}^{-y}z$ and $D_{ij}^{+y}z$ are defined in a similar manner for the y-direction.

Substituting the numerical approximation into Equation 2, we get the discrete equation:

$$\left(\max\{D_{ij}^{-x}z, -D_{ij}^{+x}z, 0\}\right)^2 + \left(\max\{D_{ij}^{-y}z, -D_{ij}^{+y}z, 0\}\right)^2 = \tilde{F}_{ij}^2 \qquad (3)$$

where $\tilde{F}_{ij} \stackrel{\text{def}}{=} \tilde{F}(i\Delta x, j\Delta y)$. The solution of this equation at point (i,j), assuming depth is known at neighboring pixels, is:

$$z_{ij} = \begin{cases} \min\{z_1, z_2\} + \tilde{F}_{ij}, & \text{if } \mid z_2 - z_1 \mid \geq \tilde{F}_{ij} \\ \frac{1}{2}\left(z_1 + z_2 \pm \sqrt{2\tilde{F}_{ij}^2 - (z_1 - z_2)^2}\right), & \text{if } \mid z_2 - z_1 \mid < \tilde{F}_{ij} \end{cases} \qquad (4)$$

where $z_1 \stackrel{\text{def}}{=} \min\{z_{i-1,j}, z_{i+1,j}\}$ and $z_2 \stackrel{\text{def}}{=} \min\{z_{i,j-1}, z_{i,j+1}\}$.

13.3.3. FAST MARCHING IN LIGHT SOURCE COORDINATES

For oblique light sources (i.e., $\vec{L} \neq (0, 0, -1)$), (Kimmel and Sethian, 2001) adopted the idea of (Lee and Rosenfeld, 1985) to rotate the brightness image to light source coordinates. This yields an 'almost' Eikonal equation (as (Kimmel and Sethian, 2001) called it), which is solved in a manner similar to the vertical case, but in the new coordinate system.

Rotation to the light source coordinate system is, however, nontrivial. The image irradiance equation (Equation 1) is invariant to depth translation. That is, $z(x,y)$ and $z(x,y) + c$ (for a constant c) generate identical irradiance. This occurs at coordinates (x,y) which are the camera coordinates (See Figure 13.1a). Following the rotation, one solves the vertical light source case of the image irradiance equation (Equation 2), which is also invariant to depth translation. However, it is now solved in a different coordinate system, so the direction of invariance is the direction of the new z-axis (i.e., the light source direction). Figure 13.1b demonstrates the invariance following the rotation.

Because of its dependence on surface depth, the rotation of the two surfaces: $z(x,y)$ and $z(x,y) + c$ to light source coordinates differ in the general case. In particular, the new (x,y) coordinates, which we denote:

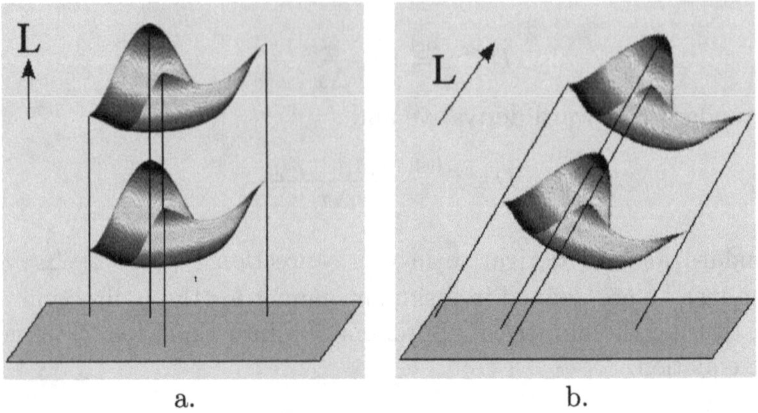

Figure 13.1. Demonstration of the invariance properties of the orthographic image irradiance equation (a) and the equation used after rotation of the image to the light source coordinates (b).

(x', y'), may be different. But these coordinates are also the new image coordinates. Thus, the image pixels used for the computation are different: $I(x', y')$ vs. $I(x' + cl_1, y' + cl_2)$, respectively, where $l_1 = p_s/\|\vec{L}\|$ and $l_2 = q_s/\|\vec{L}\|$ (the proof is omitted for brevity). Thus, a depth translation, which should preserve the irradiance under the orthographic model, requires a translation of the (x, y) coordinates as well, due to the rotation to light source coordinates. For works on the perspective model, see: (Tankus *et al.*,2003), (Tankus *et al.*, 2004a), (Tankus *et al.*, 2004b). This would be further demonstrated by experimental results (Section 13.6.2).

As a result of the dependence of image coordinates on surface depth, the new image coordinates may lie outside image boundaries. No doubt, this results in loss of information. In our implementation, for pixels outside image boundaries, we duplicated the intensity of the nearest pixel on the boundary.

Another source of error in the calculation of the rotation is the use of an approximation for the depth of the surface. An approximation is necessary because the true depth is yet unknown when the rotation to light source coordinates takes place. (Kimmel and Sethian, 2001) suggested to approximate depth as the minimal depth of neighboring pixels.

The use of the approximated depth for the rotation to light source coordinates results in an inaccurate rotation. Following that, the algorithm solves the vertical light source problem in light source coordinates, and rotates the resultant surface back to the original coordinates. The "inverse" rotation, however, is not exactly inverse to the first rotation, as it uses a more accurate depth map. An inaccurate rotation affects the shape of the $[xy]$ domain of the reconstructed surface. Consequently, a rectangular image is not necessarily reconstructed in a rectangular $[xy]$ domain, even though the projection model is orthographic.

13.4. A Rotation Iterative Solution

One way to improve the results of the Fast Marching Method for oblique light sources is to try and reduce the approximation error of the rotation to the light source coordinates.

The suggested method is iterative. It uses the depth recovered by the Fast Marching Method to recalculate the rotation. With this new rotation, rotate the image once again to the light source coordinates, and solve the vertical problem in the new coordinate system. From the depth so obtained, recalculate the rotation, solve the vertical problem and so forth. We would call this method: *Iterative Rotation.*

13.5. The Equation Iterative Solution

A different modus operandi to overcome the aforementioned flaws of the Fast Marching Method is to avoid any rotation to light source coordinates at all. Instead, we solve a series of Eikonal equations which are approximations to the image irradiance equation. Each equation should refine the approximation of its predecessor.

To formulate the approximate equations, we transform the image irradiance equation for an oblique light source (Equation 1) into the form:

$$p^2 + q^2 = F^2(p, q) \tag{5}$$

where:

$$F(p, q) \stackrel{\text{def}}{=} \sqrt{1 - \left(\frac{p_s p + q_s q + 1}{\| \vec{L} \| \, I(x, y)} \right)^2}$$

A significant difference between the vertical and oblique cases is the dependence of F on p and q.

An important observation described in (Kimmel and Sethian, 2001) is that when updating the depth values according to the discrete equation (Equation 3), information always flows from small to large values. Based on this, the Fast Marching Method reconstructs depth in an "upwind" fashion. It first sets all z values to the correct height values at local minima and to infinity elsewhere. Then, every step extends the reconstruction to higher depths. Reconstruction is thus achieved in one pass.

Nevertheless, a single pass may not be enough to solve the aforementioned formulation of the oblique problem (Equation 5), because the approximate solution (the right-hand side of Equation 4) depends on F,which depends on both p and q. Hence, we suggest another iterative method. At each iteration, F is calculated using the depth recovered at the preceding iteration:

$$p_{n+1}^2 + q_{n+1}^2 = F^2(p_n, q_n) \tag{6}$$

where p_n and q_n are the values of p and q at the $n^{\underline{\text{th}}}$ iteration. The algorithm is thus:

1. Step 0: Initialize (p_0, q_0) by the Fast Marching Method of (Kimmel and Sethian, 2001).
2. Step n:
   a) Based on the approximation (p_n, q_n), calculate the right-hand side of Equation 6, namely, evaluate $F^2(p_n, q_n)$.
   b) As we computed $F^2(p_n, q_n)$, Equation 6 is now eikonal. Use Equation 4 to obtain a solution: (p_{n+1}, q_{n+1}).
   c) Following each iteration we normalize the depth function $z(x, y)$ (divide by the mean z value) to compensate for the lack of knowledge of grid size $(\Delta x, \Delta y)$.
3. Let $n := n + 1$, and repeat Step n.

We call this method: *Iterative Equation.*

This iterative process results in a series of Eikonal equations, each solved by the Fast Marching Method. (Sethian, 1996a) showed that the Fast Marching Method produces a solution that everywhere satisfies the discrete version of the Eikonal equation. Therefore, the Fast Marching solution of each of the equations in the series satisfies the discrete version of that equation. As a result, when the series of solutions to the Eikonal equations converges, the limit is the correct solution of the discrete version of the original equation (i.e., the solution of the image irradiance equation with an oblique light source).

Empirically, in all experiments the series of solutions converged. In fact, very few iterations were necessary to obtain this convergence (i.e., to get close enough to the limit).

13.6. Experimental Results

13.6.1. THE EXPERIMENTS

To evaluate the contribution of the proposed algorithms, we compared them with the original formulation of the Fast Marching Method (Kimmel and Sethian, 2001). The evaluation involved both synthetic images and real-life images. The synthetic images were produced from a given depth map using the image irradiance equation (Equation 1). The derivatives in the equation were calculated numerically.

The initialization of the algorithms is based on points of local minima. For synthetic images, these were extracted automatically from the true depth map. For real images, they were located visually in each photograph by a human viewer, and their depths were arbitrarily set to the same constant. To demonstrate the aforementioned undesired features of the Fast Marching Method (Kimmel and Sethian, 2001) (Section 13.3.3), we ran the algorithms twice for each surface. In the second run, the depth of the original initialization (described above) was translated by a constant. Theoretically, this should merely translate the whole reconstruction along the z-axis by the same constant.

In our comparison we checked five iterations of the iterative Fast Marching Methods for each example. For the Iterative Equation method, we found out that all iterations (maybe except for the first one) yielded visually-identical images, which implies the suggested algorithm converges very fast.

To quantitatively evaluate the performance of the algorithms on synthetic data, we adopted three criteria from (Zhang *et al.*, 1999). These are: mean depth error, standard deviation of depth error, and mean gradient error. For completeness, we also supply the standard deviation of the gradient error, even though it is considered nonphysical.

13.6.2. COMPARATIVE EVALUATION

Figure 13.2 compares the original Fast Marching Method with the two iterative methods (Iterative Rotation and Iterative Equation) on the following depth map:

$$z(x, y) \stackrel{\text{def}}{=} 100 + \cos\left(\sqrt{x^2 + (y-2)^2}\right)$$

where: $x, y \in [-3.0788, 3.0788]$ (image size: 50×50 pixels). The Iterative Rotation does not improve upon the original Fast Marching; their reconstructions are very similar. The original Fast Marching Method reconstructed only a part of the cosine function (the upper right part of the surface in Figure 13.2C, iteration: 0), due to the translated $[xy]$ coordinates in the calculation. This part corresponds to the lower-left part of the original surface (Figure 13.2B). The rest of the reconstruction is a result of a calculation using pixels outside image boundaries (as described in Section 13.3.3). The Iterative Equation, on the other hand, reconstructed the right-hand side of the cosine correctly, with more noise on the left hand side (part of the elevated domain of the cosine appears almost flat there). Table 13.1 presents the error rates according to the aforementioned criteria. As expected, all error rates of the Iterative Rotation and Fast Marching methods are very close to one another. The Iterative Equation algorithm obtained lowest error rates according to all criteria.

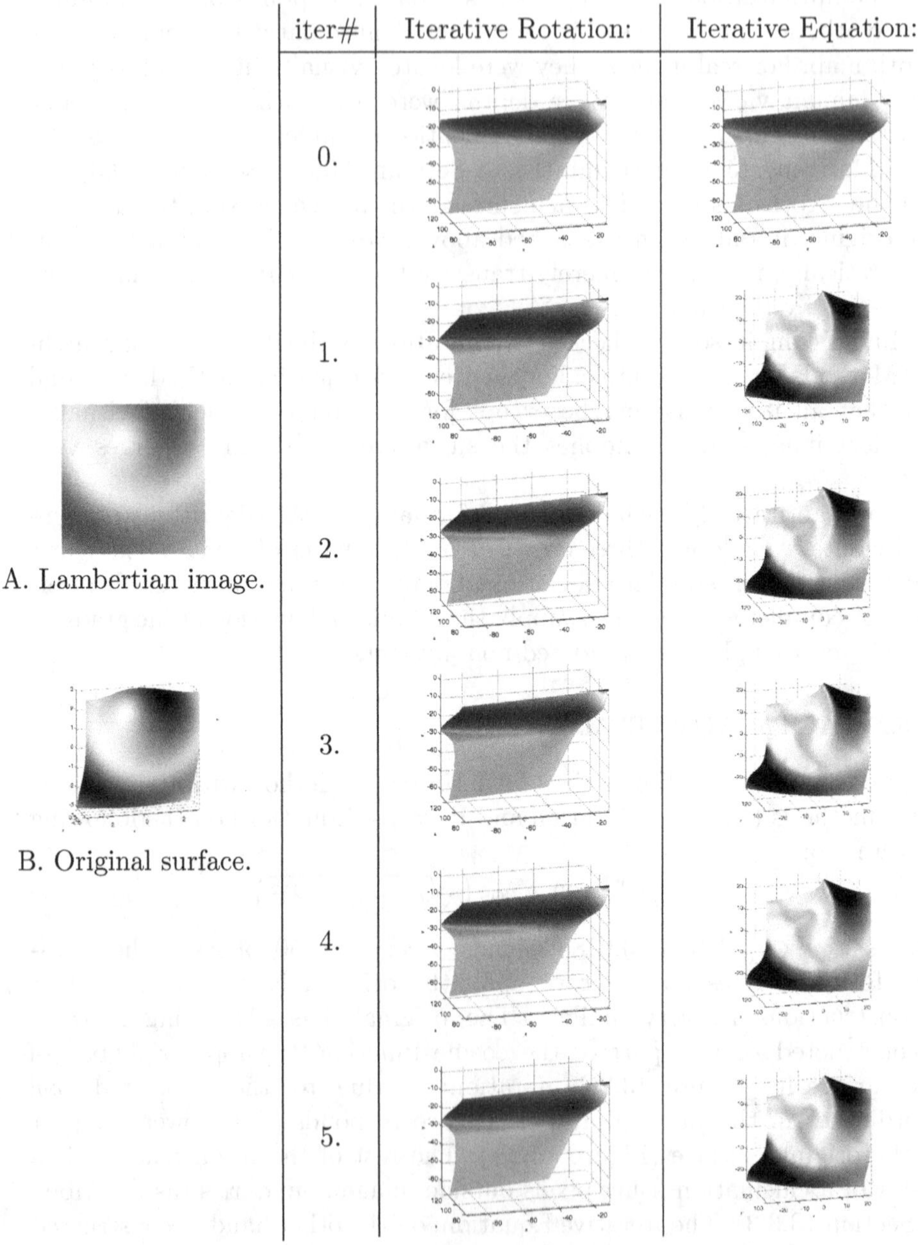

C. Reconstruction comparison.

Figure 13.2. Three variants of the Fast Marching Method for $z(x, y) = 100 + \cos\left(\sqrt{x^2 + (y-2)^2}\right)$. Each row corresponds to a different iteration (Row 0 is the original Fast Marching). Lighting is identical for all reconstructions, and is equal to that of (A).

TABLE 13.1. Error rates for the algorithms on $z(x, y) = 100 + \cos\left(\sqrt{x^2 + (y-2)^2}\right)$.

Algorithm:	No. of Iters.:	Mean Depth Error:	Std. Dev. of Depth Error:	Mean Grad. Error:	Std. Dev. of Grad. Error:
Fast Marching:	1	0.51687	0.29194	2.20442	1.01374
Iterative Rotation:	1	0.51636	0.29127	2.16523	1.01180
Iterative Rotation:	2	0.51692	0.29179	2.20046	1.01357
Iterative Rotation:	3	0.51697	0.29185	2.20573	1.01359
Iterative Rotation:	4	0.51699	0.29189	2.20869	1.01362
Iterative Rotation:	5	0.51697	0.29186	2.20626	1.01355
Fast Marching:	1	0.51687	0.29194	2.20442	1.01374
Iterative Equation:	1	0.37269	0.28148	1.05731	0.88570
Iterative Equation:	2	0.37213	0.28217	1.05174	0.88535
Iterative Equation:	3	0.37189	0.28203	1.05107	0.88494
Iterative Equation:	4	0.37188	0.28202	1.05104	0.88493
Iterative Equation:	5	0.37188	0.28202	1.05104	0.88493

Figure 13.3 shows the famous example of the Vase ($x, y \in [-63.5, 63.5]$; image size: 128×128; background depth: 100). The original Fast Marching yielded a sharp bulge at the foot of the vase. The bulge appears in a domain whose reconstruction was computed from pixels outside image boundaries. Iterative Rotation recovers similar surfaces. The Iterative Equation method, on the other hand, reconstructed a much smoother vase. Its stronger resemblance to the original is not only visible but can also be quantified by all error measures (Table 13.2). The Iterative Rotation and Fast Marching methods equate; the mean depth error of Iterative Rotation is even a little higher than that of Fast Marching.

Figure 13.4 introduces a real-world example taken by endoscopy from the gastric angulus[1] (cropped image size: 64×64). The algorithm of Kimmel and Sethian reconstructs two of the gastric folds. However, the reconstructed wall of the gastric angulus seems to consist of perpendicular planes (instead of small folds on a main low-convexity surface). On the right-hand side of the reconstruction, the surface appears "higher" (i.e., larger y-rates[2])

[1] Original is from www.gastrolab.net, courtesy of The Wasa Workgroup on Intestinal Disorders, GASTROLAB, Vasa, Finland.

[2] Recall, that the $[xy]$ domain is perpendicular to the optical axis, so the coordinate system of the reconstructed surfaces is: x – to the right; y – up; z – away from the viewer.

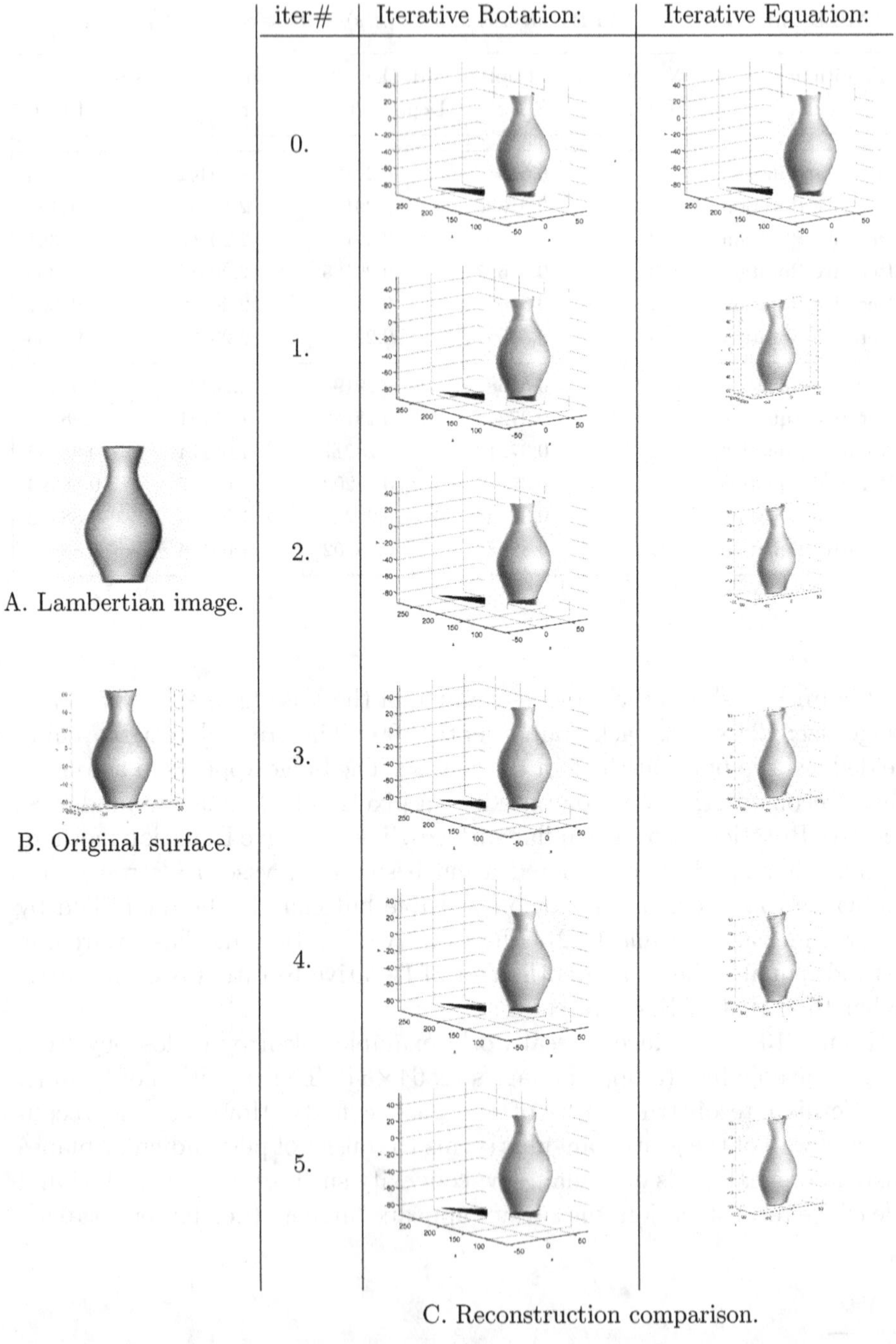

Figure 13.3. Three variants of the Fast Marching Method for the Vase example. Each row corresponds to a different iteration (Row 0 is the original Fast Marching). Lighting is identical for all reconstructions, and is equal to that of (A).

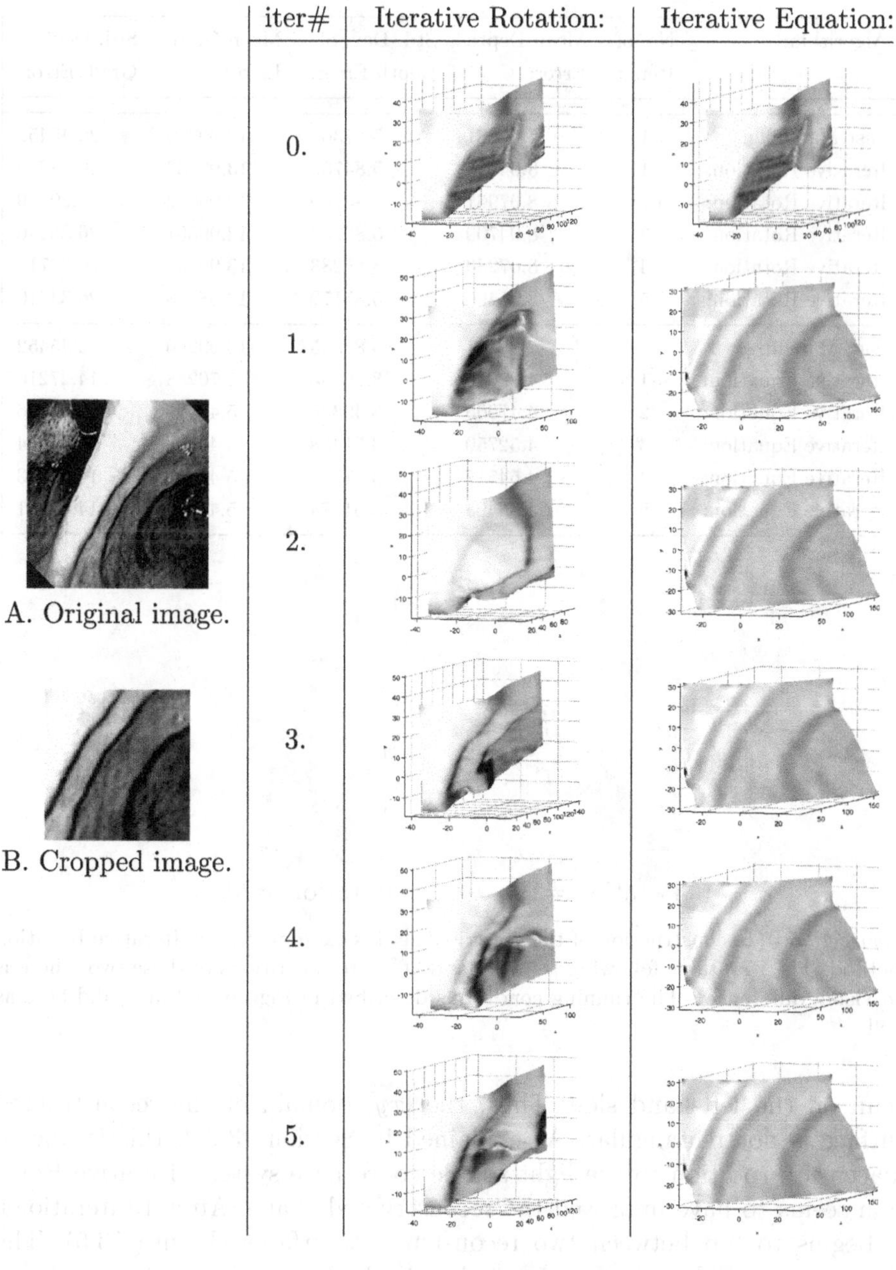
iter#
Iterative Rotation:
Iterative Equation:
0.
1.
2.
3.
4.
5.
A. Original image.
B. Cropped image.

C. Reconstruction comparison.

Figure 13.4. Three variants of the Fast Marching Method for an endoscopic image of the Gastric Angulus. Each row corresponds to a different iteration (Row 0 is the original Fast Marching). Lighting is identical for all reconstructions, and is approximately that of (B). Only (B) was used for the reconstruction.

TABLE 13.2. Error rates for the algorithms on the Vase example.

Algorithm:	No. of Iters.:	Mean Depth Error:	Std. Dev. of Depth Error:	Mean Grad. Error:	Std. Dev. of Grad. Error:
Fast Marching:	1	7.95881	5.87365	13.99250	26.45452
Iterative Rotation:	1	8.04165	5.84769	13.99933	26.28860
Iterative Rotation:	2	8.07961	5.85130	13.99862	26.29129
Iterative Rotation:	3	8.04709	5.85250	13.99551	26.36240
Iterative Rotation:	4	8.07280	5.85283	13.99254	26.34710
Iterative Rotation:	5	8.05496	5.85172	13.98958	26.34410
Fast Marching:	1	7.95881	5.87365	13.99250	26.45452
Iterative Equation:	1	4.12683	3.36455	5.70288	14.27210
Iterative Equation:	2	4.47005	3.49335	5.48740	14.09266
Iterative Equation:	3	4.52750	3.51408	5.45683	14.09894
Iterative Equation:	4	4.54517	3.51939	5.44802	14.09603
Iterative Equation:	5	4.55495	3.51974	5.44278	14.08841

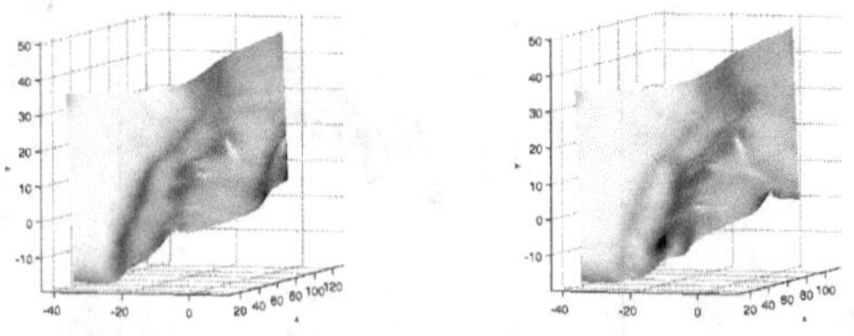

Iteration #19. Iteration #20.

Figure 13.5. Reconstruction of the Gastric Angulus example by the Iterative Rotation method. The iterations following the $19^{\underline{\text{th}}}$ and $20^{\underline{\text{th}}}$ are repetitions of these two (there is no visual difference). The running conditions described in Figure 13.4 are valid here as well.

than on the left-hand side. Thus, the $[xy]$ domain of the reconstructed surface is not rectangular. As explained in Section 13.3.3, this is due to inaccurate rotation to the light source coordinate system. Iterative Rotation seems to have improved the reconstructed shape. After 19 iterations, it begins to flip between two reconstructed surfaces (Figure 13.5). The three gastric folds are reconstructed in both states, with better results in the odd states. In the Iterative Rotation reconstruction, some cavities are present near the central fold. The upper-right part of the surface is also not faithfully recovered. The contours of the recovered folds are not as smooth as in the original image. The Iterative Equation method seems to have

Algorithm:	Reconstruction from Original Initialization:	Reconstruction from Translated Initialization:
Fast Marching:		
Iterative Rotation: (Iteration #5)		
Iterative Equation: (Iteration #5)		

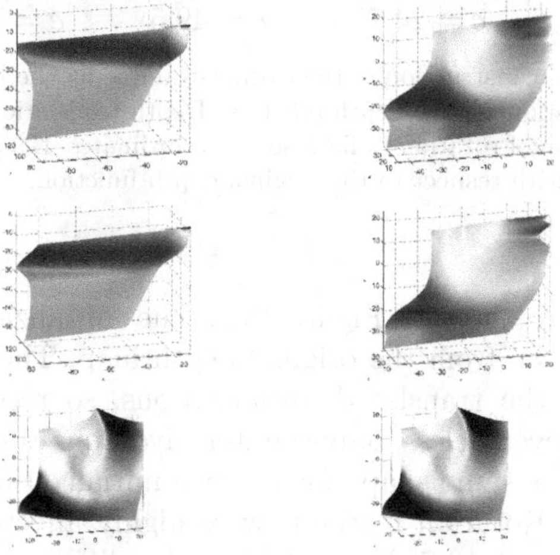

Figure 13.6. Comparison of the three algorithms on the Cosine example. Each algorithm was run with two initializations which were identical up to a constant translation along the z-direction ($c = -90$). Only the Iterative Equation algorithm remained invariant to the translation, while the two others showed a significant change between initializations. Thus, the Iterative Equation algorithm better maintains the invariance to depth translations.

reconstructed the three gastric folds in quite an accurate manner. Its accuracy appears to be higher than that of the Iterative Rotation method and in less iterations: only 1 iteration was necessary for the Iterative Equation to converge.

13.6.3. COMPARISON OF ROBUSTNESS IN DEPTH TRANSLATION

In this subsection, we would like to evaluate the robustness of the algorithms in depth translations. We therefore juxtapose the reconstructions of surfaces $z(x, y)$ and $z(x, y) + c$ (c is constant) by the three methods. To obtain reconstructions of $z(x, y) + c$, we increased the initial depth values (at minima points) by a constant with respect to the initial values employed to reconstruct $z(x, y)$. Theoretically, the reconstructions should be identical up to depth translation, due to the invariance of the orthographic image irradiance equation.

In the following examples, only one iteration of the iterative methods is displayed, for the sake of brevity.

Figure 13.6 shows the reconstructions of the Cosine example of Figure 13.2 by the three methods. The reconstructions in the middle column

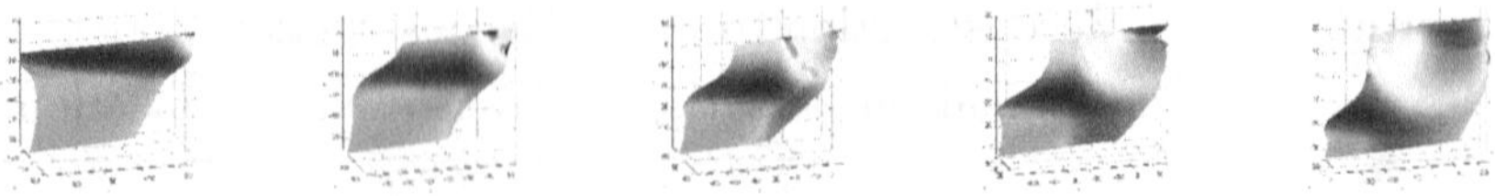

$c = 0$ $c = 24.75$ $c = 49.5$ $c = 74.25$ $c = 99$

Figure 13.7. Reconstruction of the Cosine example by the Fast Marching method with different translations of the depth function. Depth translation results in $[xy]$ translation in the calculation of rotation to light source coordinates. Below each reconstruction is its translation (c) with respect to the original depth function.

are identical to those of Figure 13.2. The initializations used for creating them were taken from the original depth map. The rightmost column was created with the translated initializations, so reconstructions should be translated. Nevertheless, only the Iterative Equation method reconstructed a surface of the same shape for the two initializations. The Fast Marching and Iterative Rotation methods were highly affected by the depth translation. As Figure 13.7 demonstrates, the difference between the surfaces reconstructed by Fast Marching with different depth translations is due to translation in the x and y coordinates in the calculation of the rotation to light source coordinates (as explained in Section 13.3.3). This translation requires some of the pixels to be taken from outside image boundaries, so in practice their values are duplicated from boundary pixels. Other pixels are simply shifted in place when the depth map is translated. The specific translation used (namely, $c = -90$) seems to have improved the reconstruction drastically. Indeed, it reduced the amount of pixels outside image boundaries prominently.

Figure 13.8 displays the reconstructed Vase (see Figure 13.3) for the original and translated initializations. For the Fast Marching method the translation was so large that the vast majority of pixels were taken from outside image boundaries. Thus, the reconstructed surface is almost planar, showing no sign of the original structure of the vase. Again, the change in reconstruction is prominent for the Fast Marching and Iterative Rotation methods, but not for Iterative Equation.

Figure 13.9 presents the original and translated reconstructions for the Gastric Angulus example of Figure 13.4. The depth-translated initialization shifts the reconstructions of the Fast Marching and Iterative Rotation methods in the $[xy]$ plane. Thus, some of the pixels are evaluated outside image boundaries. Pay attention that in this real-life example the true depth at minima points was a-priori unknown, and the algorithm was initialized based on a human guess. A different guess could result in a significant change to the reconstruction. The Iterative Equation maintained its response in spite of the depth translation.

Algorithm:	Reconstruction from Original Initialization:	Reconstruction from Translated Initialization:
Fast Marching:		
Iterative Rotation: (Iteration #5)		
Iterative Equation: (Iteration #5)		

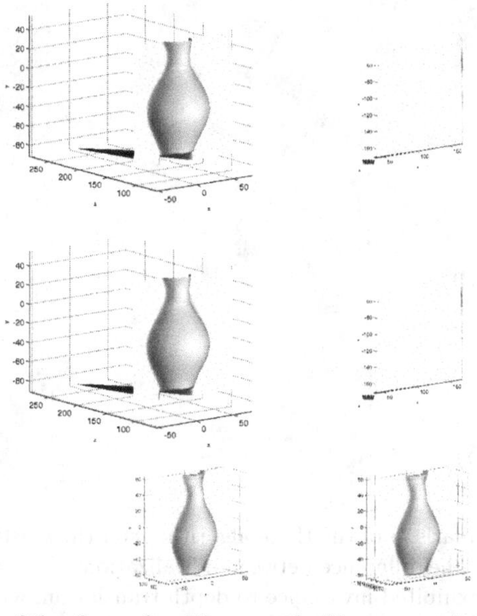

Figure 13.8. Comparison of the three algorithms on the Vase example. The initialization was translated by $c = +1000$ with respect to the original one. The Fast Marching Method and the Iterative Rotation methods yielded similar results. Both changed substantially with the change of initialization. In contrast, the Iterative Equation method maintained the invariance to depth translation.

From the figures, one can see that the Fast Marching and Iterative Rotation methods were highly affected by the translation in contrast with the theoretic invariance of the underlying equation. This demonstrates the drawbacks of rotation to light source coordinates discussed in detail in Section 13.3.3. As opposed to these two algorithms, the variation in reconstruction by the Iterative Equation method was very small. Quantification of the results in the form of depth and gradient errors appears in Tables 13.3 and 13.4 (for the synthetic examples only). Table 13.3 (the Cosine example) confirms the visual impression that reconstructions by the Fast Marching and Iterative Rotation methods were improved by the specific translation selected for this example. Indeed, due to the translation the error rates dropped significantly with respect to those of Table 13.1. However, they are still slightly higher than those of the Iterative Equation method (except for the standard deviation of the gradient error, which is considered nonphysical). In Table 13.4 (the Vase example), all error rates of the Fast Marching and Iterative Rotation methods altered with respect

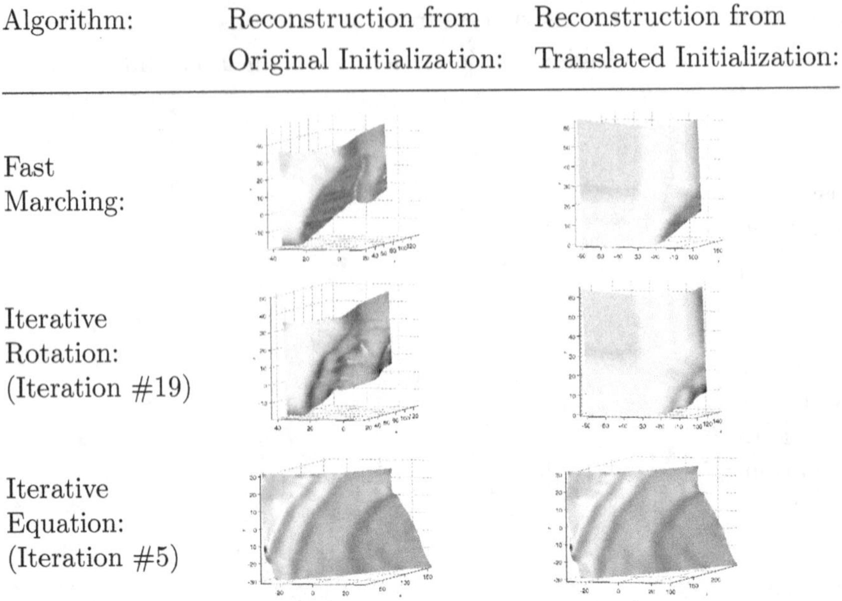

Figure 13.9. Comparison of the three algorithms on the Gastric Angulus example with two initializations. The difference between initializations was $c = +90$. Only the Iterative Equation method exhibited invariance to depth translation, while the two others showed a pronounced change between initializations.

to the corresponding values in Table 13.2. Their change is not as strong as for the Cosine example, but is still higher than that of the Iterative Equation method. In both Tables 13.3 and 13.4 variations in the error rates of Iterative Equation are only minor. Pay attention, that identical error rates do not imply identical reconstructions. Nevertheless, a significant change to these measures certainly indicates a notable change in surface shape.

We see, that in all examples, the Iterative Equation method appears to outrank the methods which rotate the image to the light source coordinate system: Fast Marching and Iterative Rotation.

When comparing the complexity of the three algorithms, no doubt the original one is the fastest, by containment. However, as the examples show, the speed in this case is at the expense of accuracy. The Iterative Equation method converges very fast and no more than 2 iterations were ever required to obtain it, so the speed difference turns out to be of secondary importance.

13.7. Conclusions

This research proposes an efficient and robust solution to the problem of Shape-from-Shading which handles both vertical and oblique light sources

TABLE 13.3. Comparison of algorithms on $z(x, y) = 100 + \cos\left(\sqrt{x^2 + (y-2)^2}\right)$, with initialization translated by -90. Pay attention to the sharp change in all measures of the Fast Marching and Iterative Rotation methods with respect to Table 13.1.

Algorithm:	No. of Iters.:	Mean Depth Error:	Std. Dev. of Depth Error:	Mean Grad. Error:	Std. Dev. of Grad. Error:
Fast Marching:	1	0.38683	0.28843	1.43113	0.80503
Iterative Rotation:	1	0.38089	0.27686	1.40019	0.75335
Iterative Rotation:	2	0.37521	0.28065	1.37679	0.77419
Iterative Rotation:	3	0.35533	0.26359	1.28275	0.72974
Iterative Rotation:	4	0.35734	0.26579	1.29597	0.74782
Iterative Rotation:	5	0.35535	0.26404	1.28020	0.73735
Fast Marching:	1	0.38683	0.28843	1.43113	0.80503
Iterative Equation:	1	0.36179	0.27772	1.02285	0.86817
Iterative Equation:	2	0.35912	0.27893	1.00283	0.85665
Iterative Equation:	3	0.35695	0.27596	1.00111	0.85411
Iterative Equation:	4	0.35850	0.27733	1.00394	0.85671
Iterative Equation:	5	0.35869	0.27748	1.00448	0.85694

under the orthographic projection model. The suggested solution is a variant of the Fast Marching Method of (Kimmel and Sethian, 2001). It employs the Fast Marching Method iteratively for oblique light sources. Each iteration solves an approximation to the image irradiance equation. The resultant solution serves for successive refinement of the approximating equation. We called this algorithm: the Iterative Equation method. When this refinement process converges, convergence is to the correct solution of the original equation.

We compared reconstruction by the original Fast Marching Method, the Iterative Rotation method (which successively refines the rotation to light source coordinates) and the Iterative Equation method on both synthetic and real-life examples (from endoscopy). We also demonstrated why rotation of the image to light source coordinates, as required by the Fast Marching and Iterative Rotation methods, is unstable. The Iterative Equation method outperformed the two other methods, and remained invariant to depth translations (due to its convergence to the correct solution).

In terms of runtime, the original Fast Marching Method is faster than the suggested ones. However, convergence of the Iterative Equation method is very fast; in all examples no more than 2 iterations were ever necessary.

TABLE 13.4. Comparison of algorithms on the Vase example with translated initialization (+1000). Note the significant change in mean gradient error of the original Fast Marching Method with respect to Table 13.2.

Algorithm:	No. of Iters.:	Mean Depth Error:	Std. Dev. of Depth Error:	Mean Grad. Error:	Std. Dev. of Grad. Error:
Fast Marching:	1	9.36239	5.74319	14.35475	18.95690
Iterative Rotation:	1	9.36239	5.74319	14.35475	18.95690
Iterative Rotation:	2	9.36239	5.74319	14.35475	18.95690
Iterative Rotation:	3	9.36239	5.74319	14.35475	18.95690
Iterative Rotation:	4	9.36239	5.74319	14.35475	18.95690
Iterative Rotation:	5	9.36239	5.74319	14.35475	18.95690
Fast Marching:	1	9.36239	5.74319	14.35475	18.95690
Iterative Equation:	1	4.60417	3.56963	5.45874	14.25520
Iterative Equation:	2	4.60904	3.57800	5.44469	14.26232
Iterative Equation:	3	4.60800	3.57882	5.44469	14.26547
Iterative Equation:	4	4.60788	3.57916	5.44462	14.26633
Iterative Equation:	5	4.60779	3.57945	5.44472	14.26703

References

Bichsel, M. and A. P. Pentland: A simple algorithm for shape from shading. In Proc. *IEEE Computer Vision and Pattern Recognition*, pages 459–465, 1992.

Brooks, M. J. and W. Chojnacki: Direct computation of shape from shading. In Proc. *IEEE International Conference on Pattern Recognition*, pages 114–119, 1994.

Brooks, M. J., W. Chojnacki, and R. Kozera: Circularly-symmetric eikonal equations and non-uniqueness in computer vision. *Journal of Mathematical Analysis and Applications*, **165**:192–215, 1992.

Brooks, M. J., W. Chojnacki, and R. Kozera: Impossible and ambiguous shading patterns. *International Journal of Computer Vision*, **7**:119–126, 1992.

Crandall, M. G. and P.-L. Lions: Viscosity solutions of Hamilton-Jacobi equations. *Transactions of the American Mathematical Society*, **277**:1–42, 1983.

Deift, P. and J. Sylvester: Some remarks on the shape-from shading problem in computer vision. *Journal of Mathematical Analysis and Applications*, **84**:235–248, 1981.

Dupuis, P. and J. Oliensis: Direct method for reconstructing shape from shading. In *IEEE Conference on Computer Vision and Pattern Recognition*, pages 453–458, 1992.

Horn, B. K. P.: Obtaining shape from shading information. In *The Psychology of Computer Vision*, (P. H. Winston, editor), pages 115–155, McGraw-Hill Book Company, 1975.

Horn, B. K. P.: Image intensity understanding. *Artificial Intelligence*, **8**:201–231, 1977.

Horn, B. K. P.: *Robot Vision.* The MIT Press/McGraw-Hill Book Company, 1986.

Horn, B. K. P. and M. J.Brooks: *Shape from Shading.* The MIT Press, 1989.

Ikeuchi, K., H. K. Nishihara, B. K. Horn, P. Sobalvarro, and S. Nagata: Determining grasp configurations using photometric stereo and the PRISM binocular stereo system. *The International Journal of Robotics Research* **5**:46–65, 1986.

Ishii, H.: A simple, direct proof of uniqueness for solutions of the Hamilton-Jacobi equations of Eikonal type. *Proceedings of the American Mathematical Society,* **100**:247–251, 1987.

Kimmel, R. and A. M. Bruckstein: Global shape from shading. *Computer Vision and Image Understanding,* **62**:360–369, 1995.

Kimmel, R. and J. A. Sethian: Optimal algorithm for shape from shading and path planning. *Journal of Mathematical Imaging and Vision,* **14**:237–244, 2001.

Kimmel, R., K. Siddiqi, B. B. Kimia, and A. M. Bruckstein: Shape from shading: Level set propagation and viscosity solutions. *International Journal of Computer Vision,* **16**:107–133, 1995.

Klette, R., R. Kozera, and K. Schlüns: Shape from shading and photometric stereo methods. In *Handbook of Computer Vision and Applications,* Vol. 2 of *Signal Processing and Pattern Recognition* (B. Jaehne, H. Haussecker, and P. Geissler, editors), pages 532–590, Academic Press Inc., San Diego, 1999.

Klette, R., K. Schlüns, and A. Koschan: *Computer Vision: Three Dimensional Data from Images.* Springer-Verlag, Singapore, 1998.

Kozera, R.: Uniqueness in shape from shading revisited. *Journal of Mathematical Imaging and Vision,* **7**:123–138, 1997.

Lee, C.-H. and A. Rosenfeld: Improved methods of estimating shape from shading using the light source coordinate system. *Artificial Intelligence,* **26**:125–143, 1985.

Lee, K. M. and C.-C. J. Kuo: Shape from shading with a linear triangular element surface model. *IEEE Transactions on Pattern Analysis and Machine Intelligence,* **15**:815–822, 1993.

Lions, P.-L.: *Generalized Solutions of Hamilton-Jacobi Equations.* Pitman, London, 1982.

Oliensis, J.: Uniqueness in shape from shading. *International Journal of Computer Vision,* **6**:75–104, 1991.

Onn, R. and A. Bruckstein: Integrability disambiguates surface recovery in two-image photometric stereo. *International Journal of Computer Vision,* **5**:105–113, 1990.

Osher, S. and J. A. Sethian: Fronts propagating with curvature dependent speed: Algorithms based on Hamilton-Jacobi formulation. *Journal of Computational Physics,* **79**:12–49, 1988.

Pentland, A. P.: Local shading analysis. *IEEE Transactions on Pattern Analysis and Machine Intelligence,* **6**:170–187, 1984.

Robles-Kelly, A. and E. R. Hancock: Model acquisition using shape-from-shading. In Proc. *The* $2^{\underline{nd}}$ *International Workshop on Articulated Motion and Deformable Objects* (F. J. Perales and E. R. Hancock, editors), pages 43–55, LNCS 2492, 2002.

Rouy, E. and A. Tourin: A viscosity solutions approach to shape-from-shading. *SIAM Journal of Numerical Analysis,* **29**:867–884, 1992.

Samaras, D. and D. Metaxas: Coupled lighting direction and shape estimation from single images. *Proceedings of the Seventh IEEE International Conference on Computer Vision,* **2**:868–874, 1999.

Sethian, J. A.: A fast marching level set method for monotonically advancing fronts. *Proceedings of the National Academy of Science of the USA* **93**:1591–1595, 1996.

Sethian, J. A.: A review of the theory, algorithms, and applications of level set methods for propagating interfaces. In *Acta Numerica,* (A. Iserles, editor), pages 309–395, Cambridge University Press, Cambridge, 1996.

Sethian, J. A.: *Level Set Methods and Fast Marching Methods: Evolving Interfaces in*

Computational Geometry, Fluid Mechanics, Computer Vision, and Materials Science. Cambridge Monograph on Applied and Computational Mathematics, Cambridge University Press, Cambridge, 1999.

Tankus, A., N. Sochen, and Y. Yeshurun: A new perspective on shape-from-shading. In Proc. *9$^{\underline{th}}$ IEEE International Conference on Computer Vision*, vol. 2, pages 862–869, 2003.

Tankus, A., N. Sochen, and Y. Yeshurun: Perspective shape-from-shading by fast marching. In Proc. *IEEE Computer Society Conference on Computer Vision and Pattern Recognition*, vol. 1, pages 43–49, 2004.

Tankus, A., N. Sochen, and Y. Yeshurun: Reconstruction of medical images by perspective shape-from-shading. In Proc. *International Conference on Pattern Recognition*, Vol. 3, pages 778–781, 2004.

Tsai, P.-S. and M. Shah: Shape from shading using linear approximation. *Image and Vision Computing,* **12**:487–498, 1994.

Woodham, R. J.: Photometric method for determining surface orientation from multiple images. In *Shape from Shading,* (B. K. P. Horn and M. J.Brooks, editors), pages 513–531, The MIT Press, 1989.

Zhang, R., P.-S. Tsai, J. E. Cryer, and M. Shah: Shape from shading: A survey. *IEEE Transactions on Pattern Analysis and Machine Intelligence*, **21**:690–705, 1999.

Zheng, Q. and R. Chellappa: Estimation of illuminant direction, albedo, and shape from shading. *IEEE Transactions on Pattern Analysis and Machine Intelligence*, **13**: 680–702, 1991.

SHAPE FROM SHADOWS

ATSUSHI IMIYA
IMIT, Chiba University
Yayoi-cho 1-33, Inage-ku, Chiba 263-8522, Japan

Abstract. This paper clarifies a sufficient condition for the reconstruction of an object from its shadows. The objects considered are finite closed convex regions in three-dimensional Euclidean space. We show a result that a series of pairs of shadows measured using a general stereo system with some geometrical assumptions is sufficient for full reconstruction of a convex object. Second, we show that a class of non-convex objects, which we define as slice convex objects, are also reconstractable from a series of shadows. Third, we clarify relations between shape reconstruction from shadows and image reconstruction from projections, which is the essential mathematical tool in medical imaging. Fourth we introduce a metric for similarity among objects from the collection of their shadows.

Key words: shape reconstruction, computerized tomography, convex object, non-convex object, voting, shape carving

14.1. Introduction

In n-dimensional Euclidean space $\mathbf{R}^n$ for $n \geq 2$, let $\boldsymbol{\omega}$ be the unit vector on S^{n-1}. A finite closed convex body K in $\mathbf{R}^n$ (Guggenheimer, 1977) is expressed as

$$K = \{\boldsymbol{x} | \boldsymbol{x}^\top \boldsymbol{\omega} \leq p(\boldsymbol{\omega}), \ \boldsymbol{\omega} \in S^{n-1}\}, \tag{1}$$

where $p(\boldsymbol{\omega})$ is the distance from the origin to a tangent plane to K. An intersection of a plane $\boldsymbol{\sigma}^\perp$, which is perpendicular to $\boldsymbol{\sigma}$ and passes through the origin, and finite closed convex body K is a shadow of K projected from the direction $\boldsymbol{\sigma}$, that is,

$$S(\boldsymbol{\sigma}) = \{\boldsymbol{x} | \boldsymbol{x}^\top \boldsymbol{\sigma} = 0\} \cap K. \tag{2}$$

Let $\partial S(\boldsymbol{\sigma})$ be the boundary curve of the shadow on plane $\boldsymbol{\sigma}^\perp$. Setting $\boldsymbol{x}^\top \boldsymbol{\omega} = p(\boldsymbol{\omega})$ to be the tangent plane to $\partial S(\boldsymbol{\sigma})$, it is possible to reconstruct a finite closed convex body K from shadows observed from all directions on S^{n-1} (Campi, 1986).

R. Klette et al. (eds.), Geometric Properties for Incomplete Data, 259-279.

For a point $\boldsymbol{a}$, a line in $\mathbf{R}^n$ is

$$l(\boldsymbol{a}, \boldsymbol{\omega}) = \{\boldsymbol{x} | \boldsymbol{x} = \boldsymbol{a} + t\boldsymbol{\omega},\ t \in \mathbf{R}\}. \tag{3}$$

For a finite closed convex body K,

$$C(\boldsymbol{a}) = \{\boldsymbol{\omega} | K \cap L(\boldsymbol{a}, \boldsymbol{\omega}) = \emptyset\}, \tag{4}$$

is the view cone with respect to the view point $\boldsymbol{a}$. A set

$$P(\boldsymbol{a}) = C(\boldsymbol{a}) \cap \boldsymbol{a}^{\perp}, \tag{5}$$

where $\boldsymbol{a}^{\perp}$ is the plane which is perpendicular to vector $\boldsymbol{a}$ and passes through the origin, is a shadow observed by the perspective projection. A set $P(\boldsymbol{a})$ corresponds the shadow observed by a pinhole camera in three-dimensional Euclidean space. In this paper, we deal with the problem of the reconstruction of K from $P(\boldsymbol{a})$ in three dimensional Euclidean space. This problem is called "Shape from shadow" (Laurentini, 1994; Laurentini, 1995; Aloimonos, 1988)

In the previous papers (Imiya and Kawamoto, 2001) and (Imiya and Kawamoto, 2002), we showed a sufficient condition for the full reconstruction of a finite closed convex body (Imiya and Kawamoto, 2001). Furthermore, we proved that a class of non-convex objects can be reconstructed fully from shadows (Imiya and Kawamoto, 2002). This paper is a sequel to these two papers. In this paper, we show the mathematical relations between "Shape from shadow" and object reconstruction from line integrals, which is the central mathematical tool for image reconstruction from projections in medical imaging. Furthermore, we deal with mathematical aspects of voting (Kawamoto and Imiya, 2001) and carving (Kutulakos and Seitz, 1999) methods for the object reconstruction.

In the classical publication (Radon, 1917), Radon introduced a problem to reconstruct a function from its line integrals and plane integrals on a plane and in a space, respectively. Nowadays, this theory is an irreplaceable mathematical tool for medical imaging (Tuy, 1983; Smith, 1985; Ludwing, 1966; Solmon, 1976; Hammaker, 1980; Kuba, 1984). Although in computer vision, a main reconstruction problem is the recovery of the object surface from photometric measurements (Zheng, 1994; Aloimonos, 1988; Vaillant and Faugeras, 1992), the reconstruction problem in medical imaging is the non-invasive visualisation of distributions in the interior of an object. In the reconstruction of a function from line integrals, we can extract the boundaries of projections and reconstructed function. Therefore, it is possible to reconstruct the object boundary from its line and plane integrals in three-dimensional Euclidean space (Stark and Peng, 1988; Prince and Willsky, 1990). This relation suggests that image reconstruction from projections and shape reconstruction from shadows have mathematically many relations (Gardner, 1995; Li, 1988; Richardson, 1996; Kölzow, 1989). In this paper, we re-summerise the results in our previous papers and show the relations between our results and mathematics in medical imaging. Furthermore, we introduce a metric for similarity among objects (Schwartz and Sharir, 1987) from the collection of their shadows. This metric measures similarity of objects from a collection of shadows without reconstructing objects from their projections.

14.2. Shape from Shadow

14.2.1. SUPPORT LINES AND PLANES

Setting $(x, y)^\top$ to be an orthogonal coordinate system on a plane, a support line of a planar convex object is expressed as

$$x\cos\theta + y\sin\theta = p(\theta), \tag{6}$$

where $\boldsymbol{\omega} = (\cos\theta, \sin\theta)^\top$ and $p(\theta)$ are the unit normal of this line and the distance from the origin to this line, respectively. The boundary of a closed convex object is reconstructed from the collection of support lines, for $0 \le \theta < 2\pi$, as

$$\begin{pmatrix} x \\ y \end{pmatrix} = \begin{pmatrix} \cos\theta & \sin\theta \\ -\sin\theta & \cos\theta \end{pmatrix} \begin{pmatrix} p(\theta) \\ \frac{d}{d\theta}p(\theta) \end{pmatrix}. \tag{7}$$

This analytical relation between a convex shape on a plane and its tangent planes was sometimes dealt with in computer vision by several authors (Zheng, 1994; Vaillant and Faugeras, 1992; Skiena, 1992; Skiena, 1991).

Setting $(x, y, z)^\top$ to be an orthogonal coordinate system on a plane, a support plane of a 3D convex object is expressed as

$$x\cos\phi\sin\theta + y\sin\phi\sin\theta + z\cos\theta = p(\theta, \phi), \tag{8}$$

where $\boldsymbol{\omega} = (\cos\phi\sin\theta, \sin\phi\sin\theta, \cos\theta)^\top$ and $p(\theta, \phi)$ are the unit normal of this plane and the distance from the origin to this plane, respectively. The boundary of a closed convex object is reconstructed from the collection of support planes, for $0 \le \theta \le \pi$ and $0 \le \phi < 2\pi$, as

$$\begin{pmatrix} x \\ y \\ z \end{pmatrix} = \begin{pmatrix} \cos\phi\sin\theta & \sin\theta\sin\phi & \cos\theta \\ \cos\phi\cos\theta & \sin\phi\cos\theta & -\sin\theta \\ -\sin\phi & \cos\phi & 0 \end{pmatrix} \begin{pmatrix} p(\theta, \phi) \\ \frac{\partial}{\partial\theta}p(\theta, \phi) \\ \frac{1}{\sin\theta}\frac{\partial}{\partial\phi}p(\theta, \phi) \end{pmatrix}. \tag{9}$$

This relation is seen in the reference (Guggenheimer, 1977) in the context of low-dimensional convex geometry.

If slices of an object along an z-axis are convex, we can reconstruct this object, for $0 \leq \theta < 2\pi$ as

$$\begin{pmatrix} x \\ y \\ z \end{pmatrix} = \begin{pmatrix} \cos\theta & \sin\theta & 0 \\ -\sin\theta & \cos\theta & 0 \\ 0 & 0 & 1 \end{pmatrix} \begin{pmatrix} p(\theta) \\ \frac{d}{d\theta}p(\theta) \\ z \end{pmatrix}. \tag{10}$$

Therefore, if slices of a non-convex object with respect to a direction are convex, we can reconstruct this non-convex object using Equation (10).

From shadows on the imaging planes, if we can construct $p(\theta)$ for $0 \leq \theta < 2\pi$ and $p(\theta, \phi)$, for $0 \leq \theta \leq \pi$ and $0 \leq \phi < 2\pi$, we can reconstruct an object. In this section, we show an orbits of the camera centre and a class of object which allow us to transform from shadows to $p(\theta)$ for $0 \leq \theta < 2\pi$ and $p(\theta, \phi)$, for $0 \leq \theta \leq \pi$ and $0 \leq \phi < 2\pi$.

14.2.2. RECONSTRUCTION OF CONVEX OBJECT

Setting $\boldsymbol{a}$ to be the origin of rays for the observation of shadow, we denote the shadow of K on an imaging plane as $\tilde{K}(\boldsymbol{a})$ and denote the boundary of shadow as $\partial\tilde{K}(\boldsymbol{a})$. Furthermore, we denote the boundary curve of $\partial\tilde{K}(\boldsymbol{a})$ as $\boldsymbol{r}(\boldsymbol{a})$. The boundary of shadow $\partial\tilde{K}(\boldsymbol{a})$ is a closed curve on an imaging plane. The vector

$$\boldsymbol{l}(\boldsymbol{a}) = \boldsymbol{r}(\boldsymbol{a})/|\boldsymbol{r}(\boldsymbol{a})|, \ \boldsymbol{r}(\boldsymbol{a}) \in \partial\tilde{K}(\boldsymbol{a}) \tag{11}$$

is the N-vector (Kanatani, 1993) of $\boldsymbol{r}(\boldsymbol{a})$ in the world coordinate system (Kanatani, 1993). $\boldsymbol{l}(t, \boldsymbol{a})$ moves on a closed curve $\partial L(\boldsymbol{a})$ on the unit sphere for each $\boldsymbol{a}$. We call these closed curve $\partial L(\boldsymbol{a})$ the N-curve of $\boldsymbol{r}(\boldsymbol{a})$.

For a point $\tilde{\boldsymbol{x}} \in \tilde{K}(\boldsymbol{a})$, setting $\boldsymbol{x}(\boldsymbol{a})$ to be the N-vector of $\tilde{\boldsymbol{x}}$, we define a cone

$$C(\boldsymbol{a}) = \{\boldsymbol{x} | \boldsymbol{x} = \lambda \boldsymbol{x}(\boldsymbol{a}), \lambda \geq 0\}. \tag{12}$$

We call $C(\boldsymbol{a})$ the view-cone at $\boldsymbol{a}$. The boundary of $C(\boldsymbol{a})$ is

$$\partial C(\boldsymbol{a}) = \{\boldsymbol{x} | \boldsymbol{x} = \lambda \boldsymbol{l}(\boldsymbol{a}), \boldsymbol{l}(\boldsymbol{a}) \subset \partial L(\boldsymbol{a}), \lambda \geq 0\}. \tag{13}$$

If a pair of view-cones which have the same vertex satisfy the relation

$$C_1(\boldsymbol{a}) \subseteq C_2(\boldsymbol{a}), \tag{14}$$

we write $\partial L_1(\boldsymbol{a}) \preceq \partial L_2(\boldsymbol{a})$, where $\partial L_i(\boldsymbol{a})$ is called the associated N-curve of a view-cone $C_i(\boldsymbol{a})$.

If $\boldsymbol{l}(t, \boldsymbol{a})$ moves on $\partial L(\boldsymbol{a})$, $\boldsymbol{n}(t, \boldsymbol{a})$ moves on the unit sphere and forms a closed curve $\partial N(\boldsymbol{a})$, which we call the orthogonal N-curve. From geometrical consideration, it is obvious that if $\partial L_1(\boldsymbol{a}) \preceq \partial L_2(\boldsymbol{a})$, then $\partial N_1(\boldsymbol{a}) \succeq \partial N_2(\boldsymbol{a})$. Furthermore, setting

$$\partial C(\boldsymbol{a})^{\perp} = \{\boldsymbol{x} | \boldsymbol{x} = \lambda \boldsymbol{n}(\boldsymbol{a}), \lambda \geq 0\} \tag{15}$$

if Equation (14) holds, then the relation $C_1(\boldsymbol{a})^{\perp} \supseteq C_2(\boldsymbol{a})^{\perp}$ holds. Setting $C_1(\boldsymbol{a})$ and $C_2(\boldsymbol{a})$ to be view-cones of K and B to be a sphere which encircles K, respectively, we obtain the following theorem, where the origin of the world coordinate system is at the centre of B.

THEOREM 14.1. *For a bounded closed set A in $\mathbf{R^3}$, if*

$$\bigcup_{\boldsymbol{a} \in \boldsymbol{A}} C_2(\boldsymbol{a})^{\perp} \supseteq \mathbf{R^3}, \tag{16}$$

then we can reconstruct K from shadows which are obtained by perspective projection.

Proof $\forall \boldsymbol{a} \in \boldsymbol{A} \subset \mathbf{R^3}$, the relation $C_1(\boldsymbol{a})^{\perp} \subseteq \mathbf{R^3}$ holds. If $\bigcup_{\boldsymbol{a} \in \boldsymbol{A}} C_2(\boldsymbol{a})^{\perp} \supseteq \mathbf{R^3}$, then we obtain

$$\mathbf{R^3} \subseteq \bigcup_{\boldsymbol{a} \in \boldsymbol{A}} \mathbf{C_2}(\boldsymbol{a})^{\perp} \subset \bigcup_{\boldsymbol{a} \in \boldsymbol{A}} \mathbf{C_1}(\boldsymbol{a})^{\perp} \subseteq \mathbf{R^3}. \tag{17}$$

This relation concludes the relation $\bigcup_{\boldsymbol{a} \in \boldsymbol{A}} C_1(\boldsymbol{a})^{\perp} = \mathbf{R^3}$. Furthermore, this equation leads to $\bigcup_{\boldsymbol{a} \in \boldsymbol{A}} n_1(t, \boldsymbol{a}) = \mathbf{S^2}$. □

In the following, we show some examples of a bounded closed set $\boldsymbol{A}$.

EXAMPLE 14.1. *Let P_1 and P_2 be a pair of perpendicular planes which pass through the centre of B. Setting $\boldsymbol{a}_1$ and $\boldsymbol{a}_2$ to be circles on P_1 and P_2 of the centre of which are at the centre of B with radii a and b, respectively, if $a^{-2} + b^{-2} > r^{-2}$, where r is the radius of B, then $\boldsymbol{a}_1 \cup \boldsymbol{a}_2$ is an example of $\boldsymbol{A}$.*

EXAMPLE 14.2. *Let P_1 and P_2 be a pair of parallel planes which touch B. Setting $\boldsymbol{a}_1$ and $\boldsymbol{a}_2$ to be circles with the radius d, the centre of which are on B, if $d > r$, $\boldsymbol{d}_1 \cup \boldsymbol{d}_2$ is an example of $\boldsymbol{A}$.*

14.2.3. RECONSTRUCTION OF NON-CONVEX OBJECTS

Equation (7) concludes that if an object is expressed as generalised cylinder of convex planar shape and if we know the axis of this object, we can reconstruct this object from a series of two-dimensional perspective projections. For the reconstruction of each slice, we are required to observe shadows from all points on a circle which encircles each slice. Figure 14.1 (a) illustrates the reconstruction of generalised cylinder from a series of perspective projections.

In the following, we assume that on each slice there exists only one closed finite region.

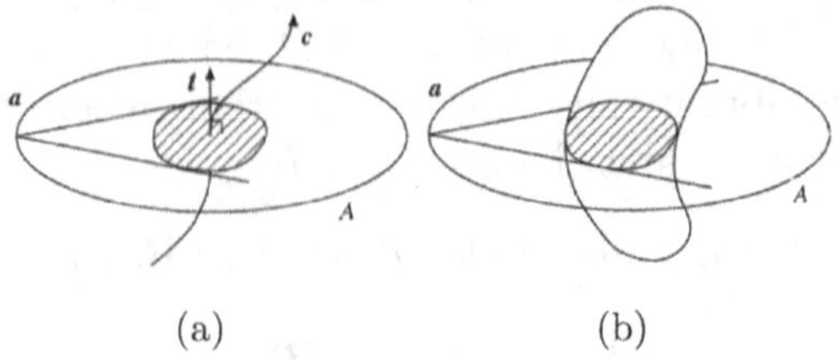

Figure 14.1. Reconstruction of Non-Convex Object: (a) A slice and its support lines of a generalised cylinder whose slices are convex, where **c**, **t**, **a**, and **A** are the axis of a generalised cylinder, the tangent vector of the axis curve at a slice, the camera centre, and a circle on which the camera centre moves, respectively. (b) A slice of non-convex object and its support lines, where ***a*** and ***A*** are the camera centre, and a circle on which the camera centre moves, respectively.

When a slice of an object is convex and we can observe this slice from all points on the circle which encircles this slice on the same plane, we can reconstruct this slice using two-dimensional method. Since all slices of a convex object are convex, we can reconstruct a convex object from a collection of slices which are perpendicular to a axis of slice.

For the reconstruction of an object from two-dimensional perspective projections, each point on the boundary is required to lie at least on a convex slice as shown in Figure 14.1 (b). A banana is an example of such a shape.

Using a geometry transform, it is possible to obtain support lines of a slice from all directions. This geometrical property implies the following lemma, when we observe a series of shadows from vertices which lie on a sphere encircling this object.

LEMMA 14.1. *From the collection of shadows which observed from vertices which lie on a sphere encircling this object, we can obtain the collection of two-dimensional perspective projections of a slice form a point which moves on a circle encircling this object.*

For any points on the boundary, if there exists at least one unique convex slice curve which contains this point, we call this object a slice convex object. A convex closed object is slice convex.

A banana-shape object in Figure 14.1 (b) is non-convex but slice convex. This geometric property and Lemma 14.1 derives the following theorem.

THEOREM 14.2. *A slice convex object is uniquely reconstructible from the collection of shadows observed from vertices which lie on the whole sphere encircling this object.*

This theorem permits us for the reconstruction of a class of non-convex objects from shadows. Furthermore, In this expression, the axis for the reconstruction is not required to be a straight line.

Next, we show an example camera motion for the measurement of two-dimensional perspective projections of a slice convex object.

EXAMPLE 14.3. *If the camera centre moves on the union of three cylinders, as shown in Figures 14.2 (a) and (b),*

$$\begin{array}{l} x^2 + y^2 = R^2,\ |Z| \leq L,\ R < L \\ y^2 + z^2 = R^2,\ |Z| \leq L,\ R < L \\ z^2 + x^2 = R^2,\ |Z| \leq L,\ R < L \end{array} \tag{18}$$

keeping the optical axis of the camera to path through the origin, we can obtain the same collection of shadows with measured from the camera whose centre moves on the sphere.

This camera motion is easier than the spherical motion for the practical instrumentation since three fixed axes of rotation are required for the measurement of shadows, since as shown in Figure 14.2 (b) this measuring system has three axes of rotation.

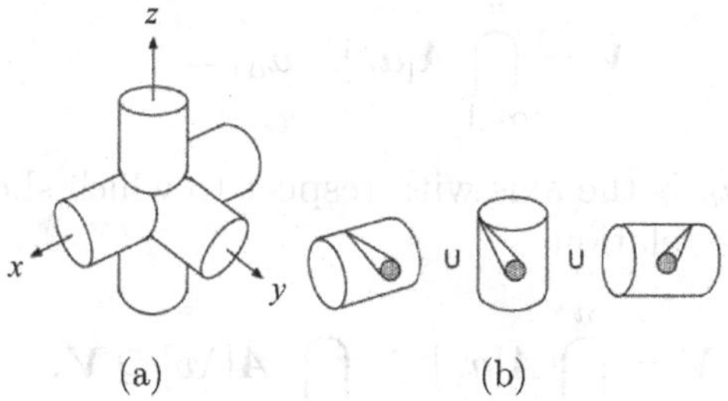

Figure 14.2. Three Orthogonal Cylinders on which the Camera Centre Moves.

14.2.4. RECONSTRUCTION OF SLICE CONVEX OBJECTS

For a slice convex object $\boldsymbol{V}$ with respect to axis $\lambda \boldsymbol{v}_0$ for $|\boldsymbol{v}_0| = 1$ and $\lambda \neq 0$, setting $\boldsymbol{A}[\boldsymbol{v}]$ to be a reconstructed object with respect to the axis $\lambda \boldsymbol{v}$, for $\lambda \neq 0$, we have the following theorem

THEOREM 14.3. *For an object* $\boldsymbol{V}$ *the relation*

$$\boldsymbol{V} = \bigcap_{\boldsymbol{v} \in \mathbf{S}^2} \boldsymbol{A}[\boldsymbol{v}] \tag{19}$$

is satisfied if $\boldsymbol{V}$ *is slice convex with respect to axis* $\lambda \boldsymbol{v}_0$.

Proof For vectors $\boldsymbol{v}_0 \neq \boldsymbol{v}$

$$\boldsymbol{V} = \boldsymbol{A}[\boldsymbol{v}_0] \supseteq \bigcap_{\boldsymbol{v}\in\mathbf{S}^2} \boldsymbol{A}[\boldsymbol{v}] \supseteq \boldsymbol{V}. \tag{20}$$

Since $\boldsymbol{V}$ is slice convex with respect to axis $\lambda\boldsymbol{v}_0$, we have the relation

$$\boldsymbol{V} = \bigcap_{\boldsymbol{v}\in\mathbf{S}^2} \boldsymbol{C}[\boldsymbol{v}]. \tag{21}$$

These two relations imply the theorem. □

In the most of previous works for the reconstruction of object from shadows, a method is applied for objects which are rotationally symmetry. and, in these works, the symmetry axis with respect to which an object is slice convex are assumed to be pre-determined. However, this theorem implies that without pre-detecting the axes for the slice convex we can reconstruct this object if we can measure shadows of slices using perspective projections from vertices over the sphere. This is an important result derived from geometric analysis in this paper.

If an object is defined as the common region of a finite number of slice convex objects, that is, object $\boldsymbol{V}$ is expressed as

$$\boldsymbol{V} = \bigcap_{\alpha=1}^{n} \boldsymbol{A}[\boldsymbol{a}_\alpha], \ |\boldsymbol{a}_\alpha| = 1, \tag{22}$$

for $\lambda \neq 0$, where $\lambda\boldsymbol{a}_\alpha$ is the axis with respect to which slices of an object is convex, we have the relation

$$\boldsymbol{V} = \bigcap_{\alpha=1}^{n} \boldsymbol{A}[\boldsymbol{a}_\alpha] \supseteq \bigcap_{\boldsymbol{v}\in S^2} \boldsymbol{A}[\lambda\boldsymbol{v}] \supseteq \boldsymbol{V}. \tag{23}$$

This relation leads to the following theorem.

THEOREM 14.4. *Object $\boldsymbol{V}$ is reconstructed as*

$$\boldsymbol{V} = \bigcap_{\boldsymbol{v}\in S^2} \boldsymbol{A}[\boldsymbol{v}]. \tag{24}$$

14.2.5. EXAMPLES OF SLICE CONVEX OBJECTS

We show some examples of slice convex objects and axes with respect of which objects are slice convex.

For a set of points $\boldsymbol{A}$ and a constant $\lambda > 0$, we define

$$\lambda\boldsymbol{A} = \{\boldsymbol{y}|\boldsymbol{y} = \lambda\boldsymbol{x},\, \boldsymbol{x} \in \boldsymbol{A}\}, \tag{25}$$

and

$$\begin{aligned} \boldsymbol{A} \oplus \{\boldsymbol{a}\} &= \boldsymbol{A_a} \\ &= \{\boldsymbol{y}|\boldsymbol{y} = \boldsymbol{x} + \boldsymbol{a},\, \boldsymbol{x} \in \boldsymbol{A}\}. \end{aligned} \tag{26}$$

Setting $\boldsymbol{B}$ to be the unit-volume cube whose centroid is at the origin of the coordinate, that is,

$$\boldsymbol{B} = \{(x, y, z)^\top | |x| \leq \frac{1}{2},\, |y| \leq \frac{1}{2},\, |z| \leq \frac{1}{2}\}, \tag{27}$$

a set

$$\boldsymbol{B_L} = \overline{\boldsymbol{A} \setminus (\frac{1}{2}\boldsymbol{B_a} \cup \frac{1}{2}\boldsymbol{B_b})} \tag{28}$$

for $\boldsymbol{a} = \frac{1}{4}(\boldsymbol{e}_1 + \boldsymbol{e}_2 + \boldsymbol{e}_3)$ and $\boldsymbol{b} = \frac{1}{4}(\boldsymbol{e}_1 - \boldsymbol{e}_2 + \boldsymbol{e}_3)$, is a L-shape polyhedron and a slice convex object with respect to the axes $\lambda\boldsymbol{e}_2$ and $\lambda\boldsymbol{e}_3$ as shown in Figure 14.3 (a).

Therefore, the shape holds the relation

$$\boldsymbol{B_L} = \bigcup_{\lambda \in \mathbf{R}} \boldsymbol{A}[\boldsymbol{e}_2],\ \boldsymbol{B_L} = \bigcup_{\lambda \in \mathbf{R}} \boldsymbol{A}[\boldsymbol{e}_3]. \tag{29}$$

It is reconstructible from perspective projections, if the camera centre moves on a sphere which encircles this object.

The L-shape block is slice convex with respect to axes $\lambda(-\boldsymbol{e}_2 + \boldsymbol{e}_3)$ and curve $c(s)$,

$$s(s) = \begin{cases} \left(\frac{1}{2} - s\right)\boldsymbol{e}_3 - \frac{1}{4}\boldsymbol{e}_2, & 0 \leq s \leq \frac{1}{2} \\ \frac{1}{\sqrt{2}}(s - \frac{1}{2})(\boldsymbol{e}_2 - \boldsymbol{e}_3) - \frac{1}{4}\boldsymbol{e}_3, & \frac{1}{2} \leq s \leq \frac{1}{2} + \frac{\sqrt{2}}{4} \\ \left\{(1 + \frac{\sqrt{2}}{4} - s\right\}\boldsymbol{e}_2 - \frac{1}{4}\boldsymbol{e}_3, & \frac{1}{2} + \frac{\sqrt{2}}{4} \leq s \leq 1 + \frac{\sqrt{2}}{4}. \end{cases} \tag{30}$$

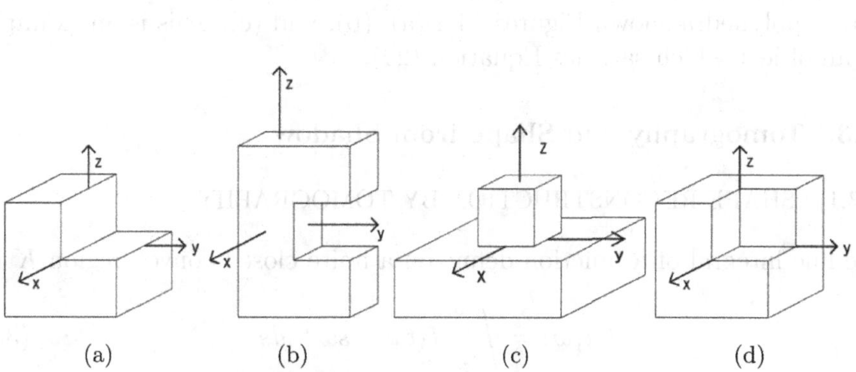

Figure 14.3. Examples of Slice Convex and Non-Slice Convex Polyhedra. (a), (b), and (c) are slice convex polyhedra and (d) is a non-slice convex polyhedron. It is possible to reconstruct polyhedra (a), (b), and (c) from shadows. (d) is non-slice-convex object.

Setting $P(s)$ to be the slice perpendicular to curve $c(s)$ the boundary of this block on the plane $P(s)$ is a rectangle for $\frac{1}{2} \leq s \leq \frac{1}{2} + \frac{\sqrt{2}}{4}$.

A polyhedron

$$\boldsymbol{B}_B = \overline{\cup_{i=0}^{4} \boldsymbol{B}_i} \tag{31}$$

such that

$$\begin{array}{ll} \boldsymbol{B}_0 = \frac{1}{2}\boldsymbol{B} & \boldsymbol{B}_1 = \frac{1}{2}\boldsymbol{B} \oplus \{\frac{1}{2}\boldsymbol{e}_3\} \\ \boldsymbol{B}_2 = \frac{1}{2}\boldsymbol{B} \oplus \{-\frac{1}{2}\boldsymbol{e}_3\} & \boldsymbol{B}_3 = \frac{1}{2}\boldsymbol{B} \oplus \{\frac{1}{2}(\boldsymbol{e}_2 + \boldsymbol{e}_3)\} \\ \boldsymbol{B}_4 = \frac{1}{2}\boldsymbol{B} \oplus \{\frac{1}{2}(-\boldsymbol{e}_2 + \boldsymbol{e}_3\} & \end{array} \tag{32}$$

is a banana shape polyhedron which is slice convex with respect to axis $\lambda \boldsymbol{e}_3$ as shown in Figure 14.3 (b). Therefore, this shape also holds the relation

$$\boldsymbol{B}_B = \overline{\cup_{\lambda \in \mathbf{R}} \boldsymbol{B}[\boldsymbol{e}_3]}, \tag{33}$$

A polyhedron

$$\boldsymbol{N} = \overline{\boldsymbol{B} \setminus \left(\frac{1}{2}\boldsymbol{C} \oplus \{\boldsymbol{v}\}\right)} \tag{34}$$

is, however, a non-convex and non-slice convex object as shown in Figure 14.4 (d) for vector $\boldsymbol{v} = \frac{1}{4}(\boldsymbol{e}_1 + \boldsymbol{e}_2 + \boldsymbol{e}_3)$

A polyhedron in Figure 14.4 (d) is the common polyhedron of three slice convex polyhedra shown Figures 14.4 (a), (b), and (c). This is an example for an object which satisfies Equation (22).

14.3. Tomography and Shape from Shadow

14.3.1. SHAPE RECONSTRUCTION BY TOMOGRAPHY

The line integral of a function define on a finite closed convex region K in $\mathbf{R}^2$ is

$$\check{f}(t, \omega) = \int_{-\infty}^{\infty} f(t\boldsymbol{\omega} + s\boldsymbol{\omega}^{\perp})ds, \tag{35}$$

where $\boldsymbol{\omega} = (\cos\theta, \sin\theta)^{\top} \in S$ and $\boldsymbol{\omega}^{\perp} = (-\sin\theta, \cos\theta)^{\top}$. If we reconstruct $f(\boldsymbol{x})$ from $\check{f}$, we can automatically reconstruct the boundary of K. Therefore, if we have the line integrals of an object on slices perpendicular to the z-axis. We can reconstruct the boundary of an object in a space. This

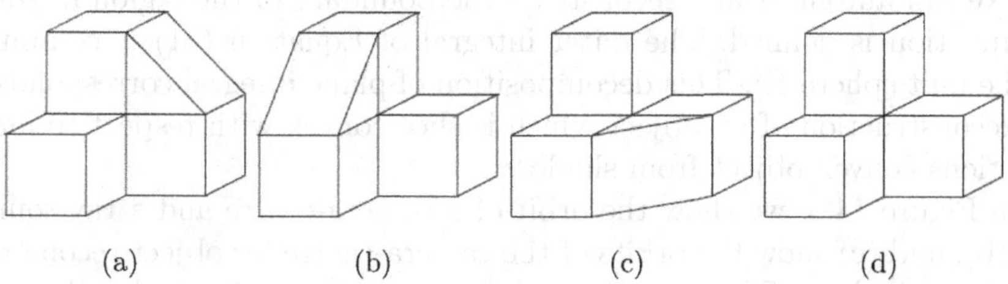

Figure 14.4. Example of a Polyhedron which is the Common Part of Slice-Convex Polyhedra. Polyhedron (d) is constructed by the common part of polyhedra (a), (b), and (c), which are slice convex, although polyhedron (d) is non-slice convex.

geometrical property allows us to reconstruct an object in a space from a collection of line integral on a series of parallel slices.

The plane integral of a function in $\mathbf{R}^3$ is

$$\check{f}(t,\omega) = \int_{\omega^\perp} f(t\boldsymbol{\omega} + x\boldsymbol{\theta} + y\boldsymbol{\sigma})dxdy, \tag{36}$$

where $\boldsymbol{\omega}^\top\boldsymbol{\theta} = 0$, and $\boldsymbol{\sigma} = \boldsymbol{\omega} \times \boldsymbol{\theta}$, for $\boldsymbol{\omega} \in S^2$ and $\boldsymbol{\theta} \in S^2$. If we reconstruct $f(\boldsymbol{x})$ from $\check{f}$,

$$f(\boldsymbol{x}) = -\frac{1}{8\pi^2}\int_{S^2} \check{f}^{(2)}(\boldsymbol{x}^\top\boldsymbol{\omega}, \boldsymbol{\omega})d\boldsymbol{\omega}, \tag{37}$$

we can automatically reconstruct the boundary of K in a space.

Setting

$$h(\boldsymbol{\omega}^\perp, \boldsymbol{\omega}) = \int_{-\infty}^{\infty} f(\boldsymbol{\omega}^\perp + x\boldsymbol{\omega})dx, \tag{38}$$

where $\boldsymbol{\omega}^\perp = (t,s)^\top$, we have the relation

$$f(\boldsymbol{x}) = \int_{S^2} \Delta^{\frac{1}{2}} h(P_{\boldsymbol{\omega}^\perp}\boldsymbol{x}, \boldsymbol{\omega})d\boldsymbol{\omega}, \tag{39}$$

for Laplacian Δ, and

$$\check{f}(t,\boldsymbol{\omega}) = \int_{-\infty}^{\infty} h(\boldsymbol{\omega}^\perp, \boldsymbol{\omega})ds. \tag{40}$$

Equation (39) reconstructs the distribution in the object and the boundary of the object. The decomposition of a plane integral to the two-step line integrals in Equation (40) means that from two-dimensional distributions on the planes $\boldsymbol{\omega}^\perp = (t,s)^\top$ computed by line integrals in a space, we can reconstruct functions as

$$f(\boldsymbol{x}) = -\frac{1}{8\pi^2}\int_{S^2} \frac{\partial^2}{\partial t^2}\left\{\int_{-\infty}^{\infty} h(\boldsymbol{\omega}^\perp, \boldsymbol{\omega})ds\right\}\Bigg|_{t=\boldsymbol{x}^\top\boldsymbol{\omega}} d\boldsymbol{\omega} \tag{41}$$

and we can automatically reconstruct the boundary of the region in which the function is defined. The outer integral of Equation (41) is computed on the unit sphere S^2. This decomposition of plane integral corresponds to the reconstruction of an object which is slice-convex with respect to many directions convex object from shadows.

In Figure 14.5, we show the orbit of a camera-centre and a ray-source. (a), (b), and (c) show the orbits of the camera-centre for object reconstruction from shadows. These configurations reconstruct a finite closed convex body fully. (c), (d), and (e) show the orbits of the ray-source for image reconstruction from projections. Configurations (d) and (e) reconstruct a function from its cone beam projections (Tuy, 1983; Smith, 1985; Axelsson, 1994). Configurations (a) and (d), and (b) and (e) correspond. However, for the reconstruction of a function from its line integrals obtained using con-beam projections, we need a line which connects two parallel circles as a part of the ray-source orbit. For both shape from shadow and image-reconstruction from projections, since a sphere is equivalent to the common region of cylinders whose axial-direction moves on the unit sphere as shown in Figure 14.6, it is possible to convert a three-dimensional problem to a collection of two-dimensional problems.

Setting $f(\boldsymbol{x})$ to be a positive, and integrable and square integrable function defined in K, for a $\boldsymbol{a} \in \mathbf{R}^3$ and $\boldsymbol{\omega} \in \mathbf{S}^2$,

$$g(\boldsymbol{a}, \boldsymbol{\omega}) = \int_0^\infty f(\boldsymbol{a} + t\boldsymbol{\omega})dt \tag{42}$$

is the divergent x-ray transform (Hammaker, 1980). Reconstruction of $f(\boldsymbol{x})$ from $g(\boldsymbol{a}, \boldsymbol{\omega})$ is the mathematical model for the reconstruction of volume distributions form cone beams for the x-ray computerised tomography. The relation

$$\check{f}(P_{\boldsymbol{\omega}}\boldsymbol{a}, \boldsymbol{\omega}) = g(\boldsymbol{a}, \boldsymbol{\omega}) + g(\boldsymbol{a}, -\boldsymbol{\omega}), \tag{43}$$

where $P_{\boldsymbol{\omega}}$ is the orthogonal projector to $\boldsymbol{\omega}^\perp$, allows us to reconstruct an object from collection of divergent projections. This relation corresponds to the reconfiguration of support lines and planes for the reconstruction of an object from view cones.

If source point $\boldsymbol{a}$ moves on the orbits which is same geometry with the view-points orbit defined in example 2 for $a = b$, it is possible to reconstruct fully $f(\boldsymbol{x})$ from $g(\boldsymbol{a}, \boldsymbol{\omega})$. Therefore, from $g(\boldsymbol{a}, \boldsymbol{\omega})$, we can reconstruct ∂K in the same condition with Example 14.1.

On the other hand our data are shadows of $f(\boldsymbol{x})$ measured by perspective projections. Therefore, denoting the characteristic function of $g(\boldsymbol{a}, \boldsymbol{\omega})$ and the ray cone as

$$\chi(\boldsymbol{a}, \boldsymbol{\omega}) = \begin{cases} 1, & \text{if } g(\boldsymbol{a}, \boldsymbol{\omega}) > 0 \\ 0, & \text{otherwise,} \end{cases} \tag{44}$$

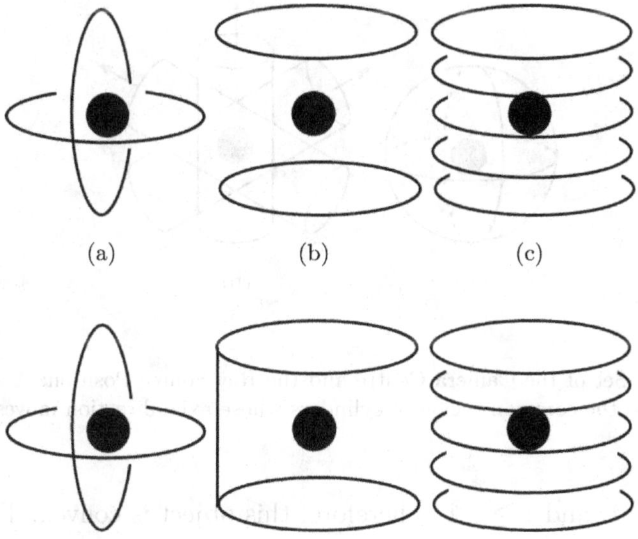

Figure 14.5. Orbit of a Camera-Centre and a Ray-Source: (a), (b), and (c) show the orbits of the camera-centre for object reconstruction from shadows. These configurations reconstruct a finite closed convex body fully. (c), (d), and (e) show the orbits of the ray-source for image reconstruction from projections. Configurations (d) and (e) reconstruct a function from its cone beam projections. Configuration (f) is for multi-slice method for the reconstruction.

and

$$C(\boldsymbol{a},\boldsymbol{\omega}) = \{(\boldsymbol{a},\boldsymbol{\omega}) \mid \chi(\boldsymbol{a},\boldsymbol{\omega}) = 1,\ \boldsymbol{a} \in \mathbf{R}^3, \boldsymbol{\omega} \in \mathbf{S}^2\}, \tag{45}$$

respectively, we obtain the following relations

$$K = \bigcap_{\boldsymbol{a}\in\boldsymbol{A}} C(\boldsymbol{a},\boldsymbol{\omega}),\ \partial K = \bigcap_{\boldsymbol{a}\in\boldsymbol{A}} \partial C(\boldsymbol{a},\boldsymbol{\omega}), \tag{46}$$

where $\partial C(\boldsymbol{a},\boldsymbol{\omega})$ is the boundary of $C(\boldsymbol{a},\boldsymbol{\omega})$.

The support plane method for shape reconstruction is an algebraic expression of the second equation of Equation (46). These geometric properties show a mathematical relationship between shape from shadows and the image reconstruction form projections of the x-ray computerised tomography since shape from shadows focuses to shape reconstruction.

For comparison of shape from shadows and and image reconstruction from projections. First we reconstruct a cube from perspective projections. Our cube $\boldsymbol{B}$ is the intersection of six half spaces $x \leq 1$, $x \geq -1$, $y \leq 1$,

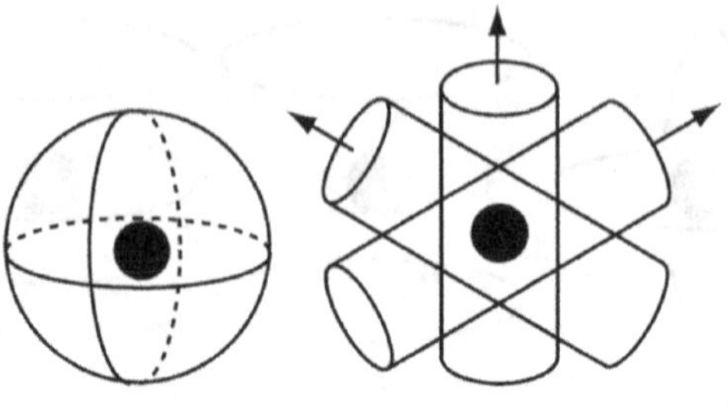

Figure 14.6. Set of the Camera Centre and the Ray-Source Position: A sphere in (a) is equivalent to the common region of cylinders whose axial-direction moves on the unit sphere (b).

$y \geq -1$, $z \leq 1$, and $z \geq -1$. Therefore, this object is convex. The vertices of cube $\boldsymbol{B}$ is

$$\begin{array}{ll} \boldsymbol{p}_1 = (1,1,1)^\top, & \boldsymbol{p}_2 = (-1,1,1)^\top, \\ \boldsymbol{p}_3 = (-1,-1,1)^\top, & \boldsymbol{p}_4 = (1,-1,1)^\top \\ \boldsymbol{q}_1 = (1,1,-1)^\top, & \boldsymbol{q}_2 = (-1,1,-1)^\top, \\ \boldsymbol{q}_3 = (-1,-1,-1)^\top, & \boldsymbol{q}_4 = (1,-1,-1)^\top. \end{array}$$

When the centre of camera moves on a circle $x^2 + y^2 = R^2, R > 1$, setting $\delta = \sin^{-1} \frac{1}{R}$, the support planes are

$$\begin{array}{lll} -\delta \leq \theta \leq \delta : & |\boldsymbol{\xi}, \boldsymbol{p}_{n-1}, \boldsymbol{p}_n, \boldsymbol{a}| = 0, & |\boldsymbol{\xi}, \boldsymbol{p}_n, \boldsymbol{q}_n, \boldsymbol{a}| = 0, \\ & |\boldsymbol{\xi}, \boldsymbol{p}_{n-1}, \boldsymbol{p}_n, \boldsymbol{a}| = 0, & |\boldsymbol{\xi}, \boldsymbol{q}_{n-1}, \boldsymbol{p}_{n-1}, \boldsymbol{a}| = 0 \\ \delta \leq \theta \leq -\delta + \frac{\pi}{2} : & |\boldsymbol{\xi}, \boldsymbol{p}_n, \boldsymbol{p}_{n+1}, \boldsymbol{a}| = 0, & |\boldsymbol{\xi}, \boldsymbol{p}_{n+1}, \boldsymbol{p}_{n+2}, \boldsymbol{a}| = 0, \\ & |\boldsymbol{\xi}, \boldsymbol{p}_{n+2}, \boldsymbol{q}_{n+2}, \boldsymbol{a}| = 0, & |\boldsymbol{\xi}, \boldsymbol{q}_{n+2}, \boldsymbol{q}_{n+1}, \boldsymbol{a}| = 0, \\ & |\boldsymbol{\xi}, \boldsymbol{q}_{n+1}, \boldsymbol{q}_n, \boldsymbol{a}| = 0, & |\boldsymbol{\xi}, \boldsymbol{q}_n, \boldsymbol{p}_n, \boldsymbol{a}| = 0, \end{array} \quad (47)$$

where $\boldsymbol{a} = (R\cos\theta, R\sin\theta, 0)^\top$ and $\boldsymbol{\xi} = (x, y, z, 1)^\top$. These support planes are periodic with respect to argument θ with period $\frac{\pi}{2}$ and do not express $z = 1$ and $z = -1$. Therefore, we cannot reconstruct cubes from perspective projections from the view cones whose vertices lie on a circle encircling this cube.

Assuming that point $\boldsymbol{b} = (0, R\sin\phi, R\cos\phi)^\top$ moves on circle $y^2 + z^2 = R^2$, the support planes of cube $\boldsymbol{B}$ which path through point $\boldsymbol{b}$ are

$$\begin{array}{lll} -\delta \leq \theta \leq \delta : & |\boldsymbol{\xi}, \boldsymbol{p}_{n-1}, \boldsymbol{p}_n, \boldsymbol{a}| = 0, & |\boldsymbol{\xi}, \boldsymbol{p}_n, \boldsymbol{q}_n, \boldsymbol{a}| = 0, \\ & |\boldsymbol{\xi}, \boldsymbol{p}_{n-1}, \boldsymbol{p}_n, \boldsymbol{a}| = 0, & |\boldsymbol{\xi}, \boldsymbol{q}_{n-1}, \boldsymbol{p}_{n-1}, \boldsymbol{a}| = 0 \\ \delta \leq \theta \leq -\delta + \frac{\pi}{2} : & |\boldsymbol{\xi}, \boldsymbol{p}_n, \boldsymbol{p}_{n+1}, \boldsymbol{a}| = 0, & |\boldsymbol{\xi}, \boldsymbol{p}_{n+1}, \boldsymbol{p}_{n+2}, \boldsymbol{a}| = 0, \\ & |\boldsymbol{\xi}, \boldsymbol{p}_{n+2}, \boldsymbol{q}_{n+2}, \boldsymbol{a}| = 0, & |\boldsymbol{\xi}, \boldsymbol{q}_{n+2}, \boldsymbol{q}_{n+1}, \boldsymbol{a}| = 0, \\ & |\boldsymbol{\xi}, \boldsymbol{q}_{n+1}, \boldsymbol{q}_n, \boldsymbol{a}| = 0, & |\boldsymbol{\xi}, \boldsymbol{q}_n, \boldsymbol{p}_n, \boldsymbol{a}| = 0, \end{array} \quad (48)$$

for

$$\begin{array}{l} \bar{\boldsymbol{p}}_1 = \boldsymbol{p}_1,\ \bar{\boldsymbol{p}}_2 = \boldsymbol{q}_1,\ \bar{\boldsymbol{p}}_3 = \boldsymbol{p}_2,\ \bar{\boldsymbol{p}}_4 = \boldsymbol{q}_2, \\ \bar{\boldsymbol{q}}_1 = \boldsymbol{p}_4,\ \bar{\boldsymbol{q}}_2 = \boldsymbol{q}_4,\ \bar{\boldsymbol{q}}_3 = \boldsymbol{q}_3,\ \bar{\boldsymbol{q}}_4 = \boldsymbol{p}_3. \end{array} \tag{49}$$

These support planes are periodic with respect to argument ϕ with period $\frac{\pi}{2}$.

This collection of planes contains planes $z = 1$ and $z = -1$. Therefore, using the orbit of the camera centre for the perspective projections, we can completely reconstruct a cube.

This collection of planes contains planes $z = 1$ and $z = -1$. Therefore, using the orbit of the camera centre for the perspective projections, we can completely reconstruct a cube. In Figure 14.7 (a), we show a shadow of a cube and a view cone of this cube and in Figure 14.7 (b), the view cones for $\theta = 0$, $\theta = \frac{\pi}{4}$, and $\theta = \frac{\pi}{2}$.

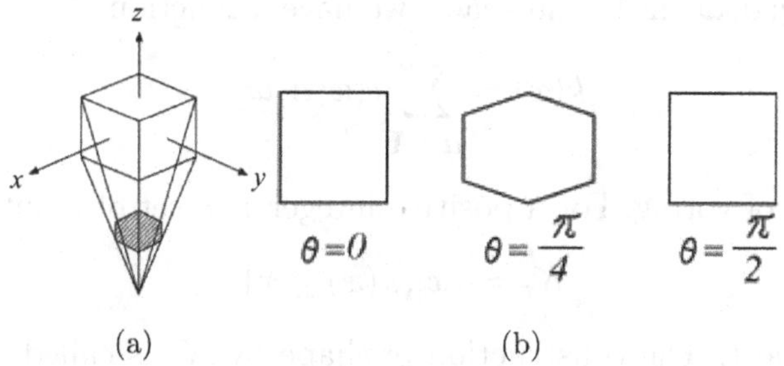

Figure 14.7. A View Cone of a Cube: (a) and the boundaries of shadows for $\theta = 0$, $\theta = \frac{\pi}{4}$, and $\theta = \frac{\pi}{2}$ (b).

Next, we show the reconstruction of a cube as the common region of view cones using the idea of image reconstruction from projections. We use same symbols with the case of the reconstruction of a cube from support planes. The view cones of cube $\boldsymbol{B}$ are

$$\begin{array}{lll} -\delta \leq \theta \leq \delta & |\boldsymbol{\eta}, \boldsymbol{p}_{n-1}, \boldsymbol{p}_n, \boldsymbol{a}| \geq 0, & |\boldsymbol{\eta}, \boldsymbol{p}_n, \boldsymbol{q}_n, \boldsymbol{a}| \geq 0, \\ & |\boldsymbol{\xi}, \boldsymbol{p}_{n-1}, \boldsymbol{p}_n, \boldsymbol{a}| \geq 0, & |\boldsymbol{\xi}, \boldsymbol{q}_{n-1}, \boldsymbol{p}_{n-1}, \boldsymbol{a}| \geq 0 \\ \delta + \frac{\pi}{2} \leq \theta \leq -\sin + \frac{\pi}{2} & |\boldsymbol{\xi}, \boldsymbol{p}_n, \boldsymbol{p}_{n+1}, \boldsymbol{a}| \geq 0, & |\boldsymbol{\xi}, \boldsymbol{p}_{n+1}, \boldsymbol{p}_{n+2}, \boldsymbol{a}| \geq 0, \\ & |\boldsymbol{\xi}, \boldsymbol{p}_{n+2}, \boldsymbol{q}_{n+2}, \boldsymbol{a}| \geq 0, & |\boldsymbol{\xi}, \boldsymbol{q}_{n+2}, \boldsymbol{q}_{n+1}, \boldsymbol{a}| \geq 0, \\ & |\boldsymbol{\xi}, \boldsymbol{q}_{n+1}, \boldsymbol{q}_n, \boldsymbol{a}| \geq 0, & |\boldsymbol{\xi}, \boldsymbol{q}_n, \boldsymbol{p}_n, \boldsymbol{a}| \geq 0, \end{array} \tag{50}$$

and

$$\begin{array}{lll} -\delta \leq \phi \leq \delta & |\boldsymbol{\xi}, \bar{\boldsymbol{p}}_{n-1}, \bar{\boldsymbol{p}}_n, \boldsymbol{b}| \geq 0, & |\boldsymbol{\xi}, \bar{\boldsymbol{p}}_n, \bar{\boldsymbol{q}}_n, \boldsymbol{b}| \geq 0, \\ & |\boldsymbol{\xi}, \bar{\boldsymbol{p}}_{n-1}, \bar{\boldsymbol{p}}_n, \boldsymbol{b}| \geq 0, & |\boldsymbol{\xi}, \bar{\boldsymbol{q}}_{n-1}, \bar{\boldsymbol{p}}_{n-1}, \boldsymbol{b}| \geq 0 \\ \delta + \frac{\pi}{2} \leq \theta \leq -\delta + \frac{\pi}{2} & |\boldsymbol{\xi}, \bar{\boldsymbol{p}}_n, \bar{\boldsymbol{p}}_{n+1}, \boldsymbol{b}| \geq 0, & |\boldsymbol{\xi}, \bar{\boldsymbol{p}}_{n+1}, \bar{\boldsymbol{p}}_{n+2}, \boldsymbol{b}| \geq 0, \\ & |\boldsymbol{\xi}, \bar{\boldsymbol{p}}_{n+2}, \bar{\boldsymbol{q}}_{n+2}, \boldsymbol{b}| \geq 0, & |\boldsymbol{\xi}, \bar{\boldsymbol{q}}_{n+2}, \bar{\boldsymbol{q}}_{n+1}, \boldsymbol{b}| \geq 0, \\ & |\boldsymbol{\xi}, \bar{\boldsymbol{q}}_{n+1}, \bar{\boldsymbol{q}}_n, \boldsymbol{b}| \geq 0, & |\boldsymbol{\xi}, \bar{\boldsymbol{q}}_n, \bar{\boldsymbol{p}}_n, \boldsymbol{b}| \geq 0, \end{array} \tag{51}$$

for $\boldsymbol{a} = (R\cos\theta, R\sin\theta, 0)^\top$ and $\boldsymbol{b} = (R\sin\theta, 0R\cos\theta)^\top$. The view cones are periodic with respect to arguments θ and ϕ with period $\frac{\pi}{2}$. The intersection of these view cones is a cube $\boldsymbol{B}$.

14.3.2. VOTING METHOD

Setting the characteristic function in the view cone to be

$$c(\boldsymbol{x}; \boldsymbol{a}, \boldsymbol{\omega}) = \begin{cases} 1, & \boldsymbol{x} \in C(\boldsymbol{a}, \boldsymbol{\omega}) \\ 0, & \text{otherwise,} \end{cases} \tag{52}$$

if we vote $c(\boldsymbol{x}; \boldsymbol{a}, \boldsymbol{\omega})$ in to the space, we have a function

$$k(\boldsymbol{x}) = \sum_{\boldsymbol{a} \in \boldsymbol{A}} c(\boldsymbol{x}; \boldsymbol{a}, \boldsymbol{\omega}). \tag{53}$$

as the results of voting. For a positive integer τ, a set of points

$$K_\tau = \{\boldsymbol{x}|, k(\boldsymbol{x}) \geq \tau\} \tag{54}$$

defines an object. The construction of shape by K_τ is called shape reconstruction by voting. Furthermore, an algorithm for the computation of K_τ is called shape carving. Equation (53) is a geometric version of backprojection in image reconstruction from projections (Solmon, 1976; Hammaker, 1980).

The voting process is the same operation with shape from shadows for slice convex objects, if we can obtain tangent lines at each point from all directions. Therefore, we have the following theorem.

THEOREM 14.5. *Voting process reconstruct slice convex objects from shadows if we have orthogonal views from all directions in* S^2.

The boolean version (Kawamoto and Imiya, 2001) of Equation (53) for lines with finite width, which is adopted in applications is

$$K = \bigcap_{\boldsymbol{a} \in A} \{ \bigcap_{\boldsymbol{l}(\boldsymbol{a}) \in C(\boldsymbol{a}, \boldsymbol{\omega})} \boldsymbol{l}(\boldsymbol{a}) \oplus \boldsymbol{B} \}, \tag{55}$$

where $\boldsymbol{B}$ is the ball with radius δ, since $\{\boldsymbol{l}(\boldsymbol{a}) \oplus \boldsymbol{B}\}$ defines a straight bar in a space whose centre line is $\boldsymbol{l}(\boldsymbol{a})$.

14.4. Metrics among Convex Objects

According to the brief survey of classical results in convex geometry in Section 14.2, parametrisation of a closed convex surface is achieved using the support plane expression. Using this parametrisation, we define a metric among convex objects.

For a vertex $\boldsymbol{x}$ on K which is parametrised by the support plane expression of a convex object, setting

$$\boldsymbol{g} = \frac{1}{4\pi}\int_0^{2\pi}\int_0^{\pi} \boldsymbol{x}\sin\theta d\theta d\phi,\ E = \sqrt{\frac{1}{4\pi}\int_0^{2\pi}\int_0^{\pi} |\boldsymbol{x}|^2 \sin\theta d\theta d\phi}, \quad (56)$$

a vector $\bar{\boldsymbol{x}} = \frac{1}{E}(\boldsymbol{x} - \boldsymbol{g})$ determines a normalised parametrisation which is invariant under scaling and translation. Then for a rotation matrix $\boldsymbol{R}$, we define a metric among convex objects K_1 and K_2 as

$$d(K_1, K_2) = \min_{\boldsymbol{R}} \sqrt{\frac{1}{4\pi}\int_0^{2\pi}\int_0^{\pi} |\bar{\boldsymbol{x}}_1 - \boldsymbol{R}\bar{\boldsymbol{y}}_2|^2 \sin\theta d\theta d\phi}, \quad (57)$$

For $\boldsymbol{x} \in \partial K_1$ and $\boldsymbol{y} \in \partial K_2$, if there is a transformation $\boldsymbol{y} = \lambda \boldsymbol{R}\boldsymbol{x} + \boldsymbol{c}$ we write $K_1 \sim K_2$ for a pair of convex objects. For this metric, we have the theorem.

THEOREM 14.6. *If and only if* $K_1 \sim K_2$, $d(K_1, K_2) = 0$ *and* $d(K_1, K_2)$.

Proof

$$\begin{aligned}
&d(K_1, K_2) + d(K_2, K_3) \\
&= \min_{\boldsymbol{R}_1} \sqrt{\frac{1}{4\pi}\int_0^{2\pi}\int_0^{\pi} |\bar{\boldsymbol{x}}_1 - \boldsymbol{R}_1\bar{\boldsymbol{x}}_2|^2 \sin\theta d\theta d\phi} \\
&\quad + \min_{\boldsymbol{R}_2} \sqrt{\frac{1}{4\pi}\int_0^{2\pi}\int_0^{\pi} |\bar{\boldsymbol{x}}_2 - \boldsymbol{R}_2\bar{\boldsymbol{x}}_3|^2 \sin\theta d\theta d\phi} \\
&\geq \min_{\boldsymbol{R}_1\boldsymbol{R}_2} \sqrt{\frac{1}{4\pi}\int_0^{2\pi}\int_0^{\pi} \left|\boldsymbol{R}_1^T\bar{\boldsymbol{x}}_1 - \boldsymbol{R}_2\bar{\boldsymbol{x}}_3\right|^2 \sin\theta d\theta d\phi} \\
&\geq \min_{\boldsymbol{R}_3} \sqrt{\frac{1}{4\pi}\int_0^{2\pi}\int_0^{\pi} |\bar{\boldsymbol{x}}_1 - \boldsymbol{R}_3\bar{\boldsymbol{x}}_3|^2 \sin\theta d\theta d\phi} \\
&= d(K_1, K_3), \qquad (58)
\end{aligned}$$

where $\boldsymbol{R}_3 = \boldsymbol{R}_1\boldsymbol{R}_2$. Thus, $d(K_1, K_2)$ satisfies the three axioms of distance. □

This theorem concludes that Equation (57) defines a metric which is invariant under rotation, translation and scaling. Furthermore, setting

$$\boldsymbol{M} = \frac{1}{4\pi}\int_0^{2\pi}\int_0^{\pi} \bar{\boldsymbol{x}}_1\bar{\boldsymbol{x}}_2^T \sin\theta d\theta d\phi, \tag{59}$$

Equation (57) becomes

$$d(K_1, K_2) = \sqrt{1 - tr(\boldsymbol{M}\boldsymbol{M}^T)^{\frac{1}{2}}}. \tag{60}$$

Thus, using cross correlation of normalised parametrisation of convex objects, we can compute the distance between two convex objects (Schwartz and Sharir, 1987).

Setting $\Delta(A, B)$ to be the symmetry difference of a pair of regions A and B in $\mathbf{R}^n$ such that

$$\Delta(A, B) = |(A \cup \overline{B}) \cap (\overline{A} \cap B)|, \tag{61}$$

we define the similarity measures of convex objects as,

$$d(K_1, K_2) = \min_{\boldsymbol{R},\boldsymbol{t}} \Delta(K_1(\boldsymbol{R}, \boldsymbol{t}), K_2), \tag{62}$$

where

$$K_i(\boldsymbol{R}, \boldsymbol{t}) = \{\boldsymbol{y}|\boldsymbol{y} = \boldsymbol{R}\boldsymbol{x} + \boldsymbol{t}, \boldsymbol{x} \in K_i\}, \tag{63}$$

for a rotation matrix $\boldsymbol{R}$ and a translation vector $\boldsymbol{t}$,

$$\tilde{d}_p(K_1, K_2) = \min_{\boldsymbol{R},\boldsymbol{t}} \int_{S^2} \left\{\Delta(\tilde{K}_1(\boldsymbol{R}\boldsymbol{\omega} \oplus \{P_{\boldsymbol{\omega}^\perp}\boldsymbol{t}\}), \tilde{K}_2(\boldsymbol{\omega}))\right\} d\boldsymbol{\omega}, \tag{64}$$

for parallel projections and

$$\tilde{d}_c(K_1, K_2) = \min_{\boldsymbol{R},\boldsymbol{t}} \int_{\boldsymbol{a}\in\{A(\boldsymbol{R},\boldsymbol{t})\}} \Delta(\tilde{K}_1(\boldsymbol{a}), \tilde{K}_2(\boldsymbol{a}))ds, \tag{65}$$

for perspective projections, which measure view corns, where s is the length on a curve $\boldsymbol{a}$ and

$$A(\boldsymbol{R}, \boldsymbol{t}) = \{\boldsymbol{y}|\boldsymbol{y} = \boldsymbol{R}\boldsymbol{a} + \boldsymbol{t}, \boldsymbol{a} \in A\}. \tag{66}$$

Equation (62) is a set theory analogous to eq. (57). Equation (64) is a similarity measure computed by shadows over all directions in S^2. Therefore, we can compute similarities of convex objects from shadows without reconstructing them. Furthermore, Equation (65) is a view-based metric which depends on the orbit of the camera for the reconstruction.

In $\mathbf{R}^2$, assuming that the origin of object lies inside of the camera orbit, we define

$$f(\theta,t) = \begin{cases} 1, & \text{if} -p(\theta+\pi) \leq t \leq p(\theta), \\ 0, & \text{otherwise.} \end{cases} \quad (67)$$

Setting $S(\theta,t) = \{(\theta,t)|f(\theta,t)=1\}$, we have

$$\begin{aligned} \tilde{d}_{p2}(K_1,K_2) &= \min_{\phi,s} \Delta(S_1(\theta-\phi,t-s), S_2(\theta,t)) \\ &= \min_{\phi,s} \int_0^{2\pi} |f_1(\theta-\phi,t-s) - f_2(\theta,\phi)| d\theta \end{aligned} \quad (68)$$

14.5. Conclusions

We summarise some open problems for shape from shadow. Analysis done in Section 14.2.2 does not depend on the dimensions of spaces. Therefore, our analysis might solve shape from shadow in the general dimensional spaces. In n-dimensional Euclidean space, k-dimensional rays observe $(n-k)$-dimensional shadows for $n > k \geq 1$. This hierarchy of the rays and shadows might derive the same mathematical properties of the X-ray transform and Radon transform (Solmon, 1976). Specially as an application, shape carving in 4-dimensional spatio-temporal space is interested in motion analysis for robot vision.

In the theory of line integrals, the Fourier-Plancherel formula, which relates the L_2 norms of an function and its transform, is well studied (Solmon, 1976). This relation allow us to evaluate the reconstruction errors from errors in the projections. The formula in shape from shadow is the relation between Equations (62) and (64). Furthermore, the construction of the robust and accurate computation of Equation (68) is also an open problem. Using this definition, machine parts are clarified from mechanically measured data (Rao and Goldberg, 1994; Rao and Goldberg, 1995).

References

Imiya, A. and K. Kawamoto: Shape reconstruction from an image sequences. In *Visual Form 2001, Proceedings of 4th International Workshop on Visual Form* (Arcelli, C., L. P. Cardella, and G. Sanniti di Baja, editors), pages 677–686, LNCS 2059, Springer, Berlin, 2001.

Imiya, A. and K. Kawamoto: Mathematical aspects of shape reconstruction from an image sequence. In *Proc. 1st Int. Symp. 3D Data Processing Visualization and Transformations*, (Cortellazzo, G. M., and C. Guerra, editors), pages 632–635, 2002.

Guggenheimer, H. W.: *Applicable Geometry*. Robert E. Kniegen Pub. Inc., New York, 1977.

Campi, S.: Reconstructing a convex surface from certain measurements of its projections. *Bollettio U.M.I.*, **6**:945–959, 1986.

Kawamoto, K. and A. Imiya: Detection of spatial points and lines by random sampling and voting process. *Pattern Recognition Letters*, **22**:199–207, 2001.

Kutulakos, K. and S. M. Seitz: A theory of shape by space carving. In Proc. *7th IEEE ICCV,* volume 1, pages 307–314, 1999.

Radon, J.: Über die Bestimmung von Functionen durch ihre Integralwerte längs gewisser Mannigfaltigkeiten, Berichte Sächsische Akademie der Wissenschaften. *Math.-Phys., Kl.,* **69**:262–267, 1917.

Tuy, H. K.: An inversion formula for cone-beam reconstruction. *SIAM J. Applied Mathematics,* **43**:546–552, 1983.

Smith, B. D.: Image reconstruction from cone-beam projections: Necessary and sufficient conditions and reconstruction methods. **IEEE Trans. Med. Imag., MI-4**:14–28, 1985.

Axelsson, C.: *Direct Fourier Methods in 3D-Reconstruction from Cone-Beam Data.* Linköping Studies in Science and Technology, Thesis No. 413, 1994.

Ludwing, D.: The Radon transform on Euclidian space. *Comm. Pure and Applied Mathematics,* **19**:49–81, 1966.

Solmon, D. C.: The X-ray transform. *Journal of Math. Anal. and Appl.*, **56**:61–83, 1976.

Hammaker, C., K. T. Smith, D. C .Solomon, and L. Wagner: The divergent beam x-ray transform. *Rocky Mountain Journal of Mathematics,* **10**:253–283, 1980.

Stark, H. and H. Peng: Shape estimation in computer tomography from minimal data. In Proc. *9th ICPR,* volume 1, pages 184–186, 1988.

Laurentini, A.: The visual hull concept for silhouette-base image understanding. *IEEE Trans. PAMI,* **16**:150–163, 1994.

Laurentini, A.: How for 3D shape can be understood from 2D silhouettes. *IEEE Trans. PAMI,* **17**:188–195, 1995.

Gardner, R. J.: *Geometric Tomography.* Cambridge University Press, Cambridge, 1995.

Kuba, A.: The reconstruction of two-dimensionally connected binary patterns from their two orthogonal projections. *Computer Vision, Graphics and Image Processing,* **27**: 249–265, 1984.

Zheng, J.-Y.: Acquiring 3-D models from sequences of contours. *IEEE Trans. PAMI,* **16**:163–178, 1994.

Vaillant, R. and O. D. Faugeras: Using external boundaries for 3-D object modeling. *IEEE Trans. PAMI,* **14**:157–173, 1992.

Aloimonos, J.: Visual shape computation. *IEEE Proceedings,* **76**:899–916, 1988.

Skiena, S. S.: Interactive reconstruction via geometric probing. *IEEE Proceedings,* **80**:1364–1383, 1992.

Skiena, S. S.: Probing convex polygon with half-planes. *Journal of Algorithms,* **12**:359–374, 1991.

Li, R. S.-Y.: Reconstruction of polygons from projections. *Information Processing Letters,* **28**: 235–240, 1988.

Prince, J. L. and A. S. Willsky: Reconstructing convex sets from support line measurements. *IEEE Trans. PAMI,* **12**:377–389, 1990.

Richardson, T. J.: Planar rectifiable curves are determined by their projections. *Discrete and Computational Geometry,* **16**:21–31, 1996.

Kölzow, D., A. Kuba, and A. Volčic: An algorithm for reconstructing convex bodies from their projections. *Discrete and Computational Geometry,* **4**:205–237, 1989.

Schwartz, J. T. and M. Sharir: Identification of partially obscured objects in two and three dimensions by matching noisy characteristic surveys. *The International Journal of Robotic Research,* **6**:29–44, 1987.

Kanatani, K.: *Geometric Computation for Machine Vision.* Oxford University Press, Oxford, 1993.

Rao, A. S. and Y. K. Goldberg: Shape from diameter: Recognizing polygonal parts with parallel-jaw gripper. *International Journal of Robotics Research,* **13**:16–37, 1994.

Rao, A. S. and Y. K. Goldberg: Manipulating algebraic parts in plane. *IEEE Trans. Robotics and Automation,* **11**:598–602, 1995.

Part III

Approximation and Regularization

A CONFIDENCE MEASURE FOR VARIATIONAL OPTIC FLOW METHODS

ANDRÉS BRUHN AND JOACHIM WEICKERT
Mathematical Image Analysis Group
Faculty of Mathematics and Computer Science
Saarland University, Building 27
66041 Saarbrücken, Germany

Abstract. In this paper we investigate the usefulness of confidence measures for variational optic flow computation. To this end we discuss the frequently used sparsification strategy based on the image gradient. Its drawbacks motivate us to propose a novel, energy-based confidence measure that is parameter-free and applicable to the entire class of energy minimising optic flow techniques. Experimental evaluations show that this confidence measure leads to excellent results, independently of the image sequence or the underlying variational approach.

Key words: optic flow, confidence measures, differential techniques, variational methods, partial differential equations, performance evaluation

15.1. Introduction

The recovery of motion fields from image sequences is one of the key problems in computer vision. Given two consecutive frames of an image sequence, one is interested in finding the projection of the 3-D motion onto the image plane: the so-called optic flow field. In order to estimate this displacement field, optic flow methods often use a constancy assumptions on image features such as the grey value.

At that point, the handling of incomplete data plays a very important role: In general, these constancy assumptions cannot provide sufficient data to determine a unique solution of the optic flow problem. For instance, in the case of the grey value constancy assumption this incompleteness manifests itself in the aperture problem. In this case not more than the flow component parallel to the image gradient can be calculated. At locations where the gradient is zero, not even this component is computable and no estimation is possible.

R. Klette et al. (eds.), Geometric Properties for Incomplete Data, 283-297.

In order to cope with these situations, variational methods regularise the problem by assuming smoothness or piecewise smoothness of the resulting flow field. At locations where the problem of incomplete data occurs, this regularisation fills in information from the neighbourhood and thus allows the estimation of a 100 % dense flow field.

However, it is clear that these estimates cannot have the same reliability at all locations. It would therefore be interesting to find a confidence measure that allows to assess the reliability of a dense optic flow field. In particular, such a measure would allow to identify locations where the problem of incomplete data has been solved successfully. Therefore, it is not surprising that (Barron *et al.*, 1994) have identified the absence of such a good measure as one of the main drawbacks of variational optic flow techniques.

In our paper we address this problem. By discussing the frequently used confidence measure based on the image gradient, we show why this method is not appropriate for sparsifying dense flow fields from variational methods. As a remedy, we propose a novel energy-based confidence measure that offers several advantages and works well over a large range of densities.

Related work. In spite of the fact that there exists a very large number of publications on variational optic flow methods (see e.g. (Horn and Schunck, 1981; Nagel and Enkelmann, 1986; Weickert and Schnörr, 2001; Brox *et al.*, 2004)) and on confidence measures for local optic flow approaches – most of them based on the evaluation of the aperture problem; see e.g. (Bigün *et al.*, 1991; Simoncelli *et al.*, 1991)) – there has been remarkably little work devoted to the application of confidence measures in the context of variational optic flow methods. First approaches go back to (Barron *et al.*, 1994) who used the magnitude of the image gradient to decide on the local reliability of a flow estimate. More recently, (Haußecker and Spies, 1999) proposed a general classification of confidence measures. However, they neither introduced any novel confidence measures for variational methods, nor performed a qualitative evaluation of existing concepts.

Organisation of the chapter. Our paper is organised as follows. In Section 15.2 we give a review on variational methods and discuss different types of regularisation strategies. The gradient-based confidence measure by (Barron *et al.*, 1994) that is widely used to sparsify the resulting flow fields is discussed in Section 15.3. Based on the results of this discussion we propose a novel energy-based confidence measure in Section 15.4 and perform a systematic experimental evaluation in Section 15.5. Finally, Section 15.6 concludes this paper with a summary.

15.2. Variational Optic Flow Computation

Let us consider some image sequence $f(x, y, t)$, where (x, y) denotes the location within a rectangular image domain Ω, and $t \in [0, T]$ denotes time. In order to retrieve objects in subsequent frames of this image sequence, many optic flow methods assume that corresponding pixels have the same grey value, i.e. that the grey value of objects remains constant over time. If we denote the movement of such an object by $(x(t), y(t))$ this assumption can be formulated as

$$0 = \frac{df(x(t), y(t), t)}{dt}. \tag{1}$$

By applying the chain rule this leads to the following *optic flow constraint (OFC)*:

$$0 = f_x u + f_y v + f_t, \tag{2}$$

where subscripts denote partial derivatives and the optic flow field satisfies $(u, v)^\top = (\partial_t x, \partial_t y)^\top$.

Evidently, this single equation is not sufficient to uniquely determine the two unknowns u and v. In particular at locations where the image gradient is zero, no estimation of the optic flow is possible. In all other cases, only the flow component parallel to $\nabla f := (f_x, f_y)^\top$ can be computed, the so-called *normal flow*:

$$w_n = -\frac{f_t}{|\nabla f|} \frac{\nabla f}{|\nabla f|}. \tag{3}$$

In the literature, this ambiguity is referred to as the *aperture problem*.

15.2.1. GENERAL STRUCTURE

Variational methods overcome the aperture problem by imposing an additional constraint on the solution: They assume that the resulting flow field is smooth or piecewise smooth. Then the optic flow can be computed as minimiser of a global energy functional, where both deviations from the data and deviations from the smoothness constraint are penalised. Let $\nabla_3 := (\partial_x, \partial_y, \partial_t)^\top$ denote the spatiotemporal gradient, let $D(u, v, \nabla_3 f)$ stand for a data term (e.g. the squared OFC) and let $S(\nabla_3 u, \nabla_3 v, \nabla_3 f)$ represent a constraint on the smoothness of the resulting flow field. Then the corresponding energy functional is given by

$$E(u, v) = \int_{\Omega \times [0,T]} \underbrace{D(u, v, \nabla_3 f)}_{\text{data term}} + \alpha \underbrace{S(\nabla_3 u, \nabla_3 v, \nabla_3 f)}_{\text{smoothness term}} \, dx\, dy\, dt, \tag{4}$$

where α serves as regularisation parameter that steers the smoothness of the estimated flow field.

15.2.2. PROTOTYPES FOR VARIATIONAL METHODS

Let us now take a closer look at the different regularisation strategies that may serve as smoothness constraints. As classified in (Weickert and Schnörr, 2001) there are basically three different types of regularisation: *Homogeneous regularisation* that assumes overall smoothness, *image-driven regularisation* that assumes piecewise smoothness and respects discontinuities in the image data, and *flow-driven regularisation* that assumes piecewise smoothness and respects discontinuities in the flow field. Moreover, when considering image and flow-driven regularisation, one can distinguish between *isotropic* and *anisotropic* smoothness terms. While isotropic regularisers do not impose any smoothness at discontinuities, anisotropic ones permit smoothing along the discontinuity but not across it.

In order to demonstrate the different regularisation concepts and to allow for a systematic experimental evaluation, we have chosen three prototypes of variational methods that cover all types of regularisation.

15.2.2.1. *The Combined Local–Global Method*

As prototype for the class of optic flow techniques with *homogeneous regularisation* we consider the so-called combined local-global (CLG) method (Bruhn *et al.*, 2002; Bruhn *et al.*, 2005). This technique combines the dense flow fields of the global approach of (Horn and Schunck, 1981) with the high noise robustness of the local method of (Lucas and Kanade, 1981).

Let $\mathbf{w} = (u, v, 1)^\top$ denote the spatiotemporal extended flow vector. Then the energy functional of the CLG method is given by

$$E(u,v) = \int_{\Omega\times[0,T]} \mathbf{w}^\top J_\rho(\nabla_3 f)\,\mathbf{w} + \alpha\left(|\nabla_3 u|^2 + |\nabla_3 v|^2\right)\,dx\,dy\,dt. \quad (5)$$

where the matrix $J_\rho(\nabla_3 f)$ is the so-called structure tensor (Bigün *et al.*, 1991; Förstner and Gülch, 1987; Rao and Schunck, 1991) given by $K_\rho * (\nabla_3 f\,\nabla_3 f^\top)$, the symbol $*$ denotes convolution in each matrix component, and K_ρ is a Gaussian with standard deviation ρ. One should note that for $\rho \to 0$ the spatial variant of the CLG approach comes down to the Horn and Schunck method, and for $\alpha \to 0$ it becomes the Lucas–Kanade algorithm.

15.2.2.2. *The Method of Nagel and Enkelmann*

For the class of optic flow methods with *image-driven* regularisation we consider the *anisotropic* technique of (Nagel and Enkelmann, 1986). This method assumes the flow field to be smooth everywhere except across discontinuities in the image data. This can be realised by penalising only the projection of the flow gradient onto the plane orthogonal to the image gradient. The corresponding energy functional for the spatiotemporal variant of the Nagel-Enkelmann algorithm is given by (Nagel, 1990)

$$E(u,v) = \int_{\Omega\times[0,T]} (f_x u + f_y v + f_t)^2 + \alpha \left(\nabla_3 u^\top D(\nabla_3 f) \nabla_3 u\right) + \alpha \left(\nabla_3 v^\top D(\nabla_3 f) \nabla_3 v\right) \, dx\, dy\, dt, \tag{6}$$

with the regularised projection matrix

$$D(\nabla_3 f) = \frac{1}{2|\nabla_3 f|^2 + 3\epsilon^2} \begin{pmatrix} f_y^2 + f_z^2 + \epsilon^2 & -f_x f_y & -f_x f_z \\ -f_x f_y & f_x^2 + f_z^2 + \epsilon^2 & -f_y f_z \\ -f_x f_z & -f_y f_z & f_x^2 + f_y^2 + \epsilon^2 \end{pmatrix} \tag{7}$$

perpendicular to $\nabla_3 f$, where ϵ serves as regularisation parameter.

15.2.2.3. *The TV-based Regularisation Method*

In contrast to image-driven regularisation methods, *flow-driven* techniques reduce smoothing where edges in the *flow* field occur during computation. Our representative for this third class of variational optic flow techniques is an *isotropic* method that penalises deviations from the smoothness assumption with the L_1 norm of the flow gradient magnitude. This corresponds to *total variation* (TV) regularisation (Rudin *et al.*, 1992). It can be related to statistically robust error penalisation (Huber, 1981), since large deviations from smoothness are penalised less severely than in the commonly used quadratic (Tikhonov) regularisation (Tikhonov and Arsenin, 1977). As a consequence, large gradient features such as edges are better preserved. Our energy functional is given by

$$E(u,v) = \int_{\Omega\times[0,T]} (f_x u + f_y v + f_t)^2 + \alpha \sqrt{|\nabla_3 u|^2 + |\nabla_3 v|^2 + \epsilon^2} \, dx\, dy\, dt, \tag{8}$$

where ϵ serves as small regularisation parameter. Related spatial energy functionals have been proposed by (Cohen, 1993; Deriche *et al.*, 1995; Kumar *et al.*, 1996), and similar spatiotemporal functionals have been investigated in (Weickert and Schnörr, 2001).

15.2.3. THE FILLING-IN EFFECT

The strategy of regularising the solution by a smoothness assumption has a useful side-effect: Variational methods always yield 100 % dense flow fields. At locations with $\nabla_3 f \approx 0$, the data term does not allow a reliable computation of a local flow estimate. However, the smoothness term fills in information from the neighbourhood. This can be explained as follows: Since the contribution of the data term to the energy functional is very small at these locations, the smoothness term becomes relatively more important. As a consequence, the local flow estimate is adjusted to its neighbourhood in accordance with the smoothness constraint. This propagation of neighbourhood information is the so called *filling-in* effect.

15.3. The Gradient-Based Confidence Measure

As we have seen in the previous section, the aperture problem does only allow the direct computation of the normal flow. Since this requires the gradient at the corresponding pixel to be different from zero, (Barron *et al.*, 1994) proposed to connect the reliability of a flow estimate to the magnitude of the underlying image gradient. Thus, the following confidence measure is obtained:

$$c_{\text{grad}} = |\nabla f|. \tag{9}$$

However, this ad-hoc criterion suffers from two drawbacks. Large gradients often result from noise and occlusions. Therefore, evaluating the magnitude of the gradient rewards exactly those locations, where the estimation of the optic flow is particularly problematic. It is not surprising, that the application of such a measure can only be of limited success. Moreover, it is clear that a-priori measures that only judge the initial situation *before* the computation, are not in the best position to decide on the reliability of a local flow estimate. They are simply not capable of considering the propagation of neighbourhood information for solving the aperture problem. This applies in particular to variational optic flow methods since they rely on the global filling-in effect of the regulariser.

15.4. A Novel Energy-Based Confidence Measure

In order to capture the filling-in effect of the regulariser in a better way, one should think of involving the computed flow in the decision process. Let us now demonstrate how this can be accomplished in a natural way. As we know from Section 15.2, variational methods are based on the minimisation of an energy functional. This energy functional penalises deviations from model assumptions by summing up the local deviations from the image domain. At locations where this deviation is small, the computed flow respects the underlying model. At locations where the deviation is large, on the other hand, the model assumptions are violated severely. In this context it appears very natural to use this indicator for assessing the local reliability of the computation. Thus, we propose a confidence measure where the reliability is inversely proportional to the local energy contribution:

$$c_{\text{ener}} = \frac{1}{D\left(u, v, \nabla_3 f\right) + \alpha\, S\left(\nabla_3 u, \nabla_3 v . \nabla_3 f\right) + \epsilon^2}, \tag{10}$$

Here ϵ serves as small regularisation parameter that prevents the denominator from becoming singular. Its actual value is not important since we are only interested in a ranking of the confidence at different locations.

Apart from its simplicity, this confidence measure has several additional advantages. Firstly, it is a consequent continuation of the concept of variational methods: *The confidence measure is based on exactly the same assumptions as the underlying energy functional.* There is no reason, why other constraints should be used for evaluating the reliability of the estimated flow field: if other constraints are considered important, they should have been taken into account earlier by incorporating them in the variational model for computing the flow field. Secondly, the energy-based confidence measure allows to consider the filling-in effect of the regulariser: in contrast to the image gradient this measure is based on the *evaluation of the flow field.* Evidently, this is the only data where the filling-in effect is present. Thirdly, it allows to detect noise and occlusions to a certain degree. Those locations have a relatively high energy and are thus easily identified. Fourthly, the proposed confidence measure can be derived in a straightforward way from any energy functional. This makes it applicable to the entire class of energy minimising optic flow techniques. And finally, it is parameter-free. Since the parameters have already been set before the computation of the flow field there is no need to readjust them afterwards.

15.5. Results

In order to be able to quantify the reliability of confidence measures in the experimental section we restrict ourselves to image sequences for which the ground truth is available. In particular, this allows us to compute error measures such as the frequently used average angular error. It is defined as the arithmetic mean of

$$\arccos\left(\frac{u_c u_e + v_c v_e + 1}{\sqrt{(u_c^2 + v_c^2 + 1)(u_e^2 + v_e^2 + 1)}}\right) \tag{11}$$

where (u_c, v_c) denotes the correct flow, and (u_e, v_e) is the estimated flow (cf. also (Barron *et al.*, 1994)).

In our first experiment we compare the performance of the gradient and the energy based confidence measures. To this end we use the spatiotemporal approach with locally integrated data term and homogeneous

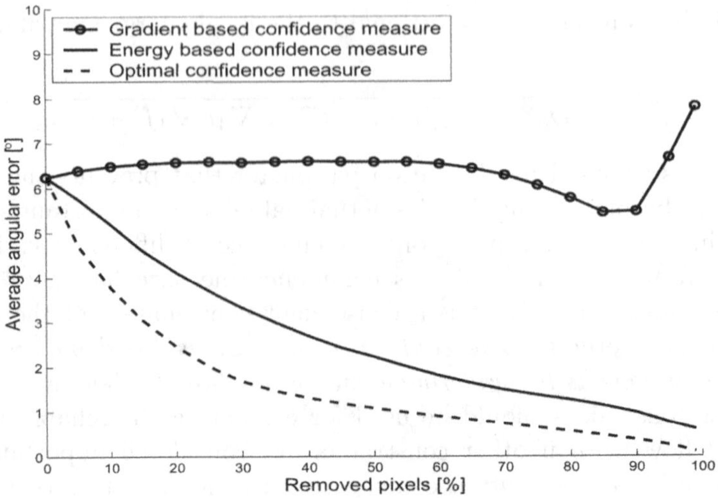

Figure 15.1. Comparison of the gradient, eigenvalue and energy-based confidence measures for the Yosemite sequence with clouds using the 3-D approach with locally integrated data term and homogeneous regularisation (CLG).

regularisation (CLG) and compute the flow field between frame 8 and 9 of the famous *Yosemite* [1] sequence with clouds. Then, we successively sparsify the estimated flow field by applying the confidence measures independently, and calculate the corresponding average angular errors within a density range from 100 % to 1 %.

The resulting graphs for both confidence measures are depicted in Figure 15.5. Moreover, a third graph is shown that illustrates the optimal sparsification performance with respect to the average angular error. By removing those locations first that contribute most to the average angular error - this requires the correct flow field - it constitutes the theoretical bound for all other confidence measures. As one can see, the proposed energy based criterion performs very favourably. It outperforms the gradient based confidence measure by far. One can also observe that the angular error decreases monotonically under sparsification over the entire range from 100 % down to 1 %. This is a clear indication for an interesting finding that may seem counterintuitive at first glance: *Regions in which the filling-in effect dominates give particularly small angular errors.* At such regions the data term vanishes and only the smoothness term contributes to the local energy. However, this contribution is often very small, since the

[1] `ftp://csd.uwo.ca/pub/vision/`

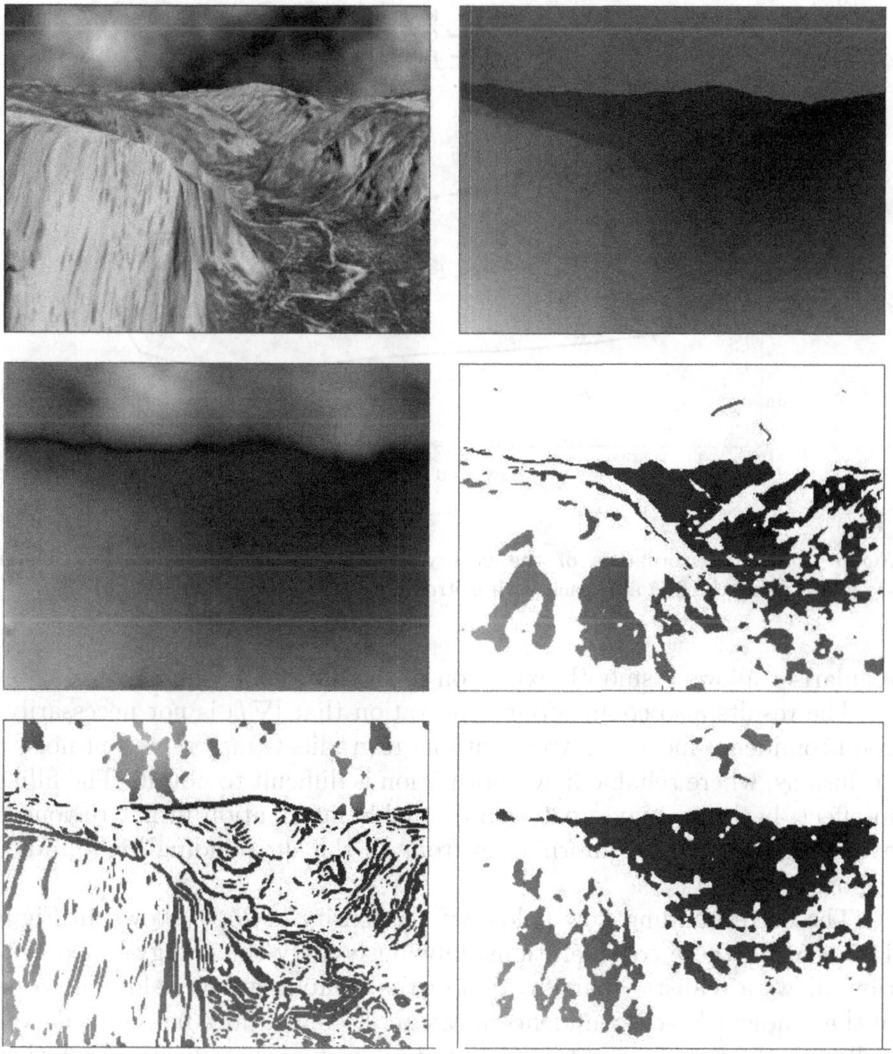

Figure 15.2. From left to right, and from top to bottom: (a) Frame 8 of the Yosemite sequence with clouds (316 × 256 pixels). *(b)* Magnitude of the ground truth. Brighter structures indicate larger values. *(c)* Magnitude of the computed field for a spatiotemporal approach with locally integrated data term and homogeneous regularisation (CLG). *(d)* 25 % quantile sparsified using the optimal confidence measure. *(e)* Ditto for the gradient-based confidence measure. *(f)* Ditto for the energy-based confidence measure.

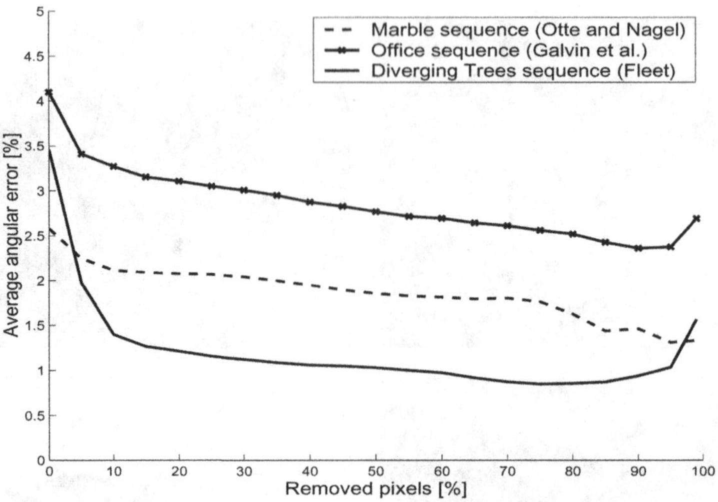

Figure 15.3. Performance of the energy-based confidence measures for different sequences using the 3-D approach with isotropic flow-driven regularisation (TV).

regulariser allows a smooth extension of the flow field in most cases.

The results also confirm our expectation that $|\nabla f|$ is not necessarily a good confidence measure: Areas with large gradients may represent noise or occlusions, where reliable flow information is difficult to obtain. The filling-in effect, however, may create more reliable information in flat regions by averaging less reliable information from all the surrounding high-gradient regions.

The corresponding flow fields with a density of 25 % shown in Figure 15.5 confirm these considerations. Obviously, only the energy-based criterion allows a realistic sparsification of the computed flow field. The result of the gradient based confidence measure, however, does not coincide very well with the flow field obtained from the optimal sparsification criterion.

Our second experiment investigates the performance of the energy-based confidence measure for a variety of image sequences. To this end we use the spatiotemporal approach with isotropic flow-driven regularisation (TV) and compute flow fields for three different image sequences. We consider the *Marble* [2] sequence by Otte and Nagel shown in Figure 15.5 (a)-(b) , the *Office* [3] sequence by (Galvin *et al.*, 1998) shown in Figure 15.5 (c)-(d) and the *Diverging Trees* sequence by Fleet shown in Figure 15.5 (e)-(f).

[2] `http://i21www.ira.uka.de/image_sequences/`
[3] `http://www.cs.otago.ac/nz/research/vision/`

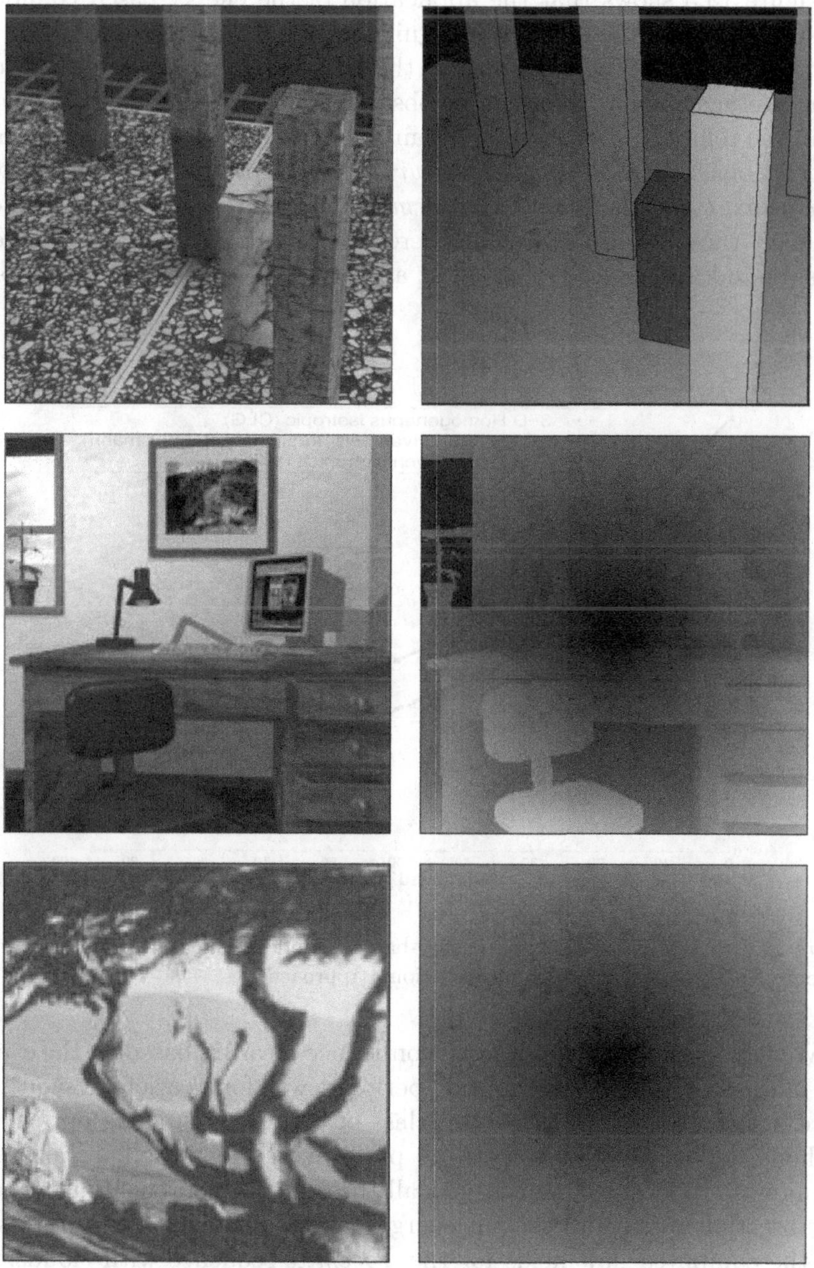

Figure 15.4. *From left to right, and from top to bottom: (a)* Frame 16 of the 512×512 Marble sequence. *(b)* Magnitude of the ground truth. *(c)* Frame 10 of the 200×200 Office sequence. *(d)* Magnitude of the ground truth. *(e)* Frame 20 of the 150×150 Translating Trees sequence. *(f)* Magnitude of the ground truth.

Figure 15.5 shows that the application of the energy-based confidence measure improves the estimation significantly in all three cases.

In particular at the beginning of the sparsification process a fast decay of the average angular error can be observed. The reason for this behaviour lies in the removal of wrong flow estimates caused by areas with high noise or occlusions. *Due to the massive occurrence of contradictory information in these areas, either the smoothness term or the data term are large.* As a consequence, these locations are considered very unreliable by the confidence measure and are already removed at an early stage of the sparsification.

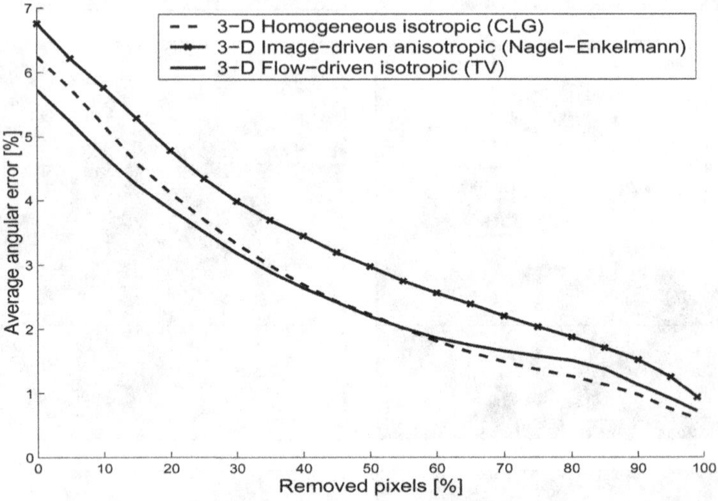

Figure 15.5. Performance of the energy-based confidence measure for the Yosemite sequence with clouds using different variational approaches.

We have seen that the proposed confidence measure based on the evaluation of the local energy contribution performs well for a variety of sequences. Since it is applicable to the entire class of energy minimising optical flow methods, let us now investigate its performance for different variational techniques. To this end we consider all three global approaches introduced in Subsection 15.2.2 and use our energy-based confidence measure to sparsify the computed flow fields for the *Yosemite* sequence with clouds. The corresponding graphs are presented in Figure 15.5. As one can see, they show once more an almost monotonic decay of the average angular error under sparsification. In particular, the observed behaviour is independent of the underlying variational approach. This is another confirmation of our findings that the evaluation of the local energy contribution is a simple

TABLE 15.1. Comparison between the "non-dense" results from (Barron *et al.*, 1994; Weber and Malik, 1995; Ong and Spann, 1999) and our results for the *Yosemite* sequence with cloudy sky (adapted from (Bruhn *et al.*, 2005). AAE = average angular error. CLG = average angular error of the spatiotemporal approach with locally integrated data term and homogeneous regularisation (CLG) with the same density. The sparse flow field has been created using our energy-based confidence criterion. The table shows that using this criterion clearly outperforms all results of non-dense methods.

Technique	Density	AAE	CLG + c_{ener}
Singh, step 2, $\lambda_1 \leq 0.1$	97.7 %	10.03°	6.04°
Ong/Spann	89.9 %	5.76°	5.26°
Heeger, level 0	64.2 %	22.82°	3.00°
Weber/Malik	64.2 %	4.31°	3.00°
Horn/Schunck, original, $\|\nabla f\| \geq 5$	59.6 %	25.33°	2.72°
Ong/Spann, tresholded	58.4 %	4.16°	2.66°
Heeger, combined	44.8 %	15.93°	2.07°
Lucas/Kanade, $\lambda_2 \geq 1.0$	35.1 %	4.28°	1.71°
Fleet/Jepson, $\tau = 2.5$	34.1 %	4.63°	1.67°
Horn/Schunck, modified, $\|\nabla f\| \geq 5$	32.9 %	5.59°	1.63°
Nagel, $\|\nabla f\| \geq 5$	32.9 %	6.06°	1.63°
Fleet/Jepson, $\tau = 1.25$	30.6 %	5.28°	1.55°
Heeger, level 1	15.2 %	9.87°	1.15°
Uras *et al.*, $\|(\| H) \geq 1$	14.7 %	7.55°	1.14°
Singh, step 1, $\lambda_1 \leq 6.5$	11.3 %	12.01°	1.07°
Waxman *et al.*, $\sigma_f = 2.0$	7.4 %	20.05°	0.95°
Heeger, level 2	2.4 %	12.93°	0.76°

confidence indicator that is efficient and widely applicable at the same time.

In our final experiment we compare our sparsified flow fields to the best non-dense results from the literature (see also (Bruhn *et al.*, 2005)). To this end we use the spatiotemporal approach with locally integrated data term and homogeneous regularisation (CLG) and compute the flow field for the *Yosemite* sequence with clouds. Using the energy-based confidence measure, the obtained flow field is then sparsified in such a way that the reduced densities coincide with the densities of other optic flow methods from the literature. The corresponding average angular errors are presented in Table 15.5. As one can see, our sparsified flow fields have a significantly lower angular error than all other methods with the same density. In this case errors down to 0.76 ° are reached for a flow density of 2.4 %. To our knowledge, these are the best values that have been obtained by non-dense methods for this sequence in the entire literature.

15.6. Summary and Conclusion

The absence of good confidence measures is regarded as one of the main drawbacks of variational optic flow methods. The goal of the present paper was to address this problem.

We have seen why the popular gradient-based measure fails: it rewards high-gradient regions where noise and occlusions dominate, and it ignores the filling-in effect of the regulariser, since it is an a-priori measure that does not take into account the estimated flow field.

As a remedy we have proposed a novel energy-based alternative that is both natural and simple: The confidence is chosen to be inversely proportional to the local energy contribution. This measure is applicable to the entire class of energy minimising optic flow techniques and it does not require additional parameters. It puts highest confidence to those locations where the model assumptions are satisfied most.

Our experiments have shown that the energy-based confidence measure performs significantly better than the gradient-based one. It may lead to excellent sparsification results, independently of the image sequence or the underlying variational approach. This is also confirmed by a final comparison to results from the literature, in which our sparsified flow fields proved to be more accurate than those of all other non-dense methods.

Acknowledgement: Our optic flow research is partly funded by the *DFG (Deutsche Forschungsgemeinschaft)* under the project WE 2602/3-1. This is gratefully acknowledged.

References

Barron, J. L., D. J. Fleet, and S. S. Beauchemin: Performance of optical flow techniques. *Int. J. Computer Vision*, **12**:43–77, 1994.

Bigün, J., G. H. Granlund, and J. Wiklund: Multidimensional orientation estimation with applications to texture analysis and optical flow. *IEEE Trans. Pattern Analysis Machine Intelligence*, **13**:775–790, 1991.

Brox, T., A. Bruhn, N. Papenberg, and J. Weickert: High accuracy optic flow estimation based on a theory for warping. In Proc. *European Conf. Computer Vision* (T. Pajdla and J. Matas, editors), pages 25–36, LNCS 3024, Springer, Berlin, 2004.

Bruhn, A., J. Weickert, and C. Schnörr: Combining the advantages of local and global optic flow methods. In *Pattern Recognition: 24th DAGM Symposium* (L. Van Gool, editor), pages 454–462, LNCS 2449, Springer, Berlin, 2002.

Bruhn, A., J. Weickert, and C. Schnörr: Lucas/Kanade meets Horn/Schunck: Combining local and global optic flow methods. *Int. J. Computer Vision*, **61**:1–21, 2005.

Cohen, I.: Nonlinear variational method for optical flow computation. In Proc. *Scandinavian Conference Image Analysis*, volume 1, pages 523–530, 1993.

Deriche, R., P. Kornprobst, and G. Aubert: Optical-flow estimation while preserving its discontinuities: A variational approach. In Proc. *Asian Conf. Computer Vision* (S. Z. Li, D. P. Mital, E. K. Teoh, and H. Wang), pages 290–295, LNCS 1035, Springer, Berlin, 1995.

Förstner, W. and E. Gülch: A fast operator for detection and precise location of distinct points, corners and centres of circular features. In Proc. *ISPRS Intercommission Conf. Fast Processing Photogrammetric Data*, pages 281–305, 1987.

Galvin, B., B. McCane, K. Novins, D. Mason, and S. Mills: Recovering motion fields: An analysis of eight optical flow algorithms. In Proc. *British Machine Vision Conf.* (J. N. Carter, M. S. Nixon, editors), pages 195–204, 1998.

Haußecker, H. and H. Spies: Motion. In *Handbook on Computer Vision and Applications. Vol. 2: Signal Processing and Pattern Recognition* (B. Jähne, H. Haußecker, and P. Geißler, editors), pages 309–396. Academic Press, San Diego, 1999.

Horn, B. and B. Schunck: Determining optical flow. *Artificial Intelligence*, **17**:185–203, 1981.

Huber, P. J.: *Robust Statistics*. Wiley, New York, 1981.

Kumar, A., A. R. Tannenbaum, and G. J. Balas: Optic flow: A curve evolution approach. *IEEE Trans. Image Processing*, **5**:598–610, 1996.

Lucas, B. and T. Kanade: An iterative image registration technique with an application to stereo vision. In Proc. *Int. Joint Conf. Artificial Intelligence*, pages 674–679, 1981.

Nagel, H.-H.: Extending the 'oriented smoothness constraint' into the temporal domain and the estimation of derivatives of optical flow. In Proc. *European Conf. Computer Vision* (O. Faugeras, editor), pages 139–148, LNCS 427, Springer, Berlin, 1990.

Nagel, H.-H. and W. Enkelmann: An investigation of smoothness constraints for the estimation of displacement vector fields from image sequences. *IEEE Trans. Pattern Analysis Machine Intelligence*, **8**:565–593, 1986.

Ong, E. P. and M. Spann: Robust optical flow computation based on least-median-of-squares regression. *Int. J. Computer Vision*, **31**:51–82, 1999.

Rao, A. R. and B. G. Schunck: Computing oriented texture fields. *CVGIP: Graphical Models Image Processing*, **53**:157–185, 1991.

Rudin, L. I., S. Osher, and E. Fatemi: Nonlinear total variation based noise removal algorithms. *Physica D*, **60**:259–268, 1992.

Simoncelli, E. P., E. H. Adelson, and D. J. Heeger: Probability distributions of optical flow. In Proc. *IEEE Conf. Computer Vision Pattern Recognition*, pages 310–315, 1991.

Tikhonov, A. N. and V. Y. Arsenin: *Solutions of Ill–Posed Problems*. Wiley, Washington, DC, 1977.

Weber, J. and J. Malik: Robust computation of optical flow in a multi-scale differential framework. *Int. J. Computer Vision*, **14**:67–81, 1995.

Weickert, J. and C. Schnörr: A theoretical framework for convex regularizers in PDE-based computation of image motion. *Int. J. Computer Vision*, **45**:245–264, 2001.

Weickert, J. and C: Schnörr: Variational optic flow computation with a spatio-temporal smoothness constraint. *J. Mathematical Imaging Vision*, **14**:245–255, 2001.

VIDEO IMAGE SEQUENCE ANALYSIS: ESTIMATING MISSING DATA AND SEGMENTING MULTIPLE MOTIONS

KENICHI KANATANI AND YASUYUKI SUGAYA
Department of Information Technology
Okayama University, Okayama 700-8530 Japan

Abstract. We discuss two issues of video processing based on our recent results: missing data estimation and multiple motion segmentation. We first show that for a rigidly moving scene we can reliably extend interrupted feature point tracking by imposing a geometric constraint based on the affine camera modeling. For scenes of multiple motions, many techniques have been proposed for segmenting moving objects into individual motions. However, many methods perform very poorly for real video sequences. We resolve this mystery by analyzing the geometric structure of the degeneracy of the motion model, which leads to a new segmentation algorithm. We demonstrate its effectiveness, using real video images.

Key words: video processing, feature tracking, missing data estimation, outlier removal, motion segmentation, affine camera model

16.1. Introduction

Video processing is one of the central topics for media technology today, and tracking feature points through the image sequence is a first step of many applications including 3-D reconstruction. Here, we discuss two issues in this respect based on our recent results (Sugaya and Kanatani, 2004a; Sugaya and Kanatani, 2004b).

The first issue is *missing data*: Feature point tracking fails when the points go out of the field of view or behind other objects. Many techniques have been proposed to estimate the missing data (Brandt, 2002; Jacobs, 2001; Saito and Kamijima, 2003; Tomasi and Kanade, 1992), but most of them are based on tentative 3-D reconstruction from sampled frames, assuming that they are correct. Here, we describe a more reliable scheme which integrates extrapolation and outlier removal. The procedure is based on (Sugaya and Kanatani, 2004a).

R. Klette et al. (eds.), Geometric Properties for Incomplete Data, 299-315.

The second issue is *multiple motion segmentation* for classifying feature point trajectories into independent motions. For this task, too, many techniques have been proposed (Chen and Suter, 2004; Costeira and Kanade, 1998; Gear, 1998; Ichimura, 1999; Ichimura, 2000; Inoue and Urahama, 2001; Kanatani, 2001; Kanatani, 2002a; Kanatani, 2002b; Park *et al.*, 2004; Vidal and Ma, 2004; Vidal and Hartley, 2004; Wu *et al.*, 2001). According to our experiments, however, many methods that exhibit high accuracy in simulations perform rather poorly for real video sequences. We show that this inconsistency is caused by the *degeneracy* of the motion model on which the segmentation is based. This finding leads to a new segmentation algorithm described in (Sugaya and Kanatani, 2004b). We demonstrate its effectiveness, using real video images.

This paper is organized as follows. Section 2 summarizes the geometric constraints. Section 3 describes our outlier removal procedure. Section 4 describes how we extend partial trajectories. In Section 5, we show real video examples of trajectory extension. In Section 6, we describe our principle of multiple-motion segmentation. In Section 7, we analyze the degeneracy of motion model. Section 8 describes our segmentation algorithm. In Section 9, we show real video examples. Section 10 concludes this paper.

16.2. Geometric Constraints

Our method is based on the geometric constraints described in (Chen and Suter, 2004; Debrunner and Ahuja, 1998; Huynh *et al.*, 2003; Irani, 2002; Kanatani, 2001; Kanatani, 2002a; Kanatani, 2002b; Kanatani and Sugaya, 2004; Sugaya and Kanatani, 2002a; Sugaya and Kanantani, 2002b; Sugaya and Kanatani, 2003; Sugaya and Kanatani, 2004a; Sugaya and Kanatani, 2004b). Suppose we track N feature points over M frames. Let $(x_{\kappa\alpha}, y_{\kappa\alpha})$ be the coordinates of the αth point in the κth frame. We stack all the coordinates vertically and represent the entire trajectory by the following $2M$-D *trajectory vector*:

$$\boldsymbol{p}_\alpha = \begin{pmatrix} x_{1\alpha} & y_{1\alpha} & x_{2\alpha} & y_{2\alpha} & \cdots & x_{M\alpha} & y_{M\alpha} \end{pmatrix}^\top . \tag{1}$$

For convenience, we identify the frame number κ with "time" and refer to the κth frame as "time κ".

We regard the XYZ camera coordinate system as a reference, relative to which the scene is moving. Consider a 3-D coordinate system fixed to the scene, and let $\boldsymbol{t}_\kappa$ and $\{\boldsymbol{i}_\kappa, \boldsymbol{j}_\kappa, \boldsymbol{k}_\kappa\}$ be, respectively, its origin and basis vectors at time κ. Let $(a_\alpha, b_\alpha, c_\alpha)$ be the coordinates of the αth point with respect to this coordinate system. Its position with respect to the reference frame at time κ is

$$\boldsymbol{r}_{\kappa\alpha} = \boldsymbol{t}_\kappa + a_\alpha \boldsymbol{i}_\kappa + b_\alpha \boldsymbol{j}_\kappa + c_\alpha \boldsymbol{k}_\kappa. \tag{2}$$

We assume an *affine camera*, which generalizes orthographic, weak perspective, and paraperspective projections (Kanatani and Sugaya, 2004; Poelman and Kanade, 1997): the 3-D point $\boldsymbol{r}_{\kappa\alpha}$ is projected onto the image position

$$\begin{pmatrix} x_{\kappa\alpha} \\ y_{\kappa\alpha} \end{pmatrix} = \boldsymbol{A}_\kappa \boldsymbol{r}_{\kappa\alpha} + \boldsymbol{b}_\kappa, \tag{3}$$

where $\boldsymbol{A}_\kappa$ and $\boldsymbol{b}_\kappa$ are, respectively a 2×3 matrix and a 2-D vector determined by the position and orientation of the camera and its internal parameters at time κ. Substituting Equation (2), we have

$$\begin{pmatrix} x_{\kappa\alpha} \\ y_{\kappa\alpha} \end{pmatrix} = \tilde{\boldsymbol{m}}_{0\kappa} + a_\alpha \tilde{\boldsymbol{m}}_{1\kappa} + b_\alpha \tilde{\boldsymbol{m}}_{2\kappa} + c_\alpha \tilde{\boldsymbol{m}}_{3\kappa}, \tag{4}$$

where $\tilde{\boldsymbol{m}}_{0\kappa}, \tilde{\boldsymbol{m}}_{1\kappa}, \tilde{\boldsymbol{m}}_{2\kappa}$, and $\tilde{\boldsymbol{m}}_{3\kappa}$ are 2-D vectors determined by the position and orientation of the camera and its internal parameters at time κ. From Equation (4), the trajectory vector $\boldsymbol{p}_\alpha$ in Equation (1) can be written in the form

$$\boldsymbol{p}_\alpha = \boldsymbol{m}_0 + a_\alpha \boldsymbol{m}_1 + b_\alpha \boldsymbol{m}_2 + c_\alpha \boldsymbol{m}_3, \tag{5}$$

where $\boldsymbol{m}_0$, $\boldsymbol{m}_1$, $\boldsymbol{m}_2$, and $\boldsymbol{m}_3$ are the $2M$-D vectors obtained by stacking $\tilde{\boldsymbol{m}}_{0\kappa}, \tilde{\boldsymbol{m}}_{1\kappa}, \tilde{\boldsymbol{m}}_{2\kappa}$, and $\tilde{\boldsymbol{m}}_{3\kappa}$ vertically over the M frames, respectively.

Equation (5) implies that all the trajectories are constrained to be in the *4-D subspace* spanned by $\{\boldsymbol{m}_0, \boldsymbol{m}_1, \boldsymbol{m}_2, \boldsymbol{m}_3\}$. In addition, the coefficient of $\boldsymbol{m}_0$ in Equation (5) is identically 1 for all α. This means that the trajectories are in the *3-D affine space* within that 4-D subspace (Kanatani, 2002b).

16.3. Outlier Removal

Before extending partial trajectories, we must remove incorrectly tracked trajectories, or "outliers", from among observed complete trajectories. For this, we adopt the method described in (Sugaya and Kanatani, 2003), which also discusses problems about the approach in (Huynh and Heyden, 2001). Let $n = 2M$, where M is the number of frames, and let $\{\boldsymbol{p}_\alpha\}$, $\alpha = 1, ..., N$, be the observed complete n-D trajectory vectors. The procedure is as follows (Sugaya and Kanatani, 2003):

1. Randomly choose four vectors $\boldsymbol{q}_1$, $\boldsymbol{q}_2$, $\boldsymbol{q}_3$, and $\boldsymbol{q}_4$ from among $\{\boldsymbol{p}_\alpha\}$.
2. Compute the $n \times n$ (second-order) moment matrix

$$\boldsymbol{M}_3 = \sum_{i=1}^{4} (\boldsymbol{q}_i - \boldsymbol{q}_C)(\boldsymbol{q}_i - \boldsymbol{q}_C)^\top, \tag{6}$$

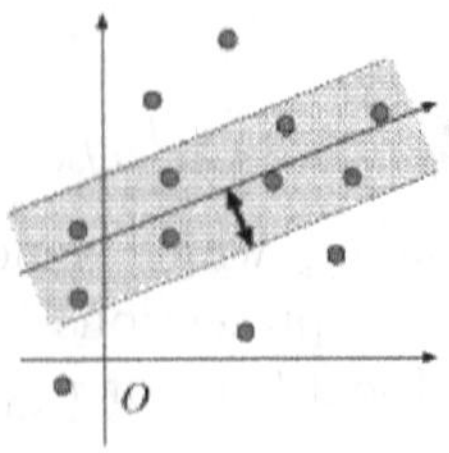

Figure 16.1. Removing outliers by fitting a 3-D affine space.

where $\boldsymbol{q}_C$ is the centroid of $\{\boldsymbol{q}_1, \boldsymbol{q}_2, \boldsymbol{q}_3, \boldsymbol{q}_4\}$.

3. Let $\lambda_1 \geq \lambda_2 \geq \lambda_3$ be the three eigenvalues of the matrix $\boldsymbol{M}_3$, and $\{\boldsymbol{u}_1, \boldsymbol{u}_2, \boldsymbol{u}_3\}$ the orthonormal system of corresponding eigenvectors.
4. Compute the following $n \times n$ projection matrix ($\boldsymbol{I}$ denotes the $n \times n$ unit matrix):

$$\boldsymbol{P}_{n-3} = \boldsymbol{I} - \sum_{i=1}^{3} \boldsymbol{u}_i \boldsymbol{u}_i^\top. \tag{7}$$

5. Let S be the number of points $\boldsymbol{p}_\alpha$ that satisfy

$$\|\boldsymbol{P}_{n-3}(\boldsymbol{p}_\alpha - \boldsymbol{q}_C)\|^2 < (n-3)\sigma^2, \tag{8}$$

where σ is an estimate of the noise standard deviation.

6. Repeat the above procedure a sufficient number of times (we stopped if S did not increase for 200 consecutive iterations), and determine the projection matrix $\boldsymbol{P}_{n-3}$ that maximizes S.
7. Remove those $\boldsymbol{p}_\alpha$ that satisfy

$$\|\boldsymbol{P}_{n-3}(\boldsymbol{p}_\alpha - \boldsymbol{q}_C)\|^2 \geq \sigma^2 \chi^2_{n-3;99}, \tag{9}$$

where $\chi^2_{r;a}$ is the ath percentile of the χ^2 distribution with r degrees of freedom.

The term $\|\boldsymbol{P}_{n-3}(\boldsymbol{p}_\alpha - \boldsymbol{q}_C)\|^2$, called the *residual*, is the squared distance of point $\boldsymbol{p}_\alpha$ from the fitted 3-D affine space. We assume that the noise in the coordinates of the feature points is an independent Gaussian random variable of mean 0 and standard deviation σ. Then, the residual $\|\boldsymbol{P}_{n-3}(\boldsymbol{p}_\alpha - \boldsymbol{q}_C)\|^2$ divided by σ^2 should be subject to a χ^2 distribution with $n-3$ degrees of freedom with expectation $(n-3)\sigma^2$. The above procedure effectively fits a 3-D affine space that maximizes the number of the trajectories whose residuals are smaller than $(n-3)\sigma^2$. Then, we remove those trajectories which cannot be regarded as inliers with significance level 1% (Figure 16.1). We have confirmed that $\sigma = 0.5$ is a reasonable value (Sugaya and Kanatani, 2003).

16.4. Trajectory Extension

After removing outlier trajectories, we optimally fit a 3-D affine space to the resulting inlier trajectories. Let $\{\boldsymbol{p}_\alpha\}$, $\alpha = 1, ..., N$, be their trajectory vectors. We first compute their centroid and the (second-order) moment matrix

$$\boldsymbol{p}_C = \frac{1}{N}\sum_{\alpha=1}^{N} \boldsymbol{p}_\alpha, \qquad \boldsymbol{M} = \sum_{\alpha=1}^{N} (\boldsymbol{p}_\alpha - \boldsymbol{p}_C)(\boldsymbol{p}_\alpha - \boldsymbol{p}_C)^\top. \tag{10}$$

Let $\lambda_1 \geq \lambda_2 \geq \lambda_3$ be the largest three eigenvalues of the matrix $\boldsymbol{M}$, and $\{\boldsymbol{u}_1, \boldsymbol{u}_2, \boldsymbol{u}_3\}$ the orthonormal system of corresponding eigenvectors. The optimally fitted 3-D affine space is spanned by the three vectors of $\boldsymbol{u}_1$, $\boldsymbol{u}_2$, and $\boldsymbol{u}_3$ starting from $\boldsymbol{p}_C$.

If the αth point can be tracked only over κ of the M frames, its trajectory vector $\boldsymbol{p}_\alpha$ has $n - k$ unknown components ($k = 2\kappa$). We partition the vector $\boldsymbol{p}_\alpha$ into the k-D part $\boldsymbol{p}_\alpha^{(0)}$ consisting of the k known components and the $(n - k)$-D part $\boldsymbol{p}_\alpha^{(1)}$ consisting of the remaining $n - k$ unknown components. Similarly, we partition the centroid $\boldsymbol{p}_C$ and the basis vectors $\{\boldsymbol{u}_1, \boldsymbol{u}_2, \boldsymbol{u}_3\}$ into the k-D parts $\boldsymbol{p}_C^{(0)}$ and $\{\boldsymbol{u}_1^{(0)}, \boldsymbol{u}_2^{(0)}, \boldsymbol{u}_3^{(0)}\}$ and the $(n - k)$-D parts $\boldsymbol{p}_C^{(1)}$ and $\{\boldsymbol{u}_1^{(1)}, \boldsymbol{u}_2^{(1)}, \boldsymbol{u}_3^{(1)}\}$ in accordance with the division of $\boldsymbol{p}_\alpha$.

We first test if each of the partial trajectories is sufficiently reliable. Let $\boldsymbol{p}_\alpha$ be a partial trajectory vector. If image noise does not exist, the deviation of $\boldsymbol{p}_\alpha$ from the centroid $\boldsymbol{p}_C$ should be expressed as a linear combination of $\boldsymbol{u}_1$, $\boldsymbol{u}_2$, and $\boldsymbol{u}_3$. Hence, there should be constants c_1, c_2, and c_3 such that

$$\boldsymbol{p}_\alpha^{(0)} - \boldsymbol{p}_C^{(0)} = c_1\boldsymbol{u}_1^{(0)} + c_2\boldsymbol{u}_2^{(0)} + c_3\boldsymbol{u}^{(0)} \tag{11}$$

for the known part. In the presence of image noise, this equality does not hold. If we let $\boldsymbol{U}^{(0)}$ be the $k \times 3$ matrix consisting of $\boldsymbol{u}_1^{(0)}$, $\boldsymbol{u}_2^{(0)}$, and $\boldsymbol{u}_3^{(0)}$ as its columns, Equation (11) is replaced by

$$\boldsymbol{p}_\alpha^{(0)} - \boldsymbol{p}_C^{(0)} \approx \boldsymbol{U}^{(0)}\boldsymbol{c}, \tag{12}$$

where $\boldsymbol{c}$ is the 3-D vector consisting of c_1, c_2, and c_3. Assuming that $k \geq 3$, we estimate the vector $\boldsymbol{c}$ by least squares in the form

$$\hat{\boldsymbol{c}} = \boldsymbol{U}^{(0)-}(\boldsymbol{p}_\alpha^{(0)} - \boldsymbol{p}_C^{(0)}), \tag{13}$$

where $\boldsymbol{U}^{(0)-}$ is the generalized inverse of $\boldsymbol{U}^{(0)}$. It is computed by

$$\boldsymbol{U}^{(0)-} = (\boldsymbol{U}^{(0)\top}\boldsymbol{U}^{(0)})^{-1}\boldsymbol{U}^{(0)\top}. \tag{14}$$

The residual, i.e., the squared distance of point $\boldsymbol{p}_\alpha^{(0)}$ from the 3-D affine space spanned by $\{\boldsymbol{u}_1^{(0)}, \boldsymbol{u}_2^{(0)}, \boldsymbol{u}_3^{(0)}\}$ is $\|\boldsymbol{p}_\alpha^{(0)} - \boldsymbol{p}_C^{(0)} - \boldsymbol{U}^{(0)}\hat{\boldsymbol{c}}\|^2$. Under our noise model, the residual $\|\boldsymbol{p}_\alpha^{(0)} - \boldsymbol{p}_C^{(0)} - \boldsymbol{U}^{(0)}\hat{\boldsymbol{c}}\|^2$ divided by σ^2 should be subject to a χ^2 distribution with $k-3$ degrees of freedom. Hence, we regard those trajectories that satisfy

$$\|\boldsymbol{p}_\alpha^{(0)} - \boldsymbol{p}_C^{(0)} - \boldsymbol{U}^{(0)}\hat{\boldsymbol{c}}\|^2 \geq \sigma^2\chi^2_{k-3;99} \tag{15}$$

as outliers with significance level 1%.

The unknown part $\boldsymbol{p}_\alpha^{(1)}$ is estimated from the constraint implied by Equation (11), namely

$$\boldsymbol{p}_\alpha^{(1)} - \boldsymbol{p}_C^{(1)} = c_1\boldsymbol{u}_1^{(1)} + c_2\boldsymbol{u}_2^{(1)} + c_3\boldsymbol{u}^{(1)} = \boldsymbol{U}^{(1)}\boldsymbol{c}, \tag{16}$$

where $\boldsymbol{U}^{(1)}$ is the $(n-k)\times 3$ matrix consisting of $\boldsymbol{u}_1^{(1)}$, $\boldsymbol{u}_2^{(1)}$, and $\boldsymbol{u}_3^{(1)}$ as its columns. Substituting Equation (13) for $\boldsymbol{c}$, we obtain

$$\hat{\boldsymbol{p}}_\alpha^{(1)} = \boldsymbol{p}_C^{(1)} + \boldsymbol{U}^{(1)}\boldsymbol{U}^{(0)-}(\boldsymbol{p}_\alpha^{(0)} - \boldsymbol{p}_C^{(0)}). \tag{17}$$

Evidently, this is an optimal estimate in the presence of Gaussian noise. However, the underlying affine space is computed only from a small number of complete trajectories; no information contained in the partial trajectories is used, irrespective of how long they are. So, we also incorporate partial trajectories in the following manner.

Note that if three components of $\boldsymbol{p}_\alpha$ are specified, one can place it, in general, in any 3-D affine space by appropriately adjusting the remaining $n-3$ components. In view of this, we introduce the "weight" of the trajectory vector $\boldsymbol{p}_\alpha$ with k known components in the form

$$W_\alpha = \frac{k-3}{n-3}. \tag{18}$$

Let N be the number of all trajectories, complete or partial, inliers or outliers. The optimization goes as follows:

1. Set the weights W_α of those trajectories, complete or partial, that are so far judged to be outliers to 0. All other weights are set to the value in Equation (18).
2. Fit a 3-D affine space to all the trajectories. The procedure is the same as before except that Equations (10) are replaced by the *weighted* centroid and the *weighted* moment matrix

$$\boldsymbol{p}_C = \frac{\sum_{\alpha=1}^N W_\alpha \boldsymbol{p}_\alpha}{\sum_{\alpha=1}^N W_\alpha}, \quad \boldsymbol{M} = \sum_{\alpha=1}^N W_\alpha(\boldsymbol{p}_\alpha - \boldsymbol{p}_C)(\boldsymbol{p}_\alpha - \boldsymbol{p}_C)^\top. \tag{19}$$

3. Test each trajectory if it is an outlier, using Equation (15).
4. Estimate the unknown parts of the inlier partial trajectories, using Equation (17).

These four steps are iterated until the fitted affine space converges. In the course of this optimization, trajectories once regarded as outliers may be judged to be inliers later, and vice versa. In the end, inlier partial trajectories are optimally extended with respect to the affine space that is optimally fitted to all the complete and partial inlier trajectories.

The iterations may not converge if the initial guess is very poor or a large proportion of the trajectories are incorrect. However, this did not happen in any of our experiments using real video sequences.

We need at least three complete trajectories for guessing the initial affine space. If no such trajectories are given, we may use the method of Jacobs (Jacobs, 2001), but it is much more practical to segment the sequence into overlapping blocks, extending partial trajectories over each block, and connecting the blocks.

16.5. Experiments

Figure 16.2(a) shows five decimated frames from a 50 frame sequence (320×240 pixels) of a static scene taken by a moving camera. We detected 200 feature points and tracked them using the Kanade-Lucas-Tomasi algorithm (Tomasi and Kanade, 1991). When tracking failed at some frame, we restarted the tracking after adding a new feature point in that frame. In the end, we obtained 29 complete trajectories, of which 11 are regarded as inliers by the procedure described in Section 3. The marks □ in Figure 16.2(a) indicate their positions; Figure 16.2(b) shows their trajectories.

Using the affine space they define, we extended the partial trajectories and optimized the affine space and the extended trajectories. The optimization converged after 11 iterations, resulting in the 560 inlier trajectories shown in Figure 16.2(c). The computation time for this optimization was 134 seconds. We used Pentium 4 2.4B GHz for the CPU with 1 GB main memory and Linux for the OS. Figure 16.2(d) is the extrapolated image of the 33th frame after missing feature positions are restored: using the 180 feature points visible in the first frame, we defined triangular patches, to which the texture in the first frame is mapped. We reconstructed the 3-D shape by factorization based on weak perspective projection (Kanatani and Sugaya, 2004) (Figure 16.2(e)); see (Sugaya and Kanatani, 2004a) for more experiment results.

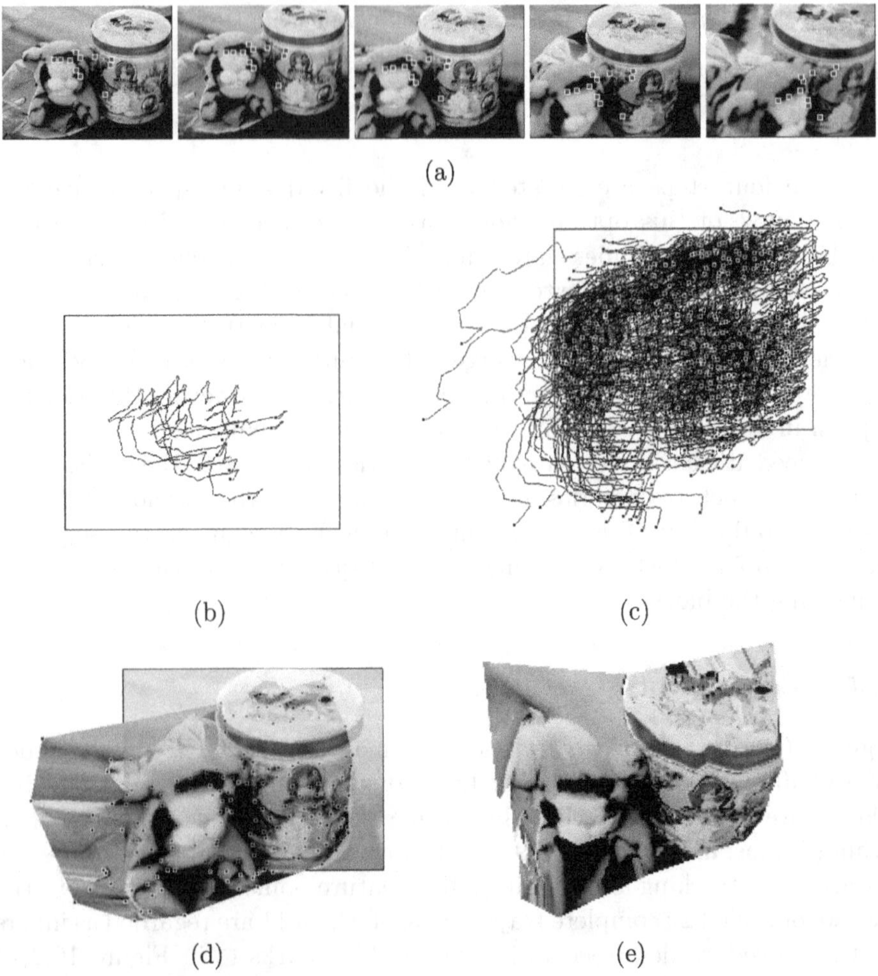

Figure 16.2. (a) Five decimated frames from a 50 frame sequence and 11 points correctly tracked throughout the sequence. (b) The 11 complete inlier trajectories. (c) The 560 optimal extensions of the trajectories. (d) The extrapolated texture-mapped image of the 33th frame. (e) The reconstructed 3-D shape.

16.6. Multiple Motion Segmentation

So far, we have regarded the observed trajectories as points undergoing a single rigid motion. We now consider the case in which multiple motions exist.

Equation (5) states that the trajectory vectors of points that belong to one object are constrained to be in the 4-D subspace spanned by $\{\boldsymbol{m}_0, \boldsymbol{m}_1, \boldsymbol{m}_2, \boldsymbol{m}_3\}$. Hence, multiple moving objects can be segmented into individual motions by separating the trajectories vectors $\{\boldsymbol{p}_\alpha\}$ into distinct

4-D subspaces. This is the principle of the method of *subspace separation* (Kanatani, 2001; Kanatani, 2002a).

Equation (5) also states that the trajectory vectors of points that belong to one object are constrained to be in a 3-D affine space within that 4-D subspace. Hence, multiple moving objects can be segmented into individual motions by separating the trajectory vectors $\{\boldsymbol{p}_\alpha\}$ into distinct 3-D affine spaces. This is the principle of the method of *affine space separation* (Kanatani, 2002b).

Theoretically, the segmentation accuracy should be higher if we use stronger constraints. For real video sequences, however, we have found that the affine space separation accuracy is often lower than that of the subspace separation (Sugaya and Kanatani, 2002a; Sugaya and Kanantani, 2002b). We will resolve this inconsistency in shortly.

As in the case of a single motion, we first need to remove outlier trajectories. If the trajectories were segmented into individual classes, we could apply the method of Section 3 to each motion separately. In the presence of outliers, however, we cannot do correct segmentation, and hence we do not know the affine spaces.

This difficulty can be resolved if we note that if the trajectory vectors $\{\boldsymbol{p}_\alpha\}$ belong to m d-D subspaces, they should be constrained to be in a dm-D subspace and if they belong to m d-D affine spaces, they should be in a $((d+1)m-1)$-D affine space. So, we robustly fit a dm-D subspace or a $((d+1)m-1)$-D affine space to $\{\boldsymbol{p}_\alpha\}$ by RANSAC and remove those that do not fit to it. We observed that all apparent outliers were removed by this method, although some inliers were also removed for safety (Sugaya and Kanatani, 2003).

16.7. Structure of Degeneracy

The motions we most frequently encounter are such that the objects and the background are translating and rotating 2-dimensionally in the image frame with varying sizes. For such a motion, we can choose the basis vector $\boldsymbol{k}_\kappa$ in Equation (2) in the Z direction (the camera optical axis is identified with the Z-axis). Under the affine camera model, motions in the Z direction do not affect the projected image except for its size. Hence, the term $c_\alpha \tilde{\boldsymbol{m}}_{3\kappa}$ in Equation (4) vanishes; the scale changes are absorbed by the scale changes of $\tilde{\boldsymbol{m}}_{1\kappa}$ and $\tilde{\boldsymbol{m}}_{2\kappa}$ over time κ. It follows that the trajectory vector $\boldsymbol{p}_\alpha$ in Equation (5) belongs to the *2-D affine space* passing through $\boldsymbol{m}_0$ and spanned by $\boldsymbol{m}_1$ and $\boldsymbol{m}_2$ (Sugaya and Kanatani, 2002a; Sugaya and Kanantani, 2002b).

If, in addition, the objects and the background do not rotate, we can fix the basis vectors $\boldsymbol{i}_\kappa$ and $\boldsymbol{j}_\kappa$ in Equation (2) to be in the X and Y directions,

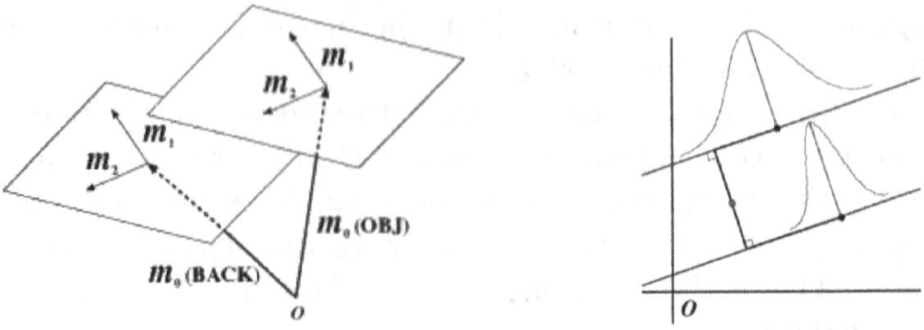

(a) (b)

Figure 16.3. (a) If the motions of the objects and the background are degenerate, their trajectory vectors belong to mutually parallel 2-D planes. (b) The data distributions inside the individual 2-D planes are modeled by Gaussian distributions.

respectively. Thus, the basis vectors $\boldsymbol{i}_\kappa$ and $\boldsymbol{j}_\kappa$ are common to all objects and the background, so the vectors $\boldsymbol{m}_1$ and $\boldsymbol{m}_2$ in Equation (5) are also common. Hence, the 2-D affine spaces, or planes, of all the motions are *parallel* (Sugaya and Kanatani, 2004b) (Figure 16.3(a)).

Note that *parallel 2-D planes can be included in a 3-D affine space.* Since the affine space separation method attempts to segment the trajectories into different 3-D affine spaces, it does not work if the objects and the background undergo this type of degenerate motions. This explains why the accuracy of the affine space separation is not as high as expected for real video sequences.

16.8. Degeneracy-tuned Learning

We now describe a learning procedure tuned to the parallel 2-D plane degeneracy (Sugaya and Kanatani, 2004b). First, we model the data distributions inside the individual 2-D planes by Gaussian distributions (Figure16.3(b)). As before, we let $n = 2M$. Suppose N n-D trajectory vectors $\{\boldsymbol{p}_\alpha\}$ are already classified into m classes by some means. Initially, we define the weight $W_\alpha^{(k)}$ of the vector $\boldsymbol{p}_\alpha$ by

$$W_\alpha^{(k)} = \begin{cases} 1 & \text{if } \boldsymbol{p}_\alpha \text{ belongs to class } k \\ 0 & \text{otherwise} \end{cases}. \tag{20}$$

Then, we iterate the following procedures A, B, and C in turn until all the weights $\{W_\alpha^{(k)}\}$ converge (we stopped the iterations when the increments in $W_\alpha^{(k)}$ are all smaller than 10^{-10}).

A. Do the following computation for each class $k = 1, ..., m$.

1. Compute the fractional size $w^{(k)}$ and the centroid $\boldsymbol{p}_C^{(k)}$ of the class k:

$$w^{(k)} = \frac{1}{N}\sum_{\alpha=1}^{N} W_\alpha^{(k)}, \qquad \boldsymbol{p}_C^{(k)} = \frac{\sum_{\alpha=1}^{N} W_\alpha^{(k)} \boldsymbol{p}_\alpha}{\sum_{\alpha=1}^{N} W_\alpha^{(k)}}. \tag{21}$$

2. Compute the $n \times n$ moment matrix $\boldsymbol{M}^{(k)}$:

$$\boldsymbol{M}^{(k)} = \frac{\sum_{\alpha=1}^{N} W_\alpha^{(k)} (\boldsymbol{p}_\alpha - \boldsymbol{p}_C^{(k)})(\boldsymbol{p}_\alpha - \boldsymbol{p}_C^{(k)})^\top}{\sum_{\alpha=1}^{N} W_\alpha^{(k)}}. \tag{22}$$

B. Do the following computation.

1. Compute the *total* $n \times n$ moment matrix

$$\boldsymbol{M} = \sum_{k=1}^{m} w^{(k)} \boldsymbol{M}^{(k)}. \tag{23}$$

2. Let $\lambda_1 \geq \lambda_2$ be the largest two eigenvalues of the matrix $\boldsymbol{M}$, and $\boldsymbol{u}_1$ and $\boldsymbol{u}_2$ the corresponding unit eigenvectors.
3. Compute the *common* $n \times n$ projection matrices:

$$\boldsymbol{P} = \sum_{i=1}^{2} \boldsymbol{u}_i \boldsymbol{u}_i^\top, \qquad \boldsymbol{P}_\perp = \boldsymbol{I} - \boldsymbol{P}. \tag{24}$$

4. Estimate the noise variance in the direction orthogonal to *all* the affine spaces by

$$\hat{\sigma}^2 = \max[\frac{\mathrm{tr}[\boldsymbol{P}_\perp \boldsymbol{M} \boldsymbol{P}_\perp]}{n-2}, \sigma^2], \tag{25}$$

where $\mathrm{tr}[\,\cdot\,]$ denotes the trace and σ is an estimate of the tracking accuracy. As before we used the value $\sigma = 0.5$ (pixels).
5. Compute the $n \times n$ covariance matrix of the class k by

$$\boldsymbol{V}^{(k)} = \boldsymbol{P}\boldsymbol{M}^{(k)}\boldsymbol{P} + \hat{\sigma}^2 \boldsymbol{P}_\perp. \tag{26}$$

C. Do the following computation for each trajectory vector $\boldsymbol{p}_\alpha$, $\alpha = 1, ..., N$.

1. Compute the conditional likelihood $P(\alpha|k)$, $k = 1, ..., m$, by

$$P(\alpha|k) = \frac{e^{-(\boldsymbol{p}_\alpha - \boldsymbol{p}_C^{(k)}, \boldsymbol{V}^{(k)-1}(\boldsymbol{p}_\alpha - \boldsymbol{p}_C^{(k)}))/2}}{\sqrt{|\boldsymbol{V}|^{(k)}}}. \tag{27}$$

2. Recompute the weights $\{W_\alpha^{(k)}\}$, $k = 1, ..., m$, by

$$W_\alpha^{(k)} = \frac{w^{(k)}P(\alpha|k)}{\sum_{l=1}^{m} w^{(l)}P(\alpha|l)}. \tag{28}$$

After the iterations of A, B, and C have converged, the αth trajectory is classified into the class k that maximizes $W_\alpha^{(k)}$, $k = 1, ..., N$.

In the above iterations, we fit 2-D planes of the same orientation to all classes by computing the common basis vectors $\boldsymbol{u}_1$ and $\boldsymbol{u}_2$ from all the data. We also estimate a common outside noise variance from all the data. Regarding the fraction $w^{(k)}$ (the first of Equations (21)) as the *a priori probability* of the class k, we compute the probability $P(\alpha|k)$ of the trajectory vector $\boldsymbol{p}_\alpha$ conditioned to be in the class k (Equation (27); common multipliers that will cancel out in Equation (28) are omitted). Then, we apply *Bayes' theorem* (Equation (28)) to compute the *a posterior probability* $W_\alpha^{(k)}$, according which all the trajectories are reclassified. Note that $W_\alpha^{(k)}$ is generally a fraction, so one trajectory belongs to multiple classes with fractional weights until the final classification is made.

This type of *unsupervised learning* (Schlesinger, 1968; Schlesinger and Hlaváč, 2002) (mathematically equivalent to the *EM algorithm* (Dempster *et al.*, 1977)) is widely used for clustering. However, the iterations are very likely to be trapped at a local maximum. So, correct segmentation cannot be obtained by this type of iterations alone unless we start from a very good initial value.

16.9. Multi-stage Learning

If we *know* that degeneracy exists, we can apply the above procedure for improving the segmentation. However, we do not know if degeneracy exists. If the trajectories were segmented into individual classes, we might detect degeneracy by checking the dimensions of the individual classes, but we cannot do correct segmentation unless we know whether or not degeneracy exists.

We resolve this difficulty by the following multi-stage learning (Sugaya and Kanatani, 2004b). First, we use the affine space separation assuming 2-D affine spaces, which effectively assumes planar motions with varying sizes. For this, we use the Kanatani's method (Kanatani, 2002b), which combines the shape interaction matrix of Costeira and Kanade (Costeira and Kanade, 1998), model selection by the geometric AIC (Kanatani, 1998), and robust estimation by LMedS (Rousseeuw and Leroy, 1987). Then, we optimize the resulting segmentation by using the parallel plane degeneracy model, as described in the preceding section.

The resulting solution should be very accurate if such a degeneracy really exists. However, rotations may exist to some extent. So, we relax the constraint and optimize the solution again by using the general 3-D motion model. This is motivated by the fact that if the motions are really degenerate, the solution optimized by the degenerate model is *not affected* by the subsequent optimization, because the degenerate constraints also satisfy the general constraints.

In sum, our scheme consists of the following three stages:

1. Initial segmentation by the affine space separation using 2-D affine spaces.
2. Unsupervised learning using the parallel 2-D plane degeneracy model.
3. Unsupervised learning using the general 3-D motion model.

The last stage is similar to the second except that 3-D affine spaces are separately fitted to individual classes. The outside noise variance is also estimated separately for each class; see (Sugaya and Kanatani, 2004b) for the actual procedure.

Here, we assume that the number m of motions is specified by the user. For example, if a single object is moving in a static background, both moving relative to the camera, we have $m = 2$. Many studies have been done for estimating the number of motions automatically (Costeira and Kanade, 1998; Gear, 1998; Inoue and Urahama, 2001), but none of them seems successful enough. This is because the number of motions is *not well-defined* (Kanatani, 2002a): one moving object can also be viewed as multiple objects moving similarly, and there is no rational way to unify similarly moving objects into one *from motion information alone*, except using heuristic thresholds or ad-hoc criteria. If model selection such as the geometric AIC (Kanatani, 1998) and the geometric MDL (Kanatani, 2004) is used, the resulting number of motions depends on criteria (Kanatani, 2002a). In order to determine the number m of motions, one needs high-level processing using color, shape, and other information.

16.10. Real Video Experiments

Figure 16.4 shows five decimated frames from three video sequences A, B, and C (320 × 240 pixels). For each sequence, we detected feature points in the initial frame and tracked them using the Kanade-Lucas-Tomasi algorithm (Tomasi and Kanade, 1991). The marks □ indicate their positions.

Table 16.1 lists the number of frames, the number of inlier trajectories, and the computation time for our multi-stage learning. The computation time is reduced by compressing the trajectory data into 8-D vectors (Sugaya and Kanatani, 2002a). We used Pentium 4 2.4GHz for the CPU with 1GB

A:

B:

C:

Figure 16.4. Three video sequences and successfully tracked feature points.

TABLE 16.1. The computation time for the multi-stage learning of the sequences in Figure 16.4.

	A	B	C
# of frames	30	17	100
# of points	136	63	73
CPU time (sec)	2.50	0.51	1.49

main memory and Linux for the OS. Table 16.2 lists the accuracies of different methods ("opt" stands for "optimized") measured by (the number of correctly classified points)/(the total number of points) in percentage.

As we can see, the Costeira-Kanade method fails to produce meaningful segmentation. Ichimura's method is effective for sequences A and B but not so effective for sequence C. For sequence A, the affine space separation is superior to the subspace separation. For sequence B, the two methods have almost the same performance. For sequence C, the subspace separation is superior to the affine space separation, suggesting that the motion in sequence C is nearly degenerate. For all the three sequences, our multi-stage learning achieves 100% accuracy.

16.11. Concluding Remarks

We discussed two issues of video processing, missing data estimation and multiple motion segmentation, based on our recent results (Sugaya and Kanatani, 2004a; Sugaya and Kanatani, 2004b).

TABLE 16.2. Segmentation accuracy (%) for the sequences in Figure 16.4.

	A	B	C
Costeira-Kanade	60.3	71.3	58.8
Ichimura	92.6	80.1	68.3
subspace separation	59.3	99.5	98.9
affine space separation	81.8	99.7	67.5
opt. subspace separation	99.0	99.6	99.6
opt. affine space separation	99.0	99.8	69.3
multi-stage learning	**100.0**	**100.0**	**100.0**

First, we described our method for extending interrupted feature point tracking (Sugaya and Kanatani, 2004a). We alternate optimal extension of the trajectories and optimal estimation of the affine space. To increase robustness, we test the reliability of the extended trajectories in every step and remove those judged to be outliers.

Next, we studied multiple motion segmentation. Our analysis of the geometric structure of the degeneracy of the motion model leads to a special type of degeneracy, which results in the multi-stage learning scheme described in (Sugaya and Kanatani, 2004b). We demonstrated its effectiveness, using real video images.

The source codes of the programs we used are available at:
`http://www.suri.it.okayama-u.ac.jp/e-program.html`.

Acknowledgements: This work was supported in part by the Ministry of Education, Culture, Sports, Science and Technology, Japan, under a Grant in Aid for Scientific Research C(2) (No. 15500113).

References

Brandt, S.: Closed-form solutions for affine reconstruction under missing data. In Proc. *Statistical Methods in Video Processing*, pages 109–114, 2002.

Chen, P. and O. Suter: Recovering the missing components in a large noisy low-rank matrix: application to SFM. *IEEE Trans. Pattern Analysis Machine Intelligence*, **26**:1051–1063, 2004.

Costeira, J. P. and T. Kanade: A multibody factorization method for independently moving objects. *Int. J. Computer Vision*, **29**:159–179, 1998.

Debrunner, C. and N. Ahuja: Segmentation and factorization-based motion and structure estimation for long image sequences. *IEEE Trans. Pattern Analysis Machine Intelligence*, **20**:206–211, 1998.

Dempster, A. P., N. M. Laird, and D. B. Rubin: Maximum likelihood from incomplete data via the EM algorithm. *J. Royal Statist. Soc.*, **B39**:1–38, 1977.

Gear, C. W.: Multibody grouping from motion images. *Int. J. Comput. Vision*, **29**:133–150, 1998.

Huynh, D. Q. and A. Heyden: Outlier detection in video sequences under affine projection. In Proc. *IEEE Conf. Computer Vision Pattern Recognition*, volume 1, pages 695–701, 2001.

Huynh, D. Q., R. Hartley, and H. Heyden: Outlier correction in image sequences for the affine camera. In Proc. *IEEE Int. Conf. Computer Vision*, volume 1, pages 585–590, 2003.

Ichimura, N.: Motion segmentation based on factorization method and discriminant criterion. In Proc. *IEEE Int. Conf. Computer Vision*, volume 1, pages 600–605, 1999.

Ichimura, N.: Motion segmentation using feature selection and subspace method based on shape space. In Proc. *IEEE Int. Conf. Pattern Recognition*, volume 3, pages 858–864, 2000.

Inoue, K. and K. Urahama: Separation of multiple objects in motion images by clustering. In Proc. *IEEE Int. Conf. Computer Vision*, volume 1, pages 219–224, 2001.

Irani, M.: Multi-frame correspondence estimation using subspace constraints. *Int. J. Comput. Vision*, **48**:173–194, 2002.

Jacobs, D. W.: Linear fitting with missing data for structure-from-motion. *Comput. Vision Image Understand.*, **82**:57–81, 2001.

Kanatani, K.: Geometric information criterion for model selection. *Int. J. Comput. Vision*, **26**:171–189, 1998.

Kanatani, K.: Motion segmentation by subspace separation and model selection. In Proc. *IEEE Int. Conf. Computer Vision*, volume 2, pages 301–306, 2001

Kanatani, K.: Motion segmentation by subspace separation: Model selection and reliability evaluation. *Int. J. Image Graphics,* **2**:179–197, 2002.

Kanatani, K.: Evaluation and selection of models for motion segmentation. In Proc. *European Conf. Computer Vision* (A.Heyden, G. Sparr, M. Nielsen, and P. Johansen, editors), pages 335–349, LNCS 2352, Springer, Berlin, 2002.

Kanatani, K.: Uncertainty modeling and model selection for geometric inference. *IEEE Trans. Pattern Analysis Machine Intelligence*, **26**:1307–1319, 2004.

Kanatani, K. and Y. Sugaya: Factorization without factorization: Complete Recipe. *Memoirs of the Faculty of Engineering, Okayama University*, **38**:61–72, 2004.

Park, J., H. Zha, and R. Kasturi: Spectral clustering for robust motion segmentation. In Proc. *European Conf. Computer Vision* (T. Pajdla and J. Matas, editors), pages 391–401, LNCS 3024, Springer, Berlin, 2004.

Poelman, C. J. and T. Kanade: A paraperspective factorization method for shape and motion recovery. *IEEE Trans. Pattern Analysis Machine Intelligence*, **19**:206–218, 1997.

Rousseeuw, P. J. and A. M. Leroy: *Robust Regression and Outlier Detection*, J. Wiley and Sons, New York, 1987.

Saito, H. and S. Kamijima: Factorization method using interpolated feature tracking via projective geometry. In Proc. *British Machine Vision Conf.* (R. Harvey and A. Bangham, editors), volume 2, pages 449–458, 2003.

Schlesinger, M. I.: A connection between supervised and unsupervised learning in pattern recognition. *Kibernetika*, **2**:81–88, 1968.

Schlesinger, M. I. and V. Hlaváč: *Ten Lectures on Statistical and Structural Pattern Recognition*, Kluwer Academic Publishers, 2002.

Sugaya, Y. and K. Kanatani: Automatic camera model selection for multibody motion segmentation. In Proc. *Workshop on Science of Computer Vision*, pages 31–39, 2002.

Sugaya, Y. and K. Kanatani: Automatic camera model selection for multibody motion segmentation. In Proc. *IAPR Workshop on Machine Vision Applications*, pages 412–415, 2002.

Sugaya, Y. and K. Kanatani: Outlier removal for motion tracking by subspace separation. *IEICE Trans. Inf. & Syst.*, **E86-D**:1095–1102, 2003.

Sugaya, Y. and K. Kanatani: Extending interrupted feature point tracking for 3-D affine reconstruction. *IEICE Trans. Inf. & Syst.*, **E87-D**:1031–1033, 2004.

Sugaya, Y. and K. Kanatani: Multi-stage optimization for multi-body motion segmentation. *IEICE Trans. Inf. & Syst.*, **E87-D**: 1935–1942, 2004.

Tomasi, C. and T. Kanade: Detection and tracking of point features. Tech. Rep. CMU-CS-91-132, Pittsburgh, 1991.

Tomasi, C. and T. Kanade: Shape and motion from image streams under orthography—A factorization method. *Int. J. Computer Vision*, **9**:137–154, 1992.

Vidal, R. and Y. Ma: A unified algebraic approach to 2-D and 3-D motion segmentation. In Proc. *European Conf. Computer Vision* (T. Pajdla and J. Matas, editors), pages 1–15, LNCS 3021, Springer, Berlin, 2004.

Vidal, R. and R. Hartley: Motion segmentation with missing data using PowerFactorization and GPCA. In Proc. *IEEE Conf. Computer Vision Pattern Recognition*, volume 2, pages 310–316, 2004.

Wu, Y., Z. Zhang, T. S. Huang, and J. Y. Lin: Multibody grouping via orthogonal subspace decomposition, sequences under affine projection. In Proc. *IEEE Conf. Computer Vision Pattern Recognition*, volume 2, pages 695–701, 2001.

ROBUST LOCAL APPROXIMATION OF SCATTERED DATA

MARKUS FENN AND GABRIELE STEIDL
Dept. of Mathematics and Computer Science,
University of Mannheim,
D–68131 Mannheim, Germany

Abstract. In this paper, we modify the robust local image estimation method of R. van den Boomgaard and J. van de Weijer for the approximation of scattered data. The derivation of our knot and data dependent approximation method is based on the relation between the Gaussian facet model in image processing and the moving least square technique known from approximation theory. Numerical examples demonstrate the advantages of our robust scattered data approximation.

Key words: moving least squares, quasi-interpolation, polynomial reproduction, robust estimators, bilateral filters

17.1. Introduction

A popular approach to scattered data approximation is the *moving least squares* (MLS) method which requires in contrast to standard interpolation methods by radial basis functions only the solution of small linear systems of equations. The size of these systems is governed by the degree of the polynomials which are reproduced by the method. The MLS approximation is theoretically well examined, see, e.g., (Fasshauer, 2003b) and the references therein. In particular, the Backus–Gilbert approach offers another way to look at the polynomial reproduction property which in turn determines the approximation order of the method.

On the other hand, there exist various local linear methods for smoothing noisy data in image processing. One example is the Gaussian facet model introduced by R. van den Boomgaard and J. van de Weijer (van den Boomgaard and van de Weijer, 2003) in the linear scale–space context. Interestingly, this method is basically the same as the MLS technique with a Gaussian weight function. The only difference consists in the fact that in scattered data approximation we know the (noisy) function only at some special, in general nonequispaced knots and no data are given within these

R. Klette et al. (eds.), Geometric Properties for Incomplete Data, 317-334.

knots, while in denoising problems in image processing the noisy function is known on the whole grid. This leads to a formulation with shifted basis functions in the MLS approach in contrast to the Gaussian facet model.

In their averaging process, the MLS method and its variants give similar weights to data within a similar distance from the evaluation point, where neighbors are heavier weighted even if these neighbors are on very different levels of the function. Consequently, edges are smoothed. This led to the development of robust estimation procedures and nonlinear filters that also data–adaptively determine the influence of each data point on the result. Among the rich variety of these methods, see, e.g., (Sochen et al., 2001) and the references therein, we focus on the robust Gaussian facet model (van den Boomgaard and van de Weijer, 2003). Having the relation between the linear approaches in image processing and scattered data approximation in mind, we modify this robust model in such a way that it can be also applied to scattered data. Moreover, we change the method slightly toward a generalized bilateral filter approach that does not only reproduce constants but also polynomials of higher degree.

This chapter is organized as follows: first, we consider the linear methods used independently in image processing and scattered data approximation, where we start with the continuous MLS method in Subsection 2.1 and move to the discrete method in Subsection 2.2. In Subsection 3.1, we use these results for introducing our robust scattered data approximation method. Its power is demonstrated by numerical examples in Subsection 3.2. The paper is concluded with a short summary.

17.2. MLS from different points of view

The aim of this section is twofold. Firstly, we want to show the relation between the well examined MLS method in approximation theory and the Gaussian facet model recently introduced in the context of linear scale–space theory by R. van den Boomgaard and J. van de Weijer (van den Boomgaard and van de Weijer, 2003). It is not hard to see that both methods differ only by the formulation with shifted basis functions such that applied to spaces of polynomials they lead to the same result. However, we find it useful to direct the attention of people from the image processing society to theoretical results from approximation theory and vice versa, to benefit from new ideas in image processing for the approximation of scattered data.

Secondly, the MLS results of this section will serve as the basis for our robust approach in Section 3. In particular, we will use the MLS approximation as initial input for our iterative algorithm.

17.2.1. CONTINUOUS MLS

Let

$$V := \text{span}\{\varphi_j : j = 1, \ldots, M\}$$

be an M-dimensional space of real–valued functions defined on $\mathbb{R}^d$. Although some results can be formulated in this general setting, we will restrict ourselves to polynomial spaces. More precisely, let $V := \Pi_s^d$ be the space of d-variate polynomials of absolute degree $\leq s$. Then V has dimension $M = \binom{s+d}{s}$. Our main reason for the restriction to polynomial spaces is that Π_s^d can be also spanned by the translates of φ_j with respect to an arbitrary fixed $x \in \mathbb{R}^d$, i.e.,

$$V = \text{span}\{\varphi_j(\cdot - x) : j = 1, \ldots, M\}. \tag{1}$$

Let w be a non–negative weight function with moments

$$\int_{\mathbb{R}^d} w(t)\, dt = 1 \qquad \text{and} \qquad \int_{\mathbb{R}^d} t^\alpha w(t)\, dt < \infty \quad \text{for all } \alpha \in \mathbb{N}_0^d, |\alpha| \leq 2s.$$

Then

$$\langle p, q \rangle_w := \int_{\mathbb{R}^d} p(t) q(t) w(t)\, dt$$

is an inner product on V with norm $\|p\|_w^2 = \int_{\mathbb{R}^d} p^2(t) w(t)\, dt$.

Now the *continuous MLS problem* can be formulated as follows, see, e.g., (Belytschko et al., 1996): for a given function $f \in L_\infty(\mathbb{R}^d)$ and $x \in \mathbb{R}^d$ find the coefficients $c_j = c_j(x)$ such that

$$u(x, t) := \sum_{j=1}^{M} c_j(x) \varphi_j(t) \tag{2}$$

minimizes the functional

$$J(x) := \int_{\mathbb{R}^d} (f(t) - u(x, t))^2 w(t - x)\, dt. \tag{3}$$

Then

$$u(x) = u(x, x) = \sum_{j=1}^{M} c_j(x) \varphi_j(x) \tag{4}$$

can be taken as an approximation of $f(x)$. Obviously, for arbitrary fixed $x \in \mathbb{R}^d$, the function $u(x, \cdot)$ is the $w(\cdot - x)$–orthogonal projection of f onto V.

On the other hand, we obtain by (1) that the polynomial $\tilde{u}(x,\cdot)$ of the form

$$\tilde{u}(x,t) := \sum_{j=1}^{M} a_j(x)\varphi_j(t-x) \tag{5}$$

which minimizes (3), i.e.,

$$\int_{\mathbb{R}^d} (f(t)-\tilde{u}(x,t))^2 w(t-x)\,\mathrm{d}t = \int_{\mathbb{R}^d} \Big(f(x+t)-\sum_{j=1}^{M} a_j(x)\varphi_j(t)\Big)^2 w(t)\,\mathrm{d}t \tag{6}$$

is also the $w(\cdot - x)$–orthogonal projection of f onto V. Consequently, $u(x,t) = \tilde{u}(x,t)$ and

$$u(x) = \tilde{u}(x,x) = \sum_{j=1}^{M} a_j(x)\varphi_j(0). \tag{7}$$

The approximation (7) of f, where the coefficients $a_j = a_j(x)$ are determined by the minimization of (6) is exactly the approximation method (van den Boomgaard and van de Weijer, 2003). considered in (van den Boomgaard and van de Weijer, 2003). In particular, they have used monomials φ_j, where $\varphi_1 \equiv 1$, as basis functions in (7), so that they have only to compute $u(x) = a_1(x)$. This simplification of MLS by using shifted monomials was also mentioned in (Fasshauer, 2003a). Having finished this paper we realized that the shifted approach (5) was also examined in detail in (Liu et al., 1997).

The minimization problem (6) can be solved for any fixed $x \in \mathbb{R}^d$ by setting the gradient with respect to $a(x) := (a_j(x))_{j=1}^M$ to zero. Using the vector notation $\varphi(t) := (\varphi_k(t))_{k=1}^M$, this leads to

$$a(x) = G^{-1}\Big(\langle f(x+\cdot), \varphi_k\rangle_w\Big)_{k=1}^M = \Big(\langle f(x+\cdot), \big(G^{-1}\varphi(\cdot)\big)_j\rangle_w\Big)_{j=1}^M, \tag{8}$$

where $\big(G^{-1}\varphi(\cdot)\big)_j$ denotes the j–th component of the vector and where the Gramian G is given by

$$G := (\langle \varphi_j, \varphi_k\rangle_w)_{j,k=1}^M.$$

In summary, we obtain by (7) and (8) that

$$\begin{aligned} u(x) &= \Big\langle f(x+\cdot), \sum_{j=1}^{M} (G^{-1}\varphi(\cdot))_j \varphi_j(0)\Big\rangle_w \\ &= \int_{\mathbb{R}^d} f(x+t)q(t)w(t)\,\mathrm{d}t = \int_{\mathbb{R}^d} f(x+t)\psi(t)\,\mathrm{d}t, \end{aligned} \tag{9}$$

where

$$q(t) := \sum_{j=1}^{M} (G^{-1}\varphi(t))_j \varphi_j(0) \qquad \text{and} \qquad \psi(t) := q(t)w(t). \tag{10}$$

In other words, u is the correlation of f with the function ψ.

R. van den Boomgaard and J. van de Weijer have used the monomials of absolute degree $\leq s$ as basis of Π_s^d. We can orthogonalize this basis with respect to $\langle \cdot, \cdot \rangle_w$ so that the new basis fulfills $\langle \varphi_j, \varphi_k \rangle_w = \|\varphi_j\|_w^2 \delta_{jk}$ $(j, k = 1, \ldots, M)$. Then $G = \operatorname{diag}(\|\varphi_j\|_w^2)_{j=1}^M$ is a diagonal matrix and the polynomial q in (10) can be represented alternatively as

$$q(t) = \sum_{j=1}^{M} \frac{\varphi_j(0)}{\|\varphi_j\|_w^2} \varphi_j(t). \tag{11}$$

The function ψ has various properties.

PROPOSITION 17.1. *The function ψ in (10) fulfills the moment condition*

$$\int_{\mathbb{R}^d} t^\alpha \psi(t) \, \mathrm{d}t = \delta_{0\alpha} \qquad (|\alpha| \leq s) \tag{12}$$

and has, for all $p \in \Pi_s^d$, the reproducing property

$$\int_{\mathbb{R}^d} p(t+x)\psi(t) \, \mathrm{d}t = p(x). \tag{13}$$

Proof Let $\{\varphi_j : j = 1, \ldots, M\}$ be w-orthogonal. Then it is easy to check that the Christoffel-Darboux kernel

$$K(t,x) = \sum_{j=1}^{M} \frac{1}{\|\varphi_j\|_w^2} \varphi_j(x) \varphi_j(t)$$

is a reproducing kernel in Π_s^d with respect to $\langle \cdot, \cdot \rangle_w$, i.e.,

$$\int_{\mathbb{R}^d} p(t) K(t,x) w(t) \, \mathrm{d}t = p(x) \qquad \text{for all } p \in \Pi_s^d.$$

In particular, we obtain for the monomials $p(t) = t^\alpha$ with $|\alpha| \leq s$ and $x = 0$ by (11) that

$$\int_{\mathbb{R}^d} t^\alpha K(t,0) w(t) \, \mathrm{d}t = \int_{\mathbb{R}^d} t^\alpha \psi(t) \, \mathrm{d}t = \delta_{0\alpha}.$$

By the binomial formula this implies for any fixed $x \in \mathbb{R}^d$ that

$$\int_{\mathbb{R}^d} (t+x)^\alpha \psi(t)\,\mathrm{d}t = x^\alpha.$$

Consequently, (13) holds true. □

In the following, we are mainly interested in radial weights w.

PROPOSITION 17.2. *Let $w(t) = \omega(\|t\|)$ be a radial weight function, where $\|\cdot\|$ denotes the Euclidian norm in $\mathbb{R}^d$. Then the function ψ in (10) is also radial.*

Proof On the one hand, the polynomial $p(y) := \sum_{k=0}^{s'} \gamma_k\, y^{2k}$ with $s' := \lfloor s/2 \rfloor$ which satisfies

$$\int_{\mathbb{R}^d} \|t\|^{2j} p(\|t\|)\,\omega(\|t\|)\,\mathrm{d}t = \delta_{0j} \qquad (j = 0, \ldots, s')$$

is uniquely determined and $p(\|t\|) \in \Pi_d^s$. Since on the other hand the polynomial $q \in \Pi_d^s$ in (10) is also uniquely determined by the moment condition (12), it suffices to show that $p(\|\cdot\|)$ actually fulfills

$$\int_{\mathbb{R}^d} t^\alpha p(\|t\|)\,\omega(\|t\|)\,\mathrm{d}t = \delta_{0\alpha}. \qquad (|\alpha| \le s) \tag{14}$$

Switching to polar coordinates, the left side of (14) reads as

$$\int_0^\infty r^{|\alpha|+d-1} p(r)\,\omega(r)\,\mathrm{d}r \int_{S^{d-1}} t^\alpha\,\mathrm{d}S,$$

where $\mathrm{d}S$ is the element of the $(d-1)$-dimensional measure on the unit sphere S^{d-1} in $\mathbb{R}^d$. If α contains any odd component, then it is easy to check by the orthogonality of sin and cos functions, that $\int_{S^{d-1}} t^\alpha\,\mathrm{d}S = 0$, cf. (Folland, 1999, p. 80). Otherwise, we have by definition of p with $|\alpha| = 2j$ that

$$\int_0^\infty r^{|\alpha|+d-1} p(r)\,\omega(r)\,\mathrm{d}r = \int_{\mathbb{R}^d} \|t\|^{2j} p(\|t\|)\,\omega(\|t\|)\,\mathrm{d}t = \delta_{0\alpha}$$

This completes the proof. □

EXAMPLE 17.1. *The most popular weight function is the Gaussian*

$$w(t) := \pi^{-d/2} \mathrm{e}^{-|t|^2}.$$

By the separability of the d-variate Gaussian, orthogonal polynomials with respect to the d-variate Gaussian weight are given by the tensor products of the univariate Hermite-polynomials

$$H_n(y) := (-1)^n e^{y^2} \frac{d^n}{dy^n} e^{-y^2}.$$

Using their three-term recurrence relation

$$H_0(y) = 1, \quad H_1(y) = 2y, \quad H_{n+1}(y) = 2yH_n(y) - 2nH_{n-1}(y),$$

we see that $H_{2n+1}(0) = 0$ *and* $H_{2n}(0) = (-1)^n \frac{(2n)!}{n!}$. *Moreover, it is well known that* $\langle H_n, H_k \rangle_w = 2^n\, n!\, \delta_{nk}$, *so that*

$$\frac{H_{2n}(0)}{\|H_{2n}\|_w^2} = \frac{(-1)^n}{4^n n!}.$$

Consequently, we obtain for even s and $t := (t_1, \ldots, t_d)$ *by (11) that*

$$\begin{aligned}
\psi(t) &= \sum_{\substack{|\alpha|\le s,\\ \alpha \text{ even}}} \prod_{j=1}^{d} H_{\alpha_j}(t_j) \frac{(-1)^{\beta_j}}{4^{\beta_j}\beta_j!} w(t) \qquad \left(\beta_j := \frac{\alpha_j}{2}\right) \\
&= \sum_{\substack{|\alpha|\le s,\\ \alpha \text{ even}}} \prod_{j=1}^{d} \frac{d^{\alpha_j}}{dt^{\alpha_j}} \omega(t_j) \frac{(-1)^{\beta_j}}{2^{\alpha_j}\beta_j!} \\
&= \sum_{\substack{|\alpha|\le s,\\ \alpha \text{ even}}} \frac{d^{\alpha}}{dt^{\alpha}} w(t) \frac{(-1)^{|\alpha|/2}}{2^{|\alpha|}\beta_1! \cdots \beta_d!} \\
&= \sum_{r=0}^{s/2} \frac{(-1)^r}{2^{2r} r!} \sum_{\substack{|\alpha|=2r,\\ \alpha \text{ even}}} \frac{r!}{\beta_1! \cdots \beta_d!} \frac{d^{\alpha}}{dt^{\alpha}} w(t) \\
&= \sum_{r=0}^{s/2} \frac{(-1)^r}{4^r r!} \Delta^r w(t),
\end{aligned}$$

where $\Delta w(t) := \sum_{j=1}^{d} \frac{\partial^2}{\partial t_j^2} w(t)$ *is the Laplacian of* w *and* $\Delta^r w(t)$ *its r-th iterate. In particular, we have for* $d = 2$ *that*

s	0	2	4
ψ	$w(t)$	$w(t) - \frac{1}{4}\Delta w(t)$	$w(t) - \frac{1}{4}\Delta w(t) + \frac{1}{32}\Delta^2 w(t)$

These special functions were also computed in (van den Boomgaard and van de Weijer, 2003) and the corresponding polynomials q in the context of the so-called approximate approximation in (Fasshauer and Zhang, 2004). For the relation of q to generalized Laguerre polynomials see (Maz'ya and Schmidt, 1996) and the references therein. Since the difference of a function f and a multiple of its convolution with the Laplacian of the Gaussian approximates a backward diffusion, the convolution with ψ for $s \geq 2$ leads to a better reproduction of f in particular at edges. This is another way of looking at the improvement of the approximation by a better polynomial reproduction with increasing s. The influence of the additional sharpening terms in ψ is illustrated in (van den Boomgaard and van de Weijer, 2003) and in our examples in Subsection 3.2.

Other weights used in the scattered data literature are the Wendland functions (Wendland, 1995). In contrast to the Gaussian these functions have a compact support. For $d = 2$ and $s = 1$ the corresponding functions ψ can be found in (Fasshauer, 2003a).

Another popular weight function in image processing is the characteristic function $w(x) := \chi_{\{x:\|x\|_\infty \leq C\}}$, which leads to the so-called Haralick facet model (Haralick et al., 1983).

REMARK 17.1. *The computation of our approximating function u of f in (9) requires the discretization of the correlation integral. If we use the rectangular quadrature rule over a grid of mesh size h and equispaced integration knots $\{x_k := hk : k \in \mathbb{Z}^d\}$, we obtain*

$$u(x) \approx h^d \sum_{k\in\mathbb{Z}^d} f(x_k)\psi(x_k - x).$$

If we replace w by its dilated version $w_\sigma = \frac{1}{\sigma^d} w(\frac{\cdot}{\sigma})$, then ψ with respect to w_σ becomes $\psi_\sigma = \frac{1}{\sigma^d}\psi(\frac{\cdot}{\sigma})$ and the discretized continuous MLS approximation of f with respect to w_σ with $\sigma = \sqrt{D}h$ is

$$u(x) \approx u_{\sqrt{D}h} = D^{-d/2} \sum_{k\in\mathbb{Z}^d} f(x_k)\psi\left(\frac{x_k - x}{h\sqrt{D}}\right). \tag{15}$$

The right-hand side of (15) is known as approximate approximation of f. In (Maz'ya and Schmidt, 2001) the authors have proved that for $f \in L_\infty(\mathbb{R}^d) \cap C^{s+1}(\mathbb{R}^d)$ and a function ψ satisfying the moment condition (12), the following error estimate holds true

$$\|f - u_{\sqrt{D}h}\|_C = \mathcal{O}(h^{s+1} + \varepsilon(\psi, D)),$$

where $\varepsilon(\psi, D)$ denotes a saturation error which can be controlled by appropriately choosing the dilation factor σ of the generating function ψ.

Note that (Maz'ya and Schmidt, 2001) contains also error estimates if nonequispaced knots x_k *are used in (15).*

17.2.2. DISCRETE MLS

In scattered data approximation, the function f is in general only known at nonequispaced knots $x_k \in \mathbb{R}^d$ $(k = 1, \ldots, N)$, where $N \geq M$. Instead of using a continuous MLS approach with a discretization of the convolution integral at these knots, we prefer a discrete MLS approach. Basically, we have the same setting as in Subsection (17.2.1), (2–4), except that we want to minimize

$$J(x) := \sum_{k=1}^{N} (f(x_k) - u(x, x_k))^2 w(x_k - x) \tag{16}$$

instead of (3). For fixed $x \in \mathbb{R}^d$, this is a weighted least squares problem for the coefficients $c_j = c_j(x)$ which has the solution

$$c(x) = (\Phi\, W(x)\, \Phi^T)^{-1} \Phi\, W(x)\, f, \tag{17}$$

where $c(x) := (c_j(x))_{j=1}^M$, $f := (f(x_k))_{k=1}^N$ and

$$\Phi := (\varphi_j(x_k))_{j,k=1}^{M,N}, \qquad W(x) := \operatorname{diag}(w(x_k - x))_{k=1}^N.$$

Here we have to assume that the points $x_k \in \mathbb{R}^d$ are distributed such that Φ has full rank, i.e., not all x_k lie on the zero set of a polynomial of degree $\leq s$. Then, by (4),

$$u(x) = \varphi(x)^T\, c(x) \tag{18}$$

is taken as approximation of $f(x)$.

REMARK 17.2. *In the case* $s = 0$, *i.e.,* $V = \{1\}$ *and* $M = 1$, *we obtain that* $\Phi = (1, \ldots, 1)$ *and consequently by (17) and (18) that*

$$u(x) = c_1(x) = \frac{\sum_{k=1}^{N} f(x_k) w(x_k - x)}{\sum_{k=1}^{N} w(x_k - x)}. \tag{19}$$

This approximation is known as Shepard's method (Shepard, 1968). The approximate value $u(x)$ *of* $f(x)$ *is the weighted average of the values* $f(x_k)$, *where the weights decrease with an increasing distance of* x_k *from* x. *We will have a look at this method again in connection with bilateral filters.*

REMARK 17.3. *From the* Backus-Gilbert approach *(Bos and Šalkauskas, 1989) it is well–known that, for an appropriate function g, the function ψ_g which solves the constrained minimization problem*

$$\frac{1}{2}\sum_{k=1}^{N}\frac{\psi_g^2(x_k,x)}{g(x_k,x)} \longrightarrow \min$$

subject to the polynomial reproducing property

$$\sum_{k=1}^{N} p(x_k)\psi_g(x_k,x) = p(x) \quad \textit{for all } p \in \Pi_s^d \tag{20}$$

is given by

$$(\psi_g(x_k,x))_{k=1}^N = \varphi(x)^T \left(\Phi\, D\, \Phi^T\right)^{-1} \Phi\, D,$$

where

$$D := \operatorname{diag}(g(x_k,x))_{k=1}^N.$$

Usually, $g(x,x_k) := w(x_k - x)$ is chosen in the literature. Then, by (17), we can rewrite (18) in the form

$$u(x) = \sum_{k=1}^{N} f(x_k)\psi_w(x_k,x). \tag{21}$$

This approach is also known as quasi-interpolation of f. If f is a polynomial of absolute degree $\leq s$, then, by the constraint (20), it is reproduced exactly, i.e., u coincides with f.

Note that on the other hand, the discrete MLS problem can be considered with the shifted formulation (5), where one has to minimize a discrete functional corresponding to (6). This leads directly to the form (21) of u.

17.3. Robust local approximation of scattered data

In (van den Boomgaard and van de Weijer, 2003), the authors suggested a robust Gaussian facet model for various applications in image processing. Robust estimators classically dealt with statistical outliers, but can be also used to better reconstruct edges. In this section, we want to use the robust facet approach in a slightly more general form for the approximation of (noisy) scattered data. Furthermore, we propose a novel method which seems to be more related to the idea of bilateral filters.

17.3.1. GENERALIZED BILATERAL FILTERS

In order to make our approximation more sensible with respect to edges we introduce a differentiable function ρ in J which punishes small differences harder but sees larger differences more gently, i.e., instead of (16) we minimize the functional

$$J_\rho(x) := \sum_{k=1}^{N} \rho\Big(\big(f(x_k) - u(x, x_k)\big)^2\Big) w(x_k - x).$$

In (16) we have simply used $\rho(r^2) = r^2$. In this section, we apply

$$\rho(r^2) := \sqrt{r^2 + \varepsilon^2} \qquad (\varepsilon \ll 1) \tag{22}$$

which results (approximately) in a weighted ℓ_1-norm of $(f(x_k) - u(x, x_k))_{k=1}^{N}$ in J_ρ, and

$$\rho(r^2) = 1 - \mathrm{e}^{-r^2/(2m^2)} \tag{23}$$

which gives an approximation of a weighted ℓ_0-norm. The function (23) was suggested in (van den Boomgaard and van de Weijer, 2003). While the function ρ in (22) and thus J_ρ are still strictly convex, ρ becomes nonconvex in (23).

Computing the gradient of $J_\rho(x)$ with respect to $c_\ell(x)$ ($\ell = 1, \ldots, M$) and setting this gradient to zero, leads to the following nonlinear system of equations

$$\Phi\, W(x) B_\rho(x)\, \Phi^T\, c(x) = \Phi\, W(x) B_\rho(x)\, f, \tag{24}$$

where

$$\begin{aligned} B_\rho(x) &:= \operatorname{diag}\Big(\rho'\big((f(x_k) - u(x, x_k))^2\big)\Big)_{k=1}^{N} \\ &= \operatorname{diag}\Big(\rho'\big((f(x_k) - \sum_{\ell=1}^{M} c_\ell \varphi_\ell(x - x_k))^2\big)\Big)_{k=1}^{N}. \end{aligned} \tag{25}$$

Note that for ρ defined by (22) or (23) the function $\rho'(r^2)$ is a monotone decreasing function in r^2. In contrast to the diagonal matrix $W(x)$ appearing in (17), we incorporate now the diagonal matrix $W(x)B_\rho(x)$ which does not only depend on the knots x_k, but also on the data $f(x_k)$. Thus, we obtain both a knot and data dependent method. We solve (24) by a fixed point iteration, i.e., we compute successively

$$c^{(i+1)}(x) = (\Phi\, W(x) B_\rho^{(i)}(x)\, \Phi^T)^{-1} \Phi\, W(x) B_\rho^{(i)}(x)\, f,$$

where

$$B_\rho^{(i)}(x) := \operatorname{diag}\Big(\rho'\big((f(x_k) - \sum_{\ell=1}^{M} c_\ell^{(i)}(x) \varphi_\ell(x_k - x))^2\big)\Big)_{k=1}^{N}$$

and set

$$u^{(i+1)}(x) := \varphi(x)^T\, c^{(i+1)}(x). \tag{26}$$

As initial vector $c^{(0)}(x)$ we use the values obtained from the discrete MLS in Subsection (17.2.2). The question of convergence of this iterative method is still open.

REMARK 17.4. *If $s = 0$, then we obtain as in Remark (17.2), that $u^{(i)}(x) = c_1^{(i)}(x)$, in particular, after one iteration,*

$$u^{(1)}(x) = \frac{\sum\limits_{k=1}^{N} f(x_k) w(x_k - x)\rho'((f(x_k) - u^{(0)}(x))^2)}{\sum\limits_{k=1}^{N} w(x_k - x)\rho'((f(x_k) - u^{(0)}(x))^2)}. \tag{27}$$

For $x := x_j$ $(j = 1, \ldots, N)$ and input $u^{(0)}(x_j) := f(x_j)$, the approximation (27) is known as bilateral filter *(Tomasi and Manduchi, 1998). In contrast to Shepard's method (19) do the weights of the values $f(x_k)$ in (27) not only decrease with an increasing distance of x_k from x, but also with an increasing distance of $f(x_k)$ from $f(x)$ (or its approximation $u^{(0)}(x)$). Thus the averaging process is reduced at edges. We remark that a generalization of bilateral filters for piecewise linear functions that completely differs from our approach was given in (Elad, 2002).*

Based on Remark (17.4) and Remark (17.3) we propose the following **new approximation method** which can be considered as a generalization of the bilateral filter. Obviously, the division by $\sum_{k=1}^{N} w(x_k - x)\rho'((f(x_k) - u^{(0)}(x))^2)$ in (27) ensures at each iteration step i that $u^{(i)}$ reproduces constants $f \equiv C$. By Remark (17.3), the idea of using bilateral filters for scattered data approximation can be generalized such that polynomials of arbitrary absolute degree $\le s$ are reproduced. We have to compute

$$u^{(i+1)}(x) := \varphi(x)^T\, (\Phi\, W(x) D_\rho^{(i)}(x)\, \Phi^T)^{-1} \Phi\, W(x) D_\rho^{(i)}(x)\, f, \tag{28}$$

where

$$D_\rho^{(i)}(x) := \operatorname{diag}\Big(\rho'((f(x_k) - u^{(i)}(x))^2)\Big)_{k=1}^{N}.$$

In contrast to $B_\rho^{(i)}$ in (25), where we find it difficult to interpret the differences $f(x_k) - u^{(i)}(x, x_k)$, our diagonal matrix $D_\rho^{(i)}$ contains the approximated differences $f(x_k) - f(x) \approx f(x_k) - u^{(i)}(x)$. The function ρ' may be any appropriate decreasing function. Moreover, as initial data $u^{(0)}$ we can take any reasonable approximation of f. Of course, for $s = 1$, both methods (26) and (28) coincide.

17.3.2. NUMERICAL EXAMPLES

In this section, we present numerical examples with the proposed algorithms in one and two dimensions. The algorithms were implemented in C. As weight function w, we have always used a dilated Gaussian function $w_\sigma(y) = e^{-y^2/(2\sigma^2)}$ which we have truncated for $|y| > 3\sigma$. In this presentation, we have restricted ourselves to the nonlinear function $\rho(r^2) = 1 - e^{-r^2/(2m^2)}$ in (23). However, we have computed various examples with the function ρ in (22) as well. In 2D, these results look very similar to those obtained by applying (23). The corresponding images can be found at our web page

`http://kiwi.math.uni-mannheim.de/~mfenn/RMLS.html`

The nonlinear methods were always performed with five iterations, since we observed reasonable convergence in all our experiments within ≤ 5 iteration steps.

Figure (17.1) shows a onedimensional example with the 'ramp'-signal. The first row contains the original 256 pixel data in (a) and 64 scattered data points (uniformly distributed random numbers) with some Gaussian noise added in (b). Here the *signal-to-noise ratio* (SNR) defined by

$$\mathrm{SNR} = 20 \log_{10} \frac{\|z - \bar{z}\|_2}{\|n\|_2}$$

with z standing for the original signal with mean $\bar{z}$, and n representing noise, is 8 dB. The following rows of Figure (17.1) show the results of the MLS approximation in (c)–(e), of iteration scheme (26) with the diagonal matrix B_ρ in (f)–(h), and of our generalized bilateral filter (28) with the diagonal matrix D_ρ in (j)–(k), where the polynomial reproduction degree increases from $s = 0$ to $s = 2$ from left to right. The parameters σ for the knot-dependent weights and the parameter m for the data-dependent weights were chosen such that the optical impression was the best. In the MLS approximations, we have taken $\sigma = 3/64$, and in the nonlinear schemes (26) and (28), the parameters $\sigma = 6/64$ and $m = 0.2$. As initial data for the iterative algorithms we have always used the results from the MLS approximation with the same degree of polynomial reproduction. However, it should be noted that our algorithm (28) has shown a quite robust behavior with respect to the choice of the initial data. Even very rough initial data approximations, e.g., a simple linear approximation, has led to nearly the same results (j)–(k).

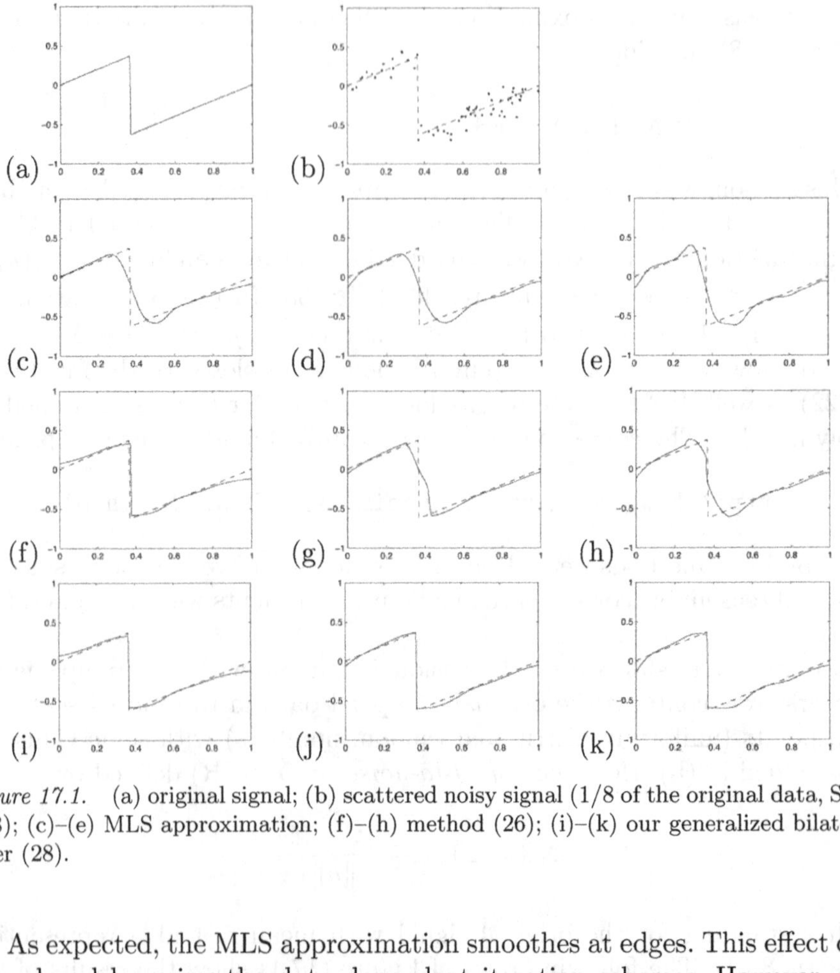

Figure 17.1. (a) original signal; (b) scattered noisy signal (1/8 of the original data, SNR = 8); (c)–(e) MLS approximation; (f)–(h) method (26); (i)–(k) our generalized bilateral filter (28).

As expected, the MLS approximation smoothes at edges. This effect can be reduced by using the data dependent iteration schemes. However, the nonlinear method (26) still introduces some artefacts at edges. The same effect can be observed in 2D.

Since the original signal is piecewise linear, the methods which reproduce quadratic polynomials (right column) do not bring some further improvements.

Figure 17.2 compares scattered data approximation in 2D. We took the 256×256 pixel image 'trui.png' in (a), added some Gaussian noise with SNR $= 16$ dB in (b). Finally, we chose randomly 1/16 of the data in (c). The images (d)–(f) in the second row of Figure 17.2 show the results of the MLS approximation for $s = 0, 1, 2$ from left to right. The parameter $\sigma = 6/256$ was chosen such that the images look best. However, we have also computed the images with respect to that parameter σ which gives the best SNR. The results are reported at our web page. The third and fourth row

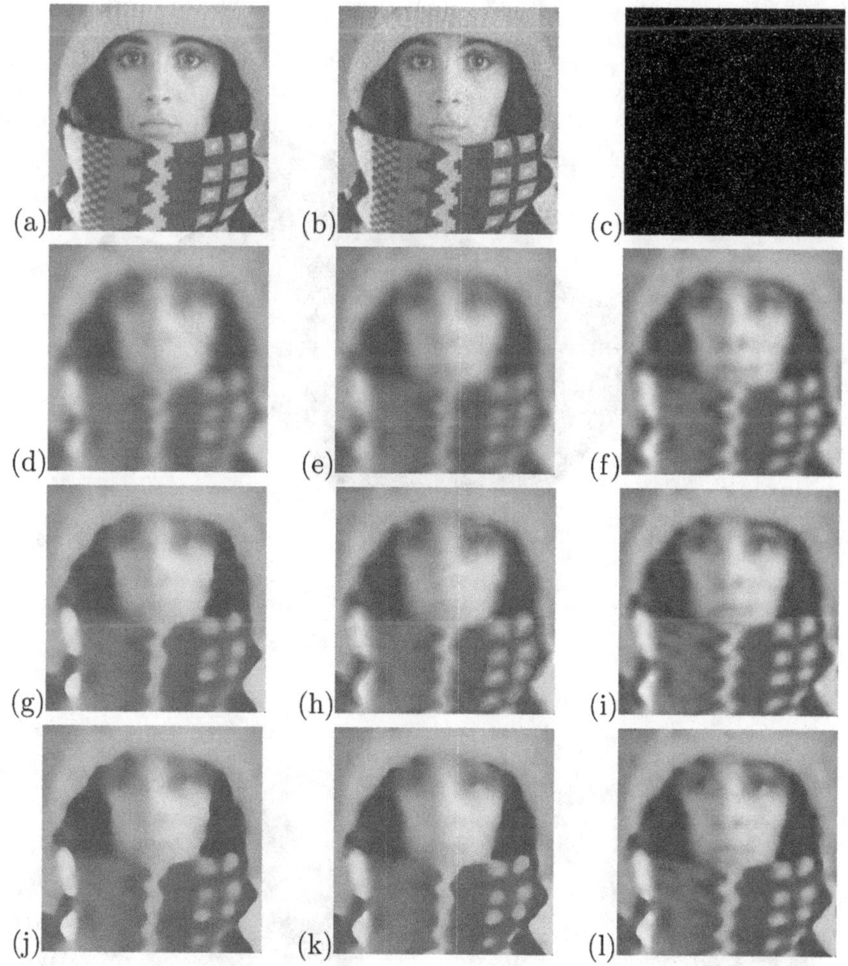

Figure 17.2. (a) original image; (b) noisy image (SNR = 16); (c) scattered noisy image (1/16 of the data); (d)–(f) MLS with $s = 0$ (SNR=7.62), $s = 1$ (SNR=7.73), $s = 2$ (SNR=9.79); (g)–(i) method (26) with $s = 0$ (SNR=8.70), $s = 1$ (SNR=8.58), $s = 2$ (SNR=10.48); (j)–(l) our generalized bilateral filter (28) with $s = 0$ (SNR=8.82), $s = 1$ (SNR=9.41), $s = 2$ (SNR=10.62). The parameter $\sigma = 6/256$ was chosen such that the images in (d)–(f) are visually best and the parameter m in (g)–(l) such that the SNR is optimal.

present the results for the nonlinear methods (26) and (28), respectively, with an increasing degree of the polynomial reproduction $s = 0, 1, 2$ from left to right and with the same parameter $\sigma = 6/256$. The parameter m in (23) was chosen such that we have obtained the best SNR. In general, we had $m \in [0.18, 0.28]$. The SNR of each image can be found in the caption of Figure 17.2. The quality of the images improves with an increasing degree of polynomial reproduction. As expected, the nonlinear methods produce

(a)

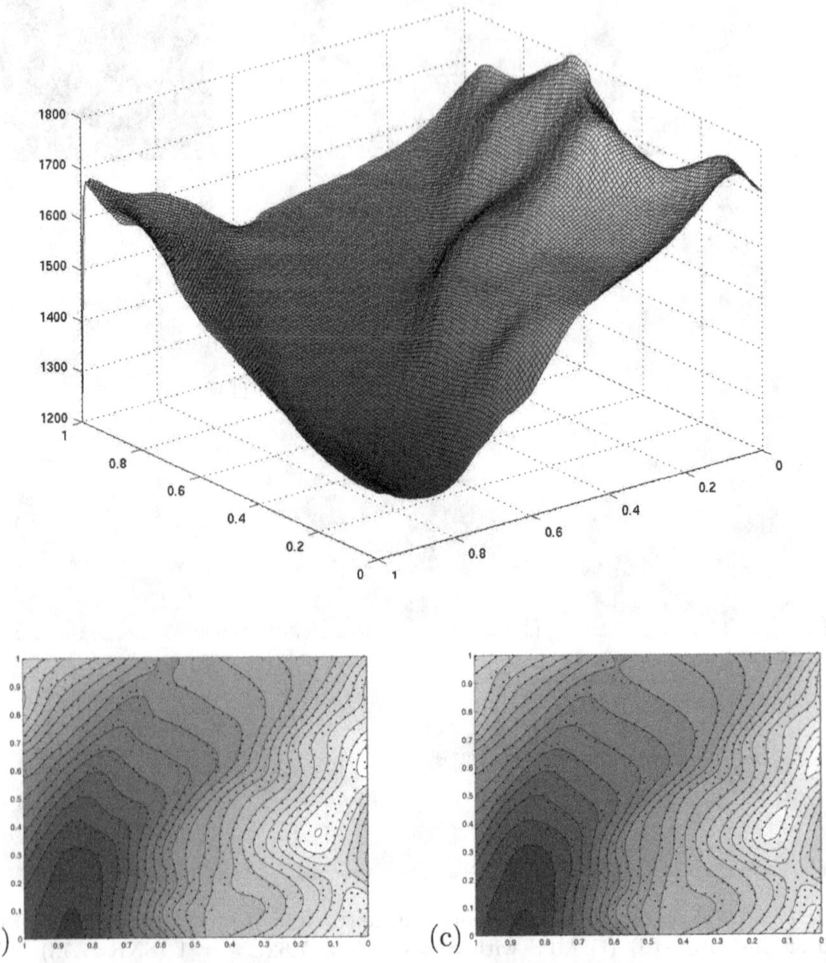

(b) (c)

Figure 17.3. Approximation of 873 scattered data points from the 'glacier' at the 128×128 grid; (a) 3-D plot of (c); (b) original data (dotted) and contour plot of the MLS approximation with $s = 2$, (c) our generalized bilateral filter (28) with $s = 2$.

somewhat sharper images. In order to observe this effect more carefully, the reader may have again a look at details of the images at our web page. The best result was obtained with our generalized bilateral filter (28) and $s = 2$. Note that one iteration step takes less than two seconds here.

Figure 17.3 is based on a data set frequently used in numerical examples for scattered data approximation: we are given 873 scattered data points representing certain contour lines of a glacier. First, we applied the MLS method with $\sigma = 6/128$ and $s = 2$. The contour plots evaluated at the 128×128 grid are presented in (b). Part (c) of the figure shows the result for our algorithm (28) applied with $\sigma = 8/128$, $m = 15$ and $s = 2$. The corresponding 3D plot can be seen in (a).

The contour plots (b), (c) reveal the differences of both methods. Although the MLS approximation (b) is quite good, our nonlinear method (c) better reconstructs smaller structures. For example, the peaks in the middle right part of the images are smoothed by the MLS, but retain by our algorithm.

17.4. Summary

We have introduced a robust local scattered data approximation method which can be considered as a generalization of bilateral filters for scattered data. In particular, the averaging process takes spatial and data values into account. Our approach provides better polynomial reproduction properties than the original bilateral filters at the cost of solving small linear systems of equations. Numerical examples have proved the advantages of the new method with respect to the reproduction of edges. However, this is our first attempt to incorporate robust estimators in scattered data approximation. A couple of theoretical questions is still open. In particular, the convergence behavior of the algorithm and its dependence on the distribution of the scattered knots as well as stability properties were not examined up to now. Furthermore, it should be possible to further speed up the performance of the algorithm by using fast Fourier transforms at nonequispaced knots (NFFT). The NFFT was considered by the authors in various papers (Fenn and Steidl, 2004; Kunis and Potts, 2002) and was recently applied by E. G. Fasshauer and J. G. Zhang (Fasshauer and Zhang, 2004) for scattered data approximation.

Acknowledgement: The basic idea of this paper goes back to a talk given by R. van den Boomgaard within the Mathematical Image Analysis Group in Saarbrücken in February 2004.

References

Belytschko, T., Y. Krongauz, D. Organ, M. Fleming, P. Krysl: Meshless methods: An overview and recent developments. *Comput. Methods Appl. Mech. Eng.*, **139**:3–47, 1996.

Bos, L. P., K. Šalkauskas: Moving least-squares are Backus-Gilbert optimal. *J. Approx. Theory*, **59**:267–275, 1989.

Elad, M.: On the origin of the bilateral filter and ways to improve it. *IEEE Trans. Image Process.*, **11**:1141–1151, 2002.

Fasshauer, G. E.: Approximate moving least-squares approximation with compactly supported radial weights. In *Meshfree methods for partial differential equations* (M. Griebel and M. A. Schweitzer, editors), pages 105–116, Lect. Notes Comput. Sci. Eng. 26, Springer, Berlin, 2003.

Fasshauer, G. E.: Multivariate Meshfree Approximation. Technical report, Dept. of Mathematics, Illinois Inst. of Technology, Chicago, 2003. `http://amadeus.math.iit.edu/~fass/603\_handouts.html`.

Fasshauer, G. E. and J. G. Zhang: Recent results for moving least squares approximation. Technical report, Dept. of Mathematics, Illinois Inst. of Technology, Chicago, 2004.

Fenn, M. and G. Steidl: Fast NFFT based summation of radial functions. *STSIP Journal*, **3**:1–28, 2004.

Folland, G. B.: *Real Analysis: Modern Techniques and Their Applications.* J. Wiley and Sons, Inc., New Nork, 1999.

Haralick, R. M., T. J. Laffey, and L. Watson: The topographic primal sketch. *The International Journal of Robotics Research*, **2**:50–72, 1983.

Kunis, S. and D. Potts: NFFT, software package, C subroutine library. `http://www.math.uni-luebeck.de/potts/nfft/`, 2002.

Liu, W.-K., S. Li, and T. Belytschko: Moving least-square reproducing kernel methods. I. Methodology and convergence. *Comput. Methods Appl. Mech. Engrg.*, **143**:113–154, 1997.

Maz′ya, V. and G. Schmidt: On approximate approximation by using Gaussian kernels. *IMA J. Numer. Anal.*, **16**:13–29, 1996.

Maz′ya, V. and G. Schmidt: On quasi-interpolation with non-uniformly distributed centers on domains and manifolds. *J. Approx. Theory*, **110**:125–145, 2001.

Shepard, D.: A two dimensional interpolation function for irregularly spaced data. In *Proc. 23rd Nat. Conf. ACM*, pages 517–523, ACM Press, 1968.

Sochen, N., R. Kimmel, and A. M. Bruckstein: Diffusions and confusions in signal and image processing. *J. Math. Imaging Vision*, **14**:195–209, 2001.

Tomasi, C. and R. Manduchi: Bilateral filtering for gray and color images. In *Proc. 6th IEEE Internat. Conf. on Computer Vision*, pages 839–846, 1998.

van den Boomgaard, R. and J. van de Weijer: Least squares and robust estimation of local image structure. In *Scale-Space 2003*(L. D. Griffin and M. Lillholm, editors), pages 237–254, LNCS 2695, Springer, Berlin, 2003.

Wendland, H.: Piecewise polynomial, positive definite and compactly supported radial functions of minimal degree. *Adv. Comput. Math.*, **4**:389–396, 1995.

ON ROBUST ESTIMATION AND SMOOTHING WITH SPATIAL AND TONAL KERNELS*

PAVEL MRÁZEK, JOACHIM WEICKERT, AND ANDRÉS BRUHN
Mathematical Image Analysis Group
Faculty of Mathematics and Computer Science, Building 27
Saarland University, 66123 Saarbrücken, Germany

Abstract. This paper deals with establishing relations between a number of widely-used nonlinear filters for digital image processing. We cover robust statistical estimation with (local) M-estimators, local mode filtering in image or histogram space, bilateral filtering, nonlinear diffusion, and regularisation approaches. Although these methods originate in different mathematical theories, we show that their implementation reveals a highly similar structure. We demonstrate that all these methods can be cast into a unified framework of functional minimisation combining nonlocal data and nonlocal smoothness terms. This unification contributes to a better understanding of the individual methods, and it opens the way to new techniques combining the advantages of known filters.

Key words: image analysis, M-estimators, mode filtering, nonlinear diffusion, bilateral filter, regularisation

18.1. Introduction

Image smoothing for the task of denoising or simplification of the visual information is a well established and thoroughly studied topic. A large number of methods have been proposed and new ones continue to appear. However, it is still not easy to see the advantages of the various approaches, and the relations between different methods are only partly understood.

This paper is intended as a contribution in this direction: by studying several methods and their relations, we end up with a better understanding

* This research was partly funded by the project *Relations between Nonlinear filters in Digital Image Processing* within the DFG Priority Program 1114: *Mathematical Methods in Time Series Analysis and Digital Image Processing*, and by the grant No. A2075302 of the Grant Agency of the Academy of Sciences of the Czech Republic. This is gratefully acknowledged.

R. Klette et al. (eds.), Geometric Properties for Incomplete Data, 335-352.

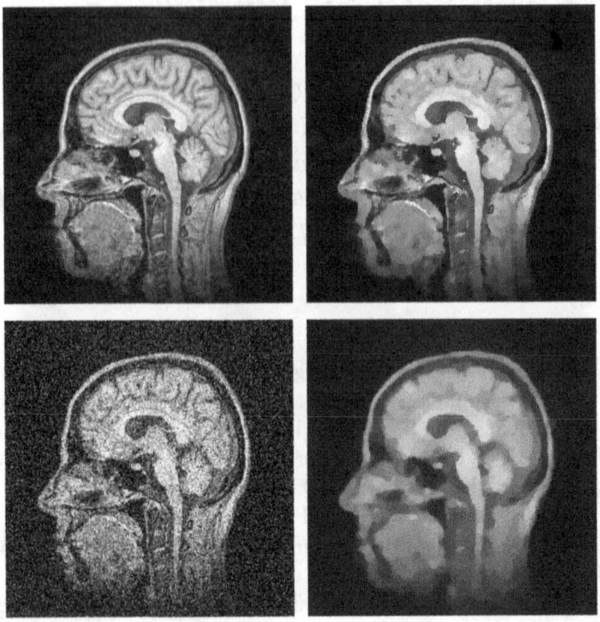

Figure 18.1. Examples of the tasks addressed by methods studied in this paper. Top: image simplification (input image on the left, processed with local mode filter on the right). Bottom: image denoising (noisy image on the left, filtered using TV flow on the right).

of each of them. We focus on M-estimators from robust statistics, median filters, mode filtering, bilateral filter, nonlinear diffusion filtering, and regularisation techniques (see Figure 18.1 for examples). Although these methods seem very different at the first glance and originate in different mathematical theories, we will show that they can lead to highly similar discrete algorithms. From there, it is not far to the observation that all these methods can be cast in a single unified framework of discrete regularisation theory: they can all be derived from minimisation of a single energy functional with (possibly nonlocal) data and (possibly nonlocal) smoothness terms.

This unifying framework has several advantages. Firstly, it explicitly shows all the freedom in selecting the penaliser type, the parameters, and the balance between smoothness and data terms; seing that freedom, it is natural to require that any reasonable smoothing methods motivates the choice of these parameters using some assumptions about the data to be reconstructed, and the noise present in the signal. Secondly, it makes explicit what assumptions are used to derive a given, previously known method from the general settings. Thirdly, after showing known methods as just special members of a whole family of discrete filters, the unifying framework allows to design novel methods tailored to the particular properties of the data and noise, and combine the advantages of known filters.

Related work. Several recent papers have offered interesting connections between different filtering strategies. In (Winkler et al., 1999) the relations between statistical methods (M-estimators) and iterative solvers are studied. The link between iterative mean shift algorithm, mode filters and clustering was analysed in (Cheng, 1995). The relations between mean, median and mode filters in the continuous settings were addressed by (Griffin, 2000). Mode filters and their connections to other approaches represented the main topic of (van de Weijer and van den Boomgaard, 2001; van den Boomgaard and van de Weijer, 2002). (Sochen et al., 2001) focused on the links between energy minimization, partial differential equations and adaptive filtering. Finally, (Elad, 2002) studied the relations between bilateral filters, robust estimation and diffusion filters. Compared to these papers, our work covers a larger number of methods and it includes them all into a single, unified framework.

Organisation of the paper. Our paper is organised as follows. Sections 18.2 to 18.7 represent a brief tour of several nonlinear filters for image processing, from robust statistical estimation and histogram operations to local M-smoothers, regularisation theory, diffusion filtering and bilateral filters. The methods differ in the use of information from local, global or windowed neighbourhood, and in computing the estimates relying either on the original data directly, or using a gradually smoothed image. Section 18.8 then proposes a unified framework which covers all of the presented methods by combining a nonlocal smoothness term and a nonlocal data term with tonal and spatial weight functions into a single functional. Section 18.9 shows some image filtering examples, and the paper is concluded with a summary in Section 18.10.

18.2. Statistical estimation

Let us assume there is an unknown (constant) signal u, and it is observed K times. We obtain the noisy samples f_i, $i = 1, \ldots, K$ according to $f_i = u + n$ where n stands for the noise. If n is a zero-mean Gaussian random variable, one can estimate u by calculating the sample mean $\overline{u} = \frac{1}{K}\sum_{j=1}^{K} f_j$. The mean $\overline{u}$ is the maximum a posteriori (MAP) estimate of u, and minimises the l_2 error $E(u) = \sum_{j=1}^{K}(u - f_j)^2$.

Complications arise if the noise n is not normally distributed, e.g. if it has heavier tails (i.e. there are more outliers, or more distant outliers in the data). This can be caused either directly by the noise properties, or when not a single constant u, but e.g. two constants get mixed in the data (we have to estimate a constant value near a discontinuity in the signal).

The classical statistical solution is to use more robust error norms, and this leads to the theory of M-estimation (Huber, 1981; Hampel et al., 1986). An M-estimate of a constant value u from noisy data f_j is found by minimising

$$E(u) = \sum_{j=1}^{K} \Psi(|u - f_j|^2) \tag{1}$$

where the error norm Ψ can attain for example one of the forms presented in Table 18.1. The right column of Table 18.1 gives an overview of what

TABLE 18.1. Examples of error norms for M-estimators. The parameter λ serves as contrast parameter.

	error norm		$\rightarrow$	*estimation result*
(a)	$\Psi(s^2) = s^2$	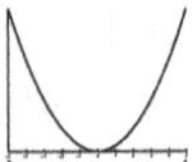	$\rightarrow$	mean
(b)	$\Psi(s^2) = \lvert s \rvert$	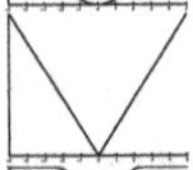	$\rightarrow$	median
(c)	$\Psi(s^2) = 1 - e^{-s^2/\lambda^2}$	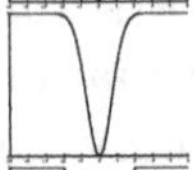	$\rightarrow$	mode approximation
(d)	$\Psi(s^2) = \min(s^2, \lambda^2)$	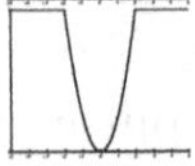	$\rightarrow$	mode approximation

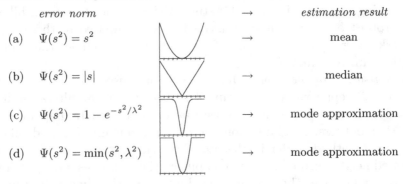

element minimises the functional (1) with the given error penaliser Ψ. For the l_2 norm (a), the solution is the mean of the noisy samples. The l_1 norm (b), the formula is minimised by the median. For the robust error norms (c) and (d), the influence of outliers is very much reduced, and the solution u minimising (1) approximates a mode (maximum) of the probability density underlying the noisy samples. Mode ideally corresponds to the most frequent value present in the data. For the discrete noisy samples, the maximum of the density can be only estimated e.g. using suitable smoothing kernels; see (Cheng, 1995) for some examples and a connection to iterative solvers. Note that while the l_2 and l_1 norms lead to a convex functional minimisation, the robust error norms (c) and (d) in Table 18.1 are nonconvex, and their corresponding functionals $E(u)$ may exhibit multiple local minima.

18.3. Histogram operations

In image analysis, the data (grey values) f_i are measured at positions (pixels) x_i, and we want to find a solution vector $\mathbf{u} = (u_i)_{i=1,\dots,N}$ where each output value u_i belongs to the position x_i. We construct the M-estimates u_i by minimising

$$E(\mathbf{u}) = \sum_{i=1}^{N} \sum_{j=1}^{N} \Psi(|u_i - f_j|^2). \tag{2}$$

$E(\mathbf{u})$ can be minimised by gradient descent (converging towards a local minimum if Ψ is nonconvex), where each element u_i may be processed independently. Initialising by $u_i^0 = f_i$, the gradient descent becomes

$$\begin{aligned} u_i^{k+1} &= u_i^k - \tau \frac{\partial E}{\partial u_i} \\ &= u_i^k - \tau \sum_{j=1}^{N} \Psi'(|u_i^k - f_j|^2)\, 2\,(u_i^k - f_j) \\ &= \left(1 - 2\tau \sum_{j=1}^{N} \Psi'(|u_i^k - f_j|^2)\right) u_i^k \\ &\quad + 2\tau \sum_{j=1}^{N} \Psi'(|u_i^k - f_j|^2)\, f_j \end{aligned} \tag{3}$$

Here τ is the step size. To speed up convergence, τ can be chosen adaptively to the data such that it is larger in plateaus and smaller in areas of large slope; see (Cheng, 1995). Setting

$$\tau := \frac{1}{2\sum_{j=1}^{N} \Psi'(|u_i^k - f_j|^2)}, \qquad g(s^2) := \Psi'(s^2), \tag{4}$$

we can rewrite (3) into the iterative formula

$$u_i^{k+1} = \frac{\sum_{j=1}^{N} g(|u_i^k - f_j|^2)\, f_j}{\sum_{j=1}^{N} g(|u_i^k - f_j|^2)}. \tag{5}$$

Note that in this formulation, the spatial distance between solution u_i and the input samples f_j is not taken into consideration since the index j runs through all pixels. This procedure is equivalent to operations with the histogram of the input image. As an example, Figure 18.2 (b) shows the steady state when iterating (5) with a weighting function g that was chosen in order to correspond to the penaliser Ψ from Table 18.1(c). This results in replacing each pixel by a local mode of the image histogram. We may thus regard it as an image adaptive quantisation strategy.

Figure 18.2. Local mode filtering. *(a) Left:* input image. *(b) Center:* iteratively smoothed using eq. (5) and the penaliser from Table 18.1(c) in a global spatial window. This approximates mode filtering of the histogram. *(c) Right:* processed with the local iterative filter from eq. (9). The tonal weight $g(s^2) = exp(-s^2/\lambda^2)$ was combined with the soft spatial window (8) with $\theta = 5$. This approximates a local mode filter.

18.4. Local M-smoothers

While the previous method was a *global*, histogram-based technique, it is often desirable to estimate a grey value of a pixel from a *local* neighbourhood only. In the framework of M-estimation, this can be achieved by introducing a second weighting term, which depends on the spatial distance between the position of restored pixel u_i and the input sample f_j. For the *local M-smoothers*, the functional to minimise has the following structure (Chu et al., 1998; Winkler et al., 1999):

$$E(\mathbf{u}) = \sum_{i=1}^{N} \sum_{j \in \mathcal{B}(i)} \Psi(|u_i - f_j|^2)\, w(|x_i - x_j|^2) \tag{6}$$

where the spatial weights w represent e.g. a hard disk-shaped window around the current position x_i,

$$w(s^2) = \begin{cases} 1 & s^2 < \theta \\ 0 & \text{otherwise,} \end{cases} \tag{7}$$

or a soft window (Chu et al., 1998),

$$w(s^2) = e^{-s^2/\theta^2}. \tag{8}$$

The local window $\mathcal{B}(i)$ is introduced in (6) for computational convenience only, to make the index j run through the neighbourhood of x_i where $w(|x_i - x_j|^2)$ exceeds some threshold of contribution importance.

In the same way as in the previous section, a minimisation of (6) with adaptive time steps leads to the iterative formula

$$u_i^{k+1} = \frac{\sum_{j \in \mathcal{B}(i)} g(|u_i^k - f_j|^2)\, w(|x_i - x_j|^2)\, f_j}{\sum_{j \in \mathcal{B}(i)} g(|u_i^k - f_j|^2)\, w(|x_i - x_j|^2)} \tag{9}$$

where the function w is called *spatial weight*, in order to distinguish it from the *tonal weight* g. The iterative process is initialised with $u_i^0 := f_i$. Note that, both in (5) and in (9), we are only interested in the *steady state* for $k \to \infty$, not in the evolution towards this minimiser.

As stated in (Winkler et al., 1999), the procedure (9) is called *W-estimator*, and represents one possibility to obtain a solution to the local M-estimation problem. It converges to a local minimum of (6) close to the input data. Depending on the penaliser Ψ, the iterations may lead e.g. to a local mode approximation as in Figure 18.2 (c) (Griffin, 2000; van de Weijer and van den Boomgaard, 2001; van den Boomgaard and van de Weijer, 2002; Comaniciu and Meer, 2002), or to an approximation of a windowed median filter or Gaussian smoothing.

18.5. Bayesian and regularisation frameworks

By taking the windowed M-estimator (6) and decreasing the spatial window size θ, we arrive at the weighting

$$w(|x_i - x_j|^2) = \begin{cases} 1 & \text{if } x_i = x_j, \\ 0 & \text{otherwise.} \end{cases} \tag{10}$$

This leads to the functional

$$E_D(\mathbf{u}) = \sum_{i=1}^{N} \Psi(|u_i - f_i|^2). \tag{11}$$

For any reasonable penaliser Ψ, (11) is minimised by $u_i = f_i$.

It is clear that such a solution is not desired: a good estimate cannot be obtained by looking at a single noisy sample. However, estimation formulated using a local neighbourhood is a highly successful practice, but it has to be combined with some assumptions about the signal to be recovered. In the Bayesian terminology such an assumptions is called *prior information*, in the framework of regularisation theory it is named *smoothness term* or *regulariser*; see e.g. (Bertero et al., 1988; Geman and Geman, 1984; Mumford, 1994).

We construct a smoothness term to express our assumptions about the signal. For the sake of convenience, let us now focus on a continuous modeling where (11) is replaced by

$$E_D(u) = \int_\Omega \Psi(|u - f|^2)\, dx \tag{12}$$

where $\Omega \subset \mathbb{R}^m$ is the image domain, and $f, u : \Omega \to \mathbb{R}$ denote the original and filtered image, respectively. In the classical example of the Mumford–Shah functional (Mumford and Shah, 1989) where the signal is assumed to be piecewise constant with step-like discontinuities, we have

$$E_S(u) = \int_{\Omega\setminus\Gamma} |\nabla u|^2 dx + \beta\,|\Gamma|$$

where Γ is the set of discontinuities and $|\Gamma|$ denotes its length (one-dimensional Hausdorff measure). The smoothness of the image is measured by the squared gradient magnitude $|\nabla u|^2$. We see that deviations from the smoothness are not penalised at the discontinuities. The parameter β balances the image smoothness against the measure of the discontinuity set Γ.

It was shown e.g. in (Winkler et al., 1999) that the explicit boundaries Γ can be expressed implicitly using a robustified prior: Let γ denote the discontinuity indicator function

$$\gamma(x) = \begin{cases} 1 & \text{on } \Gamma \text{ (edge)}, \\ 0 & \text{on } \Omega\setminus\Gamma \text{ (no edge)}, \end{cases}$$

and

$$\Psi_\beta(|\nabla u|^2) := \min\{|\nabla u|^2, \beta\} \tag{13}$$

the cup function from Table 18.1 (d). Then it follows that

$$\begin{aligned} \min_u E_S(u) &= \min_{u,\Gamma} \left(\int_{\Omega\setminus\Gamma} |\nabla u|^2 dx + \beta\,|\Gamma| \right) \\ &= \min_{u,\gamma} \int_\Omega \Big((1-\gamma)\,|\nabla u|^2 + \beta\,\gamma \Big)\, dx \\ &= \min_u \int_\Omega \min_{\gamma\in\{0,1\}} \left((1-\gamma)\,|\nabla u|^2 + \beta\,\gamma \right) dx \\ &= \min_u \int_\Omega \Psi_\beta(|\nabla u|^2)\, dx. \end{aligned} \tag{14}$$

For smoother penalisers Ψ, the discontinuity indicator γ may also attain intermediate values from the interval $[0,1]$ (Nordström, 1990).

In the Bayesian / regularisation framework, the data and smoothness terms are combined into a single functional, thus balancing the measured data against the smoothness assumptions. The resulting functional has e.g. the form

$$\begin{aligned} E(u) &= E_D(u) \;+\; \alpha\, E_S(u) \\ &= \int_\Omega \Big(\Psi_D(|u-f|^2) \;+\; \alpha\, \Psi_S(|\nabla u|^2) \Big)\, dx \end{aligned} \tag{15}$$

with some regularisation parameter $\alpha > 0$. As an example, the continuous Mumford–Shah functional fits into this framework if we choose $\Psi_D(s^2) := s^2$

and $\Psi_S(s^2) := \min(s^2, \lambda^2)$. Discrete versions of such functionals are considered e.g. in (Blake and Zisserman, 1987; Nikolova, 2000; Nikolova, 2001).

18.6. Diffusion filtering

For the sake of completeness, it should be mentioned that equation (15) also covers nonlinear diffusion filters with and without a fidelity term: Choosing $\Psi_D(s^2) := s^2$, every mininiser of (15) has to satisfy necessarily the Euler–Lagrange equation (Courant and Hilbert, 1953)

$$0 = \operatorname{div}\left(g(|\nabla u|^2)\,\nabla u\right) - \frac{u-f}{\alpha} \tag{16}$$

where $g(s^2) := \Psi_S'(s^2)$. Its solution can be regarded as the steady state of the diffusion–reaction process

$$\frac{\partial u}{\partial t} = \operatorname{div}\left(g(|\nabla u|^2)\,\nabla u\right) - \frac{u-f}{\alpha}. \tag{17}$$

where the "time" t is a purely numerical parameter. Such "biased" diffusion processes with a fidelity term have been considered by (Nordström, 1990) in the nonconvex and by (Schnörr, 1994; Stevenson et al., 1994; Charbonnier et al., 1997) in the convex case. They yield the filtered image at *infinite* time ($t \to \infty$).

Alternatively, in (Scherzer and Weickert, 2000) it has been argued that by rewriting (16) as

$$\frac{u-f}{\alpha} = \operatorname{div}\left(g(|\nabla u|^2)\,\nabla u\right) \tag{18}$$

it becomes evident that this process can be regarded as an implicit time discretisation of the diffusion process

$$\frac{\partial u}{\partial t} = \operatorname{div}\left(g(|\nabla u|^2)\,\nabla u\right), \tag{19}$$

$$u(t=0) = f \tag{20}$$

with a single time step of size α. This is a classical, "unbiased" nonlinear diffusion filter as is considered e.g. in (Perona and Malik, 1990; Weickert, 1998). Note that, in contrast to (16), such a filter gives the desired result at *finite* diffusion time $t = \alpha$.

Recently, it has also been shown (Steidl et al., 2004; Mrázek et al., 2003) that discrete diffusion filtering has close relations to wavelet shrinkage (Donoho, 1995).

18.7. Bilateral filtering

Since digital images are sampled on a quadratic pixel grid, it becomes necessary to consider discrete variants of the continuous functional (15). For the data term, this has already been discussed. Therefore let us now focus on the smoothness term $E_S(u) = \Psi_S(|\nabla u|^2)$.

One possibility is to estimate the image gradient magnitude as a sum of squared differences from a pixel to its neighbours. The discrete smoothness penaliser is then expressed in the following way:

$$E_S(\mathbf{u}) = \sum_{i=1}^{N} \Psi_S\Big(\sum_{j \in \mathcal{N}(i)} |u_i - u_j|^2 \Big) \tag{21}$$

where $\mathcal{N}(i)$ stands for the set of 4-neighbours of a pixel i. Just by exchanging the order of summation and penalisation in the last term, we can express the assumption of image smoothness in a slightly different way:

$$E_S(\mathbf{u}) = \sum_{i=1}^{N} \sum_{j \in \mathcal{N}(i)} \Psi_S(|u_i - u_j|^2). \tag{22}$$

This change of operation ordering leads to an *anisotropic* smoothness measure; see also (Weickert and Schnörr, 2001).

Let us now increase the size of the neighbourhood from which the expression (22) is estimated. Then the smoothness term becomes

$$E_S(\mathbf{u}) = \sum_{i=1}^{N} \sum_{j \in \mathcal{B}(i)} \Psi(|u_i - u_j|^2)\, w(|x_i - x_j|^2) \tag{23}$$

where $\mathcal{B}(i)$ is the larger neighbourhood set, and the summation is additionally weighted by a function w of the spatial distance between pixels.

The functional (23) can be minimised by an iterative procedure

$$u_i^{k+1} = u_i^k - \tau \frac{\partial E}{\partial u_i}, \qquad \tau := \frac{1}{2 \sum_{j \in \mathcal{B}(i)} \Psi'(|u_i^k - u_j^k|^2)}.$$

Setting $g(s^2) := \Psi'(s^2)$ leads to the weighted averaging scheme

$$u_i^{k+1} = \frac{\sum_{j \in \mathcal{B}(i)} g(|u_i^k - u_j^k|^2)\, w(|x_i - x_j|^2)\, u_j^k}{\sum_{j \in \mathcal{B}(i)} g(|u_i^k - u_j^k|^2)\, w(|x_i - x_j|^2)} \tag{24}$$

Equation (24) is exactly the *bilateral filter* (Tomasi and Manduchi, 1998); see also (Smith and Brady, 1997) for related ideas. While bilateral filtering

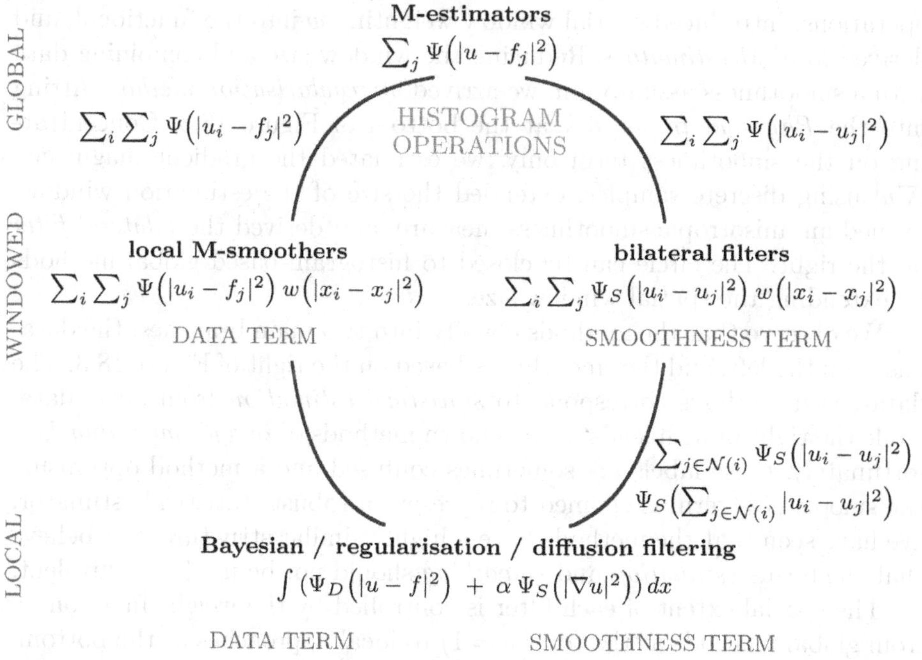

Figure 18.3. Overview of the methods studied in this paper and their corresponding energy functionals.

was originally proposed as a heuristic algorithm, we derived it here as an iterative solver to minimise the anisotropic smoothness term (23) which is evaluated in a nonlocal window.

One should observe the large amount of structural similarities between the local M-smoother (9) and the bilateral filter (24). However, there is one significant difference: Local M-smoothing uses the *initial* image in the averaging procedure and searches for the steady state, while bilateral filtering uses the *evolving* image and has to stop after a certain number of iterations in order to avoid obtaining a flat image.

Moreover, it should be noted that an alternative functional to justify the bilateral filter was proposed in (Elad, 2002). Differently to the one presented here, the functional of Elad contains a windowed smoothness term combined with a local data term. Also, bilateral filter can be viewed as an approximation to the Beltrami flow (Barash, 2002; Spira et al., 2003).

18.8. Unifying framework

Figure 18.3 presents an overview of the energy functionals minimised by all the methods discussed so far. Starting from statistical *M-estimation* at the top, we went counterclockwise down the left branch via histogram operations, introduced spatial window weighting w into the functional, and derived *local M-estimators*. Reducing the window size and combining data with a smoothness assumption, we arrived at *regularisation methods* fitting into the *Bayesian framework* at the bottom of Figure 18.3. Concentrating on the smoothness term only, we estimated the gradient magnitude $|\nabla u|$ using discrete samples, extended the size of the estimation window, formed an anisotropic smoothness measure, and derived the *bilateral filter* on the right. The circle can be closed to histogram-based global methods by extending the spatial window size.

We observe that the methods classify into two main branches: the data-based on the left, and the smoothness-based on the right of Figure 18.3. The data-based methods correspond to *statistical estimation* from noisy data, while the right branch leads to the known methods of *image smoothing*. Unfortunately, these labels are sometimes confused and a method optimising the smoothness term is claimed to represent a robust statistical estimator. We have seen that the methods have a highly similar structure, but believe that the terms *estimation* and *smoothing* should not be used as equivalent.

The spatial extent of each filter is controlled by the weight function w, from global methods at the top ($w = 1$) to local approaches at the bottom. Let us focus on the methods in the middle, where the weight w specifies a finite window. We said that the regularisation methods contain a (local) data term and a (local) smoothness term. If we understand the local M-estimators as a data term (expressed using a nonlocal window for each pixel) and the bilateral filter as a smoothness term (again using a finite window), it is natural to combine them into a single, unified functional:

$$E(\mathbf{u}) = \sum_{i=1}^{N} \sum_{j=1}^{N} \quad \sigma \, \Psi_D(|u_i - f_j|^2) \, w_D(|x_i - x_j|^2) \\ +(1-\sigma) \, \Psi_S(|u_i - u_j|^2) \, w_S(|x_i - x_j|^2). \qquad (25)$$

Depending on the choice of the penalising functions Ψ_D, Ψ_S and on the extent of the spatial weighting functions w_D and w_S, the single formulation (25) covers all the filters discussed so far. They are summarised in Table 18.2.

The unification of methods into a single framework has several advantages. It contributes to the understanding of each method as it makes explicit what parameters $\Psi_D, w_D, \Psi_S, w_S, \sigma$ are needed to derive a given filter. We can see all the freedom that this class of methods offers: four weighting functions (and their parameters) plus the parameter $\sigma \in [0, 1]$ balancing the data against the smoothness term. Obviously, to obtain a reasonable and well performing filter, the choice of parameters should be motivated by some arguments about the data and noise properties. Last

TABLE 18.2. Filtering methods structured according to the used penaliser (from Table 18.1 (a), (b), (c)), extent of the spatial weight w, and correspondence to the data or smoothness terms.

penaliser	windowed data	local smoothness	windowed smoothness
(a)	mean	linear diffusion	
(b)	median	TV diffusion	
(c)	mode approximation	nonlinear diffusion	bilateral filter

but not least, we have seen that known filters represent just several special cases in the framework of functional (25). New methods can be designed to combine the advantages of known filters. This remains the topic for further research.

Some more questions are left open, though. For example, what is the meaning of the smoothness term calculated from a larger window as in (23)? A single pixel may then be directly connected to quite distant pixels, which leads to large-scale smoothing effects of a single filter iteration, but the local topology (e.g. the classical notion of connected regions) is lost, similarly to the locally orderless images (Koenderink and van Doorn, 1999).

18.9. Experiments

In this section we present several examples in order to demonstrate the effect of individual filter components on the final results. We stress that the pictures are intended to visualise the main effects. They are not intended to claim that one method performs better than the other: optimised results in terms of image simplification or denoising can be obtained by each of them by tuning the parameters.

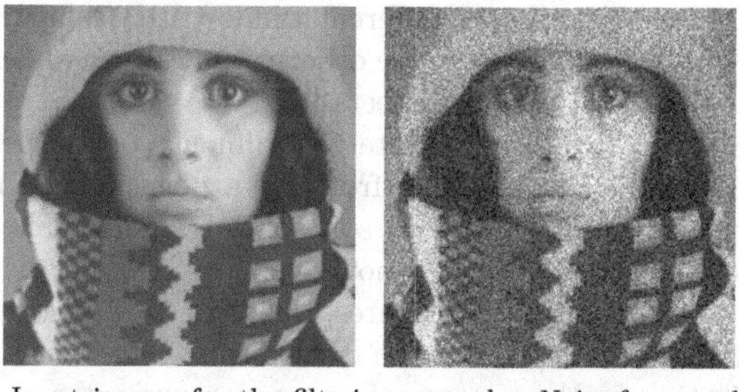

Figure 18.4. Input images for the filtering examples. Noise-free on the left (used in Figures 18.5 and 18.6), noisy on the right (Gaussian noise, SNR=4; used in Figure 18.7).

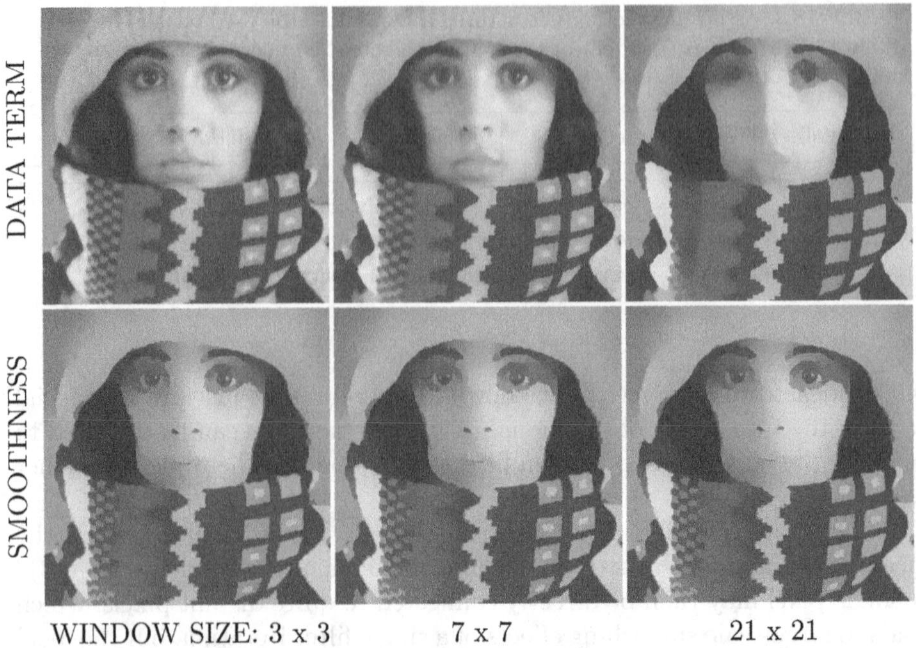

Figure 18.5. Filtering using the penaliser $\Psi(s^2) = 1 - e^{-s^2/\lambda^2}$ (Table 18.1d) with varied size of the spatial neighbourhood. Top: local M-smoothers (data term, steady state of iterating (9)). Bottom: bilateral filtering (smoothness term, 200 iterations based on (24)).

The first image simplification example is shown in Figure 18.5. The penaliser $\Psi(s^2) = 1 - \exp(-s^2/\lambda^2)$ was employed, and the soft spatial window (8) had varied size ($\theta = 1$, $\theta = 3$, $\theta = 10$, cropped circularly into windows of sizes 3×3, 7×7, and 21×21, respectively). We observe that the image filtered via the data term minimisation (i.e. local M-estimator, top row of Figure 18.5) becomes smoother as the window size increases. In this case the steady state is depicted. In the bottom row (smoothness term, bilateral filter) the effect of different window sizes is fairly small if the same number of iterations is used. We observe that already a small window applied iteratively leads to global effects.

Figures 18.6 and 18.7 demonstrate the influence of the penaliser type on the result, starting from a noise-free and a noisy image, respectively. All images were created using a 7×7 soft spatial window. The l_2 penaliser blurs the image most and removes noise very well, while the local mode approximations on the right of Figures 18.6 and 18.7 perform better at preserving the discontinuities, but the result is also more sensitive to noise. The l_1 penalisation in the center column can represent a good compromise between contrast preservation and noise removal, depending on the particular task and data properties.

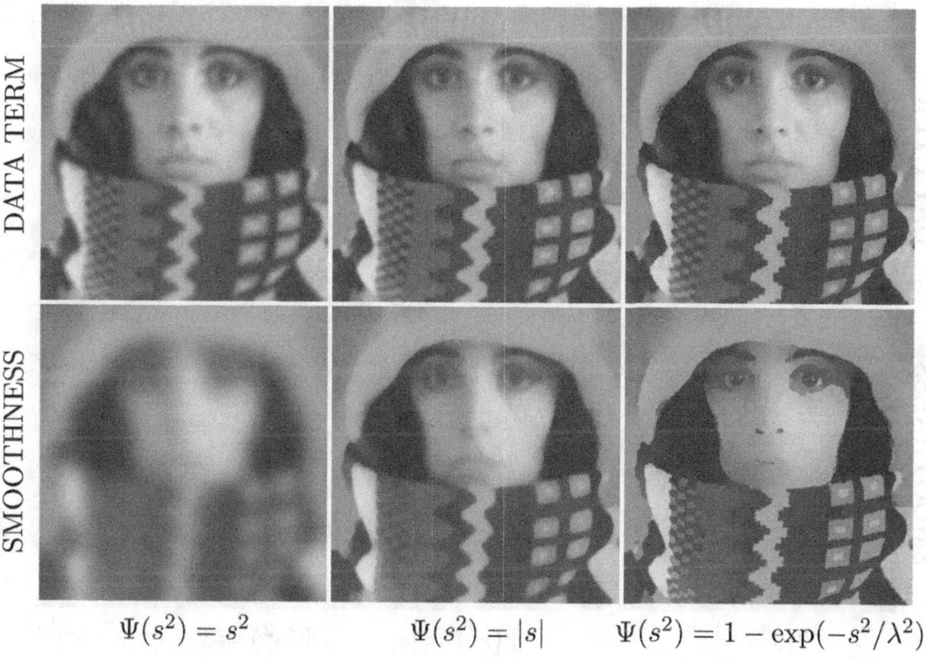

Figure 18.6. Effect of the penaliser type on the filtering result (with soft spatial weighting (8), $\theta = 3$).

18.10. Conclusion

In this paper we focused on the relations between nonlinear filters for digital image processing. We covered statistical M-estimation, mean and median filtering, mode approximation, regularisation and nonlinear diffusion approaches, and bilateral filtering. We have shown that all these methods can be cast into the unified framework of functional minimisation where the functional consists of a (possibly nonlocal) data and (possibly nonlocal) smoothness term. The mutual influence of image pixels is controlled by weighting functions depending on the spatial and tonal distances.

The unified formulation brings new insight and clarifies the relations between different methods. It makes explicit what assumptions are needed to derive known methods (often proposed ad hoc) from the general framework. Then, novel methods can be designed to combine the advantages of known filters and suit the particular data properties.

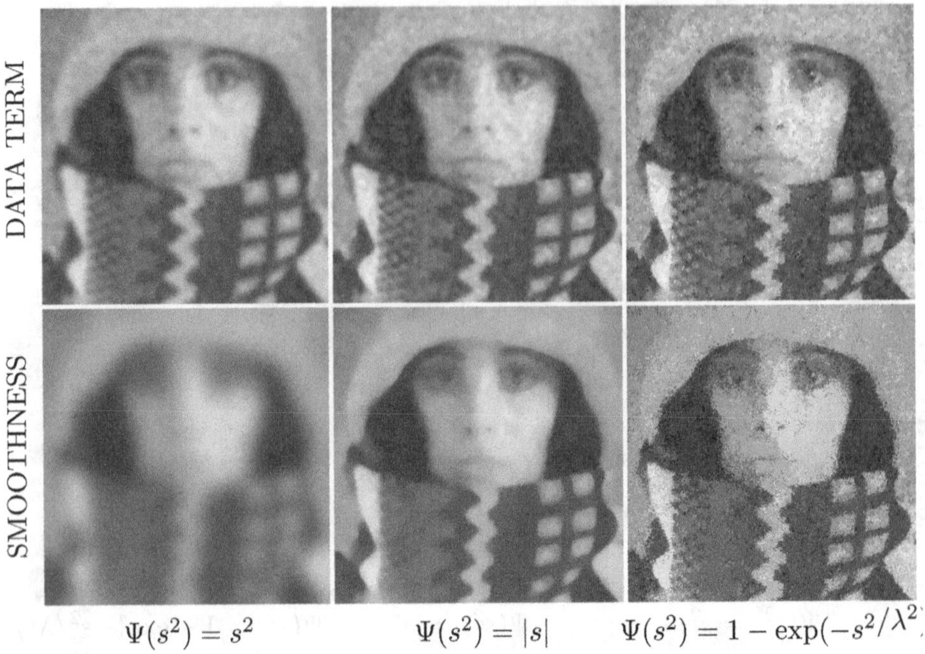

Figure 18.7. Effect of the penaliser type on the filtering result, starting from the noisy image in Fig 18.4 right (filtering in a soft window (8), $\theta = 3$).

References

Barash, D.: A fundamental relationship between bilateral filtering, adaptive smoothing and the nonlinear diffusion equation. *IEEE Trans. Pattern Analysis Machine Intelligence*, **24**:844–847, 2002.

Bertero, M., T. A. Poggio, and V. Torre: Ill-posed problems in early vision. *Proc. IEEE*, **76**:869–889, 1988.

Blake, A. and A. Zisserman: *Visual Reconstruction.* MIT Press, Cambridge, MA, 1987.

Charbonnier, P., L. Blanc-Féraud, G. Aubert, and M. Barlaud: Deterministic edge-preserving regularization in computed imaging. *IEEE Trans. Image Processing*, **6**:298–311, 1997.

Cheng, Y.: Mean shift, mode seeking, and clustering. *IEEE Trans. Pattern Analysis Machine Intelligence*, **17**:790–799, 1995.

Chu, C. K., I. Glad, F. Godtliebsen, and J. S. Marron: Edge-preserving smoothers for image processing. *J. American Statistical Association*, **93**:526–556, 1998.

Comaniciu, D. and P. Meer: Mean shift: A robust approach toward feature space analysis. *IEEE Trans. Pattern Analysis Machine Intelligence*, **24**:603–619, 2002.

Courant, R. and D. Hilbert: *Methods of Mathematical Physics*, Volume 1, Interscience, New York, 1953.

Donoho, D. L.: De-noising by soft thresholding. *IEEE Trans. Information Theory*, **41**:613–627, 1995.

Elad, M.: On the bilateral filter and ways to improve it. *IEEE Trans. Image Processing*, **11**:1141–1151, 2002.

Geman, S. and D. Geman: Stochastic relaxation, Gibbs distributions, and the Bayesian restoration of images. *IEEE Trans. Pattern Analysis Machine Intelligence*, **6**:721–741, 1984.

Griffin, L. D.: Mean, median and mode filtering of images. *Proc. Royal Society of London, Series A*, **456**:2995–3004, 2000.

Hampel, F. R., E. M. Ronchetti, P. J. Rousseeuw, and W. A. Stahel: *Robust Statistics: The Approach Based on Influence Functions.* MIT Press, Cambridge, MA, 1986.

Huber, P. J.: *Robust Statistics.* Wiley, New York, 1981.

Koenderink, J. J. and A. J. van Doorn: The structure of locally orderless images'. *Int. J. Computer Vision*, **31**:159–168, 1999.

Mrázek, P., J. Weickert, and G. Steidl: Diffusion-inspired shrinkage functions and stability results for wavelet denoising. TR 96, Dept. of Mathematics, Saarland University, Saarbrücken, Germany (to appear in *Int. J. Computer Vision*), 2003.

Mumford, D.: The Bayesian rationale for energy functionals. In *Geometry-Driven Diffusion in Computer Vision* (B. Romeny, editor), pages 141–153, Kluwer, Dordrecht, 1994.

Mumford, D. and J. Shah: Optimal approximation of piecewise smooth functions and associated variational problems. *Comm. Pure and Applied Mathematics*, **42**:577–685, 1989.

Nikolova, M.: Local strong homogeneity of a regularized estimator. *SIAM J. Applied Mathematics*, **61**:633–658, 2000.

Nikolova, M.: Image restoration by minimizing objective functions with nonsmooth data-fidelity terms. In Proc. *IEEE Workshop on Variational and Level Set Methods in Computer Vision*, pages 11–18, IEEE Computer Society Press, 2001.

Nordström, N.: Biased anisotropic diffusion – a unified regularization and diffusion approach to edge detection. *Image Vision Computing*, **8**:318–327, 1990.

Perona, P. and J. Malik: Scale space and edge detection using anisotropic diffusion. *IEEE Trans. Pattern Analysis Machine Intelligence*, **12**:629–639, 1990.

Scherzer, O. and J. Weickert: Relations between regularization and diffusion filtering. *J. Mathematical Imaging Vision*, **12**:43–63, 2000.

Schnörr, C.: Unique reconstruction of piecewise smooth images by minimizing strictly convex non-quadratic functionals. *J. Mathematical Imaging Vision* **4**:189–198, 1994.

Smith, S. M. and J. M. Brady: SUSAN: A new approach to low-level image processing. *Int. J. Computer Vision* **23**:45–78, 1997.

Sochen, N., R. Kimmel, and A. M. Bruckstein: Diffusions and confusions in signal and image processing. *J. Mathematical Imaging Vision* **14**:195–209, 2001.

Spira, A., R. Kimmel, and N. Sochen: Efficient Beltrami flow using a short time kernel. In *Scale Space Methods in Computer Vision* (L. Griffin and M. Lillholm, editors), pages 511–522, LNCS 2695, Springer, Berlin, 2003.

Steidl, G., J. Weickert, T. Brox, P. Mrázek, and M. Welk: On the equivalence of soft wavelet shrinkage, total variation diffusion, total variation regularization, and SIDEs. *SIAM J. Numerical Analysis* **42**:686–713, 2004.

Stevenson, R. L., B. E. Schmitz, and E. J. Delp: Discontinuity preserving regularization of inverse visual problems. *IEEE Trans. Systems Man Cybernetics* **24**:455–469, 1994.

Tomasi, C. and R. Manduchi: Bilateral filtering for gray and color images. In Proc. *Int. Conf. Computer Vision*, pages 839–846, Narosa Publishing House, India, 1998.

van de Weijer, J. and R. van den Boomgaard: Local mode filtering. In Proc. *Proc. IEEE Conf. Computer Vision Pattern Recognition*, volume 2, pages 428–433, IEEE Computer Society Press, 2001.

van den Boomgaard, R. and J. van de Weijer: On the equivalence of local-mode finding, robust estimation and mean-shift analysis as used in early vision tasks. In Proc. *Int. Conf. Pattern Recognition*, volume 3, pages 927–930, 2002.

Weickert, J.: *Anisotropic Diffusion in Image Processing*. Teubner, Stuttgart, 1998.

Weickert, J. and C. Schnörr: A theoretical framework for convex regularizers in PDE-based computation of image motion. *Int. J. Computer Vision* **45**:245–264, 2001.

Winkler, G., V. Aurich, K. Hahn, and A. Martin: Noise reduction in images: Some recent edge-preserving methods. *Pattern Recognition Image Analysis* **9**:749–766, 1999.

SUBSPACE ESTIMATION

WITH UNCERTAIN AND CORRELATED DATA

MATTHIAS MÜHLICH
Johann Wolfgang Goethe University
Robert-Mayer-Str. 2–4, 60054 Frankfurt, Germany

Abstract. Parameter estimation problems in computer vision can be modelled as fitting uncertain data to complex geometric manifolds. Recent research provided several new and fast approaches for these problems which allow incorporation of complex noise models, mostly in form of covariance matrices. However, most algorithms can only account for correlations within the same measurement. But many computer vision problems, e.g. gradient-based optical flow estimation, show correlations between different measurements.

In this paper, we will present a new method for improving total least squares (TLS) based estimation with suitably chosen weights and it will be shown how to compute them for general noise models. The new method is applicable to a wide class of problems which share the same mathematical core. For demonstration purposes, we included experiments for ellipse fitting from synthetic data.

Key words: total least squares based estimation, noise models, ellipse fitting

19.1. Introduction

19.1.1. THE ERRORS-IN-VARIABLES (EIV) MODEL

Parameter estimation problems of the general form

$$\varphi(\vec{x}_{i0}, \vec{p}) = 0 \qquad \forall \quad i = 1, \ldots, m \tag{1}$$

are ubiquitous in computer vision. Here $\vec{p} \in \mathbb{R}^n$ stands for the parameter vector that has to be estimated and $\vec{x}_{i0} \in \mathbb{R}^\ell$ denotes some true but unknown vectors (index i for different measurements), of which only some error-prone versions

$$\vec{x}_i = f(\vec{x}_{i0}, \vec{e}_i)$$

are available (for instance, $\vec{x}_i \in \mathbb{R}^4$ could be the stacked coordinates of corresponding points in a stereo image). Some (possibly non-linear) function f combines true values and errors; however, the assumption of additive

R. Klette et al. (eds.), Geometric Properties for Incomplete Data, 353-372.

noise $\vec{x}_i = \vec{x}_{i0} + \vec{e}_i$ is often reasonable. When $\vec{x}_{i0}$ is replaced by $\vec{x}_i$ in (1), we only achieve approximate equality: $\varphi(\vec{x}_{i0}, \vec{p}\,') \approx 0$. Usually, we have an overdetermined system, i.e. (much) more measurements than unknown parameters, or, mathematically: $m > n$. The model defined by (1) is known as errors-in-variables (EIV) model.

19.1.2. THE TOTAL LEAST SQUARES (TLS) MODEL

EIV estimation problems can be linearized to yield an equation

$$\vec{a}_i'^T \vec{p}\,' \approx b_i' \qquad (\vec{a}_i' \in \mathbb{R}^n) \tag{2}$$

for each measurement i. For many computer vision problems (most notably fundamental matrix estimation, homography estimation and camera calibration), this linearization means constructing bi-linear forms of $(\vec{x}^T, 1)^T$; in these cases, the common linearization scheme is known as direct linear transform (DLT) (Hartley and Zisserman, 2000).

Stacking these row vectors on top of each other gives $\mathbf{A}'\vec{p}\,' \approx \vec{b}'$ with $\mathbf{A}' \in \mathbb{R}^{m\times n}$. For simplicity of notation, we add $\vec{b}'$ as an additional column to $\mathbf{A}'$: $\mathbf{A} = (\mathbf{A}'|\vec{b}')$. Analogously, we append -1 to $\vec{p}\,'$ to construct $\vec{p}$. We obtain the much more convenient homogeneous form:

$$\mathbf{A}\vec{p} \approx \vec{0}\,. \tag{3}$$

Estimation problems of this type are known as total least squares (TLS) problems. [1]

19.1.3. INCLUDING ERRORS IN THE TLS CONCEPT

A general error model is defined by $\mathbf{A} = f(\mathbf{A}_0, \mathbf{D})$ with a true data matrix $\mathbf{A}_0$ and an error matrix $\mathbf{D}$, both of which being unknown. The 'true' TLS equation $\mathbf{A}_0\vec{p} = \vec{0}$ only has a non-trivial solution if $\mathbf{A}_0$ is rank-deficient. Therefore, solving a TLS problem is equivalent to estimating a rank-deficient approximation $\hat{\mathbf{A}}$ for $\mathbf{A}$ such that

$$\|\mathbf{A} - \hat{\mathbf{A}}\| = \|\hat{\mathbf{D}}\| \quad \rightarrow \quad \min \tag{4}$$

under *some appropriate norm* (or *error metric*). Throughout this paper, quantities with a hat symbol on top denote estimated values.

[1] In (1), we introduced a model with only one constraint per measurement. We set the multidimensional case aside to keep things simple here, but the extension to q constraints per measurement is straightforward: φ becomes vector-valued and the $\vec{a}_i$ and b_i in (2) have to be replaced by $q \times n$-matrices (resp. q-vectors) for each measurement. But the final result after stacking everything on top of each other is the same again: $\mathbf{A}\vec{p} \approx \vec{0}$.

The TLS solution is widely equated with the singular value of **A** corresponding to the smallest singular value. This narrows the view on the potential of TLS-based approaches considerably because nothing was said about some special error model yet and how to solve for it. With the only exception of the constraint between measurement and parameters being linearized (equation (1) vs equation (2)), this model is in no way less general than the EIV approach. Most computer vision problems can be written in linear form and even the usual assumption of additive noise in EIV problems carries over to the equally usual assumption of additive noise in TLS problems because elements of **A** are usually constructed as multi-linear forms of $\vec{x}_i$.

In this paper, the TLS notion will be restricted to the definitions given so far – no assumption on certain error metrics is made with the term 'TLS' itself. It is important to stress that taking the right singular vector is just *one variant* of TLS-based methods; this method will be denoted plain TLS or PTLS from now on. Other TLS-based approaches can differ widely in the way they are solved; for instance, constrained TLS (CTLS) (Abatzoglou *et al.*, 1991) needs iterative optimization.

PTLS estimation is widely used because it provides a closed form solution that is very easy to compute – and under certain assumptions it is indeed a statistically optimal solution. If we use an additive error model $\mathbf{A} = \mathbf{A}_0 + \mathbf{D}$ and errors of all elements of **D** are zero-mean and independent and identically distributed (iid), then taking the right singular vector corresponding to the smallest singular value as TLS solution minimizes mean squared error of the estimate. If additionally errors are Gaussian, then PTLS is even a maximum likelihood estimator (van Huffel and Vandewalle, 1991).

However, these assumption are often not very realistic; therefore, PTLS estimates can be very erroneous (e.g. highly biased in case of fundamental matrix estimation without prior data normalization). The reason is simple: PTLS implicitly takes the *Frobenius norm* as norm in equation (4) – and of course, it is not always the 'appropriate norm' indeed.

Note that in general the iid noise assumption is even violated if the underlying measurements $\vec{x}_i$ contain iid noise because of the non-linearity of the constraints in (1) with respect to the measurements (e.g. conic fitting: linear in the six homogenous conic parameters, quadratic in the measurements). The linearization of this equation usually introduces data-dependent (heteroscedastic) error terms and the data normalization (Hartley and Zisserman, 2000) that is common practice before applying PTLS effectively alleviates this effect.

19.1.4. TLS AND SUBSPACE METHODS

The TLS estimation problem can easily be embedded in the more general class of *subspace estimation problems.* In these problems, we have to estimate a rank-deficient matrix from a noisy measured matrix of higher rank. A prominent computer vision problem that belongs to this group (and which is not a TLS problem) is the factorization method in multi-view structure from motion (Irani and Anandan, 2000; Hartley and Zisserman, 2000).

Subspace problems can therefore be formulated as the problem of dividing a matrix $\mathbf{A} = \mathbf{A}_0 + \mathbf{D}$ into two parts $\mathbf{A} = \hat{\mathbf{A}} + \hat{\mathbf{D}}$ (*data subspace* and *error subspace*). The term 'subspace' refers to the row and column subspaces of the matrices $\hat{\mathbf{A}}$ and $\hat{\mathbf{D}}$.

The central point of this paper will be the presentation of the new scheme for subspace estimation problems (including TLS problems as a subset). This new approach called ETLS retains the property of providing an easy-to-compute closed-form solution. The rest of this paper is organized as follows: After a short review and comparison of different approaches to parameter estimation problems in computer vision in section 19.2, we will define a general error model for subspace problems in section 19.3. After discussion of statistical optimality for TLS estimation in section 19.4, the equilibrated TLS (ETLS) approach will be introduced (section 19.5). We continue with some experimental results in section 19.6 before we conclude with the summary.

19.2. Literature Review

Different parameter estimation approaches differ in several aspects: computational complexity, robustness against outliers, ability to include uncertainty information on the measured quantities $\vec{x}_i$, ability to account for correlations between different measurements $\vec{x}_i$ and $\vec{x}_j$. Some methods provide confidence bounds or even full posterior probability distribution function for the estimates, some do not. Additionally, for iterative methods, the convergence properties can be meaningful.

Basically, the methods range from fast suboptimal approximations to exhaustive search on complex manifolds in high dimensional space and it is everything but easy to give a fair comparison. Nevertheless, we try to summarize the basic concepts used in computer vision problems, trying to focus on more general methods instead of highly specialized ones.

19.2.1. COMPLEX ERROR MODELS AND THE TLS MODEL

The advantage of the TLS problem $\vec{a}_i^T\vec{p} \approx 0$ over the general EIV model $\varphi(\vec{x}_i, \vec{p}) \approx 0$ is that it is linear in the constraints. If we set possible problems with the linearization process aside (Matei and Meer, 2000), we can exploit the reduced mathematical complexity to allow a thorough statistical treatment of the errors, i.e. not only in terms of second order statistical moments, but compute likelihood functions $p(\vec{a}_i|\vec{x})$ that only depend on the desired parameters.

In (Nestares *et al.*, 2000), Nestares et al. showed how to compute $p(\vec{a}_i|\vec{x})$ if a probability density function for the noise $p(\vec{d}_i)$ and a conditional prior on the nuisance parameters, i.e. $p(\vec{a}_{i0}|\vec{x})$, are given. The likelihood function is defined by an integral over the nuisance parameters. For Gaussian noise and Gaussian priors with same covariance matrix (up to scale), this is possible analytically.

If noise and/or nuisance prior of measurement i are non-Gaussian (or if they are both Gaussian with different covariance matrices – the assumption of same covariances is highly restrictive!) one has to resort to numerical integration, but computation of the likelihood function still remains possible.

However, computation of the likelihood function for the whole matrix $\mathbf{A}$, i.e. $p(\mathbf{A}|\vec{x})$, is only tractable if the measurements are independent. Then $p(\mathbf{A}|\vec{x})$ can be written as a product of row-vector likelihoods $p(\vec{a}_i|\vec{x})$. Under certain assumptions, it can be shown that the negative log-likelihood function is similar to the cost function defined for the HEIV model in (Matei and Meer, 2000).

19.2.2. ITERATIVE METHODS BASED ON COST FUNCTIONS

Many algorithms essentially consist in the optimization of a suitable cost function. The proper definition and minimization of such a function can be a complicated task, especially if some information on the uncertainty of measurements $\vec{x}_i$ is to be incorporated. Most problems are only mathematically tractable if the uncertainty information is restricted to second order statistical moments, i.e. we assume that first and second order statistical moments are known. The error mean values can be subtracted in a preprocessing step and, therefore, it means no loss of generality to assume that measurements are unbiased and characterized by different covariance matrices $\mathbf{C}_i = \mathsf{E}\left[\vec{x}_i\vec{x}_i^T\right]$.

The (theoretically optimal) "gold standard" is the bundle-adjustment method derived in the photogrammetry community long ago. But the extreme computational complexity of this method is much more problematic in computer vision than in photogrammetry and this led to the development of several faster algorithms with comparable estimation quality.

An early problem of the subspace kind is ellipse fitting. For this problem, one famous early iterative algorithm was developed by Sampson (Sampson, 1982). This algorithm, however, has the major drawback of producing biased estimates. Kanatani studied this problem and developed several renormalization schemes (Kanatani, 1996) which essentially consist in removing the estimated bias in each iteration step.

Chojnacki et al. developed an new and very straightforward iterative method based on a variational approach which is called fundamental numerical scheme (FNS) (Chojnacki *et al.*, 2000). Recently, the FNS method was extended to incorporate some special kind of constraints on the estimated parameters (the constraint function $\phi(\vec{x}) = 0$ must be homogeneous of some degree κ; additionally, some conditions must apply for the measurement matrix) (Chojnacki *et al.*, 2004).

In (Matei and Meer, 2000), Matei and Meer provide an another iterative method for solving EIV problems when covariance matrices $\mathbf{C}_i$ are available for each measurement $\vec{x}_i$. This method is called heteroscedastic EIV or, in short, HEIV. This approach does not allow to handle ancillary constraints in the iterative estimation process, but the enforcement of these constraints by proper (iterative) projection on the manifolds defined by these constraints is studied.

Both HEIV and FNS do not allow the different measurements do be correlated. An important example for problems with strongly correlated measurements is gradient-based orientation estimation, e.g. the estimation of optical flow. For this problem, Ng *et al.* gave a EIV-based solution in (Ng and Solo, 2001) which assumes simple iid errors in the images but handles the resulting (much more complicated!) noise model for the gradients correctly.

Summarizing this section, one can say that research during the last few years provided fast approximations to the bundle-adjustment algorithm which can handle heteroscedastic noise and show an estimation quality comparable to bundle-adjustment. General ancillary constraints, however, are usually hard to include in the algorithm itself, and (with certain exceptions for special forms of constraints) one has to resort to two step algorithms which consist of an unconstrained estimate followed by a subsequent enforcement of the constraints.

19.3. Generalized Error Model for Subspace Problems: Covariance Tensors

In this section, we will define a generalized (additive) error model for the TLS problem $\mathbf{A} = \mathbf{A}_0 + \mathbf{D} \approx \vec{0}$ that abandons any reference to certain rows being constructed from certain independent measurements. Every element of $\mathbf{D}$ will be treated equally and arbitrary variances and covariances shall be allowed.

As a consequence, we have to transfer the concept of *covariance matrices* describing the uncertainty of *random vectors* to a higher dimension. The matrix $\mathbf{D}$ is an arbitrary $m \times n$ *random matrix* and its uncertainty has to be described with a $m \times n \times m \times n$ *covariance tensor.*

19.3.1. THE COVARIANCE TENSOR OF A RANDOM MATRIX

Let us assume that the error matrix $\mathbf{D}$ is a zero-mean random matrix and all covariances between elements ip and jq, i.e. $\mathsf{E}\left[(\mathbf{D})_{ip}(\mathbf{D})_{jq}\right]$, are known (row indices: i, j; column indices: p, q). We can now define a four-dimensional tensor $C_D \in \mathbb{R}^{m \times n \times m \times n}$

$$(C_D)_{ipjq} = \mathsf{E}\left[(\mathbf{D})_{ip}(\mathbf{D})_{jq}\right]$$

which fully decribes the error structure up to second order statistics. This model is much more general than some other methods presented in the previous section where only different covariance matrices $\mathbf{C}_i$ for each row vector were allowed – in the framework presented here, this would mean

$$(C_D)_{ipjq} = \mathsf{E}\left[(\mathbf{D})_{ip}(\mathbf{D})_{jq}\right] = (\mathbf{C}_i)_{pq}\, \delta_{ij} \ . \tag{5}$$

Sometimes this assumption is valid – but in many cases, e.g. for computation of optical flow from the structure tensor of the space-time volume, the row vectors are highly correlated.

19.3.2. COVARIANCE PROPAGATION FOR TENSORS

If we start from the EIV approach (i.e. the measurement equation is not linearized yet), all we have is a $n \times \ell$ random matrix $\mathbf{E}$ containing errors in the ℓ-dimensional measurements. Analoguously to the linearized $\mathbf{D} \in \mathbb{R}^{n \times m}$, we can define a covariance tensor of $\mathbf{E}$ (denoted as C_E).

Now the question arises of how to propagate the covariance information from C_E to C_D, the covariance tensor of the TLS error matrix $\mathbf{D}$. But the answer is simple: We have to use the Jacobian matrix of the linearization functions, i.e. $\mathbf{J}_i = \frac{\partial \vec{a}_i^T(\vec{x})}{\partial \vec{x}}$. This is best done row by row;

the cross-covariance matrices of the rows i and j transform as:[2]

$$\mathsf{E}\left[\vec{d}_i\vec{d}_j^T\right] = \mathbf{J}_i \left(\mathsf{E}\left[\vec{e}_i\vec{e}_j^T\right]\right) \mathbf{J}_j^T \tag{6}$$

where $\vec{d}_i^T$ and $\vec{e}_i^T$ are the i-th row vectors of $\mathbf{D}$ and $\mathbf{E}$, respectively. This is a straightforward extension of covariance propagation for covariance matrices. Now we know how to construct C_D if necessary; we will refer to this tensor as C (without subscript) for the rest of the paper.

19.3.3. VECTORIZING RANDOM MATRICES

In order to form the TLS error matrix $\mathbf{D} \in \mathbb{R}^{m\times n}$, we arranged m row vectors $\vec{d}_i^T$ on top of each other. Alternatively, we could have stacked m *column* vectors $\vec{d}_i$ to form a long vector $\vec{d} \in \mathbb{R}^{mn}$. The covariance matrix $\mathbf{C}_d$ of this vector would contain exactly the same elements as C_D, but arranged in different shape; both forms are different representations of the same thing.

This mapping between the 'big random vector plus covariance matrix' and the 'random matrix plus covariance tensor' representations allows to transfer certain known concepts to the random matrix world: we define the *unit covariance tensor* I as the mapping of the case $\mathsf{Cov}\left[\vec{d}\right] = \mathbf{I}_{mn}$ (the mn dimensional identity matrix):

$$(I)_{ijk\ell} = (\mathbf{I}_m)_{ik}\,(\mathbf{I}_n)_{j\ell} = \delta_{ik}\,\delta_{j\ell}\;. \tag{7}$$

In analogy to the inverse covariance matrix, we furthermore define the *inverse covariance tensor* C^{-1}. It solves $C^{-1} \cdot C = I$ when summing over the two inner indices and is given by the tensorized inverse covariance matrix $\mathbf{C}_d^{-1}$.

19.4. Solving EIV and TLS Problems

In this section, we will discuss the statistical optimality of TLS based estimation and relate it to general EIV estimation approaches.

[2] If probability density functions of the errors $\vec{e}_i$ were available, it could be advisable to carry them through the full non-linearity instead and compute the second order statistical moments (i.e. covariances) afterwards. But in general, this leads to ugly integrals which have to be solved numerically.

19.4.1. THE GENERAL APPROACH: CONSTRAINED MINIMIZATION IN MEASUREMENT SPACE

Let us now assume that the data model $\varphi(\vec{x}_{i0}, \vec{p}) = 0$ (for all $i = 1, \ldots, m$) is *non-linear* and characterized by zero-mean additive errors $\vec{x}_i = \vec{x}_{i0} + \vec{e}_i$ with known second order moments.

Stacking measurements $\vec{x}_i$ (and errors $\vec{e}_i$) to long mn-dimensional vectors with covariance matrix $\mathbf{C}_e$, the general approach for solving EIV problems is minimizing the Mahalanobis distance

$$J = (\hat{\vec{x}} - \vec{x})^T \mathbf{C}_e^{-1} (\hat{\vec{x}} - \vec{x}) \quad \rightarrow \quad \min\,, \tag{8}$$

where the individual parts of $\hat{\vec{x}}$ have to be compatible with the data model. If individual measurements are uncorrelated (equation (5) from 'random matrix perspective' or block-diagonal $\mathbf{C}_e$ from 'long random vector' perspective), this equation can be simplified further. Let $\mathsf{Cov}\,[\vec{x}_i] = \mathbf{C}_i$ be the covariance matrix of the i-th measurement. Then

$$J = \sum_{i=1}^{m} (\hat{\vec{x}}_i - \vec{x}_i)^T \mathbf{C}_i^{-1} (\hat{\vec{x}}_i - \vec{x}_i) \quad \rightarrow \quad \min \tag{9}$$

holds subject to all $\vec{x}_i$ being compatible with the data model.

In principle, we are interested in a cost function in parameter space, but all we have now is a cost function on the nuisance parameters, i.e. in the high dimensional space of measurements (which is *much* less handy). This function has to be minimized subject to non-linear constraints given by the data model.

The general solution concept for this type of problems is as follows: assuming that some approximate solutions already exist, one can linearize the constraints and iteratively compute additive correction terms for the previous approximate solutions (using the Gauß-Helmert model on the linearized equations; see e.g. (Förstner, 2001)).

19.4.2. ESTIMATION USING THE (LINEARIZED) TLS MODEL

In the TLS model, we assumed that $\varphi(\vec{x}_{i0}, \vec{p})$ depends linearly on the elements of $\vec{p}$, i.e. we consider the data model constraint

$$\varphi(\vec{x}, \vec{p}) = \mathbf{A}\vec{p} = \begin{pmatrix} \vec{a}_1^T \\ \vdots \\ \vec{a}_m^T \end{pmatrix} \vec{p} = \vec{r} \approx \vec{0} \tag{10}$$

Let $r_i = \vec{a}_i^T \vec{p}$ be the residual (deviation from data model) for the i-th row and $\vec{r} = (r_1, \ldots, r_M)^T$ be the vector of all residuals.

Estimation now leads to minimizing the Mahalanobis norm of the residual vector, i.e. minimizing the cost function

$$J = \vec{r}^T \mathbf{C}_r^{-1} \vec{r} \tag{11}$$

where $\mathbf{C}_r$ stands for the covariance matrix of the residual vector. If the assumption of uncorrelated measurements (and therefore uncorrelated residuals) is reasonable, this covariance matrix is a diagonal matrix:

$$\mathbf{C}_r = \mathsf{Cov}\left[\vec{r}\right] = \mathsf{diag}\left\{\sigma_i^2\right\} \,. \tag{12}$$

The variance σ_i^2 of the residual r_i is given by (in first order approximation):

$$\sigma_i^2 = \left(\frac{\partial(\vec{p}^T\vec{a}_i)}{\partial\vec{x}_i}\right) \mathsf{Cov}\left[\vec{x}_i\right] \left(\frac{\partial(\vec{a}_i^T\vec{p})}{\partial\vec{x}_i}\right)^T = \vec{p}^T \mathbf{J}_i \mathsf{Cov}\left[\vec{x}_i\right] \mathbf{J}_i^T \vec{p} = \vec{p}^T \mathsf{Cov}\left[\vec{a}_i\right] \vec{p} \,. \tag{13}$$

The key point is that the matrix inversion $\mathbf{C}_r^{-1}$ in (11) now transforms to a sum of fractions:

$$J = \vec{r}^T \mathbf{C}_r^{-1} \vec{r} = \sum_i (\vec{p}^T\vec{a}_i)\sigma_i^{-2}(\vec{a}_i^T\vec{p}) = \sum_i \frac{\vec{p}^T(\vec{a}_i\vec{a}_i^T)\vec{p}}{\vec{p}^T \mathsf{Cov}\left[\vec{a}_i\right]\vec{p}} \tag{14}$$

and the parameter value which minimizes this cost function can be termed estimate:

$$\hat{\vec{p}} = \arg\min_{\vec{p}} J \,. \tag{15}$$

Let us define $\mathbf{S}_i = \vec{a}_i\vec{a}_i^T$ and $\mathbf{C}_i = \mathsf{Cov}\left[\vec{a}_i\right]$. Then (11) can finally be written as

$$J = \sum_i \frac{\vec{p}^T \mathbf{S}_i \vec{p}}{\vec{p}^T \mathbf{C}_i \vec{p}} \tag{16}$$

This is exactly the function J_{AML} (AML for 'approximated maximum likelihood' (in case of Gaussian errors)) which is minimized in the FNS approach (Chojnacki *et al.*, 2000) (but here, we presented a much shorter derivation). However, it is important to stress that this derivation is restricted to a certain assumption: it holds if and only if the individual measurements are uncorrelated.

19.4.3. A GENERAL COST FUNCTION FOR SUBSPACE ESTIMATION PROBLEMS

In the last section, we started from minimizing the Mahalanobis distance in measurement space. Basically the same can be done with the linearized measurements.

The covariance tensor C can be rearranged to a symmetric and positive definite $(mn) \times (mn)$-matrix $\mathbf{C}$; this is exactly the covariance matrix of the 'vectorized' version $\vec{d} \in \mathbb{R}^{mn}$ of the $m \times n$ error matrix $\mathbf{D}$ which can be constructed by stacking all column vectors $\vec{d_i}$ on top of each other. Estimating the error matrix $\mathbf{D}$ is then equivalent to minimizing the Mahalanobis distance of the estimated $\hat{\vec{d}}$:

$$J_{opt} = \hat{\vec{d}}^T \, \mathbf{C}^{-1} \, \hat{\vec{d}} \qquad \text{subject to } \mathsf{rank}\,\{\mathbf{A} - \mathbf{D}\} = r \,.$$

We can also write down J_{opt} in matrix space (C^{-1} being the 'tensorized', i.e. simply rearranged[3], inverse covariance matrix $\mathbf{C}^{-1}$):

$$J_{opt} = \left|\hat{\mathbf{D}}\right|_C^2 = \sum_{ipjq} (\mathbf{D})_{ip} (C^{-1})_{ipjq} (\mathbf{D})_{jq} \quad \text{subject to } \mathsf{rank}\,\{\mathbf{A} - \mathbf{D}\} = r \,. \tag{17}$$

In case of Gaussian iid errors in $\vec{d}$ (or in $\mathbf{D}$, respectively), this defines a maximum likelihood estimate. Note that the optimality of PTLS for iid noise can be seen easily from (17): For iid noise, $(C^{-1})_{ipjq} = (C)_{ipjq} = \delta_{ij}\delta_{pq}$ and $J_{opt}^{(\text{iid-noise})} = \sum_{ip} (\mathbf{D})_{ip}^2 = \|\mathbf{D}\|_F^2$ becomes identical to the Frobenius norm – which is exactly the norm under which rank approximations using SVD are optimal.

The criterion J_{opt} seems obvious, but nevertheless, we have to be careful: the used error metrics is statistically justified, but what we minimize here is the distance between the measured matrix $\mathbf{A}$ and the estimate $\hat{\mathbf{A}}$, i.e. the distance between $\mathbf{A}$ and the *nearest point on the manifold of rank-deficient matrices.*

But this is not necessarily identical to the expected distance between the estimate and the underlying *true* matrix $\mathbf{A}_0$. Having a predefined rank is an inherently nonlinear constraint for a matrix and the *nearest* matrix (in terms of the matrix norm defined in (17)) on the manifold of rank r matrices is not necessarily the *true* matrix $\mathbf{A}_0$ in expectation.

The criterion J_{opt} is valid under the additional constraint of *unbiased estimation*, i.e. $\mathsf{E}\left[\hat{\mathbf{A}}\right]$ must not deviate from $\mathbf{A}_0$. But under this constraint, the cost function J_{opt} can be used for evaluating the performance of different algorithms—including possible correlations between different measurements.

[3] In MATLAB, these rearrangement operations can easily be done by defining some tensor `T` and matrix `M` and setting `T(:) = M(:)`; no special function is required.

19.4.4. A NEW DECOMPOSITION OF A COVARIANCE TENSOR

We will now show how to achieve unbiased subspace estimation. Let $C \in \mathbb{R}^{m \times n \times m \times n}$ be the covariance tensor of the random matrix $\mathbf{D}$. By permuting second and third indices we can construct a $m \times m \times n \times n$ tensor that can be mapped on a matrix $\mathbf{M} \in \mathbb{R}^{m^2 \times n^2}$. This matrix $\mathbf{M}$ can be decomposed using the singular value decomposition, i.e.

$$\mathbf{M} = \sum_{p=1}^{n^2} \alpha_p \, \vec{x}_p \, \vec{y}_p^T$$

where $\vec{x}_p \in \mathbb{R}^{m^2}$ and $\vec{y}_p \in \mathbb{R}^{n^2}$ are the left resp. right singular vectors and σ_p are the singular values. Both $\vec{x}_p$ and $\vec{y}_p$ can be re-arranged to square matrices $\mathbf{X}_p \in \mathbb{R}^{m \times m}$ and $\mathbf{Y}_p \in \mathbb{R}^{n \times n}$. Doing this, the tensor C is decomposed into the following sum of basic tensors T_p:

$$(C)_{ijk\ell} = \sum_{p=1}^{n^2} \underbrace{\alpha_p \, (\mathbf{X}_p)_{ik} \, (\mathbf{Y}_p)_{j\ell}}_{(T_p)_{ijk\ell}} = \sum_{p=1}^{n^2} (T_p)_{ijk\ell} \, . \tag{18}$$

An iid random matrix is defined by $(C)_{ijk\ell} = \delta_{ik}\delta_{j\ell}$. If this special covariance tensor is fed into the tensor decomposition algorithm described above, the result is

$$(C)_{ijk\ell} = (\mathbf{I}_m)_{ik} \, (\mathbf{I}_n)_{j\ell} \, , \tag{19}$$

i.e. the sum disappears and the only left and right 'singular matrices' are identity matrices. We can exploit the tensor decomposition to define a transformation rule for covariance tensors. If a random matrix $\mathbf{D}$ is transformed according to $\tilde{\mathbf{D}} = \mathbf{W}_L \mathbf{D} \mathbf{W}_R{}^T$, then its (decomposed) covariance tensor is transformed in the following way:

$$(\tilde{C})_{ijk\ell} = \sum_p \alpha_p \, (\mathbf{W}_L \mathbf{X}_p \mathbf{W}_L{}^T)_{ik} \, (\mathbf{W}_R \mathbf{Y}_p \mathbf{W}_R{}^T)_{j\ell} \, . \tag{20}$$

In (Stewart, 1990), Stewart introduced *cross-correlated matrices*. These are random matrices that can be constructed from an iid random matrix $\mathbf{D}$ by applying a transformation with arbitrary non-singular matrices $\mathbf{W}_L$ and $\mathbf{W}_R$. However, no method was provided to determine whether a given random matrix is cross-correlated or not. Using the calculus developed above, this is simple: Applying transformation rule (20) on (19) yields

$$(\tilde{C})_{ijk\ell} = (\mathbf{W}_L \mathbf{W}_L{}^T)_{ik} \, (\mathbf{W}_R \mathbf{W}_R{}^T)_{j\ell} \, ,$$

i.e. cross-correlated matrices are exactly those matrices that are defined by one base tensor only (no summation anymore). If the random matrix $\mathbf{D}$

given for a TLS problem is cross-correlated, then a whitening transformation can be computed; one only has to invert the Cholesky factors of $\mathbf{X}_1$ and $\mathbf{Y}_1$.

Some generalizations of basic subspace estimation algorithms for non-iid cases are in fact generalizations to cross-correlated noise. This applies to 2D homography estimation (Mühlich and Mester, 2001) and factorization with uncertainty information (Irani and Anandan, 2000).

weighted sum. If the inverse covariance tensor is now decomposed into base tensors, we can rewrite J_{opt} as is a sum of weighted expressions that are quadratic in the elements of $\hat{\mathbf{D}}$.

19.5. Equilibrated TLS/Subspace Estimation

Under the error model $\mathbf{A} = \mathbf{A}_0 + \mathbf{D}$ with $\mathsf{E}[\mathbf{D}] = \mathbf{0}$, achieving unbiased subspace estimation boils down to two simple requirements that have to be fullfilled; this can be done with appropriate weighting transformations.

19.5.1. UNBIASED SUBSPACE ESTIMATION USING EQUILIBRATION

In (Mühlich and Mester, 1998), TLS estimation is examined and it has been shown that unbiased estimates of right singular vectors of $\mathbf{A}$ (the smallest of which being the TLS solution vector) require that $\mathsf{E}\left[\mathbf{D}^T\mathbf{D}\right]$ is proportional to the identity matrix:

$$\mathsf{E}\left[\mathbf{D}^T\mathbf{D}\right] \propto \mathbf{I}_n \tag{21}$$

($\mathbf{I}_M$ denoting the $M \times M$ identity matrix). The reason for this is simple: right singular vectors are eigenvectors of $\mathbf{A}^T\mathbf{A} = \mathbf{A}_0{}^T\mathbf{A}_0 + \mathbf{A}_0{}^T\mathbf{D} + \mathbf{D}^T\mathbf{A}_0 + \mathbf{D}^T\mathbf{D}$. In expectation, the second and third term vanish and the eigenvectors of $\mathbf{A}^T\mathbf{A}$ are only identical to those of $\mathbf{A}_0{}^T\mathbf{A}_0$ (i.e. the true ones) if (21) holds; adding a multiple of the identity matrix only increases eigen*values*, but does not change eigen*vectors*.

If (21) does not hold, in general[4] the errors will introduce a bias in the right singular vectors. However, there is a very simple solution: transform the matrix to another (reweighted) space ($\tilde{\mathbf{A}} = \mathbf{W}_L\mathbf{A}\mathbf{W}_R^T$), do the rank reduction there, and transform back. This technique is called equilibration (Mühlich and Mester, 1999).

$$\mathbf{A} = \mathbf{W}_L^{-1} \underbrace{\mathbf{W}_L\mathbf{A}\mathbf{W}_R{}^T}_{\tilde{\mathbf{A}}=\mathbf{USV}^T} \mathbf{W}_R^{-T} \tag{22}$$

[4] The degenerate case which preserves the correct eigenvectors is: $\mathbf{A}_0$ and $\mathbf{D}$ share the same right singular vectors and the adding of $\mathbf{D}$ does not change the order of singular values. In this very special case, (21) is not a necessary condition. Otherwise, it is.

Equilibration means that we approximate $\tilde{\mathbf{A}}$ with a rank-deficient matrix $\hat{\tilde{\mathbf{A}}}$. As long as both equilibration matrices are non-singular, they are guaranteed to preserve the rank; therefore transforming back gives an approximation $\hat{\mathbf{A}}$ for $\mathbf{A}$ with the desired lower rank.

It is obvious that a weighting transformation changes error metrics – but it was not clear *how* to choose weights in general, especially for the left equilibration before.

Fulfilling (21) in the transformed space is sufficient for unbiased TLS estimates. But in case of general subspace estimation it is not enough. SVD is defined by *both* the left and right singular vectors and not right singular vectors only. Even for TLS type estimates, a well chosen left equilibration is important because it reduces the variance of the estimate.

An obvious extension of (21) is a second analogous requirement: $\mathsf{E}\left[\mathbf{D}\mathbf{D}^T\right] \propto \mathbf{I}_m$. In general (see last footnote again, now for both left and right singular vectors), fulfilling these two requirements *is the only way to get unbiased estimates for all singular vectors*, which is required for unbiased subspace estimation using the SVD for matrix approximation.

Both requirements can be fulfilled in a transformed space, but the problem is the coupling we get for left and right equilibration matrices: the coupled equation system

$$\mathsf{E}\left[\tilde{\mathbf{D}}^T\tilde{\mathbf{D}}\right] = \mathbf{W}_R \mathsf{E}\left[\mathbf{D}^T(\mathbf{W}_L{}^T\mathbf{W}_L)\mathbf{D}\right]\mathbf{W}_R{}^T \propto \mathbf{I}_n \tag{23}$$

$$\mathsf{E}\left[\tilde{\mathbf{D}}\tilde{\mathbf{D}}^T\right] = \mathbf{W}_L \mathsf{E}\left[\mathbf{D}(\mathbf{W}_R{}^T\mathbf{W}_R)\mathbf{D}^T\right]\mathbf{W}_L{}^T \propto \mathbf{I}_m \tag{24}$$

cannot be solved for both $\mathbf{W}_L$ and $\mathbf{W}_R$ easily. But it becomes tractable with covariance tensor decomposition.

A reweighting of the measurement matrix according to (22), i.e. $\tilde{\mathbf{A}} = \mathbf{W}_L\mathbf{A}\mathbf{W}_R^T$, changes the covariance tensor C to

$$(\tilde{C})_{ijk\ell} = \sum_p \alpha_p (\mathbf{W}_L\mathbf{X}_p\mathbf{W}_L{}^T)_{ik}(\mathbf{W}_R\mathbf{Y}_p\mathbf{W}_R{}^T)_{j\ell}$$

Here we see the advantage of the new tensor decomposition: The left (resp. right) hand equilibration matrix only affects the matrices $\mathbf{X}_p$ (resp. $\mathbf{Y}_p$).

The two requirements $\mathsf{E}\left[\tilde{\mathbf{D}}^T\tilde{\mathbf{D}}\right] \propto \mathbf{I}_m$ and $\mathsf{E}\left[\tilde{\mathbf{D}}\tilde{\mathbf{D}}^T\right] \propto \mathbf{I}_n$ now transform to

$$\sum_p \alpha_p \mathsf{Tr}\left\{\mathbf{W}_L\mathbf{X}_p\mathbf{W}_L{}^T\right\}(\mathbf{W}_R\mathbf{Y}_p\mathbf{W}_R{}^T) \propto \mathbf{I}_n \tag{25}$$

$$\sum_p \alpha_p (\mathbf{W}_L\mathbf{X}_p\mathbf{W}_L{}^T)\mathsf{Tr}\left\{\mathbf{W}_R\mathbf{Y}_p\mathbf{W}_R{}^T\right\} \propto \mathbf{I}_m\,. \tag{26}$$

This equation system (25) and (26) can be solved iteratively for $\mathbf{W}_L$ and $\mathbf{W}_R$, i.e. we

1. set $\mathbf{R}_{(0)} = 1/N\ \mathbf{I}_N$; $k = 1$
2. use (26) to compute $\mathbf{L}_{(k)}$ from $\mathbf{R}_{(k-1)}$ and scale it to Frobenius norm 1.
3. use (25) to compute $\mathbf{R}_{(k)}$ from $\mathbf{L}_{(k)}$ and scale it to Frobenius norm 1.
4. Terminate if $\mathbf{R}_{(k)}$ and $\mathbf{R}_{(k-1)}$ do not differ much; otherwise $k = k+1$ and continue at step 2.

This procedure converges very fast and we never had any convergence problems in various different experiments. In case of cross-correlated matrices, both $\mathbf{W}_L$ and $\mathbf{W}_R$ are fully determined after the first iteration.

19.5.2. MODIFICATIONS FOR RANK-DEFICIENT COVARIANCE MATRICES

In general, *all* the (cross-)covariance matrices $\mathbf{C}_{ij} = \mathsf{Cov}\left[\vec{a}_i, \vec{a}_j\right]$ are singular if (linear) covariance propagation (equation (6)) is used; then their maximum rank is the maximum rank of the covariance matrices of measurement vectors $\vec{x}_i$), i.e. ℓ at most. We cannot avoid this effect. But this is not a problem. Only if all these matrices share a *common* null space, then problems can arise because loss functions (see section 19.4.2) are undefined in this part of parameter space then.

Applications exist where some columns of the TLS measurement matrix $\mathbf{A}$ are free of errors (ellipse fitting in the experimental section will be a good example: the last column contains only 1s). Assuming that the last columns of $\mathbf{A}$ are error-free, equation (21) has to be replaced by $\mathsf{E}\left[\mathbf{D}^T\mathbf{D}\right] \propto \mathsf{diag}\left\{1, \ldots, 1, 0, \ldots, 0\right\}$. If k is the dimensionality of the erroneous column subspace, then the right equilibration matrices only have rank k as well (if the last columns of $\mathbf{A}$ are error-free, then only the first k rows and columns of $\mathbf{W}_R$ are non-zero).

In the rank reduction step (approximation of rank n erroneous matrix $\mathbf{A}$ by a matrix of lower rank; rank $n-1$ in case of TLS estimation), error-free columns simply mean that these columns must not be modified. The concept presented here goes back to Demmel (Demmel, 1987) who studied the problem of approximating a matrix with another matrix of lower rank when only a submatrix is allowed to be modified. We adapted this concept to TLS estimation in the equilibration context.

So let us assume we have to solve the (already equilibrated, just to avoid tilde symbols everywhere) TLS problem $\mathbf{A}\vec{p} \approx \vec{0}$ and we know that the last $n-k$ columns of $\mathbf{A}$ are free of errors. Let $\mathbf{A} \in \mathrm{I\!R}^{m \times n}$ be partitioned into

$\mathbf{A} = (\mathbf{A}_1|\mathbf{A}_2)$: erroneous $\mathbf{A}_1 \in \mathbb{R}^{m\times k}$ and error-free $\mathbf{A}_2 \in \mathbb{R}^{m\times(n-k)}$. Let the SVD of $\mathbf{A}_2$ be:[5]

$$\mathbf{A}_2 = \mathbf{USV}^T = (\mathbf{U}_\perp|\mathbf{U}_\parallel)\left(\frac{\mathbf{0}}{\mathbf{S}_\parallel}\right)\mathbf{V}^T . \tag{27}$$

Actually, we are mainly interested in the orthogonal matrix $\mathbf{U} = (\mathbf{U}_\perp|\mathbf{U}_\parallel)$. The columns of $\mathbf{U}$ are a basis for the vector space spanned by the column vectors of $\mathbf{A}$, and more specifically, $\mathbf{U}_\parallel \in \mathbb{R}^{m\times k}$ is a basis for the span of $\mathbf{A}_2$, while $\mathbf{U}_\perp \in \mathbb{R}^{m\times(m-(n-k))}$ is orthogonal to it.

Then $\mathbf{A}$ can be transformed to

$$\tilde{\mathbf{A}} = \mathbf{U}^T\mathbf{A} = \left(\begin{array}{c|c} \mathbf{U}_\perp^T\mathbf{A}_1 & \mathbf{0} \\ \hline \mathbf{U}_\parallel^T\mathbf{A}_1 & \mathbf{U}_\parallel^T\mathbf{A}_2 \cdot \end{array}\right) \tag{28}$$

In the context of equilibration, this can be seen as an additional left equilibration; therefore, the solution of $\tilde{\mathbf{A}}\vec{p} \approx \vec{0}$ is also the solution of $\mathbf{A}\vec{p} \approx \vec{0}$ and no back transformation for the estimate $\hat{\vec{p}}$ is necessary.

In the upper right part of $\tilde{\mathbf{A}}$, we get $\mathbf{0}$ because $\mathbf{U}_\perp$ is orthogonal to $\mathbf{A}_2$. The lower $(n-k)$ rows $(\mathbf{U}_\parallel^T\mathbf{A}_1|\mathbf{U}_\parallel^T\mathbf{A}_2) = \mathbf{U}_\parallel^T\mathbf{A}$ have rank $(n-k)$, i.e. full rank, and therefore, the TLS problem can be reduced to

$$\mathbf{U}_\perp^T\mathbf{A}_1\vec{p}_1 \approx \vec{0} . \tag{29}$$

Let $\hat{\vec{p}}_1$ be the PTLS solution of this problem (i.e. plain TLS solution in an appropriately equilibrated space). We now define

$$\hat{\vec{p}}_2 = -\underbrace{(\mathbf{U}_\parallel^T\mathbf{A}_2)^{-1}\mathbf{U}_\parallel^T\mathbf{A}_1}_{\mathbf{T}}\hat{\vec{p}}_1 \tag{30}$$

and $\hat{\vec{p}}^T = (\hat{\vec{p}}_1^T, \hat{\vec{p}}_2^T)$. Then $\hat{\vec{p}}$ is the optimal solution of $\tilde{\mathbf{A}}\vec{p} \approx \vec{0}$. Proof:

$$\begin{aligned} \tilde{\mathbf{A}}\hat{\vec{p}} &= \begin{pmatrix} \mathbf{U}_\perp^T\mathbf{A}_1\hat{\vec{p}}_1 \\ \mathbf{U}_\parallel^T\mathbf{A}_1\hat{\vec{p}}_1 - (\mathbf{U}_\parallel^T\mathbf{A}_2)(\mathbf{U}_\parallel^T\mathbf{A}_2)^{-1}\mathbf{U}_\parallel^T\mathbf{A}_1\hat{\vec{p}}_1 \end{pmatrix} \\ &= \begin{pmatrix} \mathbf{U}_\perp^T\mathbf{A}_1\vec{p}_1 \\ \vec{0} \end{pmatrix} \approx \vec{0} . \end{aligned} \tag{31}$$

This means that the estimation is carried out in a lower dimensional space (and thus gets faster). Note that $\tilde{\mathbf{A}}_1$ has $m-(n-k)$ rows (and not m like $\mathbf{A}$

[5] Attention: we need the singular values to come *last* in $\mathbf{S}$. This reflects the fact that we modelled the error-free columns in $\mathbf{A}$ to come last – which is the usual way of e.g. modelling offset parameters. Given the usual output of SVD procedures, the role of parallel and perpendicular parts is *exchanged*.

or $\mathbf{A}_1$); in some sense, we concentrated the relevant information contained within $\mathbf{A}_1$ in fewer row vectors.

Our presentation was optimized for easier understanding; we silently skipped one import aspect and we have to make up for this now. The term $(\mathbf{U}_{\|}^T\mathbf{A}_2)^{-1}$ could raise suspicion. Is this inverse defined? But looking at (27), we see that

$$\begin{aligned}\mathbf{T} &= (\mathbf{U}_{\|}^T\mathbf{A}_2)^{-1}\mathbf{U}_{\|}^T\mathbf{A}_1 = (\mathbf{S}_{\|}\mathbf{V}_{\|}^T)^{-1}\mathbf{U}_{\|}^T\mathbf{A}_1 \\ &= \mathbf{V}_{\|}\mathbf{S}_{\|}^{-1}\mathbf{U}_{\|}^T\mathbf{A}_1 = \mathbf{A}_2^{\dagger}\mathbf{A}_1 \end{aligned} \tag{32}$$

where $\mathbf{A}_2^{\dagger}$ denotes the (Moore-Penrose-)pseudoinverse of $\mathbf{A}_2$ and everything is indeed well defined.

Finally, we have to consider the equilibration weights again. In equilibrated space (characterized by $\mathbf{W}_L$ and $\mathbf{W}_R$), the same derivation hold if $\mathbf{A}_1$ and $\mathbf{A}_2$ are replaced by their equilibrated versions $\tilde{\mathbf{A}}_1 = \mathbf{W}_L\mathbf{A}_1\mathbf{W}_R^T$ and $\tilde{\mathbf{A}}_2 = \mathbf{W}_L\mathbf{A}_2$ (right equilibration only for erroneous part!). For instance, the reduced TLS problem in equilibrated space is defined by

$$\tilde{\mathbf{A}}_1\tilde{\vec{p}}_1 = \mathbf{U}_{\perp}^T\mathbf{W}_L\mathbf{A}_1\mathbf{W}_R\tilde{\vec{p}}_1 \approx \vec{0} \tag{33}$$

where $\mathbf{U}_{\perp}$ is constructed from the SVD of $\tilde{\mathbf{A}}_2$.

Let us summarize this subsection: exploiting fixed columns in $\mathbf{A}$ allows to reduce the dimensionality of the estimation problem to estimating a k-dimensional vector $\hat{\vec{p}}_1$ only and reconstructing the remaining elements with (32) and (30) at the end.

19.6. Experimental Results

The proposed method is applicable to a wide class of problems. We picked a very illustrative one for our experimental analysis: conic/ellipse fitting. This problem is important for many applications, known to be highly sensitive to errors and biased for PTLS estimation.

A conic in $\mathbb{R}^2$ is determined by a homogeneous 6-vector:

$$p_1\,x^2 + p_2\,y^2 + p_3\,xy + p_4\,x + p_5\,y + p_6 = 0$$

The measured data points are two-dimensional and the usual plain TLS approach suffers from the effect that a non-linear problem is embedded in a higher dimensional space which 'disturbs' the error metric. The result is well-known: if data points are only available from a certain arc of the ellipse, the fitted ellipses tend to be too small, i.e. the estimate is biased. We will show that our algorithm produces unbiased estimates.

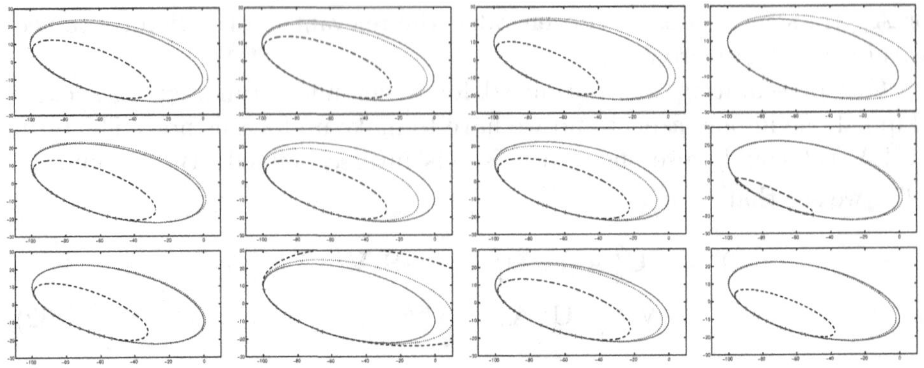

Figure 19.1. 12 runs of our ellipse fitting algorithm. True ellipse (solid; identical in all images), PTLS estimate (dashed) and ETLS estimate (dotted). The bias of PTLS (ellipses tend to be too small) is clearly visible. Data points are always chosen in lower left part of ellipse.

We have defined a test ellipse (parameter vector $\vec{p} = (.1, .2, .5, 10, 10, 50)^T$) and have taken 10 randomly chosen 2D points in the lower left arc of the ellipse. All points were disturbed by zero-mean additive Gaussian noise. The noise in the 10 measurements was *assumed to be correlated*, i.e. we used a random $10 \times 2 \times 10 \times 2$ covariance tensor (here: 190 degrees of freedom). Note that previously known ellipse fitting schemes like FNS (Chojnacki *et al.*, 2000) only allow individually different *but uncorrelated* covariance matrices for each measurement (here: $10 \cdot 3 = 30$ degrees of freedom).

From our perturbed data, we estimated the conic parameters (here: a unit vector in $\mathbb{R}^6$). One big advantage of ellipse fitting over other multidimensional parameter estimation problems is that the estimation quality can be visualized easily by plotting an ellipse with reconstructed parameters. Figure 19.6 shows 12 consecutive runs of our program. The solid light line is the true ellipse, the dashed line is the PTLS estimate and the dotted ellipse represents the ETLS estimate.

the measured points would be useless because they would appear perfectly on the ellipse. The dots indicating the data points were therefore displaced 50 times the chosen error from their true location (the corresponding plus signs). The bias of the PTLS estimates is clearly visible in figure 19.6. In these 12 consecutive runs, the 'PTLS-ellipses' were too small 10 times, too large once, and in upper right image, the PTLS estimated conic was no ellipse at all.

We also tested FNS with the same data (as covariance input, we extracted the row vector covariance matrices from the covariance tensor—

which means ignoring all correlation between different measurements) and experienced major convergence problems; in many situations, this method failed to estimate an ellipse. Good estimates were only possible for low noise levels.

We used a general conic estimation scheme here and no estimator which enforces an ellipse solution (e.g. (Halíř and Flusser, 1998)). Firstly, this leaves this example simple and easy to understand and secondly, it demonstrates that the necessity to prevent a wrong type of conic mainly arises by the usage of a wrong error metric. For the given example, our algorithm hardly ever estimates non-ellipses although it could in principle do so. Nevertheless, a combination of our approach and (Halíř and Flusser, 1998) is possible and should improve estimates in close situations.

19.7. Conclusion

Eigensystem based estimation schemes are ubiquitous in computer vision. Their estimation quality can often be improved by some previous data normalization, but still there is a desire to do further iterative optimization. The new method presented here provides a closed-form solution and might eliminate this need for many computer vision problems (or at least alleviate convergence for subsequent optimization).

In some sense, we generalized data normalization (which is a small subset of possible equilibration transformations) to more complex error models, and most importantly, we can *derive* and *justify* the statistically correct equilibration transformation, given the covariance information on the input data. Some predefined data transformation strategy (e.g. shifting to center of mass and isotropic scaling) is the first step in the correct direction, but not more than that.

Additionally, our approach applicable in cases where correlation between different measurements has to be modelled; these problems cannot be treated properly with many other approaches which only account for covariances in the same measurement.

Ongoing research will focus on the development of a unified framework that combines the method presented here with approaches like HEIV and FNS; the mathematical core of most algorithms is very similar.

References

Abatzoglou, T. J., J. M. Mendel, and G. A. Harada: The constrained total least squares technique and its applications to harmonic superresolution. *IEEE Trans. Signal Processing*, **3**:1070–1087, 1991.

Chojnacki, W., M. J. Brooks, A. van den Hengel, and D. Gawley: A new constrained parameter estimator for computer vision applications. *Image Vision Computing*, **22**: 85–91, 2004.

Chojnacki, W., M. J. Brooks, A. van den Hengel, and D. Gawley: On the fitting of surfaces to data with covariances. *IEEE Trans. Pattern Analysis Machine Intelligence*, **22**:1294–1303, 2000.

Demmel, J. W.: The smallest perturbation of a submatrix which lowers the rank and constrained total least squares problems. *SIAM J. Numerical Analysis*, **24**:199–206, 1987.

Förstner, W.: On estimating 2D points and lines from 2D points and lines. In *Festschrift anläßlich des 60. Geburtstages von Prof. Dr.-Ing. Bernhard Wrobel*, pages 69–87. Technische Universität Darmstadt, 2001.

Halíř, R. and J. Flusser: Numerically stable direct least squares fitting of ellipses. In Proc. *Int. Conf. in Central Europe on Computer Graphics Visualization Interactive Digital Media* (V. Skala, editor), pages 125–132, 1998.

Hartley, R. I. and A. Zisserman: *Multiple View Geometry in Computer Vision*. Cambridge University Press, Cambridge, 2000.

Irani, M. and P. Anandan: Factorization with uncertainty. In *Proc. ECCV 2000* (D.Vernon, editor), pages 539–553, LNCS 1862(1), Springer, Berlin, 2000.

Kanatani, K.: *Statistical Optimization for Geometric Computation: Theory and Practice.* Elsevier, Amsterdam, 1996.

Matei, B. and P. Meer: A general method for errors-in-variables problems in computer vision. In *IEEE Conference on Computer Vision and Pattern Recognition*, pages 2018–2025, 2000.

Mühlich, M. and R. Mester: The role of total least squares in motion analysis. In *Proc. ECCV* (H. Burkhardth and B. Neuman, editors), pages 305–321, LNCS 1407(2), Springer, Berlin, 1998.

Mühlich, M. and R. Mester: Subspace methods and equilibration in computer vision. Technical Report XP-TR-C-21, J. W. G. University Frankfurt, 1999.

Mühlich, M. and R. Mester: A considerable improvement in non-iterative homography estimation using TLS and equilibration. *Pattern Recognition Letters*, **22**:1181–1189, 2001.

Nestares, O., D. J. Fleet, and D. J. Heeger: Likelihood functions and confidence bounds for Total Least Squares estimation. In *Proc. IEEE Conf. on Computer Vision and Pattern Recognition (CVPR'2000)*, pages 523–530, Hilton Head, 2000.

Ng, L. and V. Solo: Errors-in-variables modeling in optical flow estimation. *IEEE Transactions on Image Processing*, **10**(10):1528–1540, 2001.

Sampson, P. D.: Fitting conic sections to 'very scattered' data: An iterative refinement of the bookstein algorithm. *Computer Vision Graphics and Image Processing*, **18**:97–108, 1982.

Stewart, G. W.: Stochastic perturbation theory. *SIAM Review*, **32**:576–610, 1990.

van Huffel, S. and J. Vandewalle: *The Total Least Squares problem: Computational aspects and analysis.* SIAM, Philadelphia, 1991.

ON THE USE OF DUAL NORMS

IN BOUNDED VARIATION TYPE REGULARIZATION

ANDREAS OBEREDER
Department of Computer Science, University of Innsbruck
Technikerstr. 25, A–6020 Innsbruck, Austria
STANLEY OSHER
Mathematics Department,
Los Angeles, CA 90095-1555, USA

OTMAR SCHERZER
Department of Computer Science, University of Innsbruck
Technikerstr. 25, A–6020 Innsbruck, Austria

Abstract. Recently Y. Meyer gave a characterization of the minimizer of the Rudin-Osher-Fatemi functional in terms of the G-norm. In this work we generalize this result to regularization models with second order derivatives of bounded variation. This requires us to define generalized G-norms. We present some numerical experiments to support the theoretical considerations.

Key words: bounded variation regularization, contact problems, Rudin-Osher-Fatemi functional

20.1. Introduction

In this paper we are concerned with minimization of *bounded variation type regularization functionals* of the form

$$F(u) := \frac{1}{2}\int (u-f)^2 + \alpha p(u) \quad (\alpha > 0)\,,$$

with

$$p(u) = \|D^k u\| \text{ for } k = 1, 2 \text{ and } p(u) = \int |\Delta u|\,.$$

Here $\|D^k u\|$ denotes the *total variation* semi-norm of the $(k-1)$-th derivative of u and $\int |\Delta u|$ denotes the *variation measure* of Δu. The results of

R. Klette et al. (eds.), Geometric Properties for Incomplete Data, 373-390.

this paper can be generalized to higher order derivatives (i.e., for functionals with regularization terms $\|D^k u\|, k = 3, \ldots$) but it is omitted due to the notational complexity.

The special case $k = 1$ is the *Rudin-Osher-Fatemi* (ROF) functional (Rudin *et al.*, 1992) (see also (Osher and Fedkiw, 2003; Osher and Paragios, 2003)) - the minimizer is called *bounded variation regularized* solution. Since the invention of the ROF-model several results for characterizing properties of the minimizer have been derived. Moreover, in special situations of data f the minimizer could be calculated analytically: Strong & Chan (Strong and Chan, 1996) characterized the minimizer of the ROF-model for 1 dimensional data and for spherically symmetric data f (see also Ring (Ring, 2000)). Nikolova (Nikolova, 2000; Nikolova, 2004a) analyzed the ROF-model in a discretized setting; in the latter paper also higher order derivatives of bounded variation have been used. Osher & Esedoglu (Osher and Esedoglu, 2004) analyzed generalized ROF-models. Y. Meyer (Meyer, 2001) gave a characterization of properties of the minimizer of the ROF-functional in terms of the G-norm. These results will be generalized to characterize minimizers of the functional F, i.e., involving regularization functionals with second order derivatives of bounded variation. Motivated from the taut-string algorithm commonly used in statistics (cf. Mammen & Geer (Mammen and Geer, 1997), Davies & Kovac (Davies and Kovac, 2001), and Dümbgen & Kovac (Dümbgen and Kovac, 2004)) and Y. Meyer's (Meyer, 2001) characterization of the minimizer we are able to reformulate bounded variation regularization as a bilateral contact problem. The well-known (undesirable) effect of stair casing of the bounded variation regularized solution can be limited by smoothing in contact zones with the tube.

In a discrete setting, for analyzing one dimensional data f, there exist various ways for calculating minimizers of the ROF-model: Mammen & Geer (Mammen and Geer, 1997) showed that the *taut-string* algorithm, commonly used in statistics, minimizes the ROF-model. Brox & Mrázek & Steidl & Weickert & Welk (Brox *et al.*, 2003; Mrázek *et al.*, 2003; Steidl *et al.*, 2004) proved that wavelet thresholding based on the Haar wavelet is equivalent to minimizing the discretized ROF-model, which in turn is *equivalent* to solving the discretized total variation flow equation

$$\frac{\partial u}{\partial t} = \left(\frac{u_x}{|u_x|}\right)_x$$

at time α. Note that the ROF model can be interpreted as a fully implicit time step of the total variation flow equation with step length α. In higher space dimensions properties of the total variation flow equation have been derived by Belletini & Caselles & Novaga (Bellettini *et al.*, 2002), Andreu

& Ballester & Caselles & Diaz & Mazön (Andreu *et al.*, 2000; Andreu *et al.*, 2001; Andreu *et al.*, 2001a; Andreu *et al.*, 2002) and Alter & Caselles & Chambolle (Alter *et al.*, 2003). Note however, that the equivalence relations do not hold in higher space dimensions.

The outline of this work is as follows: In Section 20.2 we recall some basic facts on G-norms and bounded variation regularization. In Section 20.3 we recall tube methods. Finally in Section 20.4 we present some numerical experiments.

20.2. Higher Order G-Norms

In this section we introduce generalized G-norms. We give a quite general definition, although in the subsequent sections (for notational convenience) only the cases $k = 1, 2$ and $s = 2$ (see below) are used. This section is central to prove G-norm properties of minimizers of regularization functionals and thus presented in great generality.

For $k = 1, 2, \ldots$ and $s \in [1, \infty]$ we denote by

$$G^{k,s}(\alpha) := \{v = (\nabla\cdot)^k \vec{v} : \| |\vec{v}|_{l^s} \|_{L^\infty} \le \alpha\} = \{v : \|v\|_{G^{k,s}} \le \alpha\},$$

where $\vec{v} : \mathbb{R}^n \to \mathbb{R}^{n \times k}$. Here

$$(\nabla\cdot)^k \vec{v} = \sum_{\substack{i_l = 1, \ldots, n \\ l = 1, \ldots, k}} \frac{\partial^k v_{i_1,\ldots,i_k}}{\partial x_{i_1} \ldots \partial x_{i_k}}$$

denotes the k-th divergence and

$$|\vec{v}|_{l^s} = \left(\sum_{\substack{i_l = 1, \ldots, n \\ l = 1, \ldots, k}} |v_{i_1,\ldots,i_k}|^s \right)^{1/s} .$$

Moreover

$$\nabla^k v = \left[\frac{\partial^k v}{\partial x_{i_1} \ldots \partial x_{i_k}} \right]_{\substack{i_l = 1, \ldots, n \\ l = 1, \ldots, k}}$$

denotes the k-th derivative.

If not specified otherwise we denote by $|\cdot| = |\cdot|_{l^2}$ the Euclidean norm (respectively Frobenius norm for matrices and tensors). We call

$$\|v\|_{G^{k,s}} := \inf\{\| |\vec{v}|_{l^s} \|_{L^\infty} : v = (\nabla\cdot)^k \vec{v}\}$$

generalized G-norm. For $k = 1$ and $s = 2$ the generalized G-norm corresponds to the classical definition (see (Meyer, 2001)). We denote by

$$G := \{v : \|v\|_{G^{k,s}} < \infty\}$$

the set of all distributions v which have finite G-norm. Analogously to Meyer (Meyer, 2001) (see also (Scherzer, 2004)) it can be shown that for every $v \in G$, there exists $\vec{v}$ satisfying $v = (\nabla\cdot)^k\vec{v}$ with $\|v\|_{G^{k,s}} = \||\vec{v}|_{l^s}\|_{L^\infty} < \infty$.

In the following, for notational convenience, we omit in the case $k = 1$ the first index in the pair of subscripts, i.e., $G^s := G^{1,s}$. For $s \in [1, \infty]$ let $s_* \in [1, \infty]$ satisfy $1 = 1/s + 1/s_*$.

In the following we make use of results from functional analysis and measure theory. Appropriate references, where the necessary results can be found are Yosida (Yosida, 1995), Zeidler (Zeidler, 1993), and Evans & Gariepy (Evans and Gariepy, 1992).

Below we show that G is Banachspace, by showing that it is a dual of a Sobolev space. In the case $k = 1$ and $s = 2$ this result is stated in (Meyer, 2001).

THEOREM 20.1. *The set G associated with the norm $\|\cdot\|_{G^{k,s}}$ is a Banach space, which is the dual of the Sobolev space*

$$\tilde{W}^{k,1} := \overline{C_0^\infty},$$

where the closure is taken with respect to the norm

$$\|w\|_{W^{k,1,s_*}} := \int |\nabla^k w|_{l^{s_*}}.$$

By "G is the dual space of $\tilde{W}^{k,1}$" we mean not only that the sets are identical but also the associated norms are identical, i.e., $\|\cdot\|_{G^{k,s}} = \|\cdot\|_{(W^{k,1})^*}$.

Proof On the linear space C_0^∞, $\|\cdot\|_{W^{k,1,s_*}}$ is a norm. Therefore the completion, $\tilde{W}^{k,1}$ is a Banach space (see e.g. (Yosida, 1995)).

In the following we denote by $(\tilde{W}^{k,1})^*$ the dual of $\tilde{W}^{k,1}$ and by

$$\begin{aligned} \nabla^k : \tilde{W}^{k,1} &\to L^1(\mathbb{R}^n, \mathbb{R}^{n\times k}). \\ u &\to \nabla^k u \end{aligned}$$

With the space of absolutely integrable functions from $\mathbb{R}^n$ into $\mathbb{R}^{n\times k}$, $L(\mathbb{R}^n, \mathbb{R}^{n\times k})$, we use the product space norm $\int |\vec{v}|_{l^{s_*}}$. The operator is injective. With respect to the specified norms the operator ∇^k is an isometrical isomorphism between $\tilde{W}^{k,1}$ and the range of ∇^k. Thus the range of ∇^k is closed.

Denoting by $(\nabla^k)^*$ the dual operator and by X^* the dual space of a Banach space X, it follows from Banach's closed range theorem (see e.g.

(Zeidler, 1993, p777)) and the fact that the dual of L^1 is isometrically isomorph to L^∞ (in signs $\hat{=}$) that the adjoint operator

$$(\nabla^k)^* : L^\infty(\mathbb{R}^n, \mathbb{R}^{n\times k}) \hat{=} \left(L^1(\mathbb{R}^n, \mathbb{R}^{n\times k})\right)^* \to (\tilde{W}^{k,1})^* .$$
$$\vec{w} \to (\nabla^k)^* \vec{w}$$

satisfies that the range of $(\nabla^k)^*$ equals the complement of the kernel of ∇^k. Since ∇^k is injective the kernel is trivial and thus the complement is the whole space $(\tilde{W}^{k,1})^*$.

In particular, since $(\nabla^k)^* = (\nabla\cdot)^k$, this shows that any element of $(\tilde{W}^{k,1})^*$ can be written as $(\nabla\cdot)^k \vec{w}$.

From the definition of the space G we have the characterization

$$G \hat{=} \frac{L^\infty(\mathbb{R}^n, \mathbb{R}^{n\times k})}{N((\nabla\cdot)^k)} .$$

Here N denotes the kernel of a linear operator and $\frac{X}{Y}$ denotes the factorization space of X with respect to Y. Using that $(\nabla\cdot)^k$ is the adjoint of ∇^k we see that

$$\frac{L^\infty(\mathbb{R}^n, \mathbb{R}^{n\times k})}{N((\nabla\cdot)^k)} = \frac{L^\infty(\mathbb{R}^n, \mathbb{R}^{n\times k})}{N((\nabla^k)^*)} .$$

Using that L^∞ and L^1 are isometrically isomorph and that $N((\nabla^k)^*) = (\mathrm{Range}(\nabla^k))^\perp$, we get

$$\frac{L^\infty(\mathbb{R}^n, \mathbb{R}^{n\times k})}{N((\nabla^k)^*)} \hat{=} \frac{(L^1(\mathbb{R}^n, \mathbb{R}^{n\times k}))^*}{(\mathrm{Range}(\nabla^k))^\perp} .$$

Since $\mathrm{Range}(\nabla^k)$ is closed in $L^1(\mathbb{R}^n, \mathbb{R}^{n\times k})$, we have

$$\frac{(L^1(\mathbb{R}^n, \mathbb{R}^{n\times k}))^*}{(\mathrm{Range}(\nabla^k))^\perp} \hat{=} \mathrm{Range}((\nabla^k)^*) = (\tilde{W}^{k,1})^* .$$

Combination of the identities gives the desired result. □

The above proof applies to any space which is constructed as the completion of C_0^∞. The characterization with the G-norm relies on the dual operator. For the k-th derivative it is the k-th divergence operator. The adjoint of the Laplacian is the Laplacian and so on.

We note that $\tilde{W}^{k,1}$ is independent of $1 \leq s_* \leq \infty$. Moreover, $\tilde{W}^{k,1}$ is *not* the standard Sobolev space $W^{k,1}$, where in its definition the closure of C_0^∞ is taken with respect to the norm

$$\|w\|_{W^{k,1}} = \sum_{l=0}^{k} \| |\nabla^l w|_{l^2} \|_{L^1} .$$

In fact from the Gagliardo-Nirenberg-Sobolev inequality (see e.g. (Evans and Gariepy, 1992)) it follows that

$$\| |\nabla^{k-1} w|_{l^r} \|_{L^{p_n}} \leq C \int |\nabla^k w|_{l^r} \text{ for every } w \in \tilde{W}^{k,1}\,, \tag{1}$$

where

$$p_n := \frac{n}{n-1} \text{ for space dimension } n \geq 2 \text{ and } p_n := \infty \text{ for } n = 1\,. \tag{2}$$

LEMMA 20.1. *Assume that there exists $\alpha > 0$ such that for every $v \in C_0^\infty$*

$$\left| \int wv \right| \leq \alpha \int |\nabla^k v|_{l^{s*}} \tag{3}$$

holds, then $\|w\|_{G^{k,s}} \leq \alpha$.

Proof The linear operator

$$L\colon C_0^\infty \to \mathbb{R}\,, \quad v \to \int wv$$

can be extended to a linear bounded operator on $\tilde{W}^{k,1}$. Note that by (3) for a sequence $\{v_n\}_{n\in\mathbb{N}}$ converging to v, $\{Lv_n\}_{n\in\mathbb{N}}$ is a Cauchy sequence and thus convergent with limit Lv.

Therefore, from (3) it follows that $w \in (\tilde{W}^{k,1})^*$, with dual norm (which equals the G norm) is less than α. □

Let

$$G^\Delta(\alpha) := \{v = \Delta\tilde{v} : \|\tilde{v}\|_{L^\infty} \leq \alpha\} = \{v : \|v\|_{G^\Delta} \leq \alpha\}\,.$$

We define

$$\|v\|_{G^\Delta} := \inf\{\|\tilde{v}\|_{L^\infty} : v = \Delta\tilde{v}\}\,.$$

The proof of the following theorem is analogous to the proof of Theorem 20.1 and thus omitted.

THEOREM 20.2.

$$G^\Delta := \{v : \|v\|_{G^\Delta} < \infty\}$$

associated with the norm $\|\cdot\|_{G^\Delta}$ is a Banach space, which is the dual of

$$W^\Delta := \overline{C_0^\infty}\,,$$

where the closure is taken with respect to the norm

$$\|w\|_{W^\Delta} := \int |\Delta w|\,.$$

Moreover,

$$\|\cdot\|_{G^\Delta} = \|\cdot\|_{(W^\Delta)^*}\,.$$

The proof of the following lemma is analogous to the proof of Lemma 20.1 and thus omitted.

LEMMA 20.2. *Assume that there exists* $\alpha > 0$ *such that for every* $v \in C_0^\infty$

$$\left| \int wv \right| \leq \alpha \int |\Delta v| \tag{4}$$

holds, then $\|w\|_{G^\Delta} \leq \alpha$.

In the following we highlight some properties of functions of bounded variation. For more background on this subject we refer to Evans & Gariepy (Evans and Gariepy, 1992).

DEFINITION 20.1. *The space of* functions of bounded variation (BV) *consists of functions* $u \in L^{p_n}$ *satisfying*

$$\|Du\|_{s_*} := \sup \left\{ \int u(\nabla \cdot)\vec{\varphi} \, : \vec{\varphi} \in C_0^1(\mathbb{R}^n; \mathbb{R}^n) \, , |\vec{\varphi}(x)|_{l^s} \leq 1 \right\} < \infty \, .$$

The standard definition of BV requires $u \in L^1$ (cf. (Evans and Gariepy, 1992)). Actually the assumption $u \in L^{p_n}$ is less restrictive, since any function $u \in L^1$ satisfying $\|Du\|_{s_*} < \infty$ is in L^{p_n} (which follows from the Gagliardo-Nirenberg-Sobolev inequality). Note that for $u \in \tilde{W}^{1,1}$, $\|Du\|_{s_*} = \int |\nabla u|_{l^{s_*}}$.

In this paper we further consider functions with derivatives of bounded variation and functions of bounded Laplacian. To avoid notational difficulties we just consider the case $s = s_* = 2$. The generalization to the case $s \neq 2$ is obvious.

DEFINITION 20.2.

- *We define the set of* functions with derivatives of bounded variation ($\mathtt{BV}^k$) *as functions* $u \in L^2$ *satisfying*

$$\|D^k u\| := \sup \left\{ \int u(\nabla \cdot)^k \vec{\varphi} \, : \vec{\varphi} \in C_0^k(\mathbb{R}^n; \underbrace{(\mathbb{R}^n \times \mathbb{R}^n \ldots \mathbb{R}^n)}_{k\times}) \, , |\vec{\varphi}(x)| \leq 1 \right\} < \infty \, .$$

- *The space of* functions of bounded Laplacian *(*$\mathtt{BV}^\Delta$*) consists of functions* $u \in L^2$ *satisfying*

$$\|\Delta u\| := \sup \left\{ \int u \Delta \varphi : \varphi \in C_0^2(\mathbb{R}^n; \mathbb{R}) \, , |\varphi(x)| \leq 1 \right\} < \infty \, .$$

Note that for $u \in C_0^\infty$, $\|\Delta u\| = \int |\Delta u|$.

We consider $\mathtt{BV}^k$ and $\mathtt{BV}^\Delta$ as subsets of L^2. This is not standard, but simplifies the notation considerably. We could actually proceed iteratively (analogously to Definition 20.1) and define $\mathtt{BV}^k$ as a subset of W^{k-1,p_n} which can be embedded in L^q with $p_n \le q \le \frac{n}{n-k}$ for $1 < k < n$ and in L^∞ if $k \ge n$. We avoid distinguishing between the different cases, by defining $\mathtt{BV}^k$ and $\mathtt{BV}^\Delta$ as subsets of L^2. The space $\mathtt{BV}^2$ is commonly denoted as the space of bounded Hessian (see e.g. Demengel (Demengel, 1984)).

We have the following Lemma:

LEMMA 20.3. *Assume $w \in L^2$.*

1. *Let $\|w\|_{G^{2,2}} \le \alpha$, then for any $h \in \mathtt{BV}^2$*

$$\left|\int wh\right| \le \alpha \|D^2 h\| . \tag{5}$$

2. *Let $\|w\|_{G^\Delta} \le \alpha$, then for any $h \in \mathtt{BV}^\Delta$*

$$\left|\int wh\right| \le \alpha \|\Delta h\| . \tag{6}$$

Proof The proof of the second item is analogous and thus omitted. For $h \in C_0^\infty$ and $w = (\nabla\cdot)^2 \vec{w}$ satisfying $\||\vec{w}|\|_{L^\infty} = \|w\|_{G^{2,2}}$ it follows that

$$\left|\int wh\right| = \left|\int (\nabla\cdot)^2 \vec{w} h\right| = \left|\int \vec{w}\nabla^2 h\right| \le \|w\|_{G^{2,2}} \|D^2 h\| . \tag{7}$$

Let $h \in \mathtt{BV}^2$, then there exists a sequence $\{h_l\}_{l\in\mathbb{N}}$ in C_0^∞ such that $h_l \to h$ in L^2 and $\|D^2 h_l\| \to \|D^2 h\|$. Consequently, from (7) it follows that

$$\begin{aligned} \left|\int wh\right| &= \lim_{l\to\infty} \left|\textstyle\int wh_l\right| \\ &\le \liminf_{l\to\infty} \|w\|_{G^{2,2}} \|D^2 h_l\| \\ &= \|w\|_{G^{2,2}} \|D^2 h\| . \end{aligned}$$

□

20.3. Tube Methods

In this section we recall some basic facts on the minimizer of the ROF-functional:

THEOREM 20.3. *Let $f \in L^2$. Then the minimizer u_α of the generalized ROF functional*

$$\frac{1}{2}\int (u-f)^2 + \alpha \|Du\|_{s*}$$

satisfies:

1. $u_\alpha \in L^2 \cap \mathrm{BV}$;
2. $u_\alpha \equiv 0$ *if and only if* $\|f\|_{G^s} \leq \alpha$;
3. *If* $\|f\|_{G^s} > \alpha$, u_α *is characterized by*

   a) $\|u_\alpha - f\|_{G^s} = \alpha$ *and*

   b) $\int (f - u_\alpha)u_\alpha = \alpha \|Du_\alpha\|_{s_*}$.

For $s = 2$ this results has been given in (Meyer, 2001) and for general s it has been presented in (Osher and Scherzer, 2004).

Let Φ be measurable and satisfy $\Delta\Phi = f$ with $F_f := \nabla\Phi \in L^\infty_{\text{loc}}$. All along this paper we have been considering data filtering on $\mathbb{R}^n$. If we consider data smoothing on a bounded, smooth domain Ω, the existence of a solution of Laplace's equation $\Delta\Phi = f$ with Neumann boundary data is guaranteed if $\int f = 0$. For $\mathbb{R}^n$ we assume the existence of a solution of this equation, which imposes further requirements on the data f. By definition $\|\rho - f\|_{G^s} \leq \alpha$ if and only if $\rho - f = (\nabla\cdot)\vec{v}$ and $\||\vec{v}|_{l^s}\|_{L^\infty} \leq \alpha$. This is equivalent to

$$\rho = (\nabla\cdot)(\vec{v} + F_f) \text{ and } \||\vec{v}|_{l^s}\|_{L^\infty} \leq \alpha \,.$$

Or in other words, ρ is the divergence of a vector valued function $\vec{\rho}$ which is in a tube around the "primitive" of f (to be precise, we solve Laplace's equation and differentiate). The tube is a subset of $\mathbb{R}^{2n}$ around the vector valued function F_f. We recall that u_α is the divergence of a vector valued function $\vec{u}_\alpha$ and the distance between $\vec{u}_\alpha$ and F_f is less than α, i.e., $\||F_f - \vec{u}_\alpha|_{l^s}\|_{L^\infty} \leq \alpha$. Note that the tube geometry varies with s and has an impact on the solution (cf. (Osher and Esedoglu, 2004)). For $s = 2$ the tube has a cylindrical shape and for $s = 1$ or ∞ the tube is a slot.

The following geometric interpretations of the bounded variation regularized solutions u_α are immediate: the associated vector field $\vec{u}_\alpha$ does *not* have contact with the tube if and only if $\|f\|_{G^s} \leq \alpha$. For more background on the concept of tube methods we refer to (Hinterberger *et al.*, 2003; Scherzer, 2004).

20.3.1. HIGHER ORDER DERIVATIVES OF FUNCTIONS OF BOUNDED VARIATION

To our knowledge Chambolle & Lions (Chambolle and Lions, 1997) first studied BV-models with second order derivatives for denoising. Their approach consists in minimization of the functional

$$F_{C-L}(u_1, u_2) := \frac{1}{2} \int (u_1 + u_2 - f)^2 + \beta \|Du_1\| + \alpha \|D^2 u_2\|$$

with $0 < \alpha, \beta$. The asymptotic model, for $\beta \to +\infty$, for *denoising* has been introduced in (Scherzer, 1998): the noisy function f is approximated by the minimizer of the functional

$$F_D(u) := \frac{1}{2} \int (u - f)^2 + \alpha \|D^2 u\| \ . \tag{8}$$

The motivation for studying this type of regularization arises from *nondestructive evaluation* to recover discontinuities of a derivative of a potential u in impedance problems. The discontinuities of u are locations of *material defects* (see e.g. Isakov (Isakov, 1990; Isakov, 1998)). Later on second order models for denoising have been considered by Chan & Marquina & Mulet (Chan *et al.*, 2000) and Lysaker & Lundervold & Tai (Lysaker *et al.*, 2003). It can also be used for segmentation of *low contrast data* (Hinterberger and Scherzer, 2003).

20.3.2. CHARACTERIZATION OF MINIMIZERS OF REGULARIZATION FUNCTIONALS WITH HIGHER ORDER DERIVATIVES

In the following we summarize some basic characterization for the minimizer of the functional F_D and F_Δ. The following results can be generalized in a straight forward manner to higher order derivatives: the following proofs of the results require only elementary calculations and references to the general lemmas in Section 20.2, which can be generalized for higher order derivatives. However, for the sake of simplicity of notation we restrict attention to the case $k = 2$ and $s = 2$. The case $s \neq 2$ can be treated by following the proofs in (Osher and Scherzer, 2004).

THEOREM 20.4. *Assume $f \in L^2$. Then,*

1. *the functional F_D attains a unique minimizer $u_\alpha \in L^2$ with $\nabla u_\alpha \in L^{p_n}$.*
2. *$\|f\|_{G^{2,2}} \leq \alpha$ if and only if u_α is zero.*

Proof

1. The existence of minimizer of the functional F_D follows from its weak lower semi continuity and coercivity in L^2. The weak lower semi continuity of F_D follows from the weak lower semi continuity of $\int (u - f)^2$ in L^2 and the weak lower semi continuity of the Radon measure $\|D^2 u\|$. The functional $\int (u - f)^2$ is strictly convex and the regularization functional $\|D^2 u\|$ is convex. Thus F_D is strictly convex and attains a unique minimizer. The assertion $\nabla u \in L^{p_n}$ follows from the Gagliardo-Nirenberg-Sobolev inequality.
2. We have $u \in \mathtt{BV}^2$ if and only if $F_D(u) < +\infty$. From the definition of a minimizer u_α of F_D it follows that for every $h \in \mathtt{BV}^2$, $\varepsilon \neq 0$

$$\tfrac{1}{2} \int (u_\alpha - f)^2 \ + \ \alpha \|D^2 u_\alpha\|$$

$$\begin{aligned} &\leq \frac{1}{2}\int (u_\alpha + \varepsilon h - f)^2 + \alpha\|D^2(u_\alpha + \varepsilon h)\| \\ &\leq \frac{1}{2}\int (u_\alpha - f)^2 + \varepsilon \int (u_\alpha - f)h + \frac{\varepsilon^2}{2}\int h^2 \\ &\quad + \alpha\left(\|D^2 u_\alpha\| + |\varepsilon|\|D^2 h\|\right) \end{aligned}$$

Consequently, it follows by dividing the terms in the inequality by $|\varepsilon|$ and taking $\varepsilon \to 0^\pm$ afterward that

$$\left|\int (u_\alpha - f)h\right| \leq \alpha\|D^2 h\| \text{ for } h \in \mathtt{BV}^2 . \tag{9}$$

If $u_\alpha \equiv 0$, then from (9) it follows that for every $h \in \mathtt{BV}^2$

$$\left|\int fh\right| \leq \alpha\|D^2 h\| . \tag{10}$$

If (10) holds, then for every $h \in \mathtt{BV}^2$ we have

$$\begin{aligned} F_D(h) - F_D(0) = &\ \frac{1}{2}\int ((h - f)^2 - f^2) + \alpha\|D^2 h\| \\ \geq &\ \int -fh + \alpha\|D^2 h\| \\ \geq &\ 0 . \end{aligned}$$

Or in other words $u_\alpha \equiv 0$ is a global minimizer. That is, we have shown that $u_\alpha \equiv 0$ if and only if (10) holds for every $h \in \mathtt{BV}^2$.
From (10), the assumption $f \in L^2$, and Lemma 20.1 it follows that $\|f\|_{G^{2,2}} \leq \alpha$. Conversely, if $\|f\|_{G^{2,2}} \leq \alpha$, then from Lemma 20.3 it follows that for any $h \in \mathtt{BV}^2$

$$\left|\int fh\right| \leq \|D^2 h\| \cdot \|f\|_{G^{2,2}} \leq \alpha\|D^2 h\| .$$

That is, we have shown that (10) holds for every $h \in \mathtt{BV}^2$ if and only if $\|f\|_{G^{2,2}} \leq \alpha$ and referring back to the above equivalence relation the assertion follows.

□

THEOREM 20.5. *Let $f \in L^2$ satisfy $\|f\|_{G^{2,2}} > \alpha$. Then $u = u_\alpha$ minimizes F_D if and only if*

1. $u \in \mathtt{BV}^2$,

2.

$$\|u - f\|_{G^{2,2}} = \alpha \,, \tag{11}$$

3. and

$$-\int (u - f)u = \alpha \|D^2 u\| \,. \tag{12}$$

Proof From the assumption $\|f\|_{G^{2,2}} > \alpha$ it follows from Theorem 20.4 that $u_\alpha \neq 0$.

From the definition of a minimizer u_α of F_D it follows that for every $0 \neq |\varepsilon| < 1$

$$\frac{1}{2}\int (u_\alpha - f)^2 \, + \, \alpha\|D^2 u_\alpha\| \leq \frac{1}{2}\int ((1+\varepsilon)u_\alpha - f)^2 \, + \, \alpha(1+\varepsilon)\|D^2 u_\alpha\| \,,$$

showing that

$$-\varepsilon \int (u_\alpha - f)u_\alpha - \frac{\varepsilon^2}{2}\int u_\alpha^2 \leq \alpha\varepsilon \|D^2 u_\alpha\| \,.$$

Dividing the inequality by $|\varepsilon|$ and taking $\varepsilon \to 0\pm$ shows (12). Since $\|D^2 u_\alpha\| \neq 0$, it follows from (9) that $\|u_\alpha - f\|_{G^{2,2}} = \alpha$. To prove the converse direction we note that for $u \in \mathtt{BV}^2$ satisfying $\|u - f\|_{G^{2,2}} = \alpha$ it follows from Lemma 20.3 that for any function $h \in \mathtt{BV}^2$

$$\|D^2(u+h)\| \geq -\frac{1}{\alpha}\int (u+h)(u-f) \,. \tag{13}$$

From (13), and (12) it follows that for any function $h \in \mathtt{BV}^2$

$$\begin{aligned} & \frac{1}{2}\int (u+h-f)^2 \, + \, \alpha\|D^2(u+h)\| \\ \geq \; & \frac{1}{2}\int (u-f)^2 \, + \, \int h(u-f) - \int (u+h)(u-f) \\ = \; & \frac{1}{2}\int (u-f)^2 + \alpha\|D^2 u\| \,. \end{aligned}$$

This shows that u is a global minimizer. □

In the following we consider the problem of minimization of the functional

$$F_\Delta := \frac{1}{2}\int (u-f)^2 + \alpha \int |\Delta u| \,.$$

Analogously as for F_D it can be proven that the functional F_Δ attains a unique minimizer in $\mathtt{BV}_\Delta$. For $f \in L^2$ it follows from the general results above that $\|f\|_{G^\Delta} \leq \alpha$ if and only if u_α is zero.

The minimizer u_α of F_Δ is in the tube G^Δ. Geometrically this means that the second primitive of u_α (to be precise the second primitive is the solution of Laplace's equation $\Delta u = u_\alpha$) is in a tube of radius α around the second primitive of f.

20.4. Numerical results

In this section we present some numerical results for minimizing the functionals F_{ROF} and F_Δ or F_D, respectively.

We concentrate on minimization of the functionals for one dimensional input data f. For higher dimensional data the tube properties cannot be visualized easily and are not as illustrative.

The first examples is discrete bounded variation minimization for analyzing one dimensional data. There the Rudin-Osher-Fatemi functional is discretized with a one-sided difference operator. The according optimality condition has been solved with a fixed point iteration. In the case of 1-dimensional data one might alternatively use the taut-string algorithm (cf. (Mammen and Geer, 1997)) for calculating the BV-minimizer. For one dimensional discrete data, a simple method to minimize the ROF model has been proposed in (Nikolova, 2004). In Figure 20.4 (left row) we have plotted synthetic data f with different noise levels, and the discrete BV-minimizer. In the case of noise free data (top figure) significant stair casing occurs in the inclined part of the function.

The stair casing effect is inherent to the minimizer of the ROF-model (see (Ring, 2000)). At least in the discrete setting the BV-minimizer can be exactly calculated with the taut-string algorithm (Mammen and Geer, 1997). The taut-string algorithm consists in integration of the discrete data, constructing a tube around the (discrete) primitive, and finding a string of minimal length in the tube. The finite difference quotient of the taut-string is the BV-minimizer. Since the primitive of the piecewise linear function $\hat{f}$ according to the sample data f is quadratic in the inclined region, here the taut-string approximates the inclined region by a piecewise linear function, resulting in a significant stair-casing. In the inclined region the stair casing can be removed by taking into account the information that the primitive is quadratic in the inclined region, and therefore the BV-minimizer must be linear. Thus in these regions we can use linear approximations and prevent some stair casing. Locally quadratic regions of the discrete BV-minimizer are detected when two consecutive samples have contact with the tube. In this case we approximate the BV-minimizer by a piecewise linear function instead of a piecewise constant. This visual correction has been performed in the right row of Figure 20.4. The bottom of Figure 20.4 provides a zoom into the BV-minimizer and the correction. In the noise free case the inclined part of the reconstruction is perfect linear.

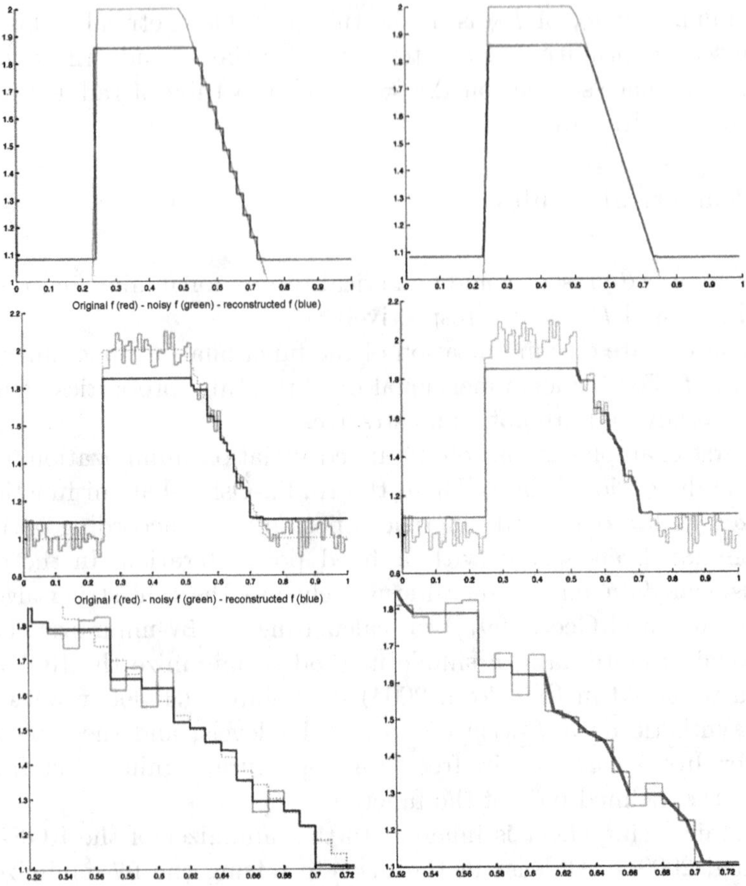

Figure 20.1. *Left:* Bounded variation regularization. *Right:* Smoothing in contact zones, using the characterization of minimizers with the G-norm. *Top:* Exact Data. *Middle* Noisy Data. *Bottom:* Zoom in the bounded variation regularized solution and the smoothing in contact zones.

In the next example (cf. Figure 20.4) we present some experiments for minimization of F_Δ. From the analytical results we know that $u_\alpha \in G_\Delta(\alpha)$. The numerical results show that the reconstruction has contact at locations, where the data f has discontinuities. Discontinuities in the derivative show up significantly in the tube, but do not have contact with the tube boundary. In comparison bounded variation regularization does not reveal

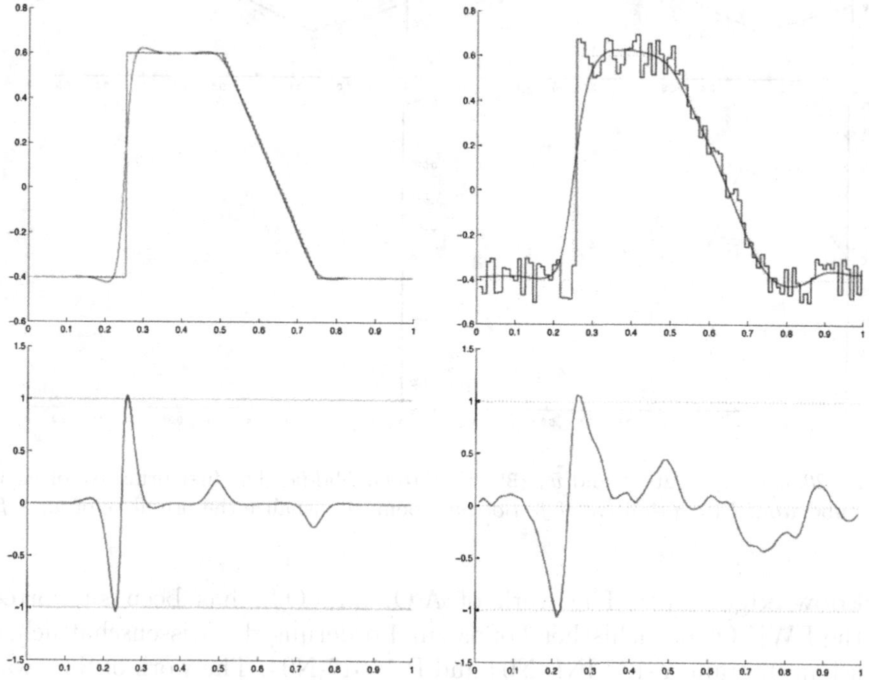

Figure 20.2. *Top:* Synthetic data and minimizer of F_Δ *Bottom:* Second primitive of the minimizer reveals features at multiple scales.

a separation in multiple scales for discontinuities in the function and derivative (cf. Figure 20.4). The bottom left image of Figure 20.4 has four pumps, which indicate the discontinuities (large pumps with contact to the tube) and small pumps for the discontinuities of the derivatives. This behavior is also clearly visible for the noisy data (cf. bottom right image of Figure 20.4). For purely denoising this model has the disadvantage that the reconstruction reveals a "ramp-shape" effect and to an imprecise edge localization, but on the other hand also discontinuities in the first derivative can be recognized. Bounded variation regularization has either contact with the tube or is linear in between. Note that for bounded variation regularization the tube is around the first primitive, while it is around the second in the second order model.

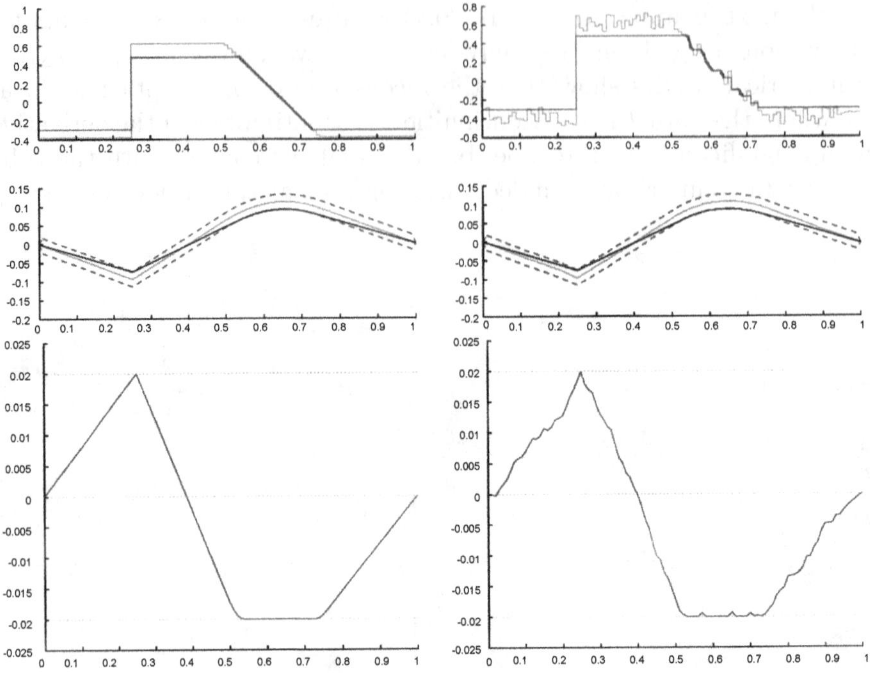

Figure 20.3. *Top:* Data f and u_α (BV minimizer). *Middle:* The first primitive of u_α is in a tube around the primitive of f. *Bottom* Zoom, to visualize the primitive of $u_\alpha - f$.

Acknowledgement: The work of A.O. and O.S. has been supported by the FWF Österreichischer Fonds zur Förderung der wissenschaftlichen Forschung), grant Y-123 INF-N04 and P-15617-N04. The work of S.O. was supported by NSF grants ACI-0321917 and DMS 0312222. O.S. would like to thank Markus Grasmair for stimulating discussions on dual spaces.

References

Alter, F., V. Caselles, and A. Chambolle: Evolution of convex sets in the plane by the minimizing total variation flow. *Preprint*, University Paris-Dauphine, 29 pages, 2003.

Andreu, F., C. Ballester, V. Caselles, and J. M. Mazön: Minimizing total variation flow. *C. R. Acad. Sci. Paris Ser. I Math.*, **331**:867–872, 2000.

Andreu, F., C. Ballester, V. Caselles, and J. M. Mazön: The Dirichlet problem for the total variation flow. *J. Funct. Anal.*, **180**:347–403, 2001.

Andreu, F., C. Ballester, V. Caselles, and J. M. Mazön: Minimizing total variation flow. *Differential Integral Equations*, **14**:321–360, 2001.

Andreu, F., V. Caselles, J. I. Diaz, and J. M. Mazön: Some qualitative properties for the total variation flow. *J. Funct. Anal.*, **188**:516–547, 2002.

Bellettini, G., V. Caselles, and M. Novaga: The total variation flow in $\mathbb{R}^N$. *J. Differential Equations*, **184**:475–525, 2002.

Brox, T., M. Welk, G. Steidl, and J. Weickert: Equivalence results for TV diffusion and TV regularisation. In Proc. *Scale-Space Methods in Computer Vision*(Griffin, L.D. and M. Lillholm, editors), pages 86–100, LNCS 2695, Springer, Berlin, 2003.

Chambolle, A. and P. L. Lions: Image recovery via total variation minimization and related problems. *Numer. Math.*, **76**:167–188, 1997.

Chan, T., A. Marquina, and P. Mulet: High-order total variation-based image restoration. *SIAM J. Sci. Comput.*, **22**:503–516 (electronic), 2000.

Davies, P. L. and A. Kovac: Local extremes, runs, strings and multiresolution. *Ann. Statist.*, **29**:1–65, 2001.

Demengel, F.: Fonctions a Hessien bornë. *Ann. Inst. Fourier (Grenoble)*, **34**:155–190, 1984.

Dümbgen, L. and A. Kovac: Extensions of smoothing via taut strings. TR, Universität Duisburg, Essen, 2004.

Evans, L. C. and R. F. Gariepy: Measure Theory and Fine Properties of Functions. *CRC–Press*, Boca Raton, 1992.

Griffin, L. D. and M. Lillholm: *Scale-Space Methods in Computer Vision.* LNCS 2695, Springer, Berlin, 2003.

Hinterberger, W., M. Hintermüller, K. Kunisch, M. von Oehsen and O. Scherzer: Tube Methods for BV regularization. *J. Mathematical Imaging Vision*, **19**:223–238, 2003.

Hinterberger, W. and O. Scherzer: Variational methods on the space of functions of bounded Hessian for convexification and denoising. TR, University Innsbruck, 2003.

Isakov, V.: *Inverse Source Problems. American Mathematical Society*, Providence, Rhode Island, 1990.

Isakov V.: *Inverse Problems for Partial Differential Equations. Applied Mathematical Sciences (127)*, Springer, New York, 1998.

Lysaker, M., A. Lundervold and X. Tai: Noise removal using fourth order partial differential equations with applications to medical magnetic resonance imaging in space and time. *IEEE Trans. Image Processing*, **12**:1579–1590, 2003.

Mammen, E. and S. van de Geer: Locally adaptive regression splines. *Ann. Statist.*, **25**:387–413, 1997.

Meyer, Y.: *Oscillating Patterns in Image Processing and Nonlinear Evolution Equations*, University Lecture Series, volume 22, American Mathematical Society, Providence, RI, 2001.

Mrázek, P., J. Weickert, and G. Steidl: Correspondences between wavelet shrinkage and nonlinear diffusion. In In Proc. *Scale-Space Methods in Computer Vision*(Griffin, L.D. and M. Lillholm, editors), pages 101–116, LNCS 2695, Springer, Berlin, 2003.

Nikolova, M.: Local strong homogeneity of a regularized estimator. *SIAM J. Appl. Math.*, **61**:633–658, 2000.

Nikolova, M.: A variational approach to remove outliers and impulse noise. *J. Mathematical Imaging Vision*, **20**:99–120, 2004.

Nikolova, M.: Weakly constraint minimization. Application to the estimation of images and signals involving constant regions. *J. Mathematical Imaging Vision*, **21**:155–175, 2004.

Osher, S. and S. Esedoglu: Decomposition of images by the anistropic Rudin-Osher-Fatemi model. *Comm. Pure Appl. Math.*, to appear, 2004.

Osher, S. and R. Fedkiw: *Level Set Methods and Dynamic Implicit Surfaces*, Applied Mathematical Sciences, volume 153, Springer, New York, 2003.

Osher, S. and N. Paragios: *Geometric Level Set Methods in Imaging, Vision, and Graphics.* Springer, New York, 2003.

Osher, S. and O. Scherzer: G-norm properties of bounded variation regularization. *Comm. Math. Sci*, **2**:237–254, 2004.

Ring, W.: Structural properties of solutions of total variation regularization problems. *M2AN Math. Model. Numer. Anal.*, **34**:799–810, 2000.

Rudin, L. I., S. Osher, and E. Fatemi: Nonlinear total variation based noise removal algorithms. *Physica D*, **60**:259–268, 1992.

Scherzer, O.: Denoising with higher order derivatives of bounded variation and an application to parameter estimation. *Computing*, **60**:1–27, 1998.

Scherzer, O.: Taut-string algorithm and regularization programs with G-norm data fit. *J. Mathematical Imaging Vision*, to appear.

Steidl, G., J. Weickert, T. Brox, P. Mrázek, and M. Welk: On the equivalence of soft wavelet shrinkage, total variation diffusion, total variation regularization, and SIDes. *SIAM J. Num. Anal*, to appear.

Strong, D. and T. F. Chan: Exact solutions to the total variation regularization problem. TR CAM 96 - 41, University of California, Los Angeles, 1996.

Yosida, K.: *Functional analysis.* Springer, New York, 1995.

Zeidler, E.: *Nonlinear Functional Analysis and its Applications*, part I, Springer, New York, 1993 (corrected printing).

Index